Instrumentation for Combustion and Flow in Engines

NATO ASI Series

Advanced Science Institutes Series

A Series presenting the results of activities sponsored by the NATO Science Committee, which aims at the dissemination of advanced scientific and technological knowledge, with a view to strengthening links between scientific communities.

The Series is published by an international board of publishers in conjunction with the NATO Scientific Affairs Division

A Life Sciences **B Physics**	Plenum Publishing Corporation London and New York
C Mathematical and Physical Sciences **D Behavioural and Social Sciences** **E Applied Sciences**	Kluwer Academic Publishers Dordrecht, Boston and London
F Computer and Systems Sciences **G Ecological Sciences** **H Cell Biology**	Springer-Verlag Berlin, Heidelberg, New York, London, Paris and Tokyo

Series E: Applied Sciences - Vol. 154

Instrumentation for Combustion and Flow in Engines

edited by

D. F. G. Durão

Instituto Superior Técnico,
Mechanical Engineering Department, Lisbon, Portugal

J. H. Whitelaw

Imperial College of Science and Technology,
London, England

and

P. O. Witze

Sandia National Laboratories,
Livermore, CA, U.S.A.

Kluwer Academic Publishers

Dordrecht / Boston / London

Published in cooperation with NATO Scientific Affairs Division

Proceedings of the NATO Advanced Study Institute on
Instrumentation for Combustion and Flow in Engines
Vimeira, Portugal
September 13–26, 1987

Library of Congress Cataloging in Publication Data

Instrumentation for combustion and flow in engines / edited by D.F.G. Durão, J.H. Whitelaw, and P.O. Witze.
p. cm. -- (NATO ASI series. Series E, Applied sciences ; no. 154)
Proceedings of the Institute on Instrumentation for Combustion and Flow in Engines, held in Vimeiro, Portugal, Sept. 13-26, 1987.
"Published in cooperation with NATO Scientific Affairs Division."
Includes index.
ISBN 0-7923-0034-3 (U.S.)
1. Internal combustion engines--Testing--Instruments--Congresses. 2. Gas-turbines--Testing--Instruments--Congresses. I. Durão, D. F. G. II. Whitelaw, James H. III. Witze, Peter O. IV. Institute on Instrumentation for Combustion and Flow in Engines (1987 : Vimeiro, Lisbon, Portugal : District) V. North Atlantic Treaty Organization. Scientific Affairs Division. VI. Series.
TJ756.I56 1988
621.43'3--dc19 88-31611
CIP

ISBN 0-7923-0034-3

Published by Kluwer Academic Publishers,
P.O. Box 17, 3300 AA Dordrecht, The Netherlands.

Kluwer Academic Publishers incorporates the publishing programmes of
D. Reidel, Martinus Nijhoff, Dr W. Junk, and MTP Press.

Sold and distributed in the U.S.A. and Canada
by Kluwer Academic Publishers,
101 Philip Drive, Norwell, MA 02061, U.S.A.

In all other countries, sold and distributed
by Kluwer Academic Publishers Group,
P.O. Box 322, 3300 AH Dordrecht, The Netherlands.

Printed in The Netherlands

CONTENTS

PREFACE

Much has been said and written about the abilities of modern instrumentation to help solve problems of combustion in engines. In the main, however, the design and fabrication of combustion chambers continues to be based on extrapolation of experience gained from use and rig tests, with little input from advanced techniques such as those based on optical diagnotics. At the same time, it has become increasingly difficult to design better combustion chambers without knowledge of the relevant flow processes. Thus, the future must involve improved understanding which, in turn, will require detailed measurements of velocity, temperature and concentration. The need to narrow the gap between current industrial practice and the acquisition and implementation of improved techniques motivated the organization of the Advanced Study Institute upon which this volume is based. This Institute on Instrumentation for Combustion and Flow in Engines was arranged to display the needs of industry and the possibilities made available by modern instrumentation and, at the same time, to make clear the relative advantages of optical and probe techniques. Held at Vimeiro during the period from 13 to 26 September, 1987, the Institute was attended by 120 participants and 16 invited lecturers.

The combustion chambers of gas turbine and internal-combustion engines have in common the feature that their efficiency is determined by the geometric and flow boundary conditions of the chamber, as is also the composition of the gases which are exhausted to the atmosphere. The two types of chamber can involve aspects of diffusion-controlled and premixed-charge combustion of both liquid and gaseous fuels, and their performance can be greatly influenced by fluid-mechanic phenomena. A major complication hindering the improved understanding of both continuous and discontinuous combustion is the time-dependent nature of the fuel vaporization and combustion processes, which involve time scales that can be of the same order as the turbulent fluid motions governing the physical phenomena that determine overall engine performance. Intrinsic to both engine types are the measurement difficulties associated with lack of convenient access and the hostile conditions brought about by high temperatures and pressures.

Because of the physical-process similarities of these two different types of engines, the Institute was arranged in two parts, but with common participation, to consider measurements needs and merits of available instrumentation. In each part, the programme included lectures describing procedures currently used in industry, and the identification of important problems which need to be solved; and presentations described instrumentation, techniques and results in a way which allowed current status to be appraised in terms of industrial needs.

An important part of the Institute was concerned with papers contributed by the international research community. Twenty-two papers of this type were presented and added greatly to the scientific content and informal discussions of the meeting. A small number of these papers are included in this Volume, chosen to complement the material provided by the invited lecturers.

The format of the Volume is similar to that of the Institute, although we have asked most of the invited lecturers to combine the material of their lectures into a single paper. Thus, there are two parts concerned with gas-turbine combustors and internal-combustion engines. We hope that the reader will find the arrangement convenient, and that the Volume will find application in industry and in research laboratories alike.

We are pleased to take this opportunity to thank the authors who have given generously of their time to prepare lectures and papers, and to make the Institute a success. It is also a considerable pleasure to acknowledge the organizational contributions of Dr. M. Heitor, who was responsible for the day-to-day running of the Institute and for communicating with the authors of this Volume. The Institute was supported by a Grant from NATO.

D.F.G. Durão
J.H. Whitelaw
P.O. Witze

Lisbon 1988

VELOCITY AND SCALAR MEASUREMENTS

IN MODEL AND REAL GAS TURBINE COMBUSTORS

M.V. Heitor
Instituto Superior Técnico
Department of Mechanical Engineering
Av. Rovisco Pais
1096 Lisbon Codex
PORTUGAL

ABSTRACT

This paper reviews recent measurements and the consequent understandings on the fundamental processes of the flow, mixing and combustion in turbulent reacting flows typical of gas-turbine combustors. The emphasis is on highly recirculating, swirling, non-premixed flames within model and real combustors but isothermal flows relevant to gas-turbine combustors are also analysed.

Finally the capabilities of optical and probe techniques for making measurements in gas-turbine combustors are discussed on the basis of the results of Heitor and Whitelaw (1986) in terms of the contributions that they can make to improve knowledge on the theory and modelling of these flows.

1. INTRODUCTION

Turbulent reacting flows such as those typical of gas-turbine combustors have attracted increased interest in recent years as a consequence of the requirements of improved efficiency, reduced emissions and use of alternative fuels. At the same time, advances in diagnostic techniques have increased the possibility of measurements with spatial and temporal resolution sufficient to allow improved understanding of turbulent combustion. Also, corresponding advances in numerical methods offer the possibility of solutions to conservation equations as a mean of interpolating and extrapolating the experimentally acquired information. This paper reviews work concerned with the application of modern experimental techniques to improve understanding of flow configurations relevant to gas-turbine combustors.

The aim of the search was to assemble available experimental results obtained with probe and optical techniques in model and real combustors. The emphasis is on local flow properties including velocity, temperature and species concentrations, but the possibility of measuring important characteristics, such as wall temperatures, is also discussed. Relevant numerical results are briefly reviewed to elucidate the needs of accurate measurements to provide guidance for the development of calculation methods.

D. F. G. Durão et al. (eds.), Instrumentation for Combustion and Flow in Engines, 1–44.

The following section considers work carried out in model and real combustors, which, in general, involve highly recirculating, swirling, non-premixed flames. The results of Heitor and Whitelaw (1986) obtained in a tuboannular combustor are particularly analysed in section 3 to indicate the needs to improve modelling techniques. Bluff-body stabilised turbulent premixed flames relevant to after-burners in gas-turbines have been reviewed by Libby et al (1986) and are not considered here. The final section summarizes the main conclusions of the present review and provides some suggestions for future research.

2. EXPERIMENTS IN GAS-TURBINE COMBUSTORS

2.1 Simple model combustors

Experiments in model combustors can enhance understanding of the flows in real combustion chambers which are extremely complex. Knowledge of the aerodynamics of these flows is incomplete and, because of its importance, experiments have been reported for isothermal and combusting flows. This subsection deals with studies which have allowed the examination of particular variables in simplified geometries relevant to gas turbine combustors.

Isothermal swirling flows relevant to the primary zone of a combustor have been analysed with hot-wire anemometry by Weske and Sturov (1974), Janjua et al (1983) and Hallet and Gunther (1983) and with laser velocimetry by Altgeld, Jones and Wilhelmi (1983). The internal structure of a recirculation zone has been determined with laser velocimetry by Faler and Leibovich (1978) and its extension and dependence upon the degree of swirl examined by Escudier and Zehnder (1982), also with laser velocimetry, and by Vu and Gouldin (1982) and Rhode, Lilley and McLaughlin (1983) with directional pitot probes and hot-wire anemometry. Similarly, the influence of an exit contraction on swirling flow behaviour has been investigated by Escudier, Bornstein and Zehnder (1980) and Escudier and Keller (1985). Leibovich (1983 and 1984) has used some of these investigations and, together with previous measurements, extended the theory of vortex stability. Escudier and Keller (1985) have, however, argued that a cold swirling flow is likely to be totally unrepresentative of a reacting flow with the same geometry, although most of the previous investigations have been performed with the aim of simulating practical combustors. In addition, they have also concluded that great care has to be taken over the boundary conditions to be imposed for the computation of swirling flows, thus confirming the need for experiments under the real combusting flow situations.

The downstream dilution zone of a combustor was initially simulated through the interaction of a row of jets in a cross flow as described, for example, by the studies of Crabb and Whitelaw (1979) and Khan and Whitelaw (1980-a) using probe techniques. Further extension to opposed rows of jets have been reported by Khan and Whitelaw (1980-b), Atkinson, Khan and Whitelaw (1983) and Wittig, Elbahar and Noll (1984). Sivasegaram and Whitelaw (1984) have suppressed the cross flow and studied, by a flow visualisation method, the effects of geometric asymmetries and misalignment of jets in a similar arrangement. A geometry closer to that found in real combustors was analysed by Chleboun et al (1982) using a pitot-static probe. Calculations with the standard '$k-\varepsilon$' model for turbulence have also been reported and show, in agreement with those of Khan, McGuirk and Whitelaw (1981), that there is a tendency to underpredict mixing and to overpredict the extent of the reverse flow zone, although the prescribed boundary conditions have significantly influenced the predicted flow field, as discussed by McGuirk and Whitelaw (1983).

Green and Whitelaw (1980 and 1983) have combined primary and dilution zones in their water models, but the upstream and downstream geometries were simplified to exclude swirl. Their results, which were based on laser velocimetry measurements, quantify the effects of the change from a two-dimensional to a three-dimensional geometry.

Qualitative and quantitative studies in water flow models closely resembling current combustors have been reported by Clarke, Gerrard and Holliday (1963) and Kennedy (1974). More recently, Koutmos and McGuirk (1985) and Bicen et al (1987) used laser velocimetry to measure the internal velocity field in a plexiglass model of the combustor studied by Heitor and Whitelaw (1986) and the measurements provide a basis for the implementation of the boundary conditions and for the development of turbulence models for calculating the flows in that combustor.

Detailed measurements of velocity and scalar characteristics of reacting flows relevant to combustors have been reported in axisymmetric geometries and the experiments and findings are summarised in tables 1.1 and 1.2 respectively for gaseous and spray flames. Swirling and non-swirling flames are included and together quantify the effects of mean droplet diameter, equivalence ratio, inlet air temperature, combustor pressure and inlet air swirl. The main conclusions are considered in the paragraphs below.

The principal differences from the equivalent isothermal flow are that the recirculation zone is shorter and wider, that the mass flow recirculated is smaller and the reverse velocities larger due to the decrease in density and that the velocity fluctuations are generally larger with the local turbulence intensities smaller, owing to thermal expansion.

The influence of droplet size has been observed in substantially different flows and it has been found that the nature of the spray can be important regardless of its characteristics, location and atomiser type. The experiments of table 1.2 and the related calculations of El Banhawy and Whitelaw (1980-a and b), Jones and McGuirk (1981) and Attya and Whitelaw (1984) suggest, however, that at least in kerosene-fuelled gas-turbine combustors with some inlet air preheat and droplets of Sauter mean diameter less than around 0.050 mm, the fuel effectively burns like a vapour and, consequently, the performance of such combustors could be analysed using results in gas flames. Similarly, the effects of soot formation and radiative heat transfer are usually small, except for purposes of calculating wall temperature, and permit the use of simple droplet and radiation models in calculation procedures as suggested by Jones and Whitelaw (1984).

Increasing equivalence ratio, inlet-air temperature, swirl or the pressure in the combustor results in an increase in temperature and NO_x and in a decrease in CO and UHC levels, and may give rise to large variations in the combusting flow field but, in general, leads to increased combustion efficiencies, as shown by Owen, Spadacini and Bowman (1977), Owen et al (1979), Nichols et al (1980), El Banhawy and Whitelaw (1981), Brum and Samuelsen (1982) and Attya and Whitelaw (1984). In particular, the influences of swirl and pressure on aerodynamic characteristics, such as the recirculation zone lenght and width, are not entirely consistent in all the investigations and are directly related to the increase in residence time within the recirculation zone, as discussed by Owen, Spadacini and Bowman (1977). The results of Attya and Whitelaw (1984) have, however, resolved some of the previous discrepancies and confirm the strong dependance of the structure and shape of the recirculation zone on the level of swirl.

In parallel to the investigations of the mean flow field summarised in the previous paragraphs, detailed measurements of velocity fluctuations have also been reported using laser velocimetry by Owen (1977), Owen et al (1979), Lightman et al (1980), Brum and Samuelsen (1982), Wilhelmi (1984) and Gouldin, Depsky and Lee (1985). In all cases the turbulent fluctuation levels are extremely high in the initial mixing region and show local maxima near the time-averaged velocity zeros associated with large-scale unsteadiness of the recirculation bubble.

Measurements of scalar fluctuations have been limited to those of Halthore and Gouldin (1984) by laser techniques and to the thermocouple measurements of LaRue, Samuelsen and Seiler (1984) which, together, support the previous aspects of the velocity field. In addition, the latter have also suggested that downstream of the recirculation zone, in the region of maximum turbulent energy, scalar transport occurs in the non-gradient sense as a result of the mean pressure field induced by swirl. Further quantification of these trends are, however, required making use of experimental methods such as those described by Heitor (1988).

Attempts to calculate the flow properties of the combustor type flows examined in the preceding paragraphs have been reported in an effort to allow the testing of calculation methods prior to their use in simulations of real arrangements. Calculations such as those of Jones and McGuirk (1981), Smith and Smoot (1981), Correa (1984) and Wilhelmi (1984) have failed to reproduce the size and strenght of the recirculation zone, the tangential velocity profiles which tend too rapidly to solid body rotation, the distribution of turbulent energy near the regions of zero mean velocity and, in general, have shown that mixing is underpredicted. Although there are some indications of failure of the 'k-ε' turbulence model, particularly as the degree of swirl increases, the major uncertainties are more likely to be associated with numerical errors and with lack of knowledge concerning the inlet conditions. The failure to predict accurately the aerodynamic field has precluded complete tests for the combustion models, although the calculations mentioned above have suggested that equilibrium assumptions are likely to be adequate as far as the calculation of major species and temperature is concerned. Nevertheless, as pointed by Jones and McGuirk (1980) and Jones (1980), the prediction of pollutant emissions, such as CO, requires at least the use of constrained models to take into account limitations in the chemical processes. Non-equilibrium models such as those reviewed by Peters (1984), although in the early stages of development, seem to be promising.

The current trend in the design of gas-turbine combustors towards increased heat flux per unit length requires improved wall cooling techniques which have been studied in slightly heated flows on simplified geometries and in model combustors. Convection, film and impingement cooling or the combination of at least two of these techniques, as studied for example by Folayan and Whitelaw (1977) and Simon, Schubert and Basler (1983), are some of the modern techniques which have been used and are reviewed by Nealy (1980) and Lefebvre (1983). The need of improved cooling has also led to the development of the laminated, pseudo-transpiration material as described by Wassel and Bhangu (1980) and used in the combustor studied by Heitor and Whitelaw (1986).

In parallel to these studies, techniques to measure wall temperature have been developed in non-reacting, heating flows, as described, for example, by Tamaru and Kurosawa (1984) using thermocouples on thin steel plates. Measurements in transpiration materials have been performed using thermal paints and embedded thermocouples by Bhanghu, Snape and Eardley (1983), but remain to be assessed and require further improvement.

2.2 Experiments in gas turbine combustors

Measurements in combustors of gas turbines have been reported by several authors and table 2 summarises the experiments and findings for a wide range of fuels. In general, they reveal the intense combustion caused by the three-dimensional nature of the primary zone which is followed by dilution by jets, where the trajectory can be a strong function of the geometry of the combustor. The results are limited in extent and subject to inaccuracies arising mainly from the use of probe techniques and to further uncertainties as to whether weighted-, or unweighted-averaged data is measured, as examined in detail by Heitor et al (1985).

The measurements have been primarily concerned with the emission of oxides of nitrogen, smoke, carbon monoxide and unburned hydrocarbons and the processes associated with their production and control have been reviewed by Sawyer, Cernansky and Oppenheim (1973), Jones (1978), Mellor (1976), Plee and Mellor (1978) and Lefebvre (1983). CO and UHC (and also aldehydes) have their peak values during idle conditions and drop to very low levels as engine power is increased, while particulates, smoke density and NO_x increase with power. Consequently, the requirements to reduce simultaneously all the pollutants are difficult to reconcile. Several approaches have however been developed towards the advanced-technology combustor concepts and include the introduction of premixed and prevaporized fuel by, for example, Parikh, Sawyer and London (1971) and Shisler, Tuttle and Mellor (1975), staged combustion by Lefebvre and Fletcher (1973) and Morr, Heywood and Fitch (1975), fluidic-controlled combustion by Rizk, Fletcher and Adkins (1981) and Nakamura, Kawaguchi and Sato (1983), multiple primary-air entries by Tilston and Hakluytt (1983) and transpiration cooling material and curved vane swirler by Bhangu, Snape and Eardley (1983).

Attempts have also been made to model the performance of gas-turbine combustion chambers. Modular techniques based on empirical approaches, such as those of Prior, Swithenbank and Felton (1978), Mellor (1980) and Edelman, Harsha and Schmotolocha (1981), have been reported together with models developed to analyse the poluttant levels as, for example, those of Fletcher, Siegel and Bastress (1971), Fletcher and Siegel (1971) and Fletcher and Heywood (1971) for NO_x and those of Sheppard (1975), Chair, Sheppard and Whittaker (1976) and Morr, Hewyood and Fitch (1975) for CO. These models are however more interpretive than predictive and a far more powerful alternative approach is one based on the numerical solution of conservation equations in differencial form. Attempts to calculate the flow properties of gas turbine combustor arrangements with this latter approach have been reported by Serag-

Eldin and Spalding (1978), Jones and McGuirk (1980), Jones and Priddin (1979) Boysan et al (1982), Coupland and Priddin (1986) and more recently by Burrus et al (1987), Wild et al (1987) and Wittig et al (1987). Among other factors, computer resources can limit the calculations as discussed, for example, by Jones and Whitelaw (1984) but, in general, there is need for new experimental data based on detailed parametric studies to assess the predictions. Swithenbank, Turan and Felton (1978) and Ewan, Boysan and Swithenbank (1984) have attempted to combine modular techniques with finite-difference solutions but, in addition to the problems associated with the latter, extrapolation to typical combustor geometries is difficult to achieve.

Table 2 reveals a deficiency of detailed information on the velocity characteristics within combustors. Jones et al (1977) measured with a 7-hole pitot probe the mean velocity distribution in the central plane of their combustor which was approximately axisymmetric and the flow pattern was not closely representative of practical combustion chambers. Toral and Whitelaw (1982) reported laser velocimeter measurements in a sector of an annular combustor which allowed access to the entire flow regime; again, only the mean axial velocity characteristics were reported. The measurements were extended by those of Bicen et al (1987), but these were obtained only for isothermal flows. Goldstein, Lau and Leung (1983) measured the mean and fluctuating axial velocity distribution, but the results are restricted to a plane downstream of the exit section of their combustor. Reliable velocity measurements within combustors are then required to improve the evaluation of current <u>turbulence</u> models in real flows, and aimed the investigations of Heitor and Whitelaw (1986), Bicen et al (1986) and Bicen and Jones (1987) summarised in the following subsection, which provided new sets of experimental data, within an arrangement similar to that of a real combustor.

The experiments of table 2 provide, however, some information to guide the development and use of <u>combustion</u> models. It is evident that equilibrium assumptions do not prevail everywhere, but further quantification is required to build and test a model. Two main zones can be distinguished within the combustors, as observed, for example, by Noyce, Sheppard and Yamba (1981) and Jones and Toral (1983). Upstream of the primary holes, fuel rich regions have been observed and the results show that, although combustion is likely to be more controlled by physical than kinetic processes, the full chemical equilibrium assumption leads to overprediction of CO and H_2 by up to four times. Downstream of the primary zone, however, combustion occurs in excess air and the results clearly indicate that both the fuel breakdown and the CO to CO_2 oxidation rates are partly controlled by finite rate chemical kinetic mechanisms.

It is also evident from the results of Toral and Whitelaw (1982) and Jones and Toral (1983) that scalar fluctuations exert a large influence on the temperature distributions in the regions around stoichiometry. Only Dils and Follansbee (1979) and Heitor and Whitelaw (1986) have reported fluctuating temperature measurements within a real combustor and show that intensities range up to 25%. Although the comparatively high operating pressure of full power conditions may suppress the chemical limitations of the combustion processes and reduce the fluctuating levels, lack of detailed experimental information at high pressures have precluded the complete evaluation of such processes and, consequently, experiments including the measurement of velocity and scalar fluctuations are required for a range of operating conditions. This implies the use of advanced optical diagnostics, which entails high costs and complexity and have only been applied to characterize jet engine exhausts as shown by Eckbreth et al (1984).

The influence of equivalence ratio (0.1< <1.0), inlet-air temperature ($300<T_o[K]<600$), combustor pressure (1<P[atm]<10), degree of swirl and type of injection on the flow within combustors have been studied with relevance to the mean flow field and, in general, the results confirm the trends observed in the axisymmetric models discussed previously, although narrow ranges for the operating conditions have been used, particularly for T_o and P, and the real situations not achieved. Bhangu, Snape and Eardley (1983) have presented one of the few studies (e.g. Suziki et al, 1978; Tilston and Hakluytt, 1983) where a whole range of engine operating conditions were used. The extension of the measurements was however rather limited and the performance of the combustor was partly determined at laboratory conditions.

The increased tendency in recent years to use low-quality and synthetic liquid fuels of relatively low hydrogen content has prompted some experimental investigations to characterise the formation of soot in arrangements similar to those of real combustors, as those of Bowden and Pearson (1983) and Naegeli, Dodge and Moses (1983), and to improve the quality of fuel preparation, for example by Rizk and Lefebvre (1982), Jasuja (1983) and Low (1983). Calculation schemes, as that of Lockwood, McGuirk and Shah (1983) and Carvalho and Coelho (1987), have also attempted to include radiation heat transfer but these problems are, as yet, far less significant than the aspects dealt with the previous paragraphs. In spite of the works listed in table 2, there is need for new experimental data to quantify the influence of geometric and flow boundary conditions such as, for example, swirler type, equivalence ratio, air splits, preheat temperature and pressure.

3. AN INVESTIGATION OF THE FLOW IN A RESEARCH GAS-TURBINE COMBUSTOR

The isothermal and combusting flow characteristics of a can-type gas-turbine combustor were reported by Heitor and Whitelaw (1986), Bicen et al (1986) and Bicen and Jones (1987) and are briefly analysed here to exemplify the use of diagnostic techniques to improve knowledge of gas-turbine combustion. In isothermal flow a laser-Doppler velocimeter was used to measure the longitudinal and tangential velocity characteristics, and the mean concentration of a tracer of helium gas, injected through the fuelling device, was used to obtain the mixture fraction distribution. In combusting flow, with propane as fuel, the velocimeter measured density weighted velocity characteristics, thin digitally compensated bare-wire thermocouples measured unweighted temperature characteristics, and a sampling probe transported gas samples to various instruments which measured near density weighted concentrations of O_2, UHC, H_2 CO_2, CO, and NO_x.

The combustion chamber was a model of the low-emissions tuboannular combustors described by Bhangu et al (1983) and is represented schematically in table 2 together with the experimental conditions used. It should be noted that the geometry of this combustor, with its resulting distribution of air, was chosen to emphasize the influence of a fuel-lean primary zone with consequent low values of NO_x

a) EXPERIMENTAL RESULTS

The velocity measurements were limited by optical access to the axis of the primary and dilution holes and at the exit plane; temperature and concentration information was provided in the same locations, in three radial planes along the combustor and in the axial vertical plane through the centre of the combustor.

The mean velocity components measured through the primary and dilution holes are shown on figure 1 (a), together with the corresponding rms of the fluctuations at the upstream location on figure 1 b), for the combustion flow (Run 1). The same mean velocities normalised by the characteristic overall mean velocity at the exit plane, U_I and U_1 for isothermal flow and Run 1 respectively, are given on figure 2: U_1 was calculated based on a mean temperature at the exit plane of 1000 K (see below). Compared with isothermal flow, the longitudinal velocity values show that the primary zone vortex increased in strength with a maximum reverse velocity of 12 m/s rather than 4 m/s: the corresponding normalized values are maintained around -0.13. The velocities between the recirculation zone and the combustor walls have also increased from 10 m/s to 14 m/s with consequently steeper radial gradients across the shear layer and a decrease in the width of the recirculation zone. This decrease was not found in simpler axisymmetric combustors, e.g. Brum and Samuelsen (1982), and is due, in part, to the penetration of the primary jets. The double cell structure characteristic of the isothermal flow has apparently been destroyed by the combustion of the fuel jets and a toroidal recirculation zone formed with a decrease in the reverse flow velocity near the centerline. The combustion has also resulted in velocities of up to 40 m/s in the plane of the secondary holes: the normalised values are similar to those in isothermal flow but reveal the absence of near zero velocity on the centerline.

The swirl (vertical) velocities are similar to those in isothermal flow. In the plane of the primary holes the magnitude was maintained and the narrower recirculation zone is associated with the inward movement of the point of maximum tangential velocity and the consequent steeper gradients. In the plane of the dilution holes, the longitudinal acceleration of the flow has reduced the magnitude and strength of the swirl.

The local turbulence intensity values are of similar magnitude of those with isothermal flow, but the local variances have increased by up to 50%. Figure 3a shows the profiles of turbulent kinetic energy, calculated on the basis of local isotropy, and the larger centerline peaks, due to the values of $\widetilde{w''^2}$, suggest that the precession of the core of the axial vortex has increased with combustion. Away from the centerline the increase in K stems mainly from $\widetilde{u''^2}$. The shear stress, $\widetilde{u''w''}$, is shown in Fig. 3b to be small and comparable to isothermal flow in the plane of the primary holes, even in the zones of higher mean radial gradient, and suggests that the increase in the turbulent kinetic energy in the recirculation zone is caused by nongradient production. Owen (1977) expalined similar increases in the turbulent kinetic energy within the recirculation zone by the presence of large turbulent structures, but the present velocity pdf results shown in figure 4 do not show the directional intermittency necessary for this assumption to hold. Further evidence is given by the skewners and flatness factors, since periodic movements of the recirculation zone would have given rise to broad pdf's characterised by flatness factors well below the Gaussian value of 3. It is noted that the

(a)

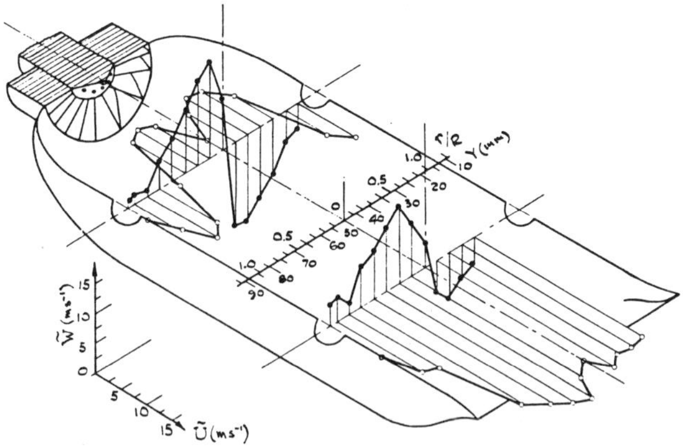

(b)

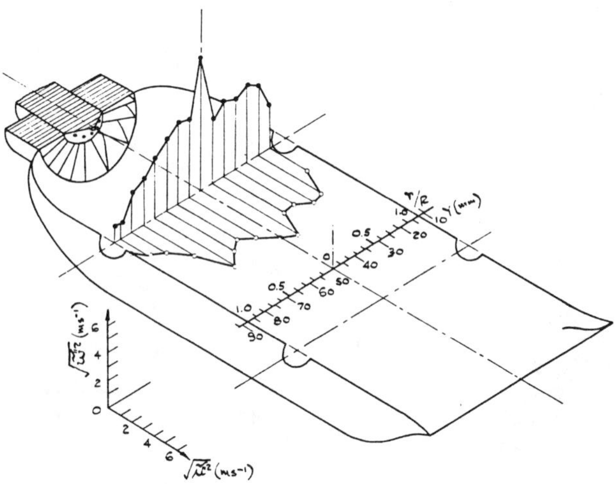

Figure 1 - Mean and turbulent velocity characteristics of the combusting flow (Run 1) within the research combustor of Heitor and Whitelaw (1986).

a) Longitudinal and tangential mean velocities along primary and dilution holes

b) Longitudinal and tangential variances of velocity fluctuations along primary hole.

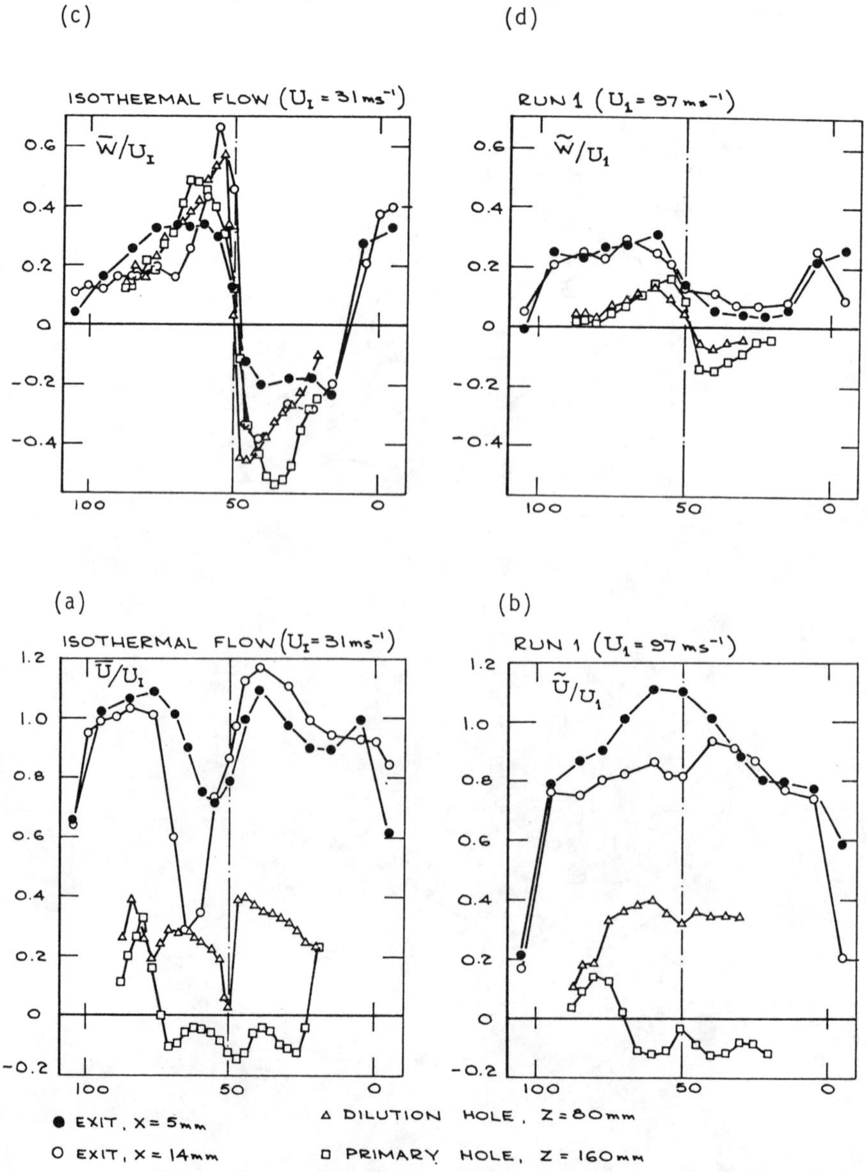

Figure 2 - Normalized mean velocity characteristics of isothermal and combusting (Run 1) flows within the research combustor of Heitor and Whitelaw (1986).
Normalization velocities: isothermal flow, U_I = 31 m/s; combusting flow, U_1 = 97 m/s
a) and b) Longitudinal velocity
c) and d) Tangential (or vertical, at the exit plane) velocity.

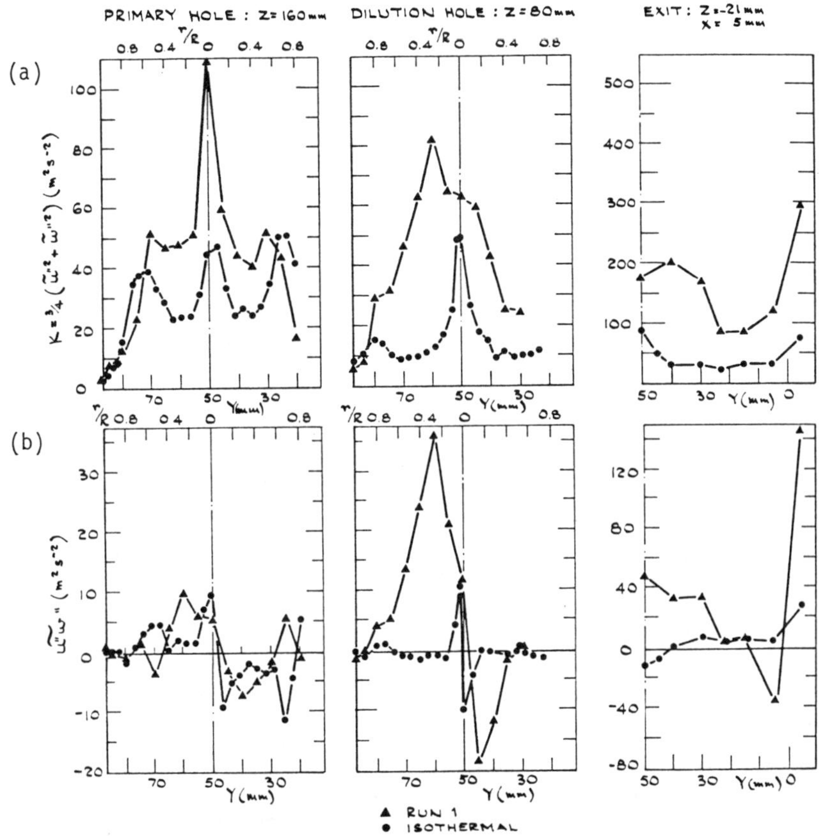

Figure 3 - Turbulent velocity characteristics of isothermal and combusting (Run 1) flows within the research combustor of Heitor and Whitelaw (1986) along primary and dilution holes and at the exit plane.
a) Turbulent kinetic energy
b) Reynolds shear stress.

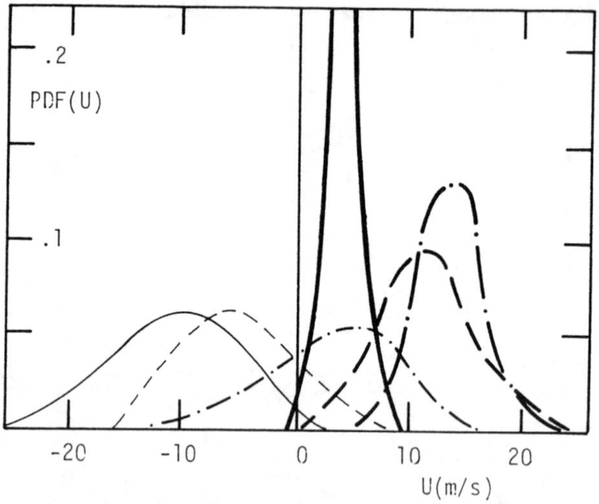

	Y (mm)	S	F
— — —	50:C_L	0.50	2.83
———	65	0.17	3.42
—·—·—	70	-0.09	2.75
— — —	75	-0.14	3.13
·———·	80	0.42	3.77
━━━	87.5	0.08	5.28

Figure 4 - Probability density distributions of the axial velocity component obtained at different radii along the axis of the primary holes for the combusting flow (Run 1) of Heitor and Whitelaw (1986). The table identifies locations and corresponding skewness and flatness factors.

positive values of the skewness factors near the centerline are associated with the shape of the toroidal recirculating zone analysed before. The increase in the turbulent kinetic energy is, therefore, likely to be related to the self-sustained mean radial pressure gradient as suggested by Jones and Whitelaw (1984). In the plane of the dilution holes, however, the shear stress is large compared with isothermal flow and its sign is consistent with mean gradient transport modelling (see Jones 1980).

Measurements in the exit plane show that the longitudinal velocity has accelerated to more than 100 m/s with a maximum in the upper half of the exit plane. The W-velocity values are less symmetric and show a significant movement upward (in the negative X-direction) as a result of the asymmetric contraction from the barrel of the combustor to the exit plane. The asymmetry stems in large part from the influence of the contraction on the swirling flow, and comparison with the isothermal-flow results show that the latter values have generaly been increased by 20 m/s, so that the negative values associated with the isothermal swirl have become positive.

The turbulence intensities are of the order of 13% across the central region of the exit plane and are not influenced by the air-fuel ratio. The intensities decreased to the exit plane, while the turbulent kinetic energy increased by some 50%. Compared with isothermal flow, the turbulent intensities have decreased slightly except in the central zone where the turbulent energy has nearly doubled. The values of $\widetilde{u''w''}$ are of the same order as in the plane of the dilution holes and correspond to around 30% of the local turbulent kinetic energy. The results suggest that the precession of the longitudinal vortex core has been partly attenuated by the acceleration of the hot gases.

Extrapolation of the results to other combustor geometries must be carried out with considerable caution because the complexity of these flows is associated with a strong geometric dependence. In the can-type combustor of Goldstein et al (1983), for example, combustion appeared to reduce the turbulence intensities by about 50% and slightly increased the turbulent kinetic energy levels.

Measured values of mean temperature and major species characteristics are shown in figure 5 along the vertical plane of simmetry. The primary air jets penetrate halfway to the centerline and the secondary jets are confined to the wall region. Partly as a consequence, a rich, high temperature, recirculation zone was formed, which appears to be controlled by the swirl-air, and a second high temperature region existed near the combustor wall and between the two sets of holes. This second region is driven by the flow between the primary jets and is annular. The influence of the swirler is stronger in the center of the combustor and the effects of the primary jets are attenuated so that high temperatures extend into the dilution zone.

Based on the mean temperature distribution measured at the exit plane, the present pattern factors are much higher (about 0.76) than those usually associated with gas-turbine practice (about 0.30) but tend to decrease with the average temperatures in the combustor. In the experiments, an increase in the preheat temperature reduces the pattern factor, whereas an increase in the air-fuel ratio increases it. In practice, the higher pressures associated with take-off and cruise conditions ensure higher temperatures and lower pattern factors.

The distribution of local values of the equivalence ratio calculated from the measured species concentrations is similar to that of mean temperature and show that the combustor can be divided into two regions by the plane of the primary holes.

Upstream of the primary holes, a significant fraction of fuel has broken down to form CO and H_2, and this is evident from the high concentrations of these species and from the low levels of O_2 and the shape of the UHC isolines. The longer residence times associated with the recirculation zone have permitted the breakdown of fuel. The absence of oxygen close to the fueling device suggests a quasi-global decomposition of fuel to CO and H_2 as described by Edelman and Fortune (1969). The increase in the air-fuel ratio resulted in the entrainment of larger concentrations of O_2 near to the center of the combustor, with consequent decrease in the UHC levels from 0.60 to 0.1%. The maximum volume concentrations of CO and H_2 were, respectively, 10 and 8%, and are not affected by the air-to-fuel ratio, indicating that, at least in the primary zone of the combustor, the oxidation reactions for CO and H_2 are likely to be more controlled by physical than kinetic processes. The values of NO_x are very low and are presented because it is unlikely that probe measurements can provide a correct relative measurement of NO/NO_2 concentrations.

Downstream of the primary holes, the effect of the primary jets is dominant. The H_2 levels have dropped to zero and are not significantly affected by the air-fuel ratio. The strong effect of the swirler is still present in the center of the combustor, and the high residence time is responsible for UHC levels below 0.3%. These values are, however, high enough to imply some

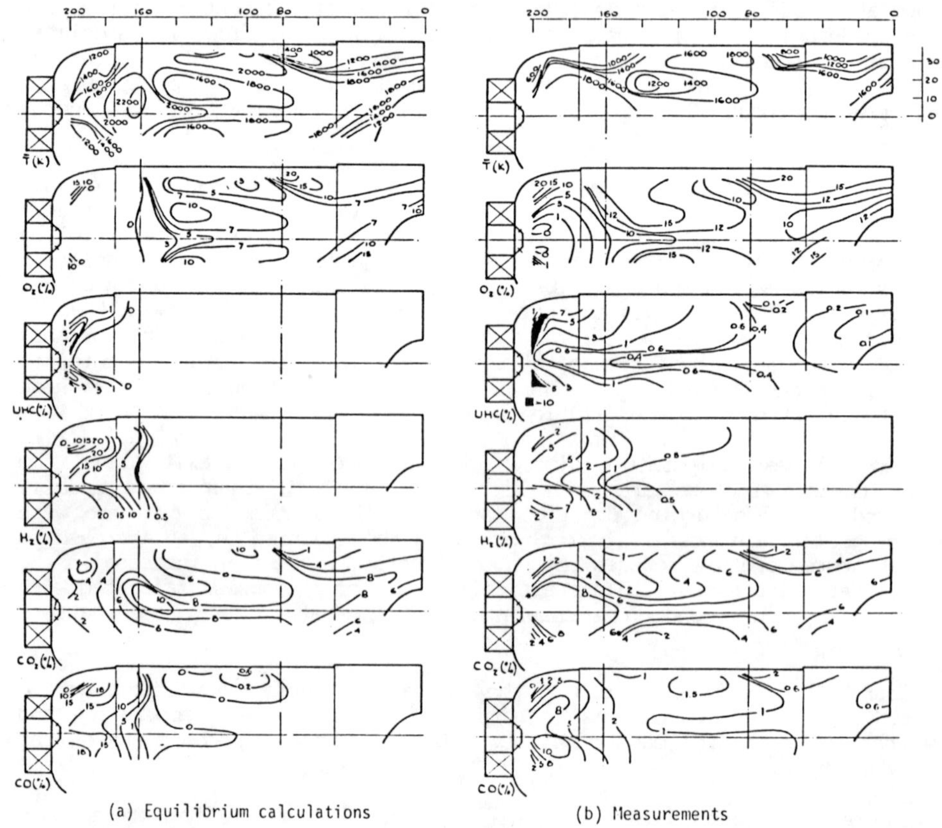

Figure 5 - Contours of mean temperatures and species concentrations (by volume) at the vertical plane of symmetry in combustor of Heitor and Whitelaw (1986). Wet analysis for all the species.
a) Equilibrium calculations (see text)
b) Measurements.

chemical kinetic control for the final stages of the fuel burn-up. Near the combustor wall and due to the peripheric fuel stream from the interprimary ports zones, the levels of UHC are around 1% and the hot region noted closed to the wall has been formed by the combustion of CO, as can be observed from the high levels of CO_2. The maximum volume concentration of CO found near the wall decreased by increasing the air-to-fuel ratio but it is relatively high, suggesting that the CO oxidation is also affected by finite rate kinetics. In the center of the combustor, the maximum concentration of CO_2 decreased with the air-fuel ratio, implying quenching of the CO oxidation by the dilution jets.

At the exit plane the results show that although the absolute concentrations of pollutants are very low, they are several orders of magnitude larger than their equilibrium values. The measurements do clearly indicate that the final stages of fuel and CO burn-up rates are limited by the reaction rates.

b) DISCUSSION

From a practical standpoint the combustion chamber has performed well in that profiles of temperature and concentrations in the exit plane are reasonably symmetric, with a tendency for the velocities to be larger at lower values of X, corresponding to larger blade radii. Operation at the absolute pressures of take-off or even cruise conditions does, of course, increase the effect of radiation, and the present experiment does not convey quantitative information of the resulting wall temperatures. The results are, however, important to assist the development of calculation methods, as summarized in the following paragraphs.

A scalar effective-viscosity turbulence model (e.g. the "k-ε" model) should be adequate for the flow in the dilution section of the combustor since the production of turbulent kinetic energy is largely due to the interaction of shear stress with shear strain. In the upstream recirculation zone, however, the flow is characterized by a large mean radial pressure gradient, and the production of turbulent kinetic energy is not accounted for by mean gradient mechanisms, so that an eddy-viscosity model would underpredict the turbulent energy levels, and is unlikely to provide correct trends for the turbulent transport of the scalar quantities. It should also be remembered that the "k-ε" model is increasingly inappropriate as the degree of swirl is increased: a second order closure is required to improve this situation but may be precluded by computer resources. The same dilemma is likely to exist with the scalar-flux model as described by Jones and Musonge (1983).

The implications of the present results for combustion modelling may be analysed on the basis of the comparisons shown in figure 5 between measurements and calculated values considering chemical equilibrium restricted by a rich flammability limit, as in Godoy (1982).

In the primary zone, the calculated species concentrations are consistent with the absence of oxygen, which is implied in fuel-rich zones by the minimization of the free energy. The estimated low values of CO_2 confirm the "fast" chemistry of the oxidation of CO and suggest that a partial equilibrium model is likely to provide a realistic prediction of the scalar quantities. A laminar flamelet model might, however, improve the calculation of the pollutant concentrations.

Downstream of the primary holes the chemical limitations imposed on the fuel and CO burn-up rates can now be quantified by the excess concentrations measured for these species, mainly in zones of low temperature. A constrained equilibrium model is likely to be required; in particular, the results clearly indicate quenching of the CO to CO_2 reaction by dilution of the secondary air.

4. CONCLUSIONS

Measurements of the velocity and scalar characteristics of turbulent reacting flows relevant to gas-turbine combustors have been reported for many years and are reviewed in the previous sections. The emphasis is on gas flames in practical combustors: the understanding of complexities associated with turbulent processes in two-phase flows undergoing chemical reaction and of those associated with radiative heat transfer will continue to require empiricism and advanced diagnostic techniques have to be further developped and applied to improve their knowledge. Difficulties in applying instrumentation and interpreting results are still experienced mainly due to the limited optical access with high heat release in practical combustors. This paper describes in detail one successful application of current diagnostic techniques to improve knowledge and to provide guidance for modelling of reacting flows typical of gas-turbine combustors.

It is clear from the experiments that practical flows include zones in which turbulence combustion interactions give rise to non-gradient production and to finite rate kinetics effects. Although some of the latter effects may be reduced at the high operating pressures associated with take-off and cruise conditions, two main zones can be distinguished within practical combustors, as follows:

a) In the primary zone the distribution of turbulent kinetic energy is influenced by the interaction between the large radial mean pressure gradient associated with the swirling flow and the density fluctuations. Consequently use of a second order closure model for turbulence may be required to predict the details of the aerodynamic field and the scalar flux. Combustion is controlled more by physical than chemical kinetic processes and the scalar quantities are likely to be represented by a partial equilibrium model: a laminar flamelet model may, however, improve the calculation of the pollutant conditions.

b) In the dilution zone, the production of turbulence relates to conventional mechanisms by the interaction of shear stress with shear strain. The fuel breakdown and the CO oxidation are reaction-rate limited and the scalar field may be described by a constrained equilibrium model.

The extension of the measurements reviewed in this paper to include two-phase flow processes and the effects of radiation characteristic of practical situations,is difficult to achieve, but experimental work should preferably include wall temperature measurements. Additional experiments are also required in simple axisymmetric model combustors to establish the nature of turbulent transport processes in swirling non-premixed flames.

General recommendations with respect to the development of calculation methods based on comparisons with the results reviewed in the previous sections, involve further work on Reynolds stress/flux description of the flow fields.

The detail of the models which can be accommodated in a calculation scheme may, however, be limited by the available computer resources and the use of current numerical features and simple assumptions to represent two-step reactions should be considered before methods based on high-order schemes to calculate gas-turbine combustor flows.

ACKNOWLEDGMENTS

The author is pleased to record many useful discussions with Prof. J.H. Whitelaw: his invaluable criticism and assistance is gratefully acknowledged.

REFERENCES

H. ALTGELD, W.P. JONES and J. WILHELMI (1983). Velocity measurements in a confined swirl driven recirculating flow. Exp. in Fluids, 1, pp.73-78.

K. ATKINSON, Z.A. KHAN and J.H. WHITELAW (1983). Experimental investigation of opposed jets discharging normally into a cross-stream. J. Fluid Mech., 115, pp. 493-504.

A.M. ATTYA and J.H. WHITELAW (1984). Measurements and calculations of preheated and unpreheated confined kerosene spray flames. Comb. Sci. and Tech., 40, pp. 139-145.

J.K. BHANGU, D.M. SNAPE and B.R. EARDLEY (1983). The design and development of a low emissions transply combustor for the civil spey engine. AGARD-CP353, paper No. 23.

A.F. BICEN, M.V. HEITOR and J.H. WHITELAW (1986). Velocity and temperature measurements in a can-type gas-turbine combustor. AGARD-CP-399, paper 14.

A. F. BICEN and W.P. JONES (1987). Velocity characteristics of isothermal and combusting flows in a model combustor. Comb. Sci. and Tech.

A.F. BICEN, J.J. MCGUIRK and J.M.L.M. PALMA (1987). Primary-zone flow characteristics of an isothermal model combustor. Proc. Nato-ASI on Instrumentation for combustion and flow in engines. September 14-26, Vimeiro, Portugal

A.F BICEN, D. TSE and J.H. WHITELAW (1987). Flow characteristics of a model annular combustor. AGARD-CPP-422, paper 14.

T.T. BOWDEN and J.H. PEARSON (1983). The influence of fuel composition upon soot emissions and flame radiation in a model gas-turbine combustor. Paper C70-83 - Inst. Mech. Eng., Oxford - 1983.

F. BOYSAN, W.H. AYERS, J. SWITHENBANK and Z. PAN (1982). Three-dimensional model of spray combustion in gas turbine combustors. J. Energy, 6, pp. 368-375.

R.D. BRUM and G.S.SAMUELSEN (1982). Two-component laser anemometry measurements in a non-reacting and reacting complex flow model combustor. Combustion Institute, Western States Section, paper WSS-C182-53.

R.D. BRUM, E.T. SEILER, J.E. LaRUE and G.S. SAMUELSEN (1983). Instantaneous two-component Laser anemometry and temperature measurements in a complex flow model combustor. AIAA Paper 83-0334.

D.L. BURRUS, W. SHYY and M.E. BRAATEN (1987). Numerical models for analytical predictions of combustor aerothermal performance characteristics. AGARD-CPP-422, paper 25.

M.G. CARVALHO and P.J.M COELHO (1987). Radiation transfer in gas turbine combustors. AGARD-CPP-422, paper 19.

C. CASCI, A. COGHE, V. GHEZZI and S. PASINI (1973). An experimental research on the behaviour of a continuous flow combustion chamber. AGARD-CP-125, paper 26.

R.S. DE CHAIR, C.G.W. SHEPPARD and M. WHITTAKER (1976). A note on carbon monoxide oxidation with particular reference to gas turbine combustion. Comb. Sci. and Tech., 12, pp. 245-247.

P. CHLEBOUN, P.J. CRAIG, F.B. SEBBOWA and C.G.W. SHEPPARD (1982). A study of transverse turbulent jets in a cross flow. Comb. Sci. and Tech., 29, pp. 107-111.

P. CHLEBOUN, S.N. NASSER, F.B. SEBBOWA and C.G.W. SHEPPARD (1983). An investigation of the interaction between multiple dilution jets and combustion products. AGARD-CP-353, Paper 26.

A.E. CLARKE, A.J. GERRARD and L.A. HOLLIDAY (1963). Some experiences in gas turbine combustion chamber practice using water flow visualisation techniques. 9th Symposium (Intl.) on Combustion, 1963, pp. 878-891.

S.M. CORREA(1984) . Prediction of an axisymmetric combusting flow. AIAA J., 22, pp. 1602-1608.

J. COUPLAND and C. H. PRIDDIN (1986). Modelling the flow and combustion in a production gas turbine combustor. Turbulent Shear Flows-5, Springer-Verlag.

D. CRABB and J.H. WHITELAW (1979). The influence of geometric asymmetry on the flow downstream of a row of jets discharging normally into a free stream. J. Heat Transfer, 10, 183-185.

R.R. DILLS, and P.S. FOLLANSBEE (1979). Use of thermocouples for gas temperature measurements in gas turbine combustor. NBS Special Publ. 561: Proc. 10th Materials Res. Symp. on Characterisation of High Temperature Vapor and Gases, Maryland - Sept., 1978.

A.C. ECKBRETH, G.M. DOBBS, J.H. STUFFLEBEAM and P.A. TELLEX (1984). CARS temperature and species measurements in augmented jet engine exhausts. Applied Optics, 23, pp. 1328-1339.

R.B. EDELMAN and O.F. FORTUNE (1969). A quasi-global chemical kinetic model for the finite rate combustion of hydrocarbon fuels with application to turbulent burning and mixing in hypersonic engines and nozzles. AIAA Paper No. 69-86.

R.B. EDELMAN, P.T. HARSHA and S.N. SCHMOTOLOCHA (1981). Modelling techniques for the analysis of ramjet combustion processes. AIAA J., 19, pp. 601-609.

Y. EL BANHAWY and J.H. WHITELAW (1980a). Assessment of an approach to the calculation of the flow properties in spray flames. Combustor Modelling. AGARD-CP-275, Paper No. 12.

Y. EL BANHAWY and J.H. WHITELAW (1980b). Calculation of flow properties of a confined kerosene spray flame. AIAA J. 18, pp. 1503-1510.

Y. EL-BANHAWY and J.H. WHITELAW (1981). Experimental study of the interaction between a fuel spray and surrounding combustion air. Comb. and Flame 42, pp. 253-275.

M.P. ESCUDIER, J. BORNSTEIN and N. ZEHNDER (1980). Observations and LDA measurements of confined turbulent vortex flow. J. Fluid Mech., 98, pp. 49-63.

M.P. ESCUDIER and J.J. KELLER (1985). Recirculation in swirling flow: a manifestation of vortex breakdown. AIAA J., 23, pp. 111-116.

M.P. ESCUDIER and N. ZEHNDER (1982). Vortex-flow regimes. J. Fluid Mech., 115, pp. 105-121.

B.C.R. EWAN, F. BOYSAN and J. SWITHENBANK (1984). Closing the gap between finite difference and stirred reactor combustor modelling procedures. 20th Symposium (Intl.) on Combustion.

J.H. FALER and S. LEIBOVICH (1978). An experimental map of the internal structure of a vortex breakdown. J. Fluid Mech., 86, pp. 313-335.

R.S. FLETCHER and J.B. HEYWOOD (1971). A model for nitric oxide emissions from aircraft gas turbine engines. AIAA Paper No. 71-123. AIAA 9th Aerospace Sciences Meeting, New York, N.W.-Jan./25-27, 1971.

S. FLETCHER and R.D. SIEGEL (1971). The control of oxides of nitrogen emissions from aircraft gas turbines. NREC Report 1162-2. Also, Federal Aviation Authority report FAA-RD-71-111-2.

S. FLETCHER, R.D. SIEGEL and E.K. BASTRESS (1971). The control of oxides of nitrogen emissions from aircraft gas turbines. NREC report 1162-1. Also, Federal Aviation Authority report FAA-RD-71-111-1.

C.O. FOLAYAN and J.H. WHITELAW (1977). Impingement cooling and its application to combustor design. Proc. Tokyo Joint Gas Turbine Congress, pp. 69-76.

N.J. FRISWELL(1979). The influence of fuel composition on smoke emissions from gas-turbine-type combustor: effects of combustor design and operating conditions. Comb. Sci. and Tech., 9, pp. 119-127.

S. GODOY (1982). Turbulent diffusion flames. Ph.D. Thesis, University of London.

R.J. GOLDSTEIN, K.Y. LAU and C.C. LEUNG (1983). Velocity and turbulence measurements in combustion systems. Exp. in Fluids, 1, pp. 93-99.

F.C. GOULDIN, J.S. DEPSKY and S.L. LEE (1985). Velocity field characteristics of a swirling flow combustor. AIAA, 23, pp. 95-102.

A.S. GREEN and J.H. WHITELAW (1980). Measurement and calculations of the isothermal flow in axisymmetric models of combustion geometries. J. Mech. Eng. Sci. 22, pp. 119-124.

A.S. GREEN and J.H. WHITELAW (1983). Isothermal models of gas-turbine combustors. J. Fluid Mech. 126, pp. 399-412.

W.L.H. HALLET and R. GUNTHER (1983). The turbulent structure of swirlling flow in a sudden expansion. Proc. 4th Symp. Turb. Shear Flows, pp. 19.25-19.31. Karlsruhe, Sept.-1983.

R.N. HALTHORE and F.C. GOULDIN (1984). Laser scattering measurements for density in a swirling flow combustor. AIAA-84-0199. AIAA: 22nd Aerospace Sciences Meeting, Jan. 9-12-Reno, Nevada.

M.V. HEITOR(1985). Experiments in turbulent reacting flows. Ph. D. Thesis, University of London.

M.V. HEITOR (1988). The simultaneous measurement of velocity and scalar properties. In "Instrumentation for combustion and flow in engines", ed. Durão et al, Martinus Nijhof Publ.

M.V. HEITOR, A.M.K.P. TAYLOR and J.H. WHITELAW (1985). Simultaneous velocity and temperature measurements in a premixed flame. Exp. in Fluids, 3. Also, Proc. ASME 105th Annual Winter Meeting, New Orleans, Dec. 9-13 (1984).

M.V. HEITOR and J.H. WHITELAW (1986). Velocity, temperature and species characteristics of the flow in a gas-turbine combustor. Comb. and Flame, 64, pp. 1-32.

T.A. JACKSON and G.S. SAMUELSEN (1984). An evaluation of fuel spray performance in a swirl stabilised combustor using optical methods for drop sizing. Paper AIAA-84-1318, AIAA-SAE-ASME, 20th Joint Propulsion Conference, June, 11-13, 1984-Cincinnati - Ohio.

S.I. JANJUA, D.K. MCLAUGHLIN, T.W. JACKSON and D.G. LILLEY (1983). Turbulence measurements in confined jets using a rotating single-wire probe technique. AIAA J., 21, pp. 1609-1610.

A.K. JASUJA (1983). Effects of airblast atomiser design upon spray quality. AGARD-CP-353, Paper 12.

R.E. JONES(1978). Gas turbine engine emissions - problems, progress and future. Prog. Energy Combust. Sci., 4, pp. 73-113.

W.P. JONES (1980). Models for turbulent flows with variable density and combustion. In: Prediction methods for turbulent flows, ed. W. Kollman, Hemisphere Pub. Corp., pp. 379-421.

W.P. JONES, W.C. CLIFFORD, C.H. PRIDDIN and R. DE CHAIR (1977). A comparison between predicted and measured species concentrations and velocities in a research combustor. Proc. AGARD 50th Meeting, Turkey. Paper 40.

W.P. JONES and J.J. MCGUIRK (1980). Mathematical modelling of gas-turbine combustion chamber. AGARD-CP-275. Paper 4.

W.P. JONES and J.J. MCGUIRK (1981). A comparison of two droplet models for gas turbine combustion chamber flows. Proc. 5th ISABE, Bangalore, India. Paper 49.

W.P. JONES and P. MUSONGE (1983). Modelling of scalar transport in homogeneous turbulent flows. 4th Symposium of Turbulent Shear Flows, pp. 17.18-17.24. Karlsruhe, Sept.-1983.

W.P. JONES and C.H. PRIDDIN (1979). Prediction of the flow field and local gas combustion in gas turbine combustors. 17th Symposium (Intl.) on Combustion, pp. 399-409.

W.P. JONES and H. TORAL (1983). Temperature and composition measurements in a research gas turbine combustor. Comb. Sci. and Tech., 31, pp. 249-275.

W.P. JONES and J.H. WHITELAW (1984). Modelling and measurements in turbulent combustion. 20th Symposium (Intl.) on Combustion. The Combustion Institute. Invited Paper.

F. JOOS and B. SIMON (1987). Comparison of the performance of a reverse flow annular combustion chamber under low and high pressure conditions. AGARD-CPP 422, paper 13.

M. KATSUKY, Y.MIZUTANI and K.SHIBUYA (1976). Emissions from gas turbine combustors. Bull. JSME, 19, pp. 1353-1359.

J.B. KENNEDY (1974). Ramburner flow visualisation studies. Proc. 11th JANNAF Combustion meeting, II, Publ. 261 - Chemical Propulsion Inf. Agency, pp. 415-440.

Z.A. KHAN, J.J. MCGUIRK and J.H. WHITELAW (1981). A row of jets in a cross flow. AGARD-CP-308, Paper 10.

Z.A. KHAN and J.H. WHITELAW (1980a). Mean velocity and concentration characteristics of rows of jets in a cross flow. J. Heat Transfer, 102, pp. 391-392.

Z.A. KHAN and J.H. WHITELAW (1980b). Vector and scalar characteristics of opposing jets discharging normally into a cross-stream. Int. J. Heat Mass Transfer, 23, pp. 1673-1680.

K. KOMIYAMA, R.C. FLAGAN and J.B. HEYWOOD (1977). The influence of droplet evaporation on fuel-air mixing rate in a burner. 16th Symposium (Intl.) on Combustion, pp. 549-560.

P. KOUTMOS and J.J. MCGUIRK (1985). Investigation of swirler-dilution jet flow split on primary zone flow patterns in a water model can-type combustor. ASME paper 85-GT-180.

D. KRETSCHNER and J. ODGERS (1987). The characterization of combustor by fuel composition-measurements in a small conventional combustor. AGARD-CPP-422, paper 10.

J.C. LaRUE, G.S. SAMUELSEN and E.T. SEILER (1984). Momentum and heat-flux in a swirl-stabilised combustor. 20th Symposium (Intl.) on Combustion. The Combustion Institute.

A.M. LEFEBVRE and R.S. FLETCHER (1973). A preliminary study on the influence of fuel staging on nitric oxide emissions from gas turbine combustors. AGARD : CP-125, paper 30.

A.H. LEFEBVRE (1983). Gas turbine combustion. McGraw-Hill, Series in energy, combustion and environment.

S. LEIBOVICH (1983). Vortex stability and breakdown. AGARD CP-342, paper 23.

S. LEIBOVICH (1984). Vortex stability and breakdown: survey and extension. AIAA J., 23, pp. 1192-1206.

M.H. LEWIS and L.D. SMOOT (1981). Turbulent gaseous combustion. Part I: local species concentrations measurements. Comb. and Flame, 42, pp. 183-196.

P.A. LIBBY, S. SIVASEGARAM and J.H. WHITELAW (1986). Premixed combustion. Prog. Energy Combust. Sci., 12, pp.393-405.

A.J. LIGHTMAN, R.D. RICHMOND, L. KRISHAMURTHY, P.D. MAGILL, W.M. ROQUEMORE, R.P. BRADLEY, J.S. STUTRUD and C.M. REEVES (1980). Velocity measurements in a bluff-body diffusion flame. Paper AIAA-80-1544: AIAA - 15th Thermophysics Conference, July 14-16, 1980 - Snowmass, Colorado.

F.C. LOCKWOOD, J.J. MCGUIRK and N.G. SHAH (1983). Radiation transfer in gas turbine combustors. AIAA Paper 83-1506.

H.C. LOW (1983). Recent research on the efflux of the ROLLS-ROYCE vaporiser fuel injector. AGARD-CP-353: paper 11.

J.J. MCGUIRK and J.H. WHITELAW (1983). Internal flows of relevance to gas-turbines. Imperial College, Mech. Eng. Dept. Report FS/83/25.

P.D. MAGILL, A.J. LIGHTMAN, C.E. ORR, R.P. BRADLEY and W.M. ROQUEMORE (1982). Simultaneous velocity and emission measurement in a bluff-body combustor. AIAA Paper 82-0883.

A.M. MELLOR (1976). Gas turbine engine pollution. Prog. Energy Combust. Sci., 1, pp. 111-133.

A.M. MELLOR (1980). Semi-empirical correlations for gas turbine emissions, ignition, and flame stabilization. Proc. Energy Combust. Sci., 6, pp. 347-358.

A.M. MELLOR, R.D. ANDERSON, R.A. ALTENKIRCH and J.H. TUTTLE (1972). Emissions from and within an Allisons J-33 combustor. Comb. Sci. and Tech., 6, pp. 169-176.

A.R. MORR, J.B. HEYWOOD and A.H. FITCH (1975). Measurements and predictions of carbon monoxide emissions from an industrial gas turbine. Comb. Sci. and Tech., 11, pp. 97-109.

D.W. NAEGELI, L.G. DODGE and C.A. MOSES (1983). Effects of flame temperature and fuel composition on soot formation in gas turbine combustors. Comb. Sci. and Tech., 35, pp. 117-131.

S. NAKAMURA, O. KAWAGUGHI and G.T. SATO (1983). A study on the combustion in the primary zone region of a gas turbine-type continuous combustion chamber. Gas Turbine Society of Japan. 1983 Tokyo Intl. Gas Turbine Congress: Paper 83-TOKYO-IGTC-30, pp. 229-236.

D.A. NEALY (1980). Combustor cooling - Old problems and new approaches. In: Gas Turbine Combustor Design Problems, (ed. A.H. Lefebvre), pp. 151-185. Hemisphere Pub. Corp.

J.A. NICHOLLS, C.W. KAUFMAN, D.G. PELACIO, D.R.GLASS and J.F. DRISCOLL (1980). The effect of fuel sprays on emissions from a research gas turbine combustor. Comb. Sci. and Tech., 23, pp. 203-213.

J.R. NOYCE and C.G.W. SHEPPARD (1982). The influence of equivalence ratio variation on pollutant formation in a gas-turbine type combustor. Comb. Sci. and Tech., 29, pp. 37-52.

J.R. NOYCE, C.G.W. SHEPPARD and F.D. YAMBA (1981). Measurements of mixing and species concentrations within a gas turbine type combustor. Comb. Sci. and Tech., 25, pp. 209-217.

Y. ONUMA and M. OGASAWARA (1975). Studies on the structure of a spray combustion flame. 15th Symposium (Intl.) on Combustion, The Combustion Institute, pp. 453-465.

Y. ONUMA, M. OGASAWARA and T. INOVE (1977). Further experiments on the structure of a spray combustion flame. 16th Symposium (Intl.) on Combustion, The Combustion Institute, pp. 561-567.

M.J. OVEN, F.C. GOULDIN and W.J. McLEAN (1979). Temperature and species concentration measurements in a swirl-stabilised combustor. 17th Symposium (Intl.) on Combustion, The Combustion Institute, pp. 363-374.

F.K. OWEN (1977). Measurements of instantaneous velocity and concentration in complex mixing flows. AIAA Progress in Astronautics and Aeronautics, Vol. 53, (ed. B.T. Zinn et al), pp. 293-303.

F.K. OWEN (1978). Measurements in combustion systems. Proc. 3rd Intl. Workshop on Laser Velocimetry, pp. 123-135, Purdue Univ., 1978.

F.K. OWEN, L.J. SPADACCINI, J.B. KENNEDY and C.T. BOWMAN (1979). Effects of inlet air swirl and fuel volatility on the structure of confined spray flames. 17th Symposium (Intl.) on Combustion, The Combustion Institute, pp. 467-473.

F.K. OWEN, L.J. SPADACCINI and C.T. BOWMAN (1977). Pollutant formation and energy release in confined turbulent diffusion flames. 16th Symposium (Intl.) on Combustion. The Combustion Institute, pp. 105-117.

P.G. PARIKH, R.F. SAWYER and A.L. LONDON (1971). Pollutants from methane fueled gas turbine combustion. Univ. California-Berkeley. Rep. No. TS-70-15, Jan. 1971.

N. PETERS (1984). Laminar diffusion flamelet models in non-premixed turbulent combustion. Prog. Energy Combust. Sci., 10, pp. 319-339.

S.L. PLEE and A.M. MELLOR (1978). Review of flashback reported in prevaporising/premixing combustors. Comb. and Flame, 32, pp. 193-203.

D.S. PRIOR, J. SWITHENBANK and P.G. FELTON (1978). Stirred reactor modelling of a low pollution liquid fuelled combustor. In Progress in Astronautics and Aeronautics, 58, Ed. L.A. Kennedy, pp. 351-372.

C.L. PROCTOR and A.M. MELLOR (1982). Numerical and experimental examination of a prevaporised-premixed combustor. Paper AIAA-82-1074. 18th Joint Propulsion Conference: AIAA SAE ASME June 21-23, 1982, Cleveland, Ohio.

D.L. RHODE, D.L. LILLEY and D.K. MCLAUGHLIN (1983). Near flow fields in axisymmetric combustor geometries with swirl. AIAA J., 21, pp. 593-600.

N.K. RIZK, R.S. FLETCHER and R.C. ADKINS (1981). The application of fluidics to control emission from gas turbine combustors. Proc. ASME Fluids Eng. Conference, pp. 233-239, Colorado, 1981.

N.K. RIZK and A.H. LEFEBVRE (1982). Air blast atomization studies on drop size atomization. J. Energy, 6, pp. 323-327.

W.M. ROQUEMORE, R.P. BRADLEY, J.S. STUTRUD, C.M. REEVES, C.A. OBRINGER and M.L. BRITTON (1983). Utilisation of laser diagnostics to evaluate combustion models. AGARD: CP-353, paper 36.

G.S. SAMUELSEN, C.P. WOOD and T.A. JACKSON (1983). Optical measurements of soot size and number density in a complex flow, swirl-stabilised combustor. AGARD-CP-353, paper 21.

R.F. SAWYER, N.P. CERNANSKY and A.K. OPPENHEIM (1973). Factors controlling pollutant emissions from gas turbine engines. AGARD: CP-125, paper 22.

R.F. SAWYER, D.P. TEIXEIRA and E.S. STARKMAN (1969). Air pollution characteristics of gas turbine engines. Trans. ASME A: J. Eng. Power, 91, pp. 290-296.

M.A. SERAG-ELDIN and D.B. SPALDING (1978). Computations of three-dimensional gas turbine combustion chamber flows. ASME 23rd Intl. Gas Turbine Conference, London.

C.G.W. SHEPPARD (1975). A simple model for carbon monoxide oxidation in gas turbine combustors. Comb. Sci. and Tech. 11, pp. 49-56.

R.A. SHISLER, J.H. TUTTLE and A.M. MELLOR (1975). Emissions from and within a film-cooled combustor. Comb. Sci. and Tech., 11, pp. 153-160.

S. SIVASEGARAM and J.H. WHITELAW (1984). Flow characteristics of opposing rows of jets in a confined space. Fluids Section Report FS/84/26, Mech. Eng. Dept., Imperial College, London.

B. SIMON, D. SCHUBERT and V. BASLER (1983). Advanced combustor liner cooling concept. AGARD-CP-353, paper 24.

P.J. SMITH and L.D. SMOOT (1981). Turbulent gaseous combustion part II: theory and evaluation for local properties. Comb. and Flame, 42, pp. 277-285.

L.J. SPADACINI, J.B. McVEY, J.B. KENNEDY, F.K. OWEN and C.T. BOWMAN (1977). Pollutant formation and energy release in liquid-fuel turbulent diffusion flames. AIAA paper 77-53, AIAA 15th Aerospace Sciences Meeting, Los Angeles, Calif-Jan. 25-26, 1977.

E.S. STARKMAN, Y. MIZUTANI, R.F. SAWYER and D.P. TEIXEIRA (1971). The role of chemistry in gas turbine emissions. Trans. ASME, Ser. A, 93, pp. 333-344.

K. SUZUKI, A. ISHII, S. HORIUCHI, T. SAITO, M. MATSUKI and T. TORISAKI (1978). Design and development of a high-pressure combustor for aero-gas-turbines. Trans. ASME: J. Eng. Power, 100, pp. 129-235.

J. SWITHENBANK, A. TURAN and P.G. FELTON (1978). Three-dimensional two-phase mathematical modelling of gas turbine combustors. In: Gas Turbine Combustor Design Problems. (ed. A.H. Lefebvre), pp. 249-314. Hemisphere Pub. Corp., 1980.

T. TAMARU and Y. KUROSAWA (1984). Evaluation of thermocouple and thermal radiation instrumentation for measuring gas turbine combustor liner wall temperature. Bull. JSME, 27, pp. 2470-2475.

J.H. TILSTON and J.F.B. HAKLUYTT (1983). The design and performance of a combustor with a multiple jet primary zone. AGARD-CP-353, paper 17.

H. TORAL and J.H. WHITELAW (1982). Velocity and scalar characteristics of the isothermal and combusting flows in a combustor sector rig. Comb. and Flame, 45, pp. 251-272.

J.H. TUTTLE, R.A. ALTENKIRCH and A.M. MELLOR (1973). Emissions from and within an Allison J-33 combustor - II: The effect on inlet temperature. Comb. Sci. and Tech., 7, pp. 125-134.

J.H. TUTTLE, R.A.SHISLER and A.M. MELLOR (1976). Investigation of liquid fueled turbulent diffusion flames. Comb. Sci. and Tech., 14, pp. 229-241.

A. VRANOS and E.F. TABACK (1976). Combustion product distributions in the primary zone of a gas turbine combustor. Comb. and Flame, 46, pp. 129-131.

B.T. VU and F.C. GOULDIN (1982). Flow measurements in a model swirl combustor. AIAA J., 20, pp. 642-651.

A.B. WASSEL and J.K. BHANGU (1980). The development and application of improved combustion wall cooling techniques. ASME 80-GT-66, March, 1980.

D.R. WESKE and G.Y. STUROV (1974). Experimental study of turbulent swirled flows in a cylindrical tube. Fluid Mech.-Sov. Res., 3, pp. 77-82.

P.N. WILD, F. BOYSAN, J. SWITHENBANK and X. LU (1987). 3-D dimensional gas turbine combustor modelling. AGARD-CPP-422, paper 27.

J. WILHELMI(1984). Axisymmetric swirl stabilised combustion. Ph.D. Thesis, University of London.

S. WITTIG, O.M.F. ELBAHAR and B.E. NOLL (1984). Temperature profile development in turbulent mixing of coolant jets with a confined hot crossflow. J. Engng. for Gas Turbines and Power, 106, 193.

S. WITTIG, H.J. BAUER and B. NOLL (1987). On the application of finite-difference techniques for the computation of the flow field in gas turbine combustors with complex geometries. AGARD-CPP-422, paper 28.

C.P. WOOD, R.A. SMITH and G.S. SAMUELSEN (1984). Spatially-resolved measurements of soot size and population in a swirl-stabilised combustor. Presented at the 20th Symposium (Intl.) on Combustion.

TABLE 1. 1 - GAS FLAMES IN SIMPLE MODEL COMBUSTORS (continued).

REFERENCE	EXPERIMENT	MEASUREMENT TECH.	FINDINGS
Owen et al (1977) Owen (1977a, 1978) 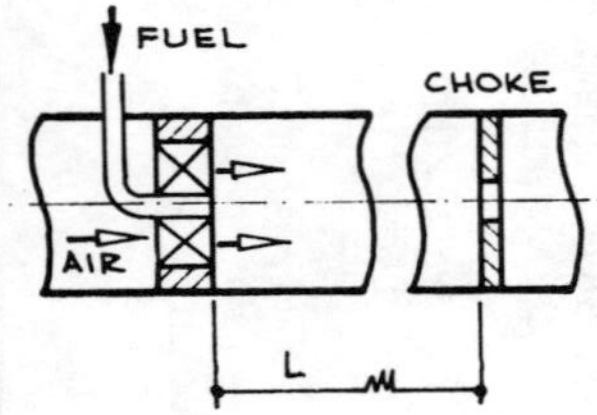 Calculations in Jones & McGuirk (1980)	D = 122 mm; L = 1878 mm T_o = 750 K P = 1 to 3.8 atm S = 0.0 to 0.6 METHANE; ϕ_m = 0.9 fuel/air velocity = 0.05 to 5.8.	U,W: LDV, 0.25x2.0 (mm); Counter processor; microballoons; d_N = 5 μm. T: thermocouple. O_2,UHC,CO_2,CO,NO_x: water-cooled probe.	Variation of air swirl, S, combustor pressure, P, and fuel/air velocity produces major changes in velocity and scalar characteristics of the combustor. High velocity fluctuations and deviations from Gaussian distributions in the initial regions are associated with large scale motions. The introduction of swirl modifies R.Z. (from ellipsoidal to toroidal), reduces reverse velocities and increases R.Z. width, stability and residence time. Consequently increases $\overline{T}$($\overline{T}_{MAX}$ < 2200 K) and NO_x(<300 ppm) and decreases UHC (always < 9%). At low swirl (0.3), increasing P increases R.Z. length, width and reverse velocities and causes a significant decrease in the turbulent energy. UHC is also decreased and NO_x significantly increased.
Brum & Samuelsen (1982) LaRue et al (1984) 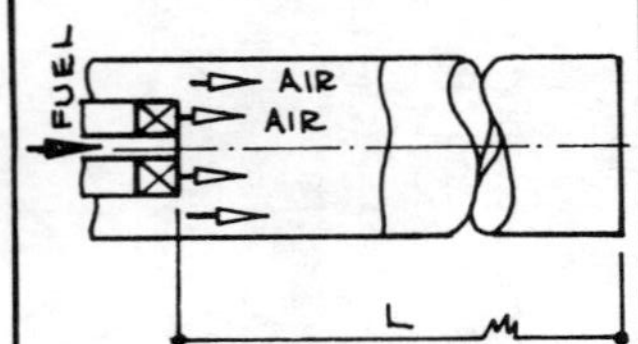 (also tables 1.7.2-A and 1.8.2)	D = 80 mm; L = 50 mm T_o = 300 K P ≈ 1 atm S = 0.8 Propane, ϕ_m = 0.1; 0.2 (Isothermal runs with CO_2)	U,W: LDV, (two-channel); d = 0.36 mm; V = 0.022 mm^3 at 20°; Alumina, d_N = 1 μm; Counter processors. T: thermocouple, Pt/Pt:10% Rd; d_W = 25; 50 μm; electrically compensated; uncoated; no radiation correction.	Combusting flows are characterised by wider and shorter R.Z. and higher reverse velocities than the isothermal flows. The r.m.s. is also increased (≈ 50%). Similar effects were observed increasing ϕ. The turbulence is isotropic inside the R.Z. with a peak near stagnation. In downstream regions $\overline{w'^2} > \overline{u'^2}$. It is suggested "large scale intermittency" and a "spiralling action of the vortex core", which reduces $\overline{u'w'}$. Temperature is uniform in R.Z. (≈ 2020 K) and about 200 K below T_{ad} (for ϕ = 1). In downstream regions near the centre line, $\partial\overline{T}/\partial x$ <0 and $\overline{u'T'}$< 0, indicating non-gradient transport, explained by the effect of swirl.

TABLE 1. 1 - GAS FLAMES IN SIMPLE MODEL COMBUSTORS (continued).

REFERENCE	EXPERIMENT	MEASUREMENT TECH.	FINDINGS
Wilhelmi (1984) (Isothermal flow: Altgeld et al, 1983) 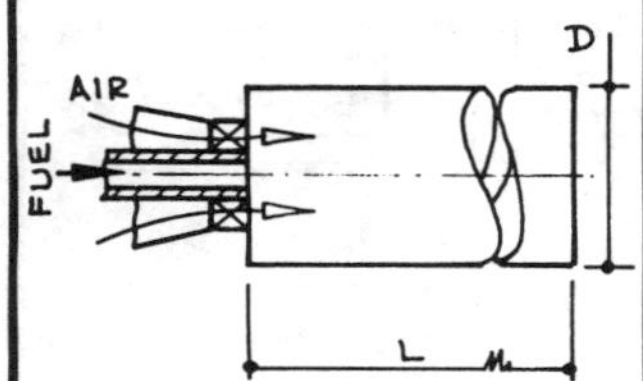 	D = 100 mm; L = 300 mm T_o = 310 K P = 1 atm S = 0.78 Propane, ϕ_m = 0.66.	U,V,W: LDV 0.17x2.84 (mm); Titanium dioxide, $d_N < 0.8$ μm; Counter processor. T: thermocouple, Pt/Pt: 13% Rd; d_W = 40 μm; uncoated; no radiation correction. O_2,UHC,H_2,CO_2,CO: water cooled probe - O.D.= 8 mm; I.D. = 1 mm.	Combustion reduces both R.Z. length (by 10%) and mass flow recirculated and increases its width and the reverse velocity (by a factor of 4). The turbulent energy levels are also increased, with maximum values in the shear layer near the stagnation zone. Within the R.Z., combustion can be assumed as complete ($O_2 \approx 2\%$; UHC < 0.1%). Maximum CO (≈ 4%) and H_2 (≈1%) were found in the shear layer. Near the walls, fuel breakdown and CO to CO_2 reaction are slower (CO reaches 1% and $T \approx 1400$ K). Calculations with the "k-ε" and equilibrium models show longer R.Z., large errors in $\overline{W}$ and that mixing is strongly underpredicted.
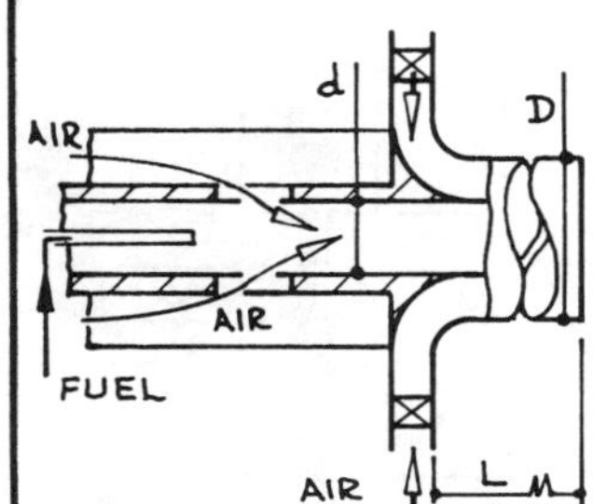 Gouldin et al (1985) Halthore & Gouldin (1984) Oven et al (1979)	d = 51 mm; D = 102 mm; L = 6.0 D T_o = 300 K P = 1 atm S = { inner - S_i = 0.50; outer - S_o = ±0.56 } Methane/Air (Premixed flame) ϕ_m = 0.23.	U,W: LDV 0.5 x 0.2 (mm); Alumina, d_N = 1 μm; Counter processor. T: thermocouple, Pt/Pt: 10% Rd; d_W = 76 μm; silica coated; no radiation correction. O_2,N_2,UHC,CO_2,CO,NO_x: water cooled probe, O.D. = 6.4 mm (Probes inserted radially). ρ: laser Rayleigh scattering 1 x 0.1 (mm); f_S = 5 kHz. $\overline{\rho}$: laser Raman scattering 1 x 0.12 x 0.18 (mm).	Swirl affects both mean and turbulent fields. For both isothermal and combusting flows, going from counter- to co-swirl decrease the length of the R.Z. and the turbulence levels. In isothermal flow, co-swirl suppresses recirculation. Combustion slightly increases the recirculation length but causes a significant increase in the reverse velocity. Velocity fluctuations are comparable in both hot and cold flows and, together with density fluctuations, have maximum values near the baffle boundary, indicating periodic oscillations. For all the flows, the scalar measurements reveal high CH_4 (6-7%), CO(11-15%) and NO_x and low combustion efficiencies. Co-swirl is associated with large gradients, low turbulent transport and increased combustion efficiencies.

TABLE 1. 1 - GAS FLAMES IN SIMPLE MODEL COMBUSTORS (concluded).

REFERENCE	EXPERIMENT	MEASUREMENT TECH.	FINDINGS
Lewis & Smoot (1981) Smith & Smoot (1981) 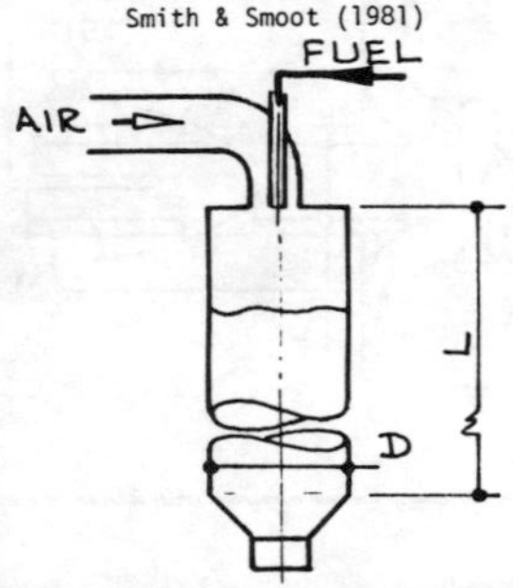	D = 203 mm; L = 1524 mm T_o = 300 to 644 K P = 1 atm METHANE ϕ_m = 1.36 Wall temperature: 540 to 1140 K	O_2,N_2,UHC,H_2,CO_2,CO: water-cooled probe and water-quenched probe T: thermocouple, Pt/Pt:10% Rd.	Direct water-quench probe is more effective in quenching reactions inside the probe. Differences in $\tilde{f}$ are, however, not significant. Mixing rates are unaffected increasing T_W, but CO_2, CO and UHC levels are slightly decreased. More complete combustion is achieved increasing T_o, but mixing is reduced due to increased air-fuel velocity ratios (constant mass flow rate). Predictions with "k-ε" and "equilibrium" models have been shown to be sensitive to grid size and boundary conditions. H_2 and CO are overpredicted by up to 50% (ϕ > 1.).
Lightman et al (1980) Magill et al (1982) Roquemore et al (1983) Correa (1984) 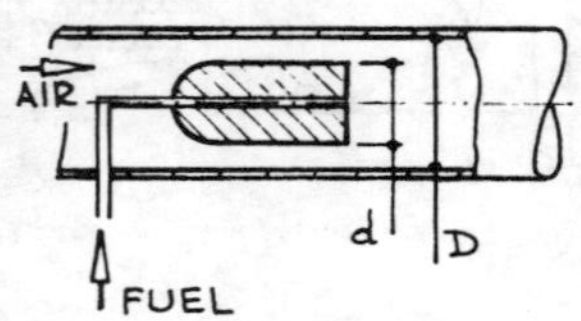	D = 254 mm; d = 140 mm; BR = 30% T_o = 300 K P = 1 atm PROPANE, ϕ_m = 0.026 to 0.035 (cold flow tests with CO_2).	U: LDV, 0.075x0.5 (mm); Titanium dioxide, d_N = 0.2 μm; Counter processor. T, N_2, O_2: CARS 1 mm^3; f_s = 10 Hz $\overline{T}$: shielded thermocouple d = 6.35 mm High speed cine pictures: up to 8000/s.	The fuel and air stagnation points move downstream for combusting flows as compared to cold flows. For a fixed fuel flow rate, the recirculation length decreases as the air flow rate is increased. Turbulence intensity peaks near the fuel stagnation point and decreases abruptly further downstream. Simultaneous LDV/luminous-emission-intensity suggest the existence of large scale "flame turbules". Calculations based on the "k-ε" model overpredict size and strength of R.Z. and do not indicate the measured peak in the turbulent energy near the first stagnation point.

TABLE 1. 2 - SPRAY FLAMES IN SIMPLE MODEL COMBUSTORS (continued).

REFERENCE	EXPERIMENT	MEASUREMENT TECH.	FINDINGS
Komiyama et al (1977) 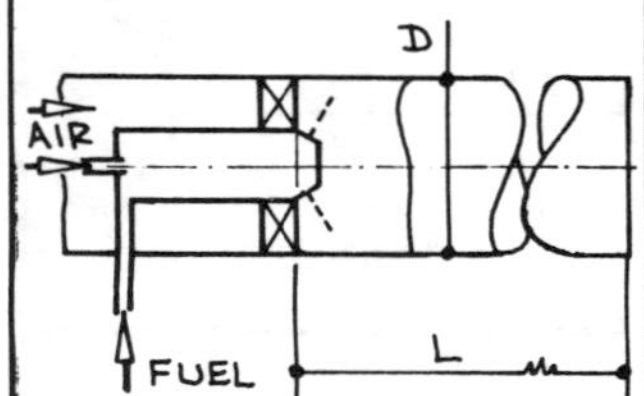 	D = 100 mm; L = 610 mm Air blast and pressure atomiser. T_o = 300 K P = 1 atm {KEROSENE, PENTANE, ISO-OCTANE PROPANE (GAS) ϕ_m = 1 SMD = 10 to 60 μm	O_2,UHC,CO_2,CO,NO_x: water cooled probe.	With air-blast atomisers (SMD = 10 to 20 μm) the droplet evaporation time is much less than the mixing time. Thus the flames are mixing-controlled. With pressure jet atomiser (SMD = 40 to 60 μm) those times are comparable and both the kinetic energy of the fuel jet and the droplet characteristics control the initial fuel-air mixing rate.
Owen et al (1979) Spadacini et al (1977) 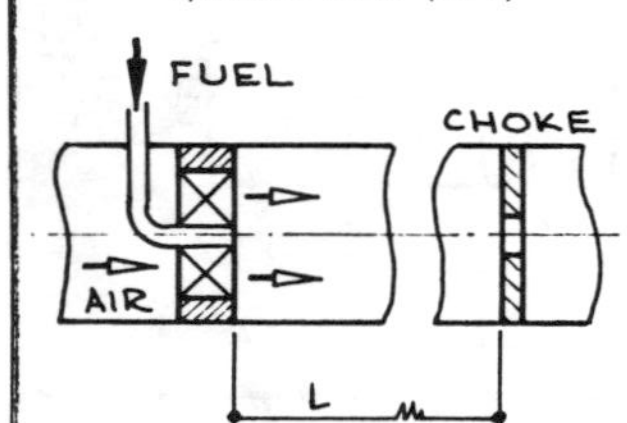 (gas flames in table 1.8.1)	D = 122 mm; L = 1878 mm Pressure-atomiser T_o = 533 K P = 1 atm S = 0.3; 0.6 ISO-OCTANE; HEAVY OIL ϕ_m = 0.65 SMD = 120 to 140 μm.	U,U_p,N_p: LDV, 0.24x2.0 (mm); resin particles:d_N =6 μm; counter processor and frequency tracker. O_2,UHC,CO_2,CO: water-cooled probe. $\overline{T}$: thermocouple, corrected for radiation.	The flame consists of a conical core comprising larger fuel droplets (whose velocities are governed primarily by the fuel nozzle characteristics), surrounded by clouds of small fuel droplets of small inertia. Variations on fuel volatility have negligible effects on the flame structure which resembles that observed for gaseous fuels (table 1.8.1), e.g. increasing swirl increases recirculation length (P = 1 atm).

TABLE 1. 2 - SPRAY FLAMES IN SIMPLE MODEL COMBUSTORS (continued).

REFERENCE	EXPERIMENT	MEASUREMENT TECH.	FINDINGS
ElBanhawy & Whitelaw (1981) 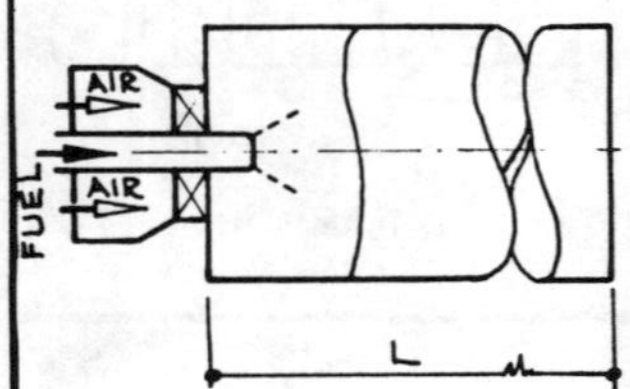	D = 157 mm; L = 758 mm Rotating cup atomiser T_o = 300 K P = 1 atm KEROSENE, ϕ_m = 0.35 Swirler vane angle: 30°; 45°; 60° SMD = 33; 47; 96 μm.	U_p,W_p: LDV spectrum analyser. $\bar{T}$: suction pyrometer, O.D. = 6 mm; thermocouple, Pt/Pt:13% Rh; d_W = 300 μm. O_2,CO_2,CO: water-cooled probe - O.D. = 6 mm; I.D. = 1.55 mm.	Increasing the mean drop size reduces combustion intensity downstream of the spray flame for large radii. The time scales of droplet evaporation and turbulent mixing are comparable and confirm that droplet characteristics contribute to the combustion process. Decreasing swirl reduces combustion intensity within the initial part of the flame close to the centre line and increases flame length.
	Calculations in ElBanhawy & Whitelaw (1980) and Jones & McGuirk (1981)		
Attya & Whitelaw (1984) 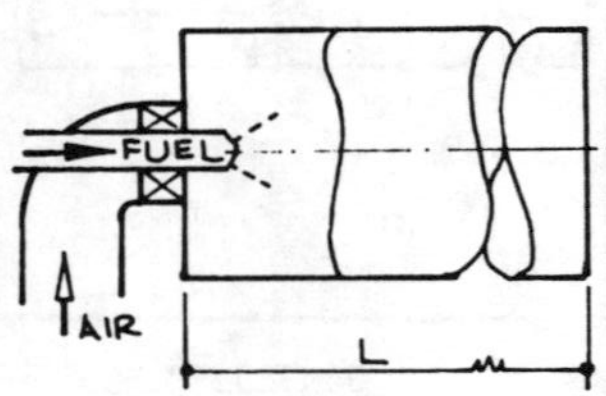	D = 206 mm, L = 900 mm Rotating cup atomiser T_o = 300 to 500 K P = 1 atm KEROSENE, ϕ_m = 0.46 S = 0.47 to 1.4 SMD = 54; 77; 115 μm.	U_p,W_p: LDV, spectrum analyser. $\bar{T}$: thermocouple, Pt/Pt:13% Rd; d_W = 80 μm; uncoated; no radiation correction. O_2,UHC,CO_2,CO: water-cooled probe - O.D. = 6 mm; I.D. = 2.05 mm.	Increasing T_o increases combustion intensity and improves mixing. Decreasing swirl decreases both combustion intensity and flame length. For low swirl the flame is stabilised near the wall. Flames for high swirl, high T_o and small droplets are characterised by complete droplet evaporation in the R.Z. and are shown to be mixing controlled. It is suggested that for $D_p < 50$ μm, and increasing with T_o, droplet models are unnecessary and that a simple model will suffice for gas turbine applications.

TABLE 1. 2 - SPRAY FLAMES IN SIMPLE MODEL COMBUSTORS (continued).

REFERENCE	EXPERIMENT	MEASUREMENT TECH.	FINDINGS
Jackson & Samuelsen (1984) Wood et al (1984) Samuelsen et al (1983) 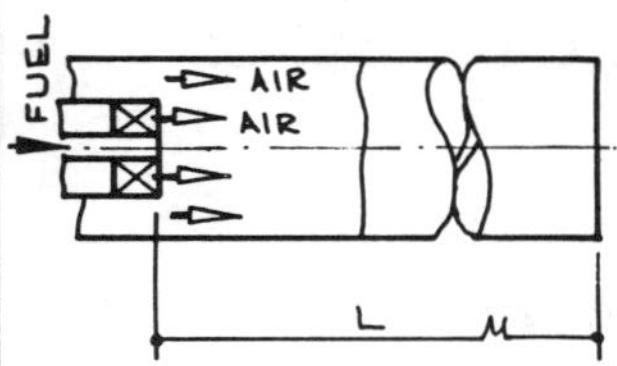 (also tables 1.7.2-A and 1.8.1)	D = 80 mm; L = 320 mm Twin-fluid atomiser T_o = 300 K P = 1 atm S = 1.3 LIQUID ISO-OCTANE and Blends with the same smoke point. ϕ_m = 0.3; 0.5 (AFR = 0.75 to 3.5) SMD = 5 to 50 μm.	U,W: LDV - 2-channel; vol: 0.022 mm^3 (at 20°); d = 0.36 mm; Alumina, d_N = 1 μm; counter processor. D_p: LD - d = 9 mm. D_p; Soot: Optical method (Mie scattering at 20° and 60°), D_p = 0.08 to 0.38 μm. $\overline{T}$: thermocouple, Pt/Pt:13% Rd. O_2,UHC,CO_2,CO: water-cooled probe.	Soot is as much dependent on the combined aerodynamics and fuel distribution in the combustor as on fuel molecular structure. Sooting is low or non-existent in high temperature regions. Adding ring compounds to iso-octane and using double ring blends increases the total soot which is reduced by decreasing the fuel loading. Increasing AFR decreases D_p and narrows size distribution.
Onuma & Ogasawara (1975) Onuma et al (1977) 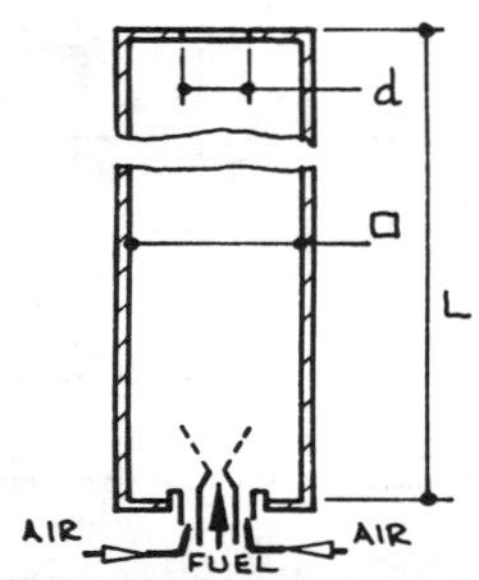 	L_{eq} = 430 mm; L = 1140 mm d = 70 mm Air atomising spray nozzle T_o = 300 K P = 1 atm {KEROSENE; HEAVY OIL; PROPANE (GAS) ϕ = 0.17 to 0.66 SMD = 4 to 80 μm.	N_p,D_p: glass plate collector and microphotographs. U: Pitot probe. $\overline{T}$: suction pyrometer; thermocouple, Pt/Pt:13% Rd; d_W = 300 μm, uncoated. O_2,UHC,CO_2,CO,NO: water-cooled probe.	The region where the droplets exist is limited to a small area above the burner nozzle. Most of the droplets in the flame do not burn individually, but fuel vapour from the droplets concentrate and burns like a gas diffusion flame. Kerosene, heavy oil and propane flames are similar in structure.

TABLE 1. 2 - SPRAY FLAMES IN SIMPLE MODEL COMBUSTORS (continued).

REFERENCES	EXPERIMENT	MEASUREMENT TECH.	FINDINGS
Tuttle et al (1976) 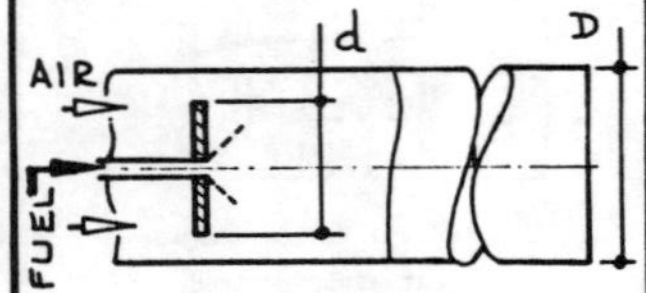 (calculations reported by ElBanhawy & Whitelaw, 1980)	D = 146 mm; d = 114 mm; BR = 61% Simple pressure-atomiser T_o = 500 K P = 5 atm LIQUID PROPANE ϕ_m = 0.1; 0.2 SMD = 130 μm	$\bar{T}$: thermocouple, Pt/Pt:10% Rd; d_W = 3 mm. UHC,CO,NO_x: water-cooled probe, O.D. = 9.5 mm; I.D. = 3.2 mm	The flame consists of a hollow reaction region (shear layer) along the boundary between the recirculation zone and the free stream. As the air flow rate increases, the relative importance of the heterogeneous flame structure increases and a flame zone following the fuel spray is also detected. CO is formed in the R.Z. with the maximum values in the shear layer. NO is formed near the stagnation zone (high T) and in the downstream shear layer.
Proctor & Mellor (1982) 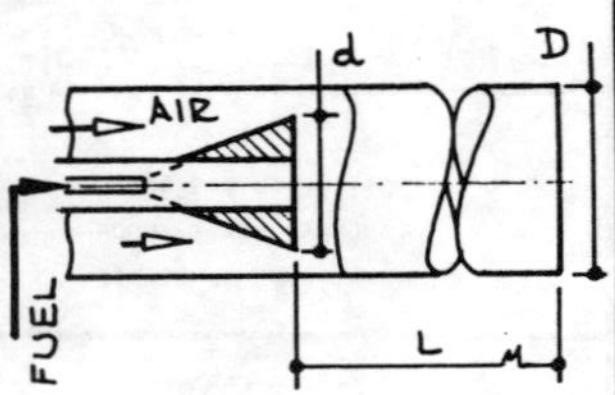	D = 146 mm; d = 127 mm; BR = 75% L = 522 Simplex pressure atomising nozzle T_o = 800 K P = 4; 8 atm LIQUID PROPANE; JET-A ϕ_m = 0.23; 0.30 SMD = 87 μm (JET-A)	Wall temperature: thermocouple. O_2,UHC,CO_2,CO: water-colled probe.	The general characteristics of the jet-A flow field are similar to those of the propane, although combustion does occur differently. Fuel droplets do not penetrate into the recirculation zone and act as fuel sources within the combustor. Consequently, less UHC and CO are observed for Jet-A. Using propane as fuel, combustion efficiency in the region of CO oxidation is better than 99%, but drops to below 20% at the centre line.

TABLE 1. 2 - SPRAY FLAMES IN SIMPLE MODEL COMBUSTORS (concluded).

REFERENCE	EXPERIMENT	MEASUREMENT TECHNIQUES	FINDINGS
Nicholls et al (1980) 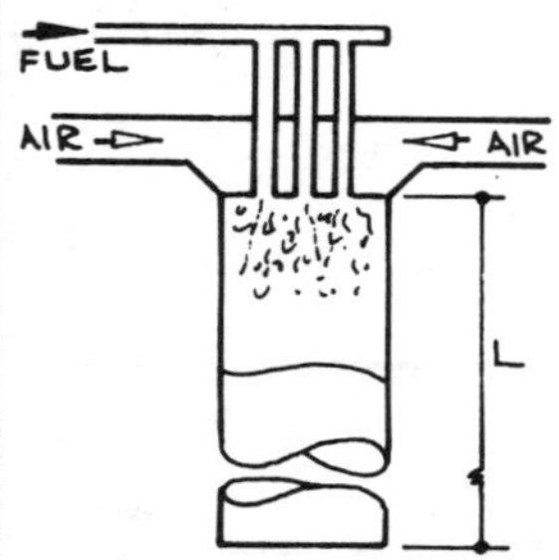 	D = 108 mm; L = 300 mm Air blast atomization of fuel (37 needles) T_o = 360 K P = 1 atm JET - A $\emptyset_m$ = 0.6 to 0.9 SMD = 10 to 58 μm	U: LDV, alumina particles; counter processor. $\overline{T}$: thermocouple, Pt/Pt:Rd; corrected for radiation (<5%). UHC, CO, NO_x: water-cooled probe.	Based on high levels of UHC and low temperatures is concluded that the spray burns as a gaseous flame. Increasing drop size increases UHC and decreases temperature gradients. Increasing residence time decreases UHC and CO and increases NO_x. Increasing $\emptyset_m$ increases flame temperature and NO_x and decreases UHC.

TABLE 2 - GAS TURBINE COMBUSTORS (continued).

REFERENCE	EXPERIMENT	MEASUREMENT TECH.	FINDINGS
Sawyer et al (1969) Starkman et al (1971) 10 sets of dilution holes 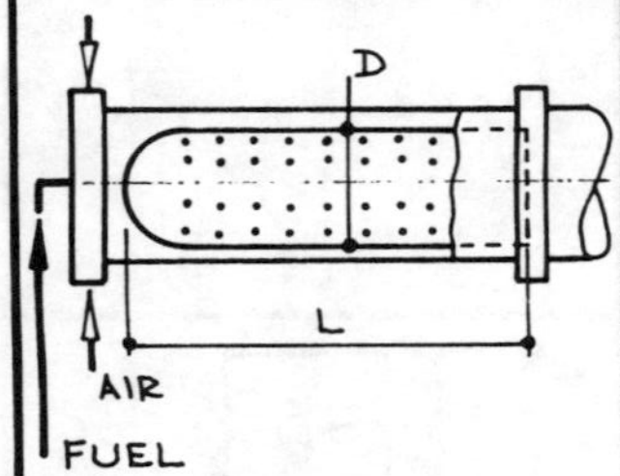 	L = 355 mm; D = 76 mm T_o = 300 K P = 1.5 atm HEPTANE (LIQUID) ϕ_m = 0.119; 0.205; 0.281 (controlled by the air flow)	UHC,CO_2,CO,NO: water-cooled probe: O.D. = 6.4 mm; d_{OR} = 0.38 mm; (inserted from exit).	CO reaches local peak values of 1 to 4% and exceeds equilibrium everywhere. CO oxidation is quenched by air addition. NO reaches local peak values of 40 to 100 ppm in high temperature zones and only exceeds equilibrium as the gases approach the exit.
Parikh et al (1971) GEOMETRY AS IN SAWYER et al (1969)	T_o = 300 K P = 1.2 atm METHANE (GAS) ϕ_m = 0.20; 0.27 (controlled by the air flow) TWO fuel injectors.	O_2,UHC,CO_2,CO,NO: water-cooled probe: O.D. = 6.4 mm; d_{OR} = 0.38 mm. $\overline{T}$: thermocouple, Pt/Pt:13% Rd; d_W = 100 μm; silica coated; corrected for radiation (< 200 K).	NO measured (< 25 ppm) is at least an order of magnitude below the equilibrium values. Its production is rate controlled and increases as T_o, ϕ_m or P are increased. Prevaporization of fuel is suggested to reduce NO. CO and UHC distributions are similar to those for liquid fuel: CO oxidation is quenched by air addition. $\overline{T}_{MAX}$ (corrected) approaches local $\overline{T}_{ad}$.

TABLE 2 - GAS TURBINE COMBUSTORS (continued).

REFERENCE	EXPERIMENT	MEASUREMENT TECHNIQUES	FINDINGS
(a) Mellor et al (1972) (b) Tuttle et al (1973) 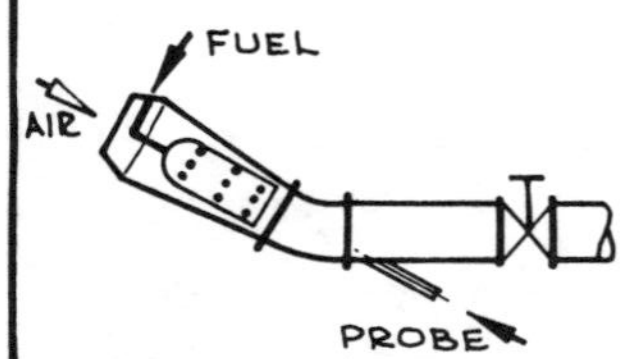 ALLISON - J33 3 sets of dilution holes + small upstream holes for P.Z. air addition.	L = 508 mm; D = 148 mm T_o = 300 K(a); 500 K(b) P = 3 to 10 atm LIQUID PROPANE ϕ_m = 0.15 to 0.345.	UHC,CO,NO: water-cooled probe: O.D. = 9.5 mm. $\bar{T}$: thermocouple, Pt/PT:10% Rd; d_W = 3 mm; MgO coated; No radiation correction.	Hollow flame in P.Z. with maximum temperatures (≈1600 K) along the walls and ≈ 400 K below the theoretical value due to liquid fuel impingement upon the thermocouple. CO and UHC increase with increasing $\dot{m}_a$ but decrease with increasing P and T_o; in contrast, NO decreases with increasing $\dot{m}_a$ but increases with increasing P and ϕ_m. In general, NO is low (< 50 ppm) but CO is high (up to 10%) due to the air distribution in the P.Z
Shrisler et al (1975) SIMILIAR ARRANGEMENT TO TUTTLE et al (1973) WITH THE FILM-COOLED T-56 COMBUSTOR	T_o = 394 to 584 K P = 3.5 to 9.7 atm LIQUID AND VAPOR PROPANE ϕ_m = 0.11 to 0.30	UHC,CO,NO: water-cooled probe: O.D. - 9.5 mm.	Average drop lifetimes are small, thus turbulent mixing controls combustion and emission of NO. In general, the P.Z. resembles a turbulent diffusion flame. Poor mixing, as a result of film cooling, causes reactions to continue beyond exit plane. Emission levels do not change with prevaporisation of propane because of its low evaporation time. The results cannot be extrapolated to practical fuels.

TABLE 2 - GAS TURBINE COMBUSTORS (continued).

REFERENCE	EXPERIMENT	MEASUREMENT TECH.	FINDINGS
Casci et al (1973) 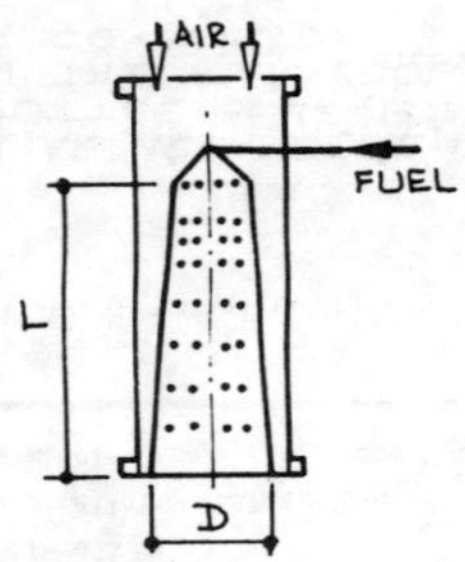	D = 110 mm; L = 345 mm T_o = 285 K P = 1 to 2.9 atm METHANE ϕ_m = 0.15; 0.22 (ϕ_m controlled by air flow rate).	UHC,CO_2,CO,NO_x: water-cooled probe: O.D. = 10 mm; I.D. = 6 mm; (inserted radially). $\bar{T}$: thermocouple within sampling probe.	CO and UHC levels decrease as P and AFR increase, except near the exit for P = 1 atm, because of quenching by air addition. Maximum values were respectively equal to 5 and 1%. NO production is controlled mainly by the temperature and both decrease as P decreases and AFR increases. Maximum NO about 100 ppm for temperatures around 2000 K.
Lefebvre & Fletcher (1973) 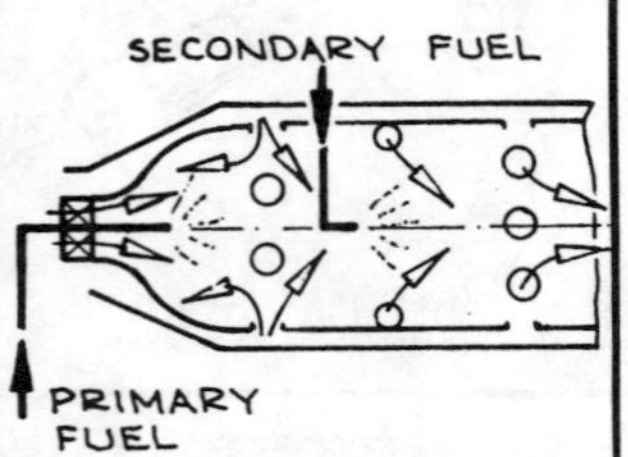 3 sets of holes and film cooling slots.	L = 500 mm; D = 250 mm T_o = 450 K P = 2.7 atm KEROSENE ϕ_m = 0.14 to 0.22 Three separate combustor builds with different air flow split.	NO: EXIT PLANE, water-cooled diameter probe with 9 holes of 1 mm.	Fuel staging has considerable potential for lowering NO emissions. Maximum reduction (from 35 to 15 ppm) is obtained when 75% of the fuel is employed in P.Z. Predictions by Fletcher and Heywood (1971) are in agreement with measurements, although the model uses adjusted parameters.

TABLE 2 - GAS TURBINE COMBUSTORS (continued).

REFERENCE	EXPERIMENT	MEASUREMENT TECHNIQUES	FINDINGS
Morr, Heywood & Fitch (1975) GEOMETRY IS NOT GIVEN.	T_o = 450 K P = 4 atm KEROSENE	O_2,UHC,CO_2,CO,NO_x: water-cooled probe (only in the exit plane).	Excessive CO emissions are caused by the rapid quenching of P.Z. combustion gas due to the dilution air jets. A simple model is formulated based on a rate limited partial equilibrium approach. Minimum CO emission would be obtained by burning a uniform mixture of fuel and air in the P.Z. at ϕ = 0.7. Staged combustion is suggested to reduce the NO_x emission.
Vranos & Taback (1976) 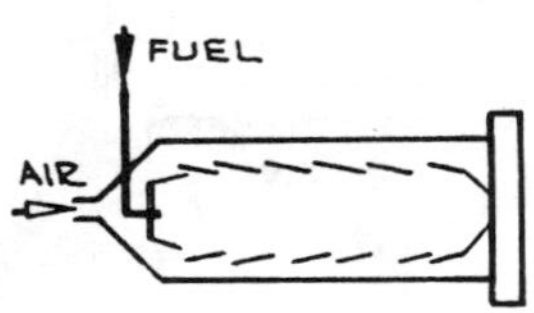 film-cooled JT8D combustor.	L = 560 mm; D = 180 mm T_o = 390 to 625 K P = 2.4 atm JP-5 FUEL ϕ_m = 0.15; 0.20.	O_2,N_2,UHC,H_2,CO_2,CO: steam-cooled probe.	Quenching of the CO oxidation is evident mainly near the walls, where CO reaches 10%.

TABLE 2 - GAS TURBINE COMBUSTORS (continued).

REFERENCE	EXPERIMENT	MEASUREMENT TECHNIQUES	FINDINGS
Katsuki et al (1976) 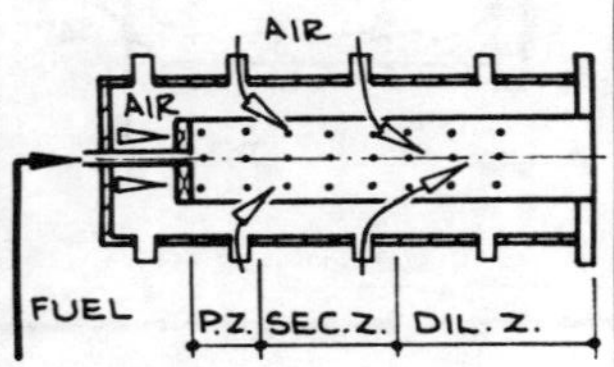 8 sets of holes. Calculation in Jones & McGuirk (1980).	L = 350 mm; D = 60 mm T_o = 320; 430 K P = 1 atm KEROSENE ϕ_m = 0.23 SWIRL VANE: 0°; 30°; 45°.	O_2,N_2,UHC,H_2,CO_2,CO,NO: water-cooled probe, I.D. = 0.8 mm. $\overline{T}$: Thermocouple, Pt:30% Rd / Pt:6% Rh; d_W = 200 μm; uncoated; no radiation correction.	CO and H_2 are produced in initial oxidation process at the end of P.Z. (T_{MAX} = 1900 K), while NO is formed by thermal process and depends on local ϕ and T. Increasing T_o, reduces CO and increases NO (maximum values are respectively 4% and 50 ppm). With excessive swirl the burning zone is transferred downstream. Optimum swirl decreases CO but does not affect NO.
Jones et al (1977) 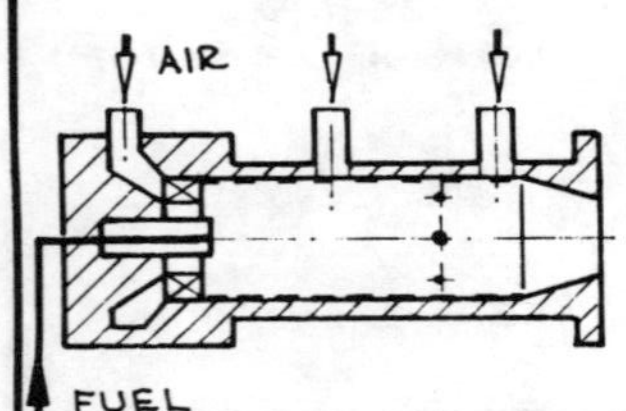 1 set of holes; porous wall.	L = 200 mm; D = 72 mm T_o = 570 K P = 2.1 atm PROPANE (GAS) ϕ_m = 0.34	O_2,N_2,UHC,H_2,CO_2,CO,NO_x: water-cooled probe. $\overline{U}$: 7-hole water cooled pitot probe, O.D. = 6 mm.	CO increases initially up to 2% and decreases slowly further downstream, indicating a chemically controlled oxidation. Predicted CO is too high everywhere but UHC is well calculated. Errors due to assumed pdf. Improved calculations of Jones and Priddin (1979) still overestimated CO and UHC due to model and numerical errors. It is concluded that detailed velocity measurements are required to improve models.

TABLE 2 - GAS TURBINE COMBUSTORS (continued).

REFERENCE	EXPERIMENT	MEASUREMENT TECHNIQUES	FINDINGS
Dils & Follansbee (1979) 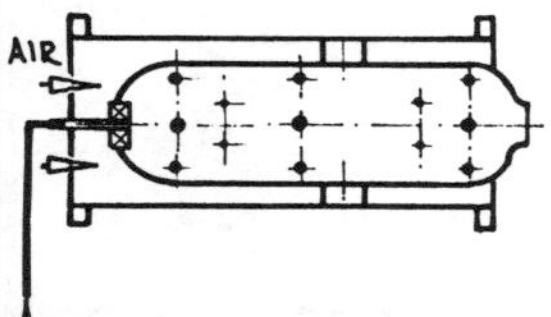 2 main sets of holes and small upstream holes for P.Z. air addition.	L = 318 mm; D = 100 mm T_o = 310 K P = 1 atm JET-A FUEL $\phi_m \simeq 0.4$	$\overline{T}, \overline{t'^2}$: thermocouple; Pt:30% Rd/ Pt:6% Rd; d_W = 76 µm; alumina coated; no radiation correction; Electrically compensated (Use of low-pass filter: f_c = 1 kHz).	Mean gas temperatures range from T_o up to about 200 K below T_{ad}. Temperature fluctuations vary from 5 to 25% of the temperature rise across the combustor. The use of a film-cooled liner decreases both the penetration of jets and the peak temperatures in the forward sections of the combustor by 260 K. Fuel nozzle spray angle affects flow pattern only in P.Z.
Noyce et al (1981) Noyce & Sheppard (1982) 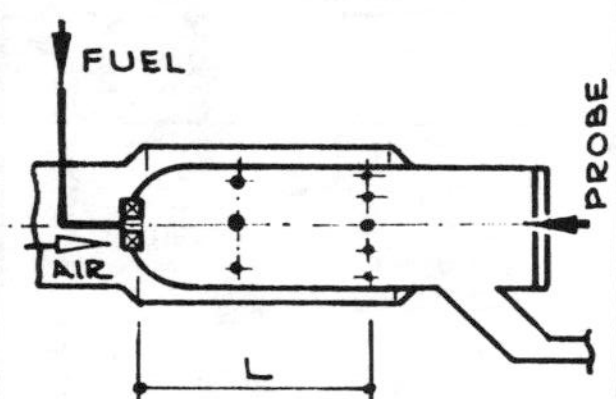 2 sets of holes	L = 444 mm; D = 168 mm T_o = 300 K P = 1 atm PROPANE (GAS) ϕ_m = 0.23 to 0.77 Two alternative fuel injectors.	O_2,UHC,H_2,CO_2,CO,NO_x: water-cooled probe: d_{OR} = 1 mm.	Combustion is determined by the interaction between the recirculation zone flow, the fuel jets (dominant within the P.Z.) and the primary air flows. A rich P.Z. is established for all the cases, with CO levels up to 8% for ϕ_m = 0.23 (i.e. below equilibrium) leading to low combustion efficiencies. Substantial super-equilibrium CO and UHC down the combustor indicate kinetic limitations for the combustion process. The EI_{CO} at the exit plane have a <u>minimum</u> (≈ 25 gCO/ g fuel) at ϕ_m = 0.35. Equilibrium is approached as ϕ_m is increased.

TABLE 2 - GAS TURBINE COMBUSTORS (continued).

REFERENCE	EXPERIMENT	MEASUREMENT TECHNIQUES	FINDINGS
Chleboun et al (1983) 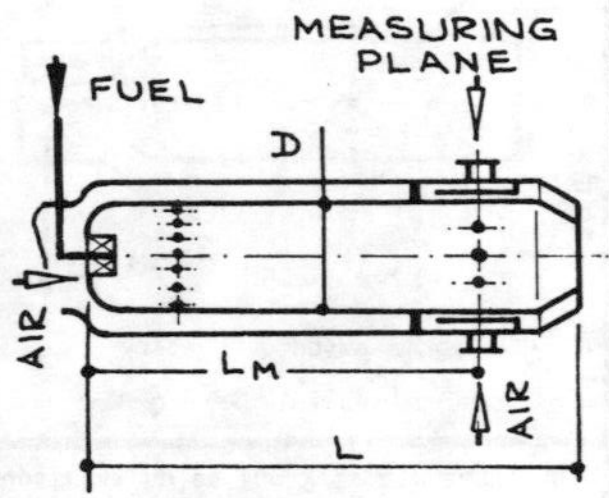 (isothermal flow: Chleboun et al, 1982)	$L = 317.5$ mm; $D = 175$ mm $L_M = 6.55$ mm $T_o = 325$ K $P = 1$ atm PROPANE (GAS) $\phi_m = 0.10;\ 0.11;\ 0.13$ (In the primary zone: $\phi = 0.19$)	UHC, O_2, CO_2, CO: water-cooled probe, I.D. = 1 mm; O.D. = 12.7 to 25.4 mm. Thermocouples: $\bar{T}$: Pt:20% Rd / Pt:40% Rd; $d_W = 127$ µm; Uncoated; No radiation correction. $\overline{t'^2}$: Pt/Pt: 10% Rd; electricity compensated $d_W = 50$ µm; Uncoated; No radiation correct.	Measurements were restricted to the mixing plane ($X = L_M$), where uniform conditions ($\bar{T} = 715$ K; $\phi < 3.2\%$) had been established. Temperature fluctuations have maxima at the leading edge of the jets and about 120 K (up to about 24%). Error analysis is not given. Fully 3-D calculations using the "k-ε" model have shown greater jet penetration and smaller rate of jet dissipation than the experiments. In general, mixing is underpredicted.
Goldstein et al (1983) 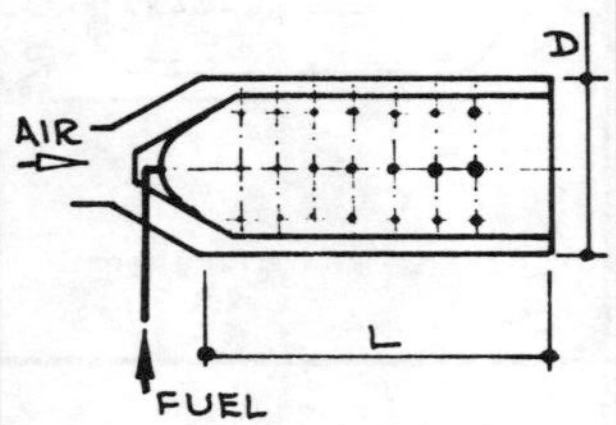 	Two different combustors: 1 { $L = 605$ mm; $D = 150$ mm; 7 sets of holes (only) 2 { $L = 387$ mm; $D = 128$ mm; 4 sets of holes and film-cooling secondary air non-symmetrical $T_o = 300$ K $P = 1$ atm JP-4 FUEL $\phi_m = 0.14$ to 0.26.	EXIT PLANE: $\bar{T}$: thermocouple, Ch/Al $d_W = 640$ µm U: LDV, M_gO particles, $d_N < 2$ µm; counter processor.	For combustor 1, the exit profiles are uniform with $\Delta T_{MAX} \approx 150^oC$. For combustor 2, the asymmetry of the secondary air causes large variations and $\Delta T_{MAX} \approx 420$ K. With combustion, the mean velocities are increased by a factor of 2-3, the turbulence intensities (which are ≈ 15%) are reduced to half the value in cold flow and the r.m.s. of the velocity fluctuations are slightly increased (by ≈ 20%).

TABLE 2 - GAS TURBINE COMBUSTORS (continued).

REFERENCE	EXPERIMENT	MEASUREMENT TECHNIQUES	FINDINGS
Friswell (1979) 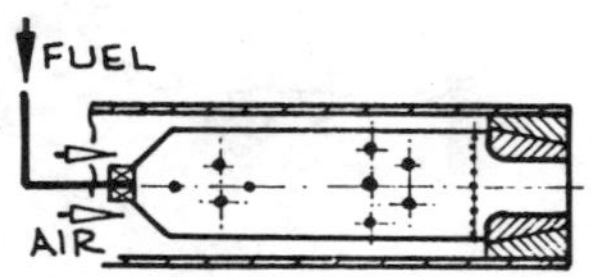 	D = 76 mm T_o = 300 to 600 K P = 2 to 22 atm Range of KEROSENES, GAS OILS, and SOLVENTS. AFR = 60/1 (constant) Three combustor designs varying length and shape of tube and holes.	Smoke Emission: heated steel probe at exit. $\bar{T}_{WALL}$ thermocouple, Ch/Al; (fitted into pockets on the outer skin of the combustor).	Fuel hydrogen content gives the best prediction of smoke emission and variations on $\bar{T}_{WALL}$ caused by changes in flame radiation. The influence of fuel composition on smoke formation/flame radiation falls to zero above P = 10 atm. In the hottest point of the combustor wall (920 K, for P = 8 atm, T_o = 300 K): ΔT_{MAX} = 50°C for 1% wt. change in hydrogen content.
Naegeli et al (1983) 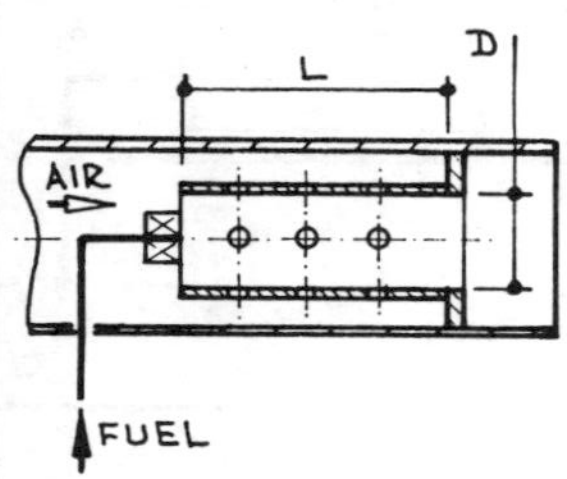 3 sets of holes.	D = 51 mm T_o = 533 to 922 K P = 7 to 13 atm KEROSENE-type fuels (Hydrogen content wt. = 11.5 to 14.3%) ϕ_m = 0.14 to 0.20.	T and Opacity (soot concentration): "Kurlbaum technique" (soot reversal method).	Soot concentration increases significantly as flame temperature increases, but the increase in soot with fuels of lower H/C ratio was much stronger than could be attributed to associated increases in the flame temperature alone.

TABLE 2 - GAS TURBINE COMBUSTORS (continued).

REFERENCE	EXPERIMENT	MEASUREMENT TECHNIQUES	FINDINGS
Bowden & Pearson (1983) 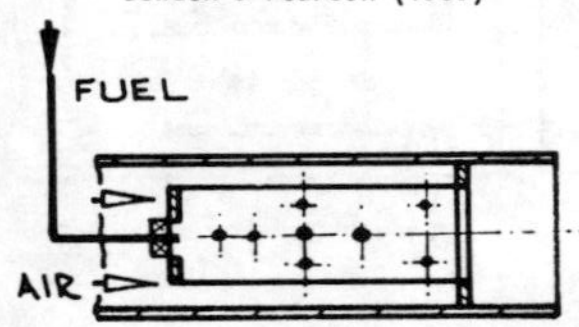 	L = 150 mm; D = 40 mm T_o = 655 to 753 K P = 7.8 to 19.6 atm Wide range of 'non-blended' and 'blended' FUELS: Hydrogen cont. = 12 to 15% wt. Aromatic cont. = 0 to 46% wt. Napht. cont. = 0 to 11% wt. AFR = 60/1 (constant)	$\bar{T}_{WALL}$ Water-cooled fibre optics connected to pyro-electric detector FLAME RADIATION: optical pyrometer. UHC, CO_2, CO, NO_x, Soot at EXIT: air cooled rake extractors.	Hydrogen content gives a better prediction of the combustion performance than aromatic content. However, for fuels with high concentrations of napthalenes or tetralins, the smoke point may provide the best prediction.
Kretschner & Odger (1987) 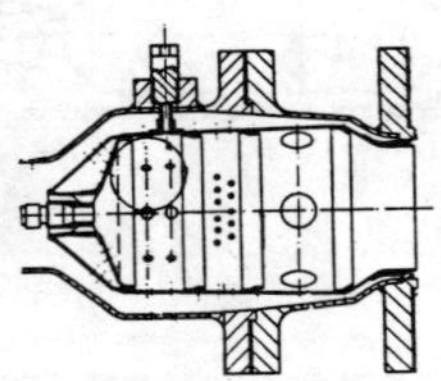	L=160 mm D=95 mm To=350 K P= 1 atm m=0.21 - 0.35 Fuel: various	CO_2, CO, HC, O_2, NO, NO_x: water cooled probe T: red/green brightness pyrometers Wall temperature: thermocouples	Flame transparency has not allowed to determine primary zone flame temperature. Exhaust temperature is not affected by fuel The results have shown the inadequacy of Mach number scaling and have proved the reaction volume scaling for small combustors.

TABLE 2 - GAS TURBINE COMBUSTORS (continued).

REFERENCE	EXPERIMENT	MEASUREMENT TECHNIQUES	FINDINGS
Toral & Whitelaw (1982) (Sector of annular combustor: rectangular cross-section. 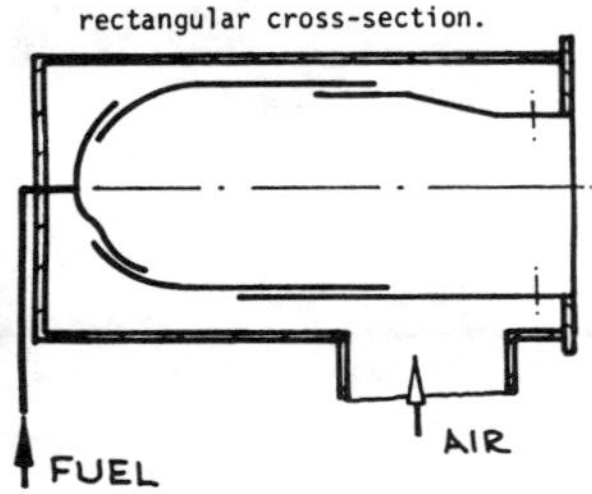 	L = 95 mm; L_y = 100 mm T_o = 300 K P = 1 atm METHANE (GAS) ϕ_m = 0.373; 0.488 Isothermal flow tests with He.	He (cold flow tests): sampling total head probe. O_2,UHC,CO_2,CO: water-cooled probe. $\overline{T}$; thermocouple, Pt/PT:13% Rh; d_W = 40 μm; $\overline{U},\overline{V},\overline{W}$: 3-hole probe; LDV, spectrum analyser.	Mean mixture fraction distributions measured with combustion are similar to those in isothermal flow. Comparison with equilibrium compositions suggests the inhibition of the global fuel breakdown in fuel rich zones. In fuel lean regions, high levels of CO (< 6%) indicate finite kinetic rate mechanisms. Measured temperatures exceed equilibrium values due to differences in type of average and to the neglect of fluctuations in the calculations.
Nakamura et al (1983) 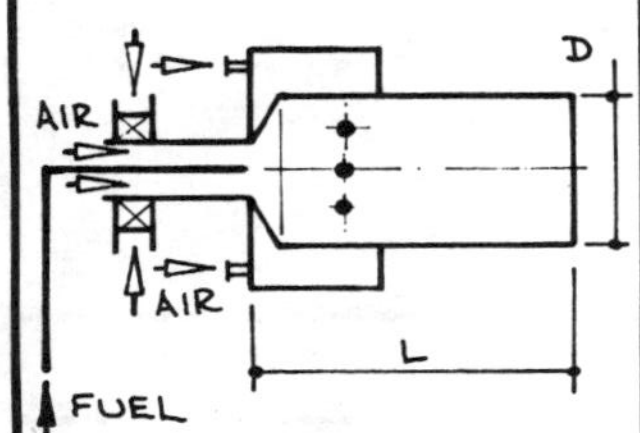 1 set of holes	L = 366 mm; D = 125 mm T_o = 300 K P = 1 atm PROPANE ϕ_m = 0.20 to 1.00 S = 0.6 to 2.0	O_2,UHC,CO_2,CO,NO_x: water-cooled probe. $\overline{T}$: thermocouple, d_W = 300 μm; uncoated; no radiation correction. Ion concentration: Langmuir probe. τ_r: Optical cell (injection of sodium salt). Isothermal flow: Ar trace method.	A stable flame is held at the P.Z. when $S > 0.6$. Optimum combustion efficiency depends on flow rate, swirl and air-fuel ratio. For S = 1.0 and ϕ_m = 1.0, the optimum secondary air flow is one third of total air flow. The effect of swirl is not monotonic, but the optimum value concerning the inflow of secondary air is about S = 1.0. Increasing the total air flow increases the penetration of secondary air into the P.Z. and improves combustion efficiency.

TABLE 2 - GAS TURBINE COMBUSTORS (continued).

REFERENCE	EXPERIMENT	MEASUREMENT TECHNIQUES	FINDINGS
Bicen, Tse and Whitelaw (1987) 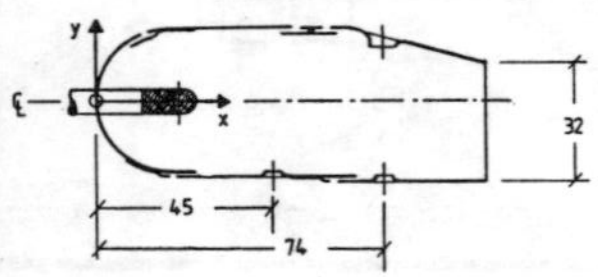 sector of Gem-60 combustor similar to that of Toral and Whitelaw (1982)	L=100 mm To=300K P=1 atm Methane (Gas) AFR=37 Isothermal flow test with helium	U; u'^2; V; v'^2 (only isothermal flow): LDV, frequency counter T: Pt/Rh thermocouples, 0.080 mm passive scalar (only isothermal flow): suction probe, katharometer	The primary zone is characterized by a Vortex driven by the film-cooling flows and limited in its downstream extent by the primary jets. There are two hot regions in the P.Z. in line with vaporizer exits. Changes in the position of the primary holes result in poorer mixing.
Joos & Simon (1987) 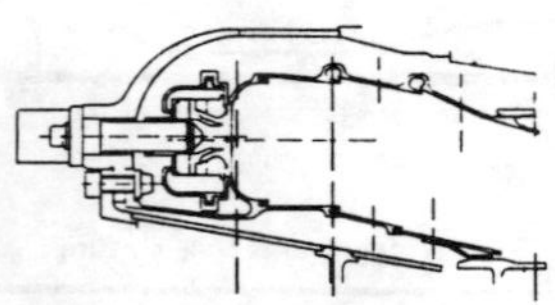 annular combustion chamber	To=670 K P=13 bar FAR=0.026	Fuel distribution: water cooled probe F:thermocouple probe	Experimental results agree with 3-D calculations, but calculation procedure has to be improved. Increase in wall temperature under elevated pressure is due to better fuel treatment under these conditions, to a larger spray-angle and to radiative heat transfer.

TABLE 2 - GAS TURBINE COMBUSTORS (continued).

REFERENCE	EXPERIMENT	MEASUREMENT TECHNIQUES	FINDINGS
Jones & Toral (1983) 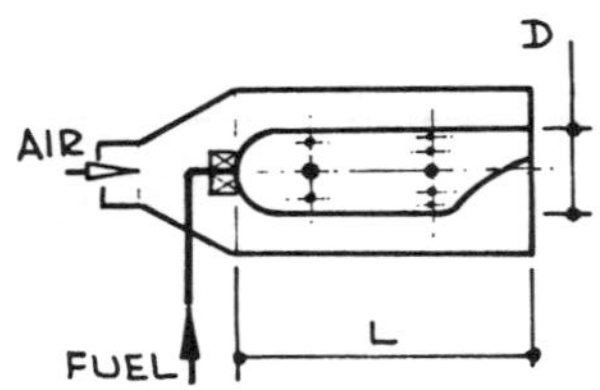 2 sets of holes; porous wall.	$L = 210$ mm; $D = 75$ mm $T_o = 313; 523$ K $P = 1$ atm PROPANE (GAS) $\phi_m = 0.30$.	O_2,UHC,H_2,CO_2,CO,NO_x: water-cooled probe, O.D.: 5 mm; I.D.: 1 mm. $\overline{T}$: thermocouple, Pt/Pt:13% Rd; $d_W = 80$ µm; uncoated; no radiation correction.	In the P.Z. UHC (≃ 5-6%) is only slightly affected by T_o, suggesting that the fuel breakdown reactions are physically controlled. The CO levels (≃ 6%) are less than the equilibrium values due to the high UHC concentrations. In the downstream region the CO and UHC burnup rates are both chemical kinetic rate limited. Measured temperatures are in agreement with equilibrium values at exit, but large differences arise in the P.Z. due to the neglect of the fluctuations in the calculations. Increasing T_o increase $\overline{T}$ by the same amount throughout the combustor.
Bhangu et al (1983) 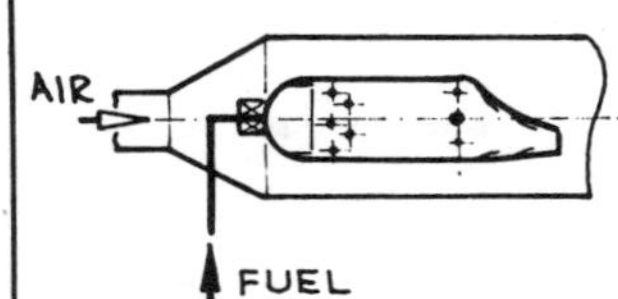 2 main sets of holes plus additional trimmer holes and cooling ring near the head; porous wall. (Fig. 4.34)	$D = 130$ mm (engine) Tests reported for both engine and rig: $T_o \simeq 750$ to 800 K $P = 1$ to 26 atm KEROSENE	UHC,CO_2,CO,NO_x,smoke: water-cooled probe. $\overline{T}$ (exit planes): thermocouple. $\overline{T}_{WALL}$: thermal paints; embedded thermocouple.	Previous examination of wiggle-strip combustor (film-cooled based) has shown high emission levels due to reduced reaction volume and increased quenching of CO reaction. A new design is studied to minimise the CO level at ground idle and comprises: - curved vane swirler to increase η and decrease smoke; - porous wall: "Transply" - Wassel and Bhangu (1980) - improved distribution of dilution holes. UHC,CO and smoke have been reduced respectively by factors of 20, 3.5 and 2.

TABLE 2 - GAS TURBINE COMBUSTORS (concluded).

REFERENCE	EXPERIMENT	MEASUREMENT TECHNIQUES	FINDINGS
Heitor (1985) Heitor & Whitelaw (1986) 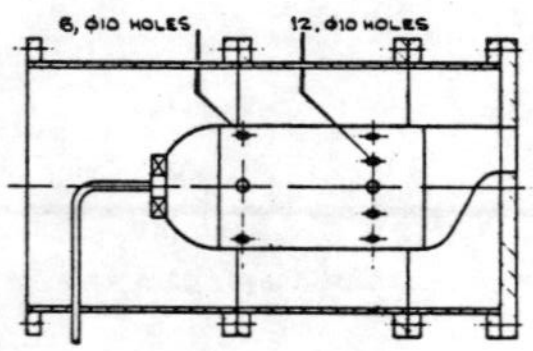 	ISOTHERMAL FLOW; simulation of: T_o = 318 K P = 1 atm Fuel: C_3H_8 AFR = 52.1 ϕ_m = 0.30 $\dot{m}_a$ = 0.085 kg/s.	Mean mixture fraction: helium trace technique, probe, O.D. = 1.47 to 2.65 mm. U,W: LDV	SEE TEXT
	T_o = 318 to 750 K P = 1 atm Fuel: C_3H_8 AFR = 52.1; 70 ϕ_m = 0.22; 0.30 $\dot{m}_a$ = 0.080; 0.085 kg/s $\dot{m}_f$ = 1.21; 1.53; 1.63 g/s	T: thermocouple, d_W = 40, 80 μm. (digitally compensated). Mean concentrations of O_2, UHC, H_2, CO_2, CO, NO_x: water-cooled probe, O.D. = 8 mm; I.D. = 1 mm. U,W: LDV	

COMBUSTION OSCILLATIONS IN DUCTS

S. SIVASEGARAM and J.H. WHITELAW

Imperial College of Science and Technology, London SW7 2BX, England

ABSTRACT

A series of measurements has been carried out with ducted flames and has quantified the relative importance of longitudinal-acoustic, Helmholtz and aerodynamic-shedding oscillations. The paper reviews the results and discusses their implications for ramjets and thrust augmentors.

INTRODUCTION

Many excellent contributions have been made to our understanding of combustion oscillations in ducts and include those of Kilham et al. [1], El Banhawy et al. [2], Putnam et al. [3], Yang and Culick [4], Crump et al. [5], Bloxidge et al. [6] and Lang et al. [7]. The motivations for the investigations include the improvement of ramjet, augmentor and furnace combustion, and all have been concerned with simplified arrangments, and some with low velocities and corresponding heat release.

In a series of related investigations, Heitor et al. [8], Katsuki and Whitelaw [9] and Sivasegaram and Whitelaw [10-16] have examined the influence of geometry, equivalence ratio of premixed reactants and velocity, and their results, together with those of earlier investigations, provide detailed understanding of many aspects of combustion oscillations. These investigations are reviewed in the following two sections which are concerned with ducted flames without and with an exit nozzle. The implications of the results for ramjets and the thrust augmentors of jet engines are clarified, though they must be regarded as tentative in view of the more stringent conditions and requirements of missiles and jet aircraft.

The experimental techniques in references [8] to [16] involved the measurement of mean, rms and frequency spectra of wall static pressure and of free-field sound intensity. The instrumentation comprised a Druck pressure transducer (PDCR 810), a Bruel and Kjaer condenser microphone (B&K 4136) and a Spectral Dynamics fast Fourier transform spectrum analyser (SD 340). In isothermal flows, used to examine the nature of aerodynamic shedding from disks and sudden expansions, hot-wire anemometry was also employed.

DUCTS WITHOUT AN EXIT NOZZLE

The flow configurations of importance here involve a duct with a bluff-body stabilised flame and with an unconstricted exit which simulates an engine configuration with an open nozzle. It can be anticipated that the open end will act as a pressure node to acoustic oscillations and that the corresponding impedance will be insufficient to reflect a pressure wave. The configurations examined are shown on Figure 1, and correspond to disk-stabilised and dump-combustor arrangements. Both are shown with a closed upstream end formed by a swirl register, but experiments were also performed with open ends and with a range of lengths and shapes. With the closed upstream end, the disk-stabilised arrangement gives rise to flammability and stability limits of the form shown on Figure 2a, and the dump-combustor with an area expansion ratio larger than 2.5 to those of Figure 2b. In both cases, there is a range of equivalence ratios which allows combustion and within its bounds, a narrower but extensive range over which the sound pressure measured 20 diameters away from the duct axis exceeds 120 dB and corresponds to rough combustion. The magnitude of the combustion

D. F. G. Durão et al. (eds.), Instrumentation for Combustion and Flow in Engines, 45–54.

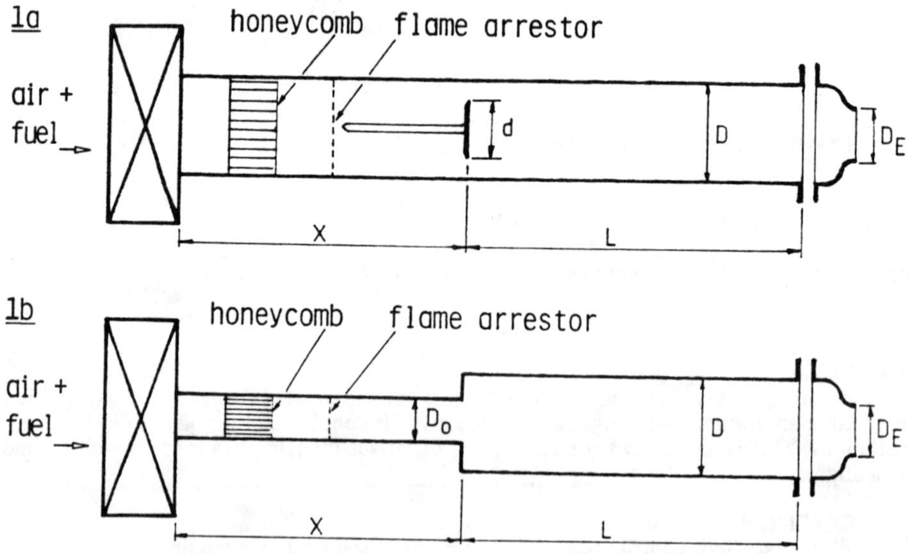

Figure 1: Burner Configurations
1a - disk-stabilized flame; 1b - dump combustor

oscillations within the bounds of rough combustion depends on the heat release and therefore upon the velocity and the diameter of the duct: it also depends on the length of duct downstream of the flame stabiliser, since a short duct can permit a fraction of the combustion to take place in a downstream jet where it will not drive oscillations within the duct.

In the régimes of equivalence ratio where the radiated sound intensities were low and the combustion smooth, it was possible to detect discrete frequencies corresponding to the longitudinal acoustic and shedding frequencies. In the spectra of Figure 3, for example, f_1 corresponds to shedding from the disk and f_2 to a quarter-wave based on the length from the swirl register to the baffle. When the duct length is such that the acoustic frequency is the same as the shedding frequency, the two sources of oscillations combine to give one larger peak and an rms signal which exceeds that of other smooth-combustion configurations by up to 5 dB. This phenomenon has been noted previously, especially by Schadow et al. [17,18], who devised non-circular arrangements of the ducts upstream of their dump combustors to reduce the contribution of the shedding frequency. Thus, even the 5 db increase over smooth combustion can be important and worthy of special efforts to suppress the shedding component, especially when the much larger heat release associated with engine configurations is considered.

Transition from smooth to rough combustion was characterised by a sharp increase in radiated sound level over a narrow range of equivalence ratios. At these equivalence ratios, the acoustic and shedding frequencies control the oscillations in a random manner with the two dominating at different times separated by tens of milliseconds.

In the range of equivalence ratios corresponding to rough combustion, only the longitudinal-wave frequency could be identified and, with the closed upstream end of the disk-stabilised flame, corresponded to a quarter-wave based on this length or to a higher harmonic, as shown on Figure 4a. Measurements of the wall static pressure confirmed an antinode in the plane of the swirl register and a node in the plane of the baffle. With the dump-combustor of large area ratio (Figure 4b), the quarter-wave was based on the

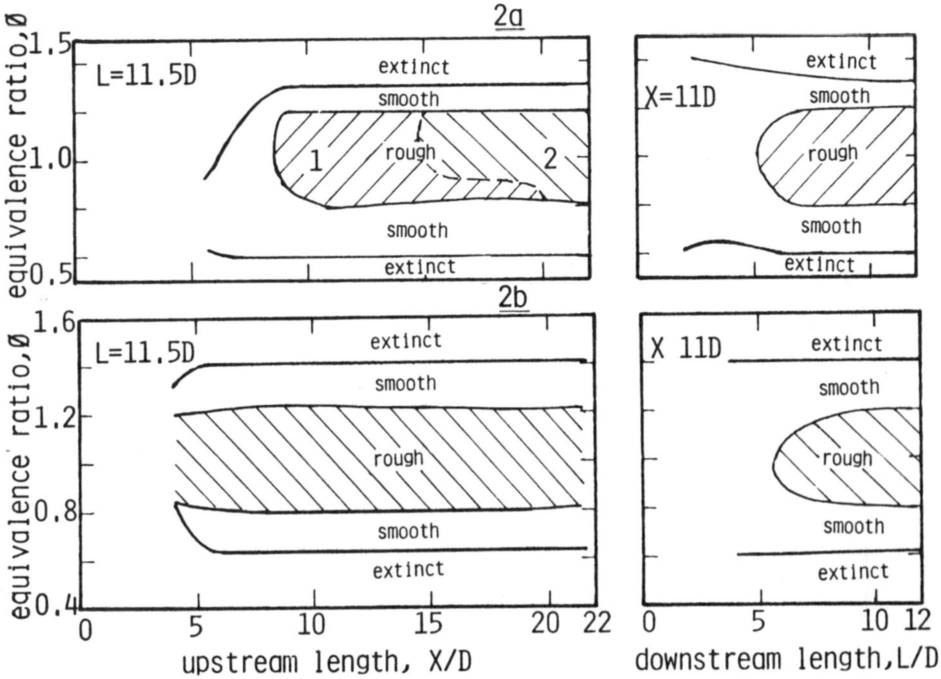

Figure 2: Influence of upstream and downstream duct length on stability and flammability limits

2a - disk-stabilized flames, D=80mm, d=40mm, Re=45 000;

2b - dump combustor, D=80mm, D_0=40mm, Re=90 000.

Labels 1 and 2 in 2a refer to rough combustion associated with 1/4-wave and 3/4-wave frequencies, respectively.

downstream length of the combustor with an antinode in the dump plane and a node in the exit plane. As might be expected, small area ratios (< 2.5) gave rise to quarter-waves (or their harmonics) based on the full length of the duct, as shown in Figure 4c.

With an open upstream end, the disk-stabilised flames gave rise to half-waves or their harmonics, as shown on Figure 5, with nodes in the plane of the upstream end and in the plane of the baffle. The dump-combustor with small expansion ratio also gave rise to half-waves based on the complete duct length. These results led to the idea of nozzles or orifices in the upstream duct so as to create two characteristic frequencies, one based on the distance from the upstream end to the orifice and the other from the orifice to the plane of the disk or the downstream open end. Figure 6a shows two arrangements, the component waves and the reduction in sound intensity which results. Similar reductions can be achieved with chambers such as those of Figure 6b which lead to the interaction of quarter-wave or bulk mode oscillations with the main longitudinal wave.

One aspect of the above results which remains to be explained concerns the existence of the dominant frequency in rough combustion within a limited range of values, as may be seen from Figures 4 and 5. Sivasegaram and Whitelaw [15] observed that this range is a function of mean velocity in the plane of the flame holder and exists for dump and disk stabilised flames and expansion ratios in the plane of the flame holder in the range between

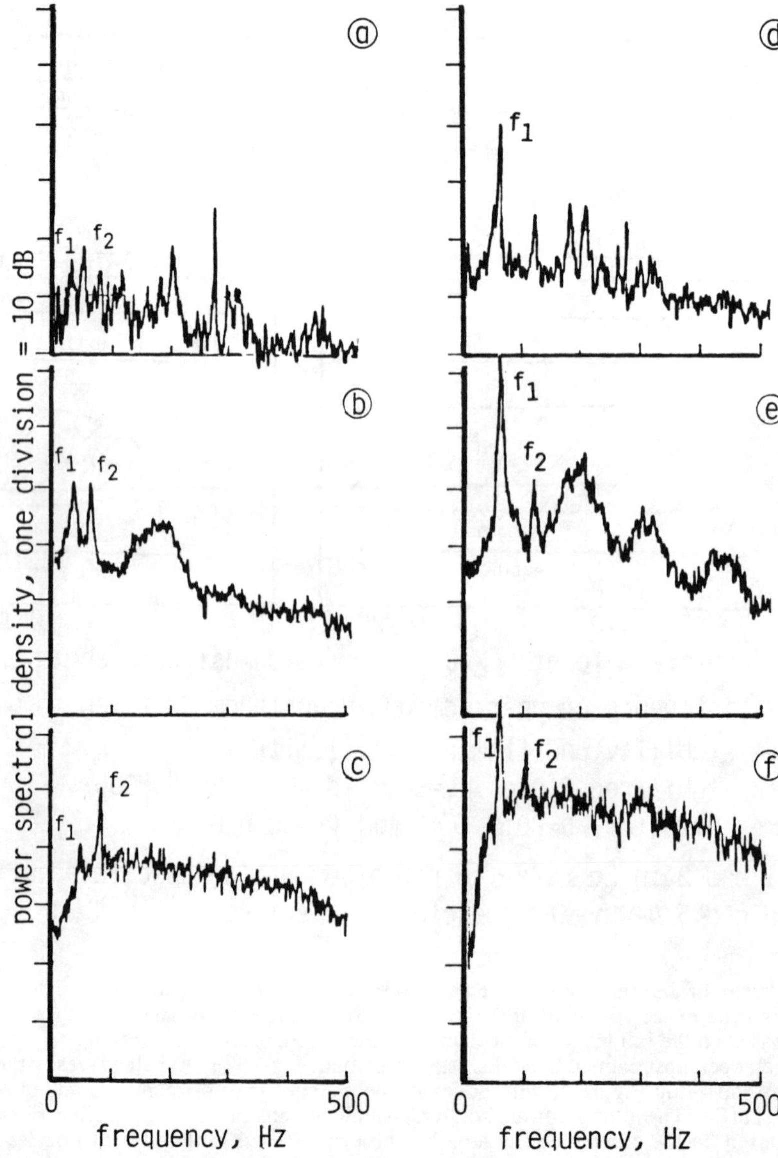

Figure 3: Power Spectra for Smooth Combustion in Disk-Stabilized Flames
D=80mm, d=40mm, L=11.5D, Re=45 000
a,b,c - Ø=0.68; d,e,f - Ø=0.71
a,d - sound spectra; b,e - pressure spectra near location of disk; c,f - pressure spectra near upstream end.

1.3 and 12, with the frequency range varying from between 75 and 170 Hz for a mean velocity of 10 m/s to between 200 and 500 Hz for a mean velocity of 100 m/s.

4a

4b

4c

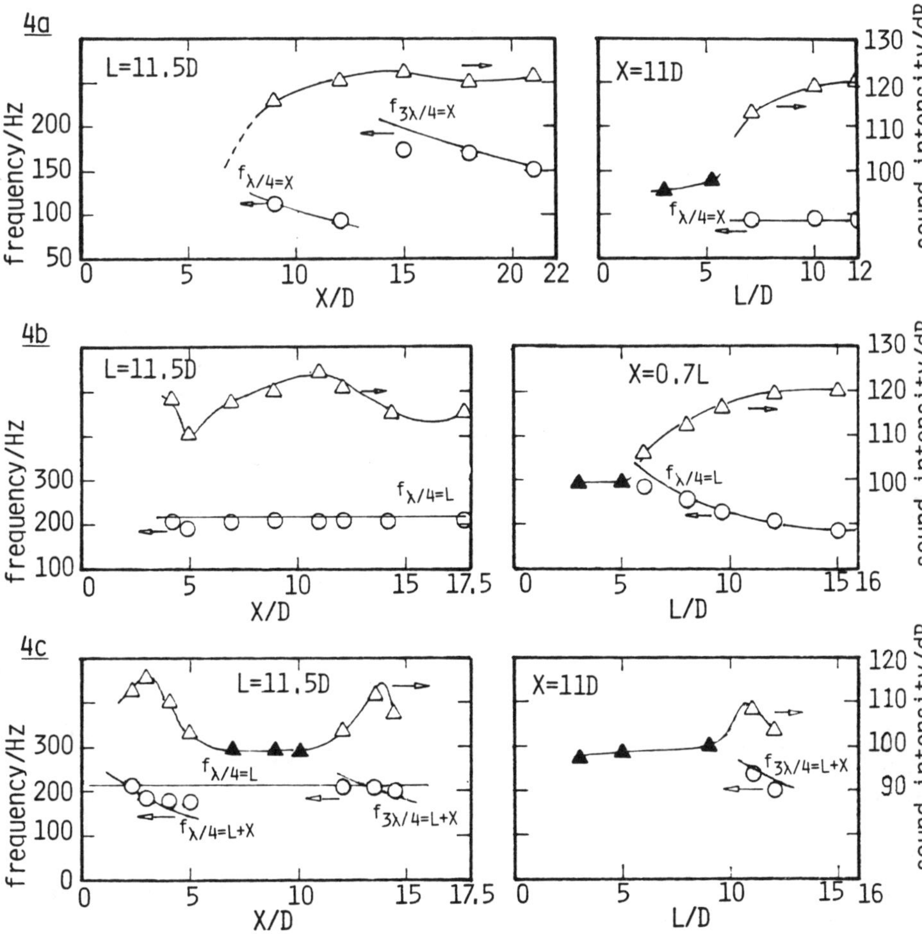

Figure 4: Influence of Upstream and Downstream Duct Lenth on Sound Intensity and Dominant Frequency at Stoichiometry

4a - disk-stabilized flame, D=80mm, d=40mm, Re=45 000;

4b - dump combustor, D=80mm, D_0=40mm, Re=90 000;

4c - dump combustor, D=65mm, D_0=51mm, Re=55 000.

▲,△ - sound intensity; ○ - frequency. Open symbols refer to rough combustion and the solid symbol to smooth combustion.

DUCTS WITH A NOZZLE

The addition of a nozzle to the duct is an essential feature of ramjets and thrust augmentors and it can be readily envisaged that, with velocities of sufficient magnitude, the nozzle will act as an antinode to the longitudinal wave so that the disk-stabilised arrangement and dump-combustors of large area ratio will give rise to half-wave acoustic oscillations. In dump combustors, however, the cavity formed by the dump plane, the cylinder and the nozzle can act as a Helmholtz oscillator so that bulk-mode oscillations can be expected to play an important rôle.

The importance of bulk-mode oscillations is made clear by Figure 7a, which shows that it dominates the flow at equivalence ratios close to the flammability limits where stabilisation is at its weakest. The corresponding oscillations do not have the sine-wave characteristics of the higher, longitudinal frequencies but rather that of a series of explosions separated by regular time intervals. With the addition of swirl, the stabilisation of the flame was improved and the bulk mode oscillations were absent and, in the case of longer ducts,

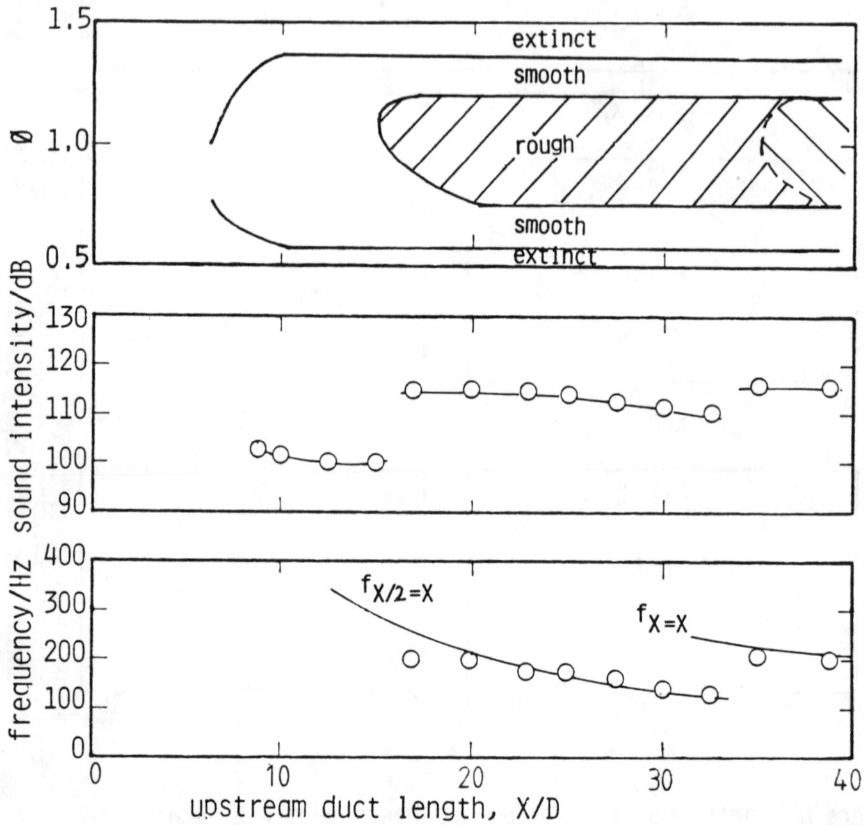

Figure 5: Flammability and Stability Limits, Radiated Sound Intensity and Dominant Frequency at Stoichiometry for Disk-Stabilized Flame with the Upstream End Acoustically Open

D=40mm, d=20mm, L=12D, Re=31 000.

6a

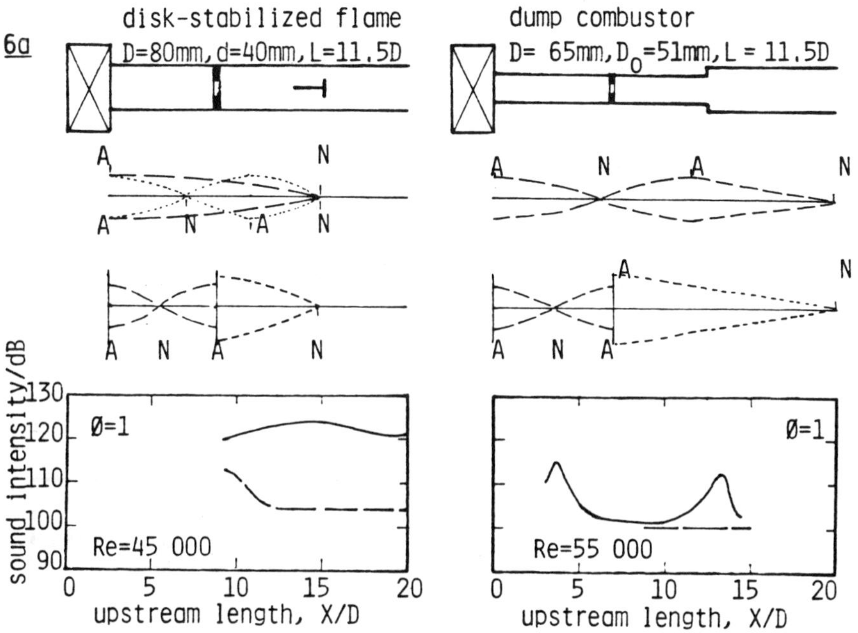

6b

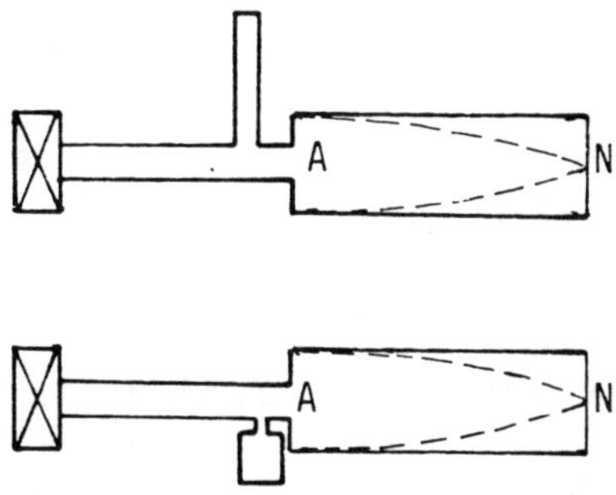

Figure 6: Methods of Suppressing Combustion Oscillations

6a - suppression of oscillations using an orifice located in the upstream duct in disk-stabilized flames and in dump combustors with a small area ratio; longitudinal pressure waves with and without orifice; ——— sound intensity without suppression, – – – sound intensity with most effective suppression.

6b - suppression using quarter-wave tube and Helmholtz resonator replaced by half-wave oscillations and their harmonics, as shown in Figure 7b. Once again, the preference of the combustion oscillations to exist in a limited range of frequencies is clear so that, with short ducts and high acoustic frequencies, either the bulk mode was preferred or combustion oscillations were absent.

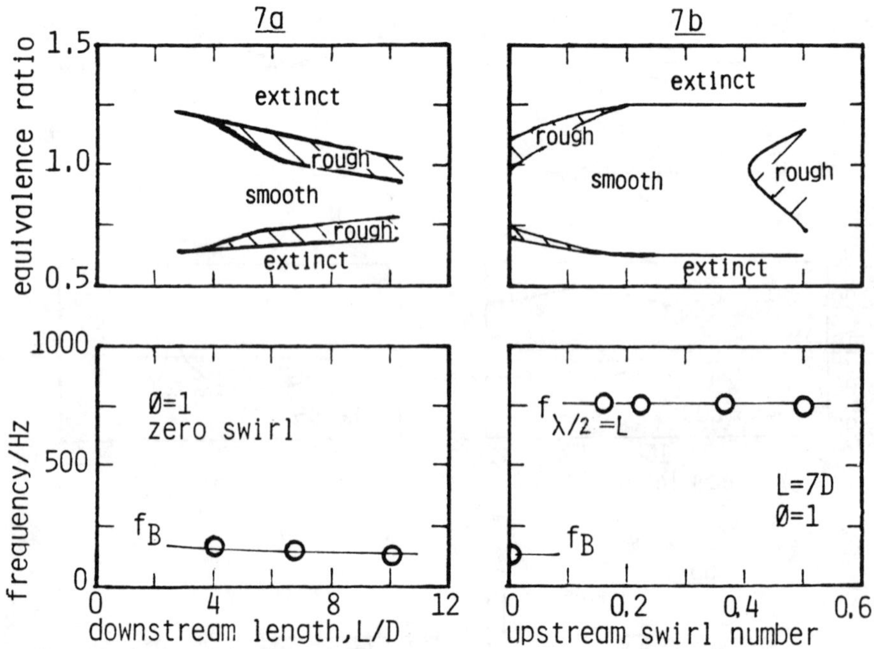

Figure 7: Flammability and Stability Limits and Dominant Frequency at Stoichiometry

D=80mm, D_0=23mm, D_E=30mm, Re=150 000

7a - influence of downstream length in nonswirling flow with a constricted exit; 7b- influence of swirl

DISCUSSION

The disk-stabilised flames of references [8] to [13] corresponded to gas turbine practice in terms of blockage ratio and, in the case of reference [13], simulated the effect of a by-pass duct. It is clear from reference [13] that, where acoustic oscillations exist, they can be ameliorated by careful arrangement of the by-pass flow. The results were obtained for comparatively low velocities, however, and it seems that they were insufficient to induce half-wave frequencies in the combustor or to offer the alternative of the bulk mode. With larger velocities and shorter length ($1 < L/D < 2$), it can be expected that the longitudinal frequencies will be high so that the bulk mode will be present and cause severe oscillations at equivalence ratios close to flammability limits. The remedy for this could lie in the generation of low frequency acoustic waves, based on the by-pass duct and out of phase with the bulk mode. Otherwise, it is necessary to avoid fuel to air ratios and geometries which give rise to poor flame stabilisation. The use of swirl from the low pressure turbine may aid the stabilisation process.

With dump-diffuser arrangements, the same principles hold but the duct upstream of the dump plane will not influence the oscillation characteristics if the area ratio is larger than 2.5. On the other hand, swirl can be introduced to assist stabilisation and to remove or to ameliorate the problem of the bulk mode frequency; and the longitudinal frequency will not occur if the duct is short and velocities high. These comments apply to rough combustion as interpreted here and not to the coupling of shedding and other frequencies in regions of smooth combustion: in this case, it is wise to avoid the influence of shedding perhaps by the procedures suggested by Schadow et al. [18].

NOMENCLATURE

D	diameter of downstream duct
D_E	diameter of exit nozzle
D_o	diameter of upstream duct in dump-combustor
d	diameter of disk
f	frequency
L	length of downstream duct
Re	Reynolds number = ρ U/ν
U	bulk mean velocity in upstream duct
X	length of upstream duct
λ	wave length
ν	kinematic viscosity
ρ	density
ϕ	equivalence ratio = (fuel to air ratio)/(fuel to air ratio at stoichiometry)

SUBSCRIPTS

B	bulk node
$\lambda/4$	quarter-wave
$\lambda/2$	half-wave
$3\lambda/4$	three quarter-wave
λ	full wave

ACKNOWLEDGEMENTS

Financial support for this work was from the US Office of Naval Research under Grant N00014-84G-0185 and is gratefully acknowledged. Useful discussions with many colleagues, particularly Dr K.C. Schadow and Dr E. Gutmark of the Naval Weapons Center at China Lake, are also acknowledged.

REFERENCES

1. Kilham JK, Jackson EG and Smith TBJ: Oscillatory Combustion in Tunnel Burners. Tenth Symposium (International) on Combustion, pp. 1231-1240, 1965.
2. El Banhawy Y, Melling A and Whitelaw JH: Combustion-Driven Oscillations in a Small Tube. Comb and Flame 33, 281, 1978.
3. Putnam AA, Murphy MJ and Locklin DW: Burner Technology Bulletin: Control of Combustion Noise from Small Gas Burners. Gas Research Institute, Chicago, Topical Report GRI-85/210, 1985.
4. Yang V and Culick FEC: Analysis of Low Frequency Combustion Instabilities in a Laboratory Ramjet Combustor. Comb Sci and Tech 45, 1, 1986.
5. Crump JE, Schadow KC, Yang V and Culick FEC: Longitudinal Combustion Instabilities in Ramjet Engines: Identification of Acoustic Modes. J Propulsion and Power 2, 105, 1986.

6. Bloxidge GJ, Dowling AP, Hooper N and Langhorne PJ: Active Control of Reheat Buzz. AIAA Paper 87-0433, 1987.
7. Lang W, Poinsot T and Candel S: Active Control of Combustion Instability. Comb and Flame (to be published), 1987.
8. Heitor MV, Taylor AMKP and Whitelaw JH: Influence of Confinement on Combustion Instabilities in Premixed Flames Stabilised on Axisymmetric Baffles. Comb and Flame 57, 109, 1984.
9. Katsuki M and Whitelaw JH: The Influence of Duct Geometry on Unsteady Premixed Flames. Comb and Flame 63, 87, 1986.
10. Sivasegaram S and Whitelaw JH: Oscillations in Confined Disk-Stabilised Flames. Comb and Flame 68, 121, 1987.
11. Sivasegaram S and Whitelaw JH: Suppression of Oscillations in Disk-Stabilised Flames. J Propulsion and Power 3, 291, 1987.
12. Sivasegaram S and Whitelaw JH: Experiments with Premixed Flames. Proc NATO Advanced Studies Workshop on Numerical Methods for Combusting Flows, 1987.
13. Sivasegaram S and Whitelaw JH: Combustion Oscillations of Relevance to Augmentors. AIAA Paper 87-2107, 1987.
14. Sivasegaram S and Whitelaw JH: Oscillations in Axisymmetric Dump Combustors. Comb Sci and Tech 52, 413, 1987.
15. Sivasegaram S and Whitelaw JH: Combustion Oscillations in Dump Combustors with a Constricted Exit. Department of Mechanical Engineering, Imperial College, London, Report FS/87/01, 1987.
16. Sivasegaram S and Whitelaw JH: Oscillations in Nonaxisymmetric Dump Combustors. Department of Mechanical Engineering, Imperial College, London, Report FS/87/06, 1987.
17. Schadow KC, Crump JE, Mahan VA, Nabity JA, Wilson KJ and Gutmark E: Large-Scale Coherent Structures as Drivers of Ramjet Combustion Instabilities. Proc JANNAF Propulsion Meeting, 1985.
18. Schadow KC, Gutmark E, Parr TP, Parr DM and Wilson KJ: Passive Shear Flow Control to Minimise Ramjet Combustion Instabilities. 23rd JANNAF Combustion Meeting, 1986.

EFFECT OF MEAN AND VARIABLE TIME CONSTANT ON COMPENSATED THERMOCOUPLE MEASUREMENTS*

K.C. HOPKINS, J.C. LARUE+, AND G.S. SAMUELSEN

UCI Combustion Laboratory
Department of Mechanical Engineering
University of California, Irvine, CA 92717

ABSTRACT

Variable and mean time constant compensation techniques applied to thermocouples are used to obtain measurements of the temperature moments, axial heat flux, time series, probability density functions, and power spectra in a heated axisymmetric jet. These results are compared to the corresponding results obtained using a cold wire which has adequate intrinsic uncompensated frequency response to measure the time resolved temperature. Comparison of corresponding results indicates that at high mean velocities and moderate velocity intensities (~10%), both compensation techniques lead to results that are similar to those obtained using the reference cold wire system. At lower mean velocities and higher velocity intensities (~20%) the variable time constant technique yields results that are in better agreement with those obtained using the reference system.

1.0 INTRODUCTION

The measurement of time resolved temperature in reacting and non-reacting turbulent flows is of fundamental importance in obtaining information concerning such properties as the progress of reaction, the geometrical characteristics of the reaction zone, the heat flux and the entrainment mechanism. Compensated fine wire thermocouples have been used in many studies to obtain the time resolved temperature.

However, a general concern associated with measurements of this type is the uncertainty due to the compensation technique. The compensation technique utilized in most studies involves the use of a mean or average value for the system response or time constant. In fact, the time constant is a function of fluid temperature, velocity and composition and thus is not a constant. Therefore, the temperature signal will at times be over- or under-compensated which leads to uncertainty in the measured temperature. The purpose of the study described herein is to determine experimentally the uncertainties in the statistical, spectral, and time resolved properties of the compensated temperature signal due to moderate to large velocity fluctuations in a flow with relatively small temperature fluctuations. The uncertainty is determined by comparing the statistical and spectral properties of the temperature field in a slightly heated axisymmetric jet obtained using the compensated thermocouple to the corresponding results obtained using a 0.625 μm diameter cold wire - a sensor that has adequate uncompensated frequency response for the accurate time resolved measurement of temperature in this flow. In addition to the

* Sponsored by NSF Grant CPE 8412073
+ Corresponding Author

D. F. G. Durão et al. (eds.), Instrumentation for Combustion and Flow in Engines, 55–68.

use of a mean or average time constant, a variable time constant, which is corrected for the instantaneous velocity is used to effect the compensation.

The test flow is a heated, axisymmetric jet where relatively high velocity and temperature intensities are produced. The velocity intensities are similar to those encountered in reacting flows and thus the results concerning the uncertainties due to the compensation technique can be applied, with some caution, to reacting flows.

2.0 BACKGROUND

The compensation technique used in the present study has its genesis in the early work of Shepard and Warshawsky (1952) who recognized that time resolved measurements of temperature could be obtained using compensated thermocouples. The compensation technique they proposed is based on the application of the transient form of the conservation of energy equation applied to the thermocouple subject to several simplifying assumptions.

Notable among those assumptions are that end conduction effects are negligible, that the temperature is uniform in both the radial and axial directions, and that radiation effects are negligible. Paranthoen et al. (1983) investigated the effect of sensor length on the measured temperature and found that end conduction effects should be negligible for probes with length to diameter ratios greater than 200. Yule et. al. (1978) have performed a careful analysis of the effect of bead size relative to wire size and report that conduction between the larger bead and the smaller wire can lead to non-first order response effects which become more significant as the ratio of bead diameter to wire diameter (d_b/d_w) increases. For $d_b/d_w < 2.5$, Paranthoen et al. (1983) find that this non-first order behavior is nearly negligible in its effect.

For all the work reported herein, the length to diameter ratio is greater than 1000, and $1.2 \leq d_b/d_w \leq 2.0$. Thus end conduction and bead-wire conduction effects should be negligible. The low temperature differences between the probe and ambient surroundings leads to negligible radiation effects.

Subject to these assumptions, the compensated temperature is related to the measured thermocouple temperatures as follows:

$$T(t) = T_m(t) + \tau(U(t), T_g(t)) \frac{dT_m(t)}{dt} \tag{1}$$

where T(t) is the compensated temperature; $T_m(t)$ is the temperature obtained directly from the thermocouple; τ is the time constant; U(t) is the velocity; $T_g(t)$ is the gas temperature; and t is time. (The time constant variation with gas composition would have to be included if there were significant changes in the gas composition.)

Using the definition of the Nusselt number it can be shown that

$$\tau = c_w m_w / Nu\ k_g \pi \tag{2}$$

where c_w is the heat capacity of the thermocouple material; m_w is the mass per unit length; Nu is the Nusselt number; and k_g is the thermal conductivity of the gas. In order to determine the functional relationship between the time constant and the velocity, the empirical relationship proposed by Collis and Williams (1959), valid for the Reynolds number range 0.02 < Re < 44, with the film temperature taken equal to the gas temperature as suggested by Bradbury and Castro (1972) is used

$$Nu = (0.24 + 0.56\ Re^{0.45}) \qquad (3)$$

where Re is the Reynolds number based on wire diameter. Substitution of the right hand side of Equations 3 into 2 leads to the explicit relationship between time constant and velocity

$$\tau = c_w m_w / \{ (0.24 + 0.56\ (\frac{Ud_w}{\nu_g})^{0.45}) \ k_g \pi \} \qquad (4)$$

where ν_g is the kinematic viscosity at the gas temperature. For the range of temperature encountered in this study, variations in ν_g and k_g are small and the gas composition is fixed. For this reason, the explicit dependences on temperature and gas composition are not shown.

In principle once the time constant is computed using Equation 4 or determined experimentally, the compensated temperature can be obtained from the measured thermocouple temperature by means of Equation 1. One apparent difficulty is that, even in the relatively simple case where variations in gas composition and temperature are not significant, as in the present study, the time constant remains a function of velocity. In the more general case of a reacting flow, the time constant is a function of velocity, gas composition and temperature. In most studies the simultaneous time resolved values of the velocity, gas composition and temperature are not available and hence neither is the instantaneous value of the time constant. In order to overcome this lack of availability, many studies make use of the mean or average value of the time constant which corresponds to the "average" flow conditions at a position of interest. In most studies of direct relevance to the study discussed herein, the time constant is determined by applying a d.c. or a.c. internal heating current to the sensor, switching off the heating current and observing the exponential decay of the thermocouple (cf. Lockwood and (1981) and LaRue et al. (1984)). With internal heating, Paranthoen et. al. (1983) have shown, due to the Peltier effect, that the correct value of the time constant determined from the initial portion of the temperature decay is in error. Heitor et. al. (1985) have also noted the same effect. Thus, the initial portion of the temperature decay (equal in time to about one-half the mean time constant) is not used in the determination of the time constant. In a turbulent flow, the mean time constant, as suggested by Yule et al. (1987) is determined as the ensemble average of the time constant determined from the appropriate portion of 50 to 100 temperature decay curves. The ensemble average is substituted for the time verying value of τ in Equation 1 and along with the instantaneous value of the temperature and temperature derivative from the thermocouple is used to evaluate the right hand side of Equation 1 and hence determine the compensated temperature. It is clear, however, as pointed out by Ballantyne and Moss (1977), that at times the thermocouple temperature will be under and over-compensated.

The uncertainty in the compensated temperature due to the use of a mean time constant provides in part the motivation for the use of a variable time constant which takes into account velocity and temperature fluctuations. Heitor et. al. (1985) appear to be the first to have used this approach but did not compare their results to those obtained using a reference system. More recently, Talby et. al. (1987) have compared spectral and statistical results obtained by means of a thermocouple compensated using a mean time constant to those obtained using a cold wire. Measurements are obtained in a heated jet and the time constant is determined using a novel technique which involves cooling the sensor with a cool air jet and then monitoring the temperature rise when the cool air is removed. Due to electronic noise, comparisons are limited to 1 kHz or less. In that range, good agreement is found between the spectra and also the probability density functions obtained using the mean compensated thermocouple and the cold wire. The present measurements extend the frequency range to 5 kHz and also include an assessment of the variable time constant.

In the next section, we discuss the experimental systems and in particular the mean and variable compensation techniques used in the present study. This section is followed by a section describing the results and then one presenting the conclusions.

3.0 EXPERIMENTAL SYSTEMS

First considered is the flow field and flow system and next a discussion of the instrumentation. Subsequently, the method used to determine the time constant is discussed which is then followed by a discussion of the signal processing and data analysis systems.

3.1 Flow Field

The test flow consists of a heated axisymmetric jet with an exit diameter, D, of 2.59 cm which is mounted at the center of a 1.4 m diameter flat plate. The jet exit is preceded by a smooth contraction section, with a 9 to 1 area ratio, a plenum, and other flow management devices. The resulting profiles of the velocity and temperature at the exit plane are "top hat" with a variation in the velocity and temperature (away from the boundary layer) of less than ±1% with corresponding intensities of less than respectively 0.4 and 0.1%.

3.2 Flow System and Conditions

The jet is supplied by house air with pressure maintained a constant by means of a single stage pressure regulator which is equipped with particle and mist filters. The air is heated by means of an insulated heat exchanger equipped with two 600 W Chromolox tubular heaters which is connected to the jet by means of insulated tubing. The jet is wrapped with heater tape and insulation to help maintain the air temperature. The jet exit velocity and temperature are monitored by means of respectively an MKS Baratron pressure transducer and an Instrulab platinum thermometer. Variations in the velocity and temperature are less than respectively ±0.1 m/sec and ±0.5°C. The ambient temperature is monitored by means of a second platinum thermometer and is found to vary by less than ±0.5°C during the course of data collection. The jet is placed away from boundaries and care is taken to minimize velocity fluctuations in the test facility.

For all the work discussed herein the jet exit velocity and temperature are respectively 26.0 m/s and about 115°C which results in a jet Reynolds number of 25,900.

3.3 Instrumentation and Traverse

Time resolved velocity is obtained by means of a CTA while time resolved temperature is obtained using both a 0.625 μm diameter cold wire and a compensated thermocouple. The hot and cold wire sensors are mounted 0.8mm apart on a TSI 1244 double sensor probe. The thermocouple is mounted on a separate shaft and is attached to a small traverse which is used to place the thermocouple about 0.7mm from the cold wire. The axes of all the sensors are aligned perpendicular to the radius along which data are obtained with the hot wire placed closest to the jet centerline. The three sensors are attached to a three dimensional traverse with a radial resolution of $\pm$0.05 mm and a vertical resolution of $\pm$0.2mm.

The hot wire is platinum, 1.2 μm in diameter, 1.25mm in length, and is operated by means of a TSI 1050A CTA bridge at overheat ratios of 1.8 to 2.1 which corresponds to wire temperatures of 250 to 350°C. While the hot wire temperature exceeds the maximum flow temperature by a factor of two or three, the wire remains sensitive to both temperature and velocity fluctuations and is therefore calibrated as a function of both temperature and velocity in a calibration jet. Calibrations of the hot wire are performed both before and after data collection. In most cases the calibrations differ only slightly though in some cases there is drift. For the data presented here, the drift is always less than 5%. The frequency response of the hot wire, determined using the square wave test, is found to exceed 20 KHz.

The 0.625 μm diameter, 0.8 mm long, platinum cold wire is operated in a dc-Wheatstone bridge with a current of 120 μa. This current is low enough that velocity sensitivity is negligible and the frequency response is estimated to exceed 6 KHz (cf. LaRue et. al. (1975)). The root mean square noise level of the cold wire is 0.01°C and the cold wire is also calibrated in the calibration jet as a function of temperature (and velocity). As expected the velocity sensitivity is negligible.

All thermocouples junctions are formed by spot welding chromel-constantan wires that are 12.7 $\pm$2 μm in diameter. At the conditions of the experiment, the wire Reynolds number is about 20. Only thermocouples with junction diameters, d_b, less than 2.5 times the wire diameter, d_w, are used. The small diameter wires are spot welded to 0.508 mm diameter support wires of the same material. The support wires are placed in insulated twin-bore tubing and are separated so that the length to wire diameter ratio is about 1200. Further details on probe construction are discussed by Seiler (1983).

3.4 Time Constant Determination

The technique used in the present study to determine the time constant is similar to that of Lockwood and Moneib (1981). The only significant difference is that a 10 KHz sinewave is used to provide the heating current. Briefly, the square wave output of a signal generator, as shown in the block diagram of Figure 1 is used to power a mercury-wetted relay (GORDOS 651C) which connects the heating current to the probe and disconnects the preamplifier. The duration of the square wave is long enough (at least 20 times the value of the time constant) so that the sensor reaches steady state. The heating current is high enough to overheat the sensor by 200 to 250°C. After about 20 τ, the square wave returns to a value of 0 volts, the heating circuit is disconnected and the thermocouple is again connected to the preamplifier which is operated at a gain of 450. The falling edge of the square wave is also used to trigger digital sampling of the temperature decay at a sample rate of 20 KHz by means of a DATA TRANSLATION DT 2801-A analog to digital converter which is connected to an IBS-2000 personal computer.

The method used to determine the time constant, which employs the IBS-2000 Personal Computer and suitable software is not straightforward in its application. For example, unavoidable noise due to relay bounce which has a duration of 0.5 to 1.0 msec is introduced into the initial portion of the temperature decay. Therefore the initial 1 ms of the decay is not used in the determination of the time constant. Consistent with the results of Paranthoen et. al. (1983) (attributed to Peltier effects) and Heitor et al. (1985), the initial portion ($\sim 0.5\,\bar{\tau}$) of the temperature decay does not have an exponential response. Therefore, the time constant is determined only for temperature decay curves in the range $0.5\bar{\tau} \le t \le 2.5\bar{\tau}$ where $\bar{\tau}$ is the average value of the time constant at a particular flow condition. In this time range, the response is reasonably well described by an exponential equation of the form

$$T_m(t) = T_i e^{-t/\tau} + T_g \tag{5}$$

where T_i is the initial temperature. The algorithm employed to determine the time constant first forms the natural log of $(T_m(t) - T_g)$. The method of least squares is next applied to this quantity as a function of time and the slope so determined is used to obtain the time constant. An advantage of this method is that T_i is not required and T_g can be obtained by averaging the temperature decay data starting at times greater than about 5 τ.

The time constant as a function of velocity is shown in Figure 2. Also shown for comparison is the time constant predicted for a 12.7 μm diameter wire using Equation 4. The predicted time constant for all velocities is less than the computed value. As noted on the figure the bead or junction diameter is about 1.2 times the wire diameter and for this reason the measured time constant is expected to be larger than the predicted value. Moffat (1958) suggests a correction factor to account for the increased bead diameter and resulting increase in the thermal inertia of the form

$$\tau_c = \tau\,(d_b/d_w)^{0.37} \tag{6}$$

where τ_c is the modified time constant. The modified time constant is also shown on Figure 2. There, similar to the results of Heitor et. al. (1985), it can be seen that the measured and modified predicted time constants are found to be in agreement. Thus using Equations 4 and 6 it is possible to obtain a "calibration equation" which provides a relationship between time constant and velocity of the form

$$\frac{1}{\tau} = \bar{A} + \bar{B}\,U^{.45} \tag{7}$$

where $\bar{A}$ and $\bar{B}$ are calibration constants which are determined from a least square fit of pairs of $\frac{1}{\tau}$ and $U^{.45}$.

In turbulent flow, since flow conditions vary as a function of time, the time constant also varies. Average or mean values of the time constant, $\bar{\tau}$, are determined, as suggested by Yule et. al. (1978), as the average of between 50 and 100 realizations obtained at the position of interest. As suggested by Gouldin (1985) the time constant is found to have negligible dependence on the temperature for the temperature range of 25 to 115°C encountered in the present study.

3.5 Signal Processing

The hot wire, cold wire, thermocouple, thermocouple derivative, and compensated thermocouple signals are recorded on an FM Instrumentation tape recorder. (The latter three signals are indicated respectively by the numbers 1, 2, and 3 on Figure 1.) The tape speed is 38.1 cm/sec which corresponds to a frequency response (3 db point) of 10 kHz. The hot wire and cold wire signals are amplified by precision buck and gain amplifiers to match the dynamic range of the tape recorder and are low passed filtered at 10 KHz by means of 8 pole Butterworth filters. As indicated on Figure 1 the signals corresponding to the temperature, temperature derivative, and compensated temperature from the thermocouple are also amplified by means of precision buck and gain amplifiers and are subsequently filtered at 5 KHz which leads to an rms noise level of less than 1°C for the temperature from the compensated thermocouple for $x/d \leq 25$ where x is the axial downstream distance measured relative to the jet exit plane. Further downstream, the velocity decreases and the time constant increases. Thus increased levels of compensation are required which in turn would lead to a noise level higher than 1°C for the same frequency range. In order to maintain the noise level at less than 1°C, the low pass filter for the compensated temperature is reduced to 2.5 kHz for $x/D > 25$.

The compensated temperature based on the mean time constant is obtained using an analogue electronic system to form the terms on the right hand side of Equation 1 and the means of implementation corresponds to the items labeled in Figure 1 as the Thermocouple Preamplifier, the Differentiator, Buck and Gain Amplifier and the Summer. The Thermocouple Preamplifier provides a signal proportional to $T_m(t)$ while the Differentiator provides a signal proportional to $\bar{\tau}\, dT_m(t)/dt$. The purpose of the Summer is clear.

Subsequently the signals from the analogue tape recorder are played back, filtered at 5 KHz, and simultaneously digitized by means of a Phoenix 16 bit analog to digital converter which is equipped with sample and hold units for each signal and controlled by a DEC 11/23 CPU. The data are digitized at a rate of 10.24 KHz and 100 records of 2048 samples per record which corresponds to 20 sec of data are stored on digital magnetic tape for analysis. Computations to obtain moments, probability density functions, power spectra and compensation are all performed on an LSI 11/73 microcomputer using appropriate software.

The algorithm used in the variable compensation method first computes the instantaneous value of the time constant by using the velocity at a particular instant in time and the inverse of Equation 7 which is expressed as

$$\tau(t) = \frac{1}{\bar{A} + \bar{B}\, U(t)^{.45}} \tag{8}$$

Combined with the thermocouple temperature and derivative obtained at the same instant in time, the right hand side of Equation 4 can be evaluated to yield the compensated temperature.

4.0 RESULTS

As an aid in the interpretation of these results it is useful to review the structure of the jet flow. On the centerline from the jet exit plane to about x/D = 3, the temperature is nearly equal to the jet exit temperature with intensities characteristic of the exit plane of the jet (0.1%). The flow becomes intermittent starting at about x/D = 3 and at about x/D = 8 is well mixed with no evidence of unmixed fluid from the potential core. Thus starting at about x/d = 3 the mean temperature begins to decrease and the root mean square (rms) begins to increase. At x/D = 3 due to the presence of only a few cool particles of fluid, the skewness will become negative and the kurtosis will increase. Downstream of this intermittent or transition region the temperature field becomes increasingly more well mixed and the skewness and kurtosis will approach values of respectively 0 and 3.

The mean velocity begins to fall starting at about x/D = 3 while the velocity intensity starts to increase starting at x/D = 1 and increases nearly linearly to a value of about 20% at x/D = 8. Between x/D = 8 and 15, the intensity increases more slowly and reaches a nearly asymptotic value of about 25% at x/D = 20.

4.1 Axial Variation of Temperature Moments and Axial Heat Flux

The axial variation of the first four moments of the temperature are shown on Figures 3 to 6. In these and other figures, $\Theta \triangleq T_g - T_\infty$, where Θ is the temperature excess; and T_∞ is the ambient temperature. The overbar on symbols in these figures corresponds to the time average and the symbol θ corresponds to the value of the temperature measured relative to the mean temperature excess, $\bar{\Theta}$. The subscripts "o" and "c" correspond respectively to the jet exit and centerline and R corresponds to the jet radius. Through the potential core and transition region, the mean, intensity, and skewness obtained from the cold wire and from the thermocouple using either the mean or variable time constant technique are essentially the same. Measurements of the kurtosis at x/d = 4 in the intermittent region differ from the cold wire reference value by about 12 and 20% for respectively the variable and mean time constant. Downstream of the intermittent region, both compensation methods yield results for the mean and kurtosis that are essentially the same as the reference cold wire results.

Further downstream, the relative intensity from the thermocouple using either compensation technique is about 10% higher than the reference value except at x/d = 10 where the differences are as high as 16%. The skewness values obtained using the variable time constant agrees with the reference values while those obtained using the mean time constant are only slightly higher. Thus, except in the intermittent or transition region temperature statistics up to fourth order obtained using either a variable or mean time constant differ by less than 10% from the reference values. Of note is that the uncertainty in the even moments appears to be somewhat higher in the intermittent region of the flow.

The axial heat flux correlation as shown in Figure 7 obtained using either compensation technique appear to lead to values less than the reference value starting at about x/d = 6. The maximum difference occurs near the end of the intermittent region at x/d = 8 where the reference value of the axial heat flux reaches a peak value of 0.644 and the results corresponding to the variable and mean compensation techniques are respectively 11 and 17% lower. The velocity intensity is about 20% at this location. It is not surprising that the heat flux determined from the compensated thermocouple is in general lower than the reference values as the thermocouple is placed further from the hot wire than is the cold wire. However, the relative differences must be related to the method of compensation.

4.2 Time Series

Temperature time series from the cold wire and the compensated and uncompensated thermocouple are shown on Figure 8. The two compensation techniques lead to signals which differ only slightly from the reference cold wire signal. The potential core intermittency is clearly visible in these time series. When fluid from the potential core traverses the measurement location, the cold wire time series displays a flat plateau while jet fluid mixed with ambient air corresponds to the more variable portion of the signals. Some oscillations due either to noise or non-first order behavior are seen in the high temperature portions of the compensated thermocouple time series. The interfaces between turbulent (low temperature) fluid and potential core (high temperature) fluid are nearly coincident for all three signals indicating negligible phase lag. Small differences are to be expected as the thermocouple sensor and the cold wire sensor are displaced by about 0.7 mm.

4.3 Probability Density Functions

Probability density functions (pdf) for the compensated and uncompensated thermocouple and the cold wire at three representative radial locations are shown in Figure 9. At the centerline, the pdf is distinguished by a very large spike approaching a delta function at the jet exit temperature. This is consistent with the intermittency noted in the temperature time series at this location. Note also the peak value in the cold wire pdf is higher by a factor of more than two than the peak value of the pdf for either of the compensated thermocouple signals. This lower value for the peak is attributable to the higher level of noise in the compensated thermocouple signal relative to that in the cold wire signal. The effect of noise is to increase the width and decrease the peak value of a spike in the pdf and also to lead to the appearance of temperature entries in the pdf at temperatures higher than the potential core temperature. The rms noise at this location is about 0.5°C. Thus entries as high as 1.5°C above the potential core temperature might be due to noise. This argument does not account for the high maximum temperatures exhibited in the pdf for the variable time constant signal at r/R = 0.00. While half of the relative shift of about 3.5°C is probably due to noise, the other half is probably due to an error in the d.c. offset introduced in the signal processing.

The probability density functions for three radial locations at x/d = 15 are shown in Figure 10. The velocity intensity is twice that at x/d = 5 and the mean velocity is about half. Thus variations in the time constant will be increased relative to those found at the five diameter station. A close inspection of the pdf at the two inner radial locations shows that the variable time constant method yields results that are in better agreement with the reference pdf that does the mean time constant method. However at the outer radial position both methods lead to overcompensation as evidenced by temperature entries in the pdf lower than those measured by the cold wire.

4.4 Power Spectra

Power spectra at positions corresponding to those of Figure 9 are shown on Figure 11. At all three positions, it is clear that both compensation techniques recover the detailed structure of the power spectra and extend the frequency response by about two orders of magnitude. At all positions there appears to be a slight (1/4 to 1/2 db) overcompensation at low frequencies. Some of these differences are due simply to uncertainty arising in the calibration and signal processing. This is consistent with the observation that at low frequencies the spectral level of the uncompensated thermocouple at three of the four radial positions is slightly higher than that of the cold wire. The moderate to high frequency portions of the spectrum from 100 up to about 2 kHz are in good agreement with the reference spectra. The effect of noise starting at about 2 KHz is clearly visible in the power spectra of the compensated thermocouple at the largest radial location where the signal level is the lowest and the time constant is relatively large.

5.0 CONCLUSIONS

At five diameters downstream of the jet exit plane, use of either the variable or mean time constant leads to results from the compensated thermocouple for the time series, temperature moments up to fourth order, probability density functions and spectra that are in generally good agreement with the corresponding results obtained from a reference system consisting of cold wire. However, fifteen diameters downstream where the mean velocity is lower and the velocity intensity is higher, a comparison of the probability density functions at several radial locations indicates that the variable time constant method yields results that are in better agreement with the reference results. Thus, in flows with high relative intensities, especially at lower values of the mean velocity, the variable time constant will yield more accurate results.

ACKNOWLEDGMENT

The authors wish to acknowledge Esteban Gonzalez, Russell B. Jones, and Hoi Hung Cheong for their help in preparing the figures, and Janice Johnson and Verna Bruce for their excellent preparation of the manuscript.

1. Ballantyne, A., Moss, J.B., Fine Wire Thermocouple Measurements of Fluctuating Temperature. Comb. Sci. Tech. 17, p. 63-72, 1977.

2. Bradbury, L.J.S., Castro, I.D., Some Comments on Heat-Transfer Laws for Fine Wires, J. Fluid Mech. 51, p. 487-495, 1972.

3. Collis, D.C. and Williams, M.J., Two Dimensional Convection from Heated Wires at Low Reynolds Number, J. Fluid Mech. 6, p. 357-384, 1959.

4. Gouldin, F.C., An Analysis of the Use of Compensated Thermocouples in Variable Temperature, Turbulent Flow. Presented at the Spring Meeting of the Western States Section of the Combustion Institute, San Antonio, TX, 1985.

5. Heitor, M.V., Taylor, A.M.K.P., and Whitelaw, J.H., Simultaneous Velocity and Temperature Measurements in a Premixed Flame, Exp. in Fluids, 3, p. 323-339, 1985.

6. LaRue, J.C., Deaton, T., and Gibson, C.H., Measurement of High Frequency Turbulent Temperature, Rev. Sci. Instrum., 46, p. 757-763, 1975.

7. LaRue, J.C., Samuelsen, G.S., Seiler, E.T., Momentum and Heat Flux in a Swirl Stabilized Combustor, Twentieth Symposium (International) on Combustion, the Combustion Institute, p. 277-285, 1984.

8. Lockwood, F.C. and Moneib, H.A., Fluctuating Temperature Measurements in a Heated Round Free Jet, Comb. Sci. and Teh., 22, p. 63-81, 1980.

9. Moffat, R.J., Designing Thermocouples for Response Rate, Trans. ASME, 80, p. 257-262, 1958.

10. Paranthoen, P., Lecordier, J.C., Petit, C., Gajan, P., Survey and Recent Developments of Frequency Response Studies of Cold Wires and Fine Wire Thermocouples in Turbulent Heated Flows, 8th Australian Fluid Mech. Conf., Univ. Newcastle, N.S.W., p. 6B.1-6B.4, 1983.

11. Seiler, E.T., Compensated Fine-Wire Thermocouple Measurements in a Complex Reacting Flow, Masters of Science Thesis, Dept. of Mech. Engr., Univ. of CA, Irvine, 1984.

12. Shepard, C.E. and Warshawsky, I., Electrical Techniques for Compensation of Thermal Time Lag of Thermocouples and Resistance Thermometer Elements, NACA 2703, 1952.

13. Talby, R., Anselmet, F., and Fulacher, L, "Temperature Fluctuation Measurements with Fine Thermocouples," Paper 7.4, Sixth Symposium on Turbulent Shear Flows, Toulouse, France, September 7-9, 1987.

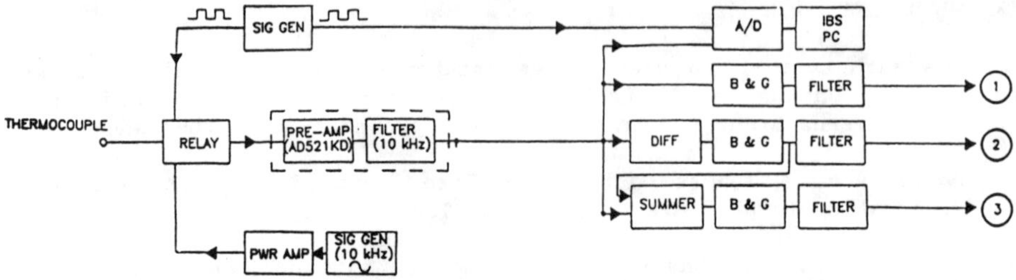

Figure 1. Block diagram of signal processing and compensation electronics for the thermocouple

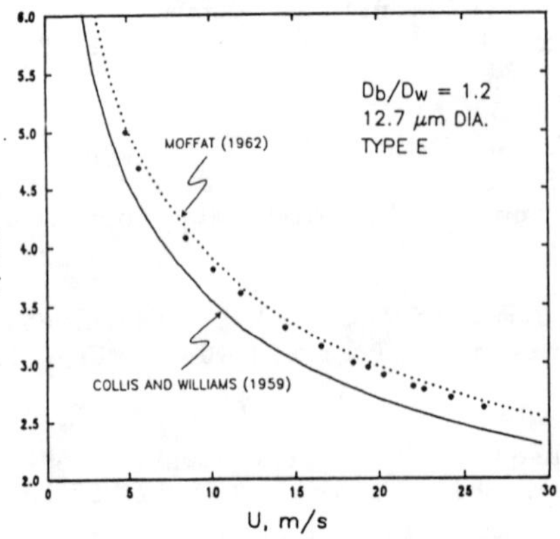

Figure 2. Time constant variation as a function of velocity

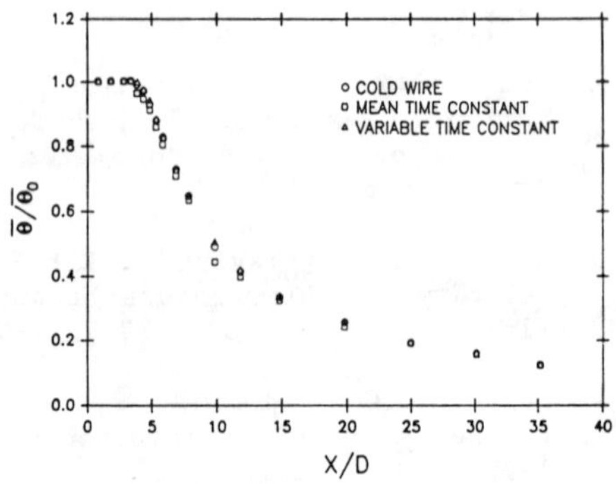

Figure 3. Axial centerline variation of mean temperature

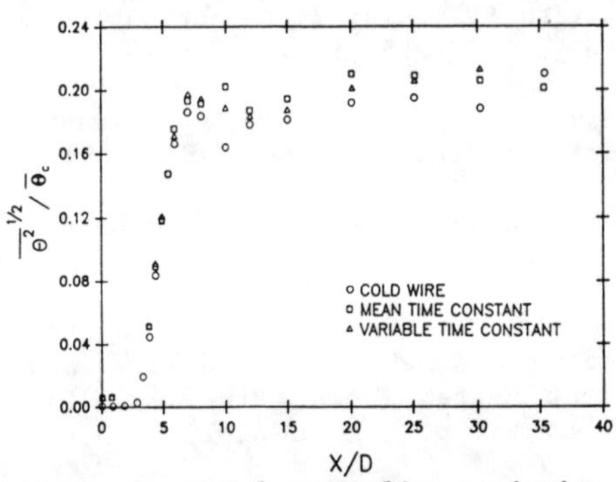

Figure 4. Axial centerline variation of temperature intensity

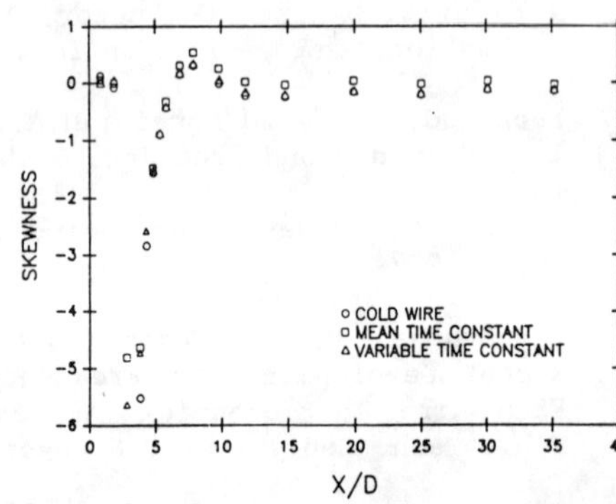

Figure 5. Axial centerline variation of skewness

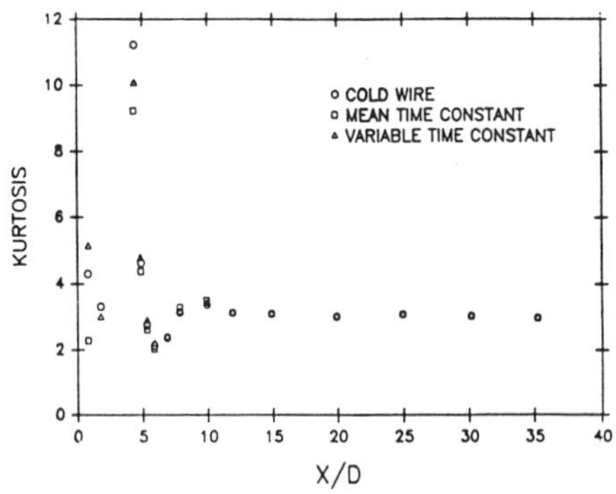

Figure 6. Axial centerline variation of kurtosis

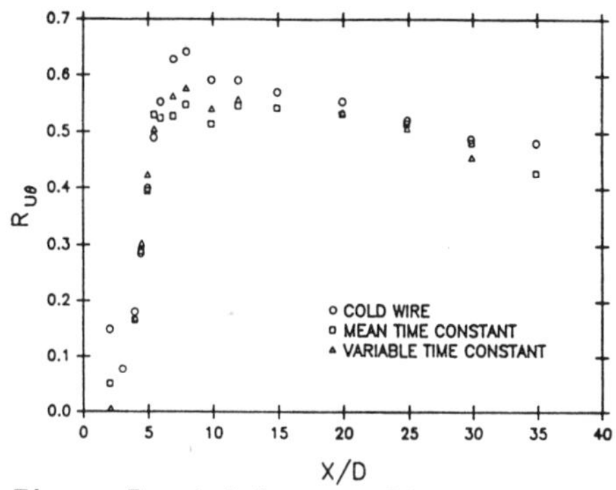

Figure 7. Axial centerline variation of heat flux

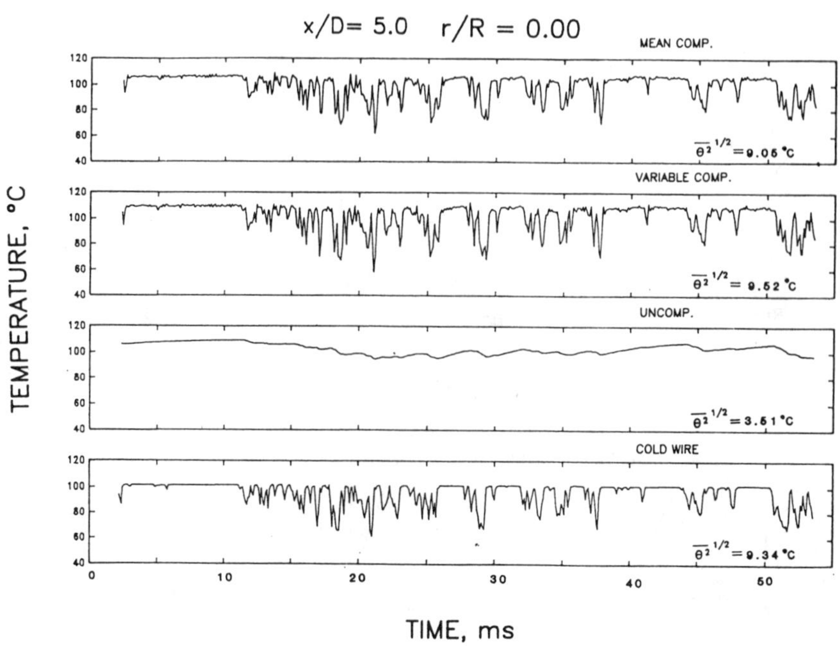

Figure 8. Time series of the temperature

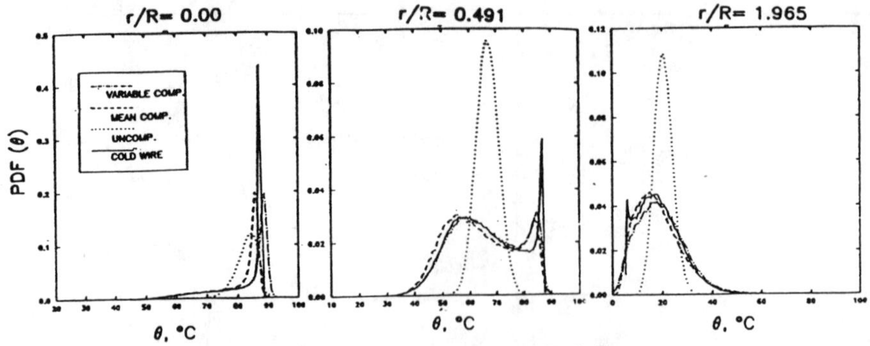

Figure 9. Probability density functions at x/D=5.0

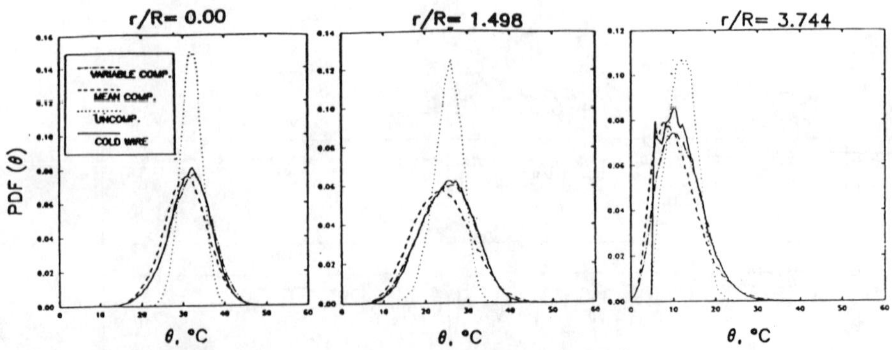

Figure 10. Probability density functions at x/D=15.0

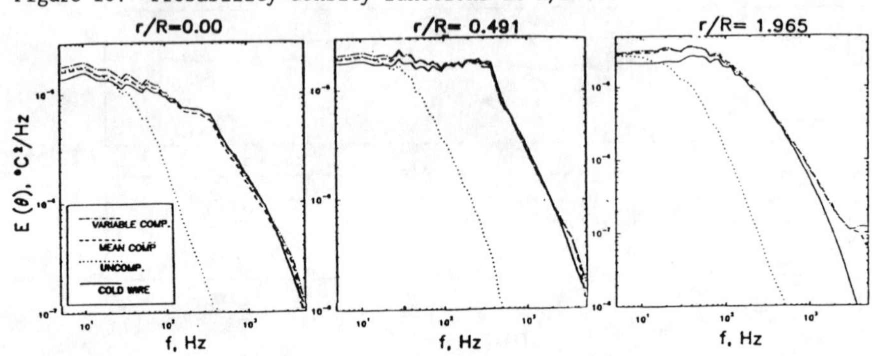

Figure 11. Power spectra at x/D=5.0

LASER DIAGNOSTICS FOR GAS TURBINE THERMOMETRY AND SPECIES MEASUREMENTS

ALAN C. ECKBRETH

United Technologies Research Center
Silver Lane, MS 90
East Hartford, Connecticut 06108 USA

1. INTRODUCTION

With the increasing availability and reliability of laser sources, laser diagnostic techniques are assuming an ever broadening role in probing the hostile, yet easily perturbed, environments characteristic of practical combustion processes. Physical probes, due to their intrusion, can seriously perturb the fundamental flame behavior being investigated, are often limited in their spatial resolution and temporal response, and, in addition, are confronted with survival at high temperatures and pressures. Laser spectroscopic methods are, by definition, nonintrusive and, thus, capable of very high temperature probing. They offer the potential for remote, nonperturbing, in-situ examination of combustion phenomena and are capable of simultaneously high spatial (10^{-3} cm^3) and temporal ($\leq 10^{-6}$ sec) resolution. In practice however, compromises in either or both scales may be sometimes necessary.

Laser probing can play two major roles in the development of gas turbine combustors. The first, which has seen the most application to date, is in the fundamental arena. Here spatially-precise and simultaneous measurements of temperature, species and/or velocity are used to gauge the validity and accuracy of numerical codes and attendant sub-models used to predict gas turbine combustor behavior. Such measurements are made in combustion tunnels, generously equipped with optical access, where practical behavior can be simulated and the various constitutive processes isolated. The second role is in empirical development and performance testing where the characteristics and capabilities of laser techniques may prove advantageous compared with traditionally employed probes. In this arena, laser techniques have not yet made much impact for several reasons. Optical access to the primary combustor in an actual gas turbine engine is a formidable problem. To date, the only applications of laser techniques have been to exhaust measurements where optical viewing is not a problem. In the future, advances in fiber optics and long-pulse laser sources may permit primary combustor diagnostics. Laser techniques are generally complex and expensive requiring highly-trained operators. In the very economically competitive gas turbine industry, it is difficult to replace fairly standardized instrumentation where large investments have already been made, e.g., gas sampling. Laser techniques provide more information,

D. F. G. Durão et al. (eds.), Instrumentation for Combustion and Flow in Engines, 69–106.

such as fluctuation magnitudes, correlations, etc., which the design system currently can't accommodate. Nevertheless, as manufacturers push to higher pressures and temperatures for increased performance, physical probes will be severely compromised and laser approaches may be anticipated to make inroads into this area also.

In this chapter, the application of laser diagnostics to temperature and species measurements in gas turbine combustion will be treated. Many of the considerations to be reviewed are equally germane to internal combustion engines as well. Other chapters in this volume deal with the use of lasers for velocimetry and droplet/particle sizing and these techniques would have to be integrated with those discussed here for simultaneous measurement of all the parameters of interest. Emphasis here will be on pointwise measurements, whereas other chapters will consider simultaneous measurements over an entire two-or three-dimensional field. In the next section, various laser diagnostic techniques, potentially applicable to gas turbine thermometry and species measurements, will be reviewed and compared. From this screening, the most appropriate techniques will emerge. Their practical applicability will be judged by considerations such as interferences, perturbations and optical access which will be discussed in the subsequent section. The most applicable laser approach, coherent anti-Stokes Raman spectroscopy (CARS) will then be reviewed. The chapter concludes with an examination of advanced CARS techniques.

2. LASER DIAGNOSTICS FOR TEMPERATURE AND SPECIES

2.1 Advantages/Disadvantages

Laser diagnostics are nonintrusive by definition, i.e., nothing is inserted other than photons, and remote; thus, they have unlimited temperature capability. Laser techniques are generally, but not always, nonperturbing which is the ultimate goal of any diagnostic approach, nonintrusive or physical. Laser techniques are clearly well suited to the delicately stabilized recirculation zones typical of gas turbine primary zones. The techniques to be considered here are fast and spatially-precise, i.e., measurements are made in a localized region and often with a single laser pulse of duration 10^{-6} sec or less. The techniques for the most part are in-situ and, thus, unambiguous. Corrections for radiation, conduction, chemical transformation, etc., are not necessary.

The requirement for optical access, particularly in a jet engine context, is a major disadvantage, but not necessarily insurmountable. The techniques require expensive laser, spectrographic and computer equipment and generally highly skilled personnel to operate them. Signal strength is not unlimited and, thus, compromises in spatial and/or temporal scales as well as detectivity limitations must be considered. Interferences can limit ranges of applicability, although the more successful approaches can be broadly utilized. Lastly, droplet and particulate loadings must be such to permit reasonable laser and signal beam transmissions. Clearly, optical techniques will not be successful if the opacity of the medium is too large.

2.2 Light Scattering and Wave Mixing

Before beginning the survey of laser diagnostics, it is instructive to review the origin of the various physical phenomena being exploited. In view of the increasing exposure that combustion researchers will likely have to laser spectroscopic techniques in the coming years, an acquaintance with the nature of the processes would seem beneficial. In this vein, a classical picture will suffice here, although detailed explanations generally need to invoke quantum mechanical treatments. All optical phenomena are governed by Maxwell's equations which can be manipulated to yield the wave equation[1]

$$\nabla^2 \vec{E} - \frac{1}{c^2}\frac{\partial^2 \vec{E}}{\partial t^2} = \mu_o \frac{\partial^2 \vec{P}}{\partial t^2} \tag{1}$$

where E is the electric field of the incident electromagnetic light wave, c is the speed of light in vacuum and P is the generalized electric polarization. Often several monochromatic or quasi-monochromatic field components are involved and E and P can be expanded into Fourier components. The induced polarization can be expressed as power series of E (ω_i), namely

$$\begin{aligned} P_i(\omega_i) &= \epsilon_o \chi^{(1)}(\omega_i)\, E(\omega_i) + \sum_{j,k} \epsilon_o \chi^{(2)}(\omega_i,\omega_j,\omega_k)\, E(\omega_j)E(\omega_k) \\ &+ \sum_{j,k,l} \epsilon_o \chi^{(3)}(\omega_i,\omega_j,\omega_k,\omega_l)\, E(\omega_j)E(\omega_k)E(\omega_l) + \ldots \end{aligned} \tag{2}$$

$\chi^{(1)}$ is the linear susceptibility of the medium, the $\chi^{(n)}$ are the nth order nonlinear susceptibilities, so named because they express how susceptible the medium is to be polarized. The polarization is given by the vector sum of the induced dipole moments per unit volume. The polarization and susceptibility are macroscopic properties of the medium. The dipole moment is related to the electric field through the molecular polarizability which is a microscopic or molecular property. The effects of the nonlinearities in the polarization become manifest only at very high laser intensities as will become clear shortly. The induced linear polarization modifies the propagation of the light wave through the medium, accounted for by introduction of the refractive index. Dispersion and absorption phenomena accompany the real and imaginary parts, respectively, of the complex refractive index usually modeled classically by an electron oscillating in a harmonic potential well. Raman and Rayleigh scattering arise from the oscillating polarization induced through the linear susceptibility. Recall that an oscillating polarization, e.g., electric dipoles, produces electromagnetic radiation. Rayleigh scattering arises from the induced polarization oscillating at the same frequency as the incident radiation. In general, the polarizability and, hence, the induced polarization, are dependent upon the nuclear positions of the molecule and are thus modulated by the rotation and/or vibration of the molecule. This leads to Raman scattering which is frequency shifted from the incident light by the rotational/vibrational frequency and which may be interpreted as the beat frequency between the incident radiation and nuclear motions.[2]

The higher order polarizations are considerably weaker and to an approximation, the succeeding polarizations may be expressed as[3]

$$P^{(n+1)}/P^{(n)} = E/E_{at} \tag{3}$$

where E_{at} is the intra-atomic electric field and typically of order $3(10^8)$ V/cm. Even at high laser intensities of 10^9 W/cm^2, the ratio of succeeding polarizations is small, about 10^{-3}. Thus, the nonlinear polarization is small, about 10^{-3}. Nonlinear phenomena were not experimentally observed until the early sixties when giant lasers were produced by Q-switching. Today such lasers are commonplace and readily available commercially. Nonlinear processes in actuality are very strong at high laser intensities due to their coherent nature and the presence of resonance denominators in the nonlinear susceptibilities. In isotropic media such as gases, there are no second order effects due to inversion symmetry. The lowest order nonlinearities in a gas are thus third order in incident electric field strength and arise through the third order nonlinear susceptibility. Important examples of third order processes for combustion diagnostics are CARS and stimulated Raman gain/loss spectroscopy.

The third order nonlinear susceptibility $\chi^{(3)}$ is complex and may be written as

$$\chi^{(3)} = \sum_j (\chi'_j + i\chi''_j) + \chi_{nr} \tag{4}$$

where χ'_j, χ''_j are the real and imaginary components of the resonant susceptibility from the species whose Raman resonances j are being interrogated. χ_{nr} is the nonresonant electronic background susceptibility contributed by the molecular electronic clouds of all of the constituents present in the gas mixture. Various nonlinear Raman processes exhibit different functional differences on $\chi^{(3)}$ or its components.

2.3 Survey of Laser Diagnostic Techniques

2.3.1 Incoherent Approaches. In these approaches, the signal induced by the exciting laser beam is scattered into essentially 4π steradians in accordance with dipole radiation physics. These techniques are implemented in the manner shown in Fig. 1. The scattered signal light is collected at some angle θ to the incident laser beam over some solid angle Ω determined by the lens and spectrographic system employed. The sampling extent, ℓ, is determined by triangulation of the detection system with the laser beam and is typically a few millimeters. The laser beam is focussed to several hundred micrometers resulting in a cylindrical sample volume. An advantage of these approaches is that they can be implemented in backscattering with the laser and optical collection axes coaxial permitting single ported optical access. In all of the spectroscopic techniques to be described, concentration information derives from the strength of the scattering and temperature from the spectral distribution, i.e., the intensity variation with frequency.

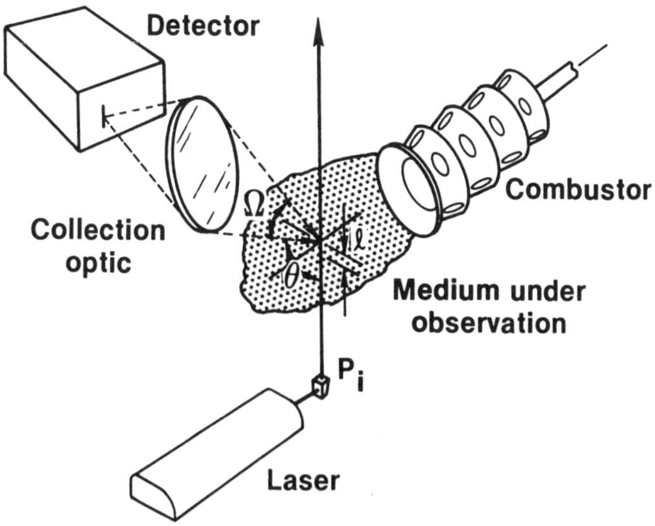

FIGURE 1. Laser light scattering optical measurement schematic.

The prominent incoherent scattering processes are depicted in Fig. 2. The parabolic shaped curves represent electronic states whose potential energy as a function of internuclear separation is shown. The horizontal lines represent allowed vibrational energy states of the molecule discretized in accordance with quantum mechanics. Rotational energy level splitting is not shown explicitly but occurs for each vibrational state. The molecules are partitioned among the various energy states in accordance with Boltzmann statistics. By interrogating the population distribution, temperature information is extracted. Knowing this and the number density in any one state from the signal magnitude permits the total species concentration to be ascertained. Strictly speaking, vibrational-rotational temperatures are measured. In high pressure combustion situations, these temperatures are equilibrated with the translational temperature of measurement interest.

Rayleigh scattering is the elastic scattering of light quanta or photons from molecules and is the phenomenon giving rise to the blue appearance of the sky. Because the scattering is elastic, i.e., no energy exchange between the photons and molecules, the scattered signal light is unshifted in frequency and, hence, not specific to the molecule causing the scattering. Thus the technique cannot be used for individual species measurements. It can be used for total density measurements if the composition is known. In low Mach number situations, temperature can be obtained from the total density through the gas equation of state assuming the pressure is known. Because of its signal strength, Rayleigh scattering has found considerable use in fundamental turbulent combustion studies.[4] From a practical viewpoint, Rayleigh diagnostics suffer from Mie scattering interferences (from particles) and spuriously scattered laser light and can only be employed in essentially particle-free situations. For this reason, it is not suitable for gas turbine combustor probing.

Raman scattering is the inelastic scattering of light from molecules as illustrated in Fig. 2 and is termed rotational, vibrational or electronic depending on the nature of the energy exchange which occurs between the molecule and photons.[2] The process is essentially instantaneous occurring within a time of 10^{-12} sec or less. The molecule may either become excited or deexcited depending on its original state prior to the interaction. Raman scattering is ideally suited to combustion diagnostics and has been widely applied in clean flames.[1,5] Visible wavelength lasers are employed since the strength of the scattering scales as the fourth power of the Raman frequency, but no specific wavelength is required. Due to the quantization of the molecular energy states, the Raman spectrum resides at fixed frequency separations from the laser line characteristic of the molecule from which the scattering emanates. Thus the Raman scattering is species specific and linearly proportional to species number density. Furthermore, spectral interferences between vibrational Raman bands in gases are rare. With proper detection, many species can be monitored simultaneously since the scattering occurs concurrently from all of the constituents present. Temperature measurements are readily made from the distribution of the scattering. Unfortunately, Raman scattering is very weak with cross sections typically around 10^{-30} cm^2/sr, resulting in a collected Raman to laser energy ratio of approximately 10^{-14} in flames. In practical combustion situations, Raman scattering diagnostics are typically plagued by low signal to interference (S/I) ratio. Despite all of its advantages, the weakness of Raman scattering limits its applicability and leads to the utilization of stronger processes, e.g., CARS, for probing gas turbine type flames.

Fluorescence is the emission of light from an atom or molecule following promotion to an excited state by various means: electron bombardment, heating, chemical reaction (chemiluminescence) or photon absorption (Fig. 2). Here only the last means will be considered. The precise definition of fluorescence requires that emission occur between electronic energy states of the same multiplicity, i.e., same electronic spin states. Emission between states of different electronic spin is termed phosphorescence. In general, fluorescent lifetimes vary between 10^{-10} and 10^{-5}sec, much shorter than

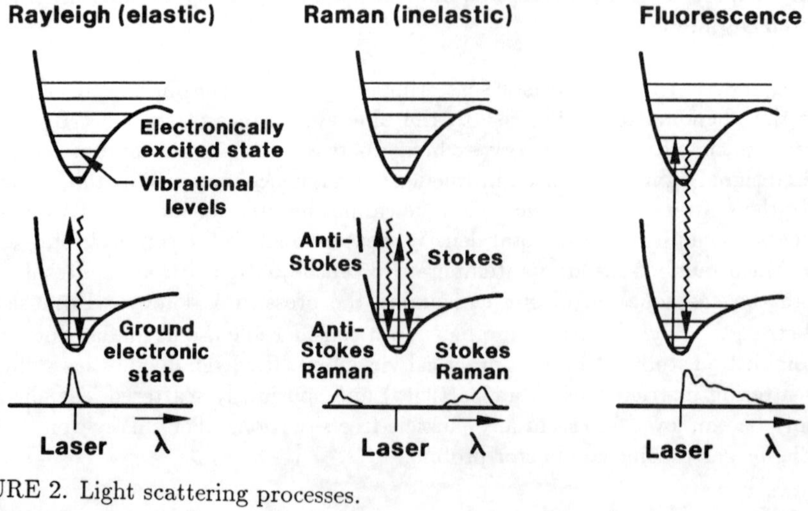

FIGURE 2. Light scattering processes.

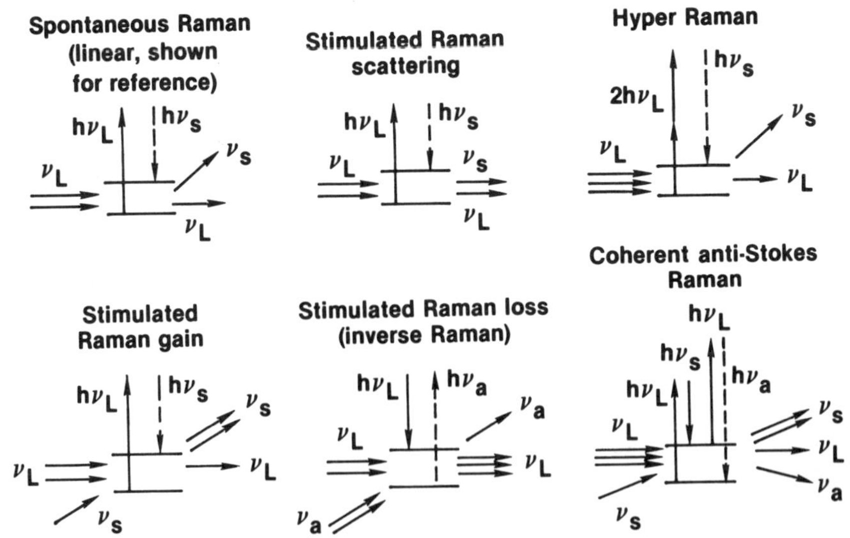

FIGURE 3. Nonlinear optical effects illustrated.

the phosphorescent lifetimes of 10^{-4}sec to seconds. The light emission may be shifted in wavelength from the incident light, fluorescence, or occur at the same wavelength, resonance fluorescence. In general it is desirable to examine shifted emission to avoid potential interferences from particle (Mie) or spurious laser scattering. Fluorescence is of diagnostic interest since it is species specific, and the cross sections for fluorescence are generally many orders of magnitude stronger than those for Raman scattering. Fluorescence detection is of course possible only if the molecule has an absorption wavelength accessible to a laser source, i.e., > 190 nm. Most of the major constituents in airfed hydrocarbon-fueled combustion possess absorptions in the vacuum uv and are not amenable to fluorescence detection. Fluorescence is typically employed to detect flame radicals at ppm levels and is of considerable importance in chemical kinetics and flame structure studies. In order to focus this paper, and since temperature and major species are most critical to characterizing the performance of gas turbine configurations, fluorescence will not be considered further. Rather the emphasis will be on Raman processes, both incoherent and coherent, which can provide temperature and major species concentration measurements.

2.3.2 Coherent Methods. This class of methods produces signals encoded in coherent, or laserlike, signal beams either through modulation of a probe laser beam or by generation of a nascent beam. The coherent methods to be considered here are all third-order nonlinear processes involving at least two laser beams which can be crossed to achieve high spatial resolution. Absorption spectroscopy can be considered a coherent technique if a laser source is employed. It is not included here since it is line-of-sight and is not spatially precise unless a sophisticated crossed-beam variant is employed. Some of the more common nonlinear optical techniques are displayed in Fig. 3. It should be pointed out that not all nonlinear techniques are coherent.

Stimulated Raman scattering is compared to spontaneous Raman scattering in Fig. 3. Under intense laser radiation, the spontaneously generated Stokes photons experience exponential growth in the direction of laser propagation and emerge as a coherent beam. The technique, despite high conversion efficiencies in certain circumstances, is not promising for gas phase diagnostics for several reasons. It can only be generated in certain selected gases, e.g., N_2, H_2 and then only at many atmospheres of pressure with very high intensity laser pulses. Only selected lines of the Raman spectrum emerge which make data interpretation ambiguous. The effect to date has not been applied for the above reasons to gas phase diagnosis.

Stimulated Raman gain/loss spectroscopy employs two laser beams whose frequencies differ by the Raman shift of the molecule being probed. In the gain variant, a probe laser Stokes-shifted to lower frequencies from an intense pump laser experiences gain. In the analogous loss process, an anti-Stokes-shifted probe beam experiences pump-induced absorption; this process is also termed inverse Raman scattering and was demonstrated first. In the original experiments,[6] two collinear beams, one a high intensity monochromatic pump source, the other a broadband continuum, traversed the medium under observation. Under high enough pump intensity, absorption of the continuum was observed on photographic plates at the anti-Stokes frequencies. Intensities just short of producing stimulated effects were required to produce darkening of the plates since the induced absorptions in low density gases are very weak. In either case, the signal originates through the imaginary component of the third order nonlinear susceptibility, and, at small gains, is linearly proportional to gas density. In contrast to CARS, the technique requires no phase matching, i.e., the beams can be arbitrarily oriented, and does not contain contributions from the nonresonant background susceptibility. It has been extensively utilized for high resolution spectroscopic studies,[7] but has seen relatively little diagnostic use relative to CARS for several reasons. Because stimulated Raman gain/loss spectroscopy is a beam modulation approach, it is difficult to perform in a broadband or multiplex fashion on optical multichannel detectors at the small gains, i.e., $< 10^{-2}$, typical of gas phase work. Such approaches are required for single pulse measurements which are typically necessary in turbulent combustion investigations. It is also difficult to reference since it contains no dependence on the nonresonant background susceptibility. Referencing is necessary in turbulence work since refractive index gradients may cause laser beam steering and uncrossing which can contribute or mask the probe beam modulation.

Raman-induced Kerr effect spectroscopy (RIKES) is somewhat similar to the inverse Raman effect in that an intense monochromatic pump source interacts with a broadband probe source to produce a birefringence at the Raman frequencies. The effect is resonantly enhanced when the frequency difference between the pump and the probe beams is tuned to the vibrational Raman resonances of the molecule being probed. To avoid interferences from the optical Kerr effect, the pump beam is circularly polarized. The probe beam is linearly polarized and blocked by a crossed polarizer. In the presence of the pump source, the probe beam experiences birefringence at the Raman frequencies which are transmitted through the crossed polarizer to a suitable detector. Practically, RIKES suffers from the rejection achievable with crossed polarizers which, at best, is about 10^{-6}. Due to slight birefringence in intervening optical elements, e.g., lenses, windows, practically achievable rejection is less. Turbulent fluctuations in combustion media may also produce undesired anisotropies. Clearly if the effect is not stronger than the best rejection achievable, no signal is observed which may well be the case in gases particularly at low concentrations. Until demonstrated to the contrary, this phenomenon does not appear promising for gas-phase combustion diagnostics.

Hyper Raman processes are illustrated in Fig. 3 and result in Raman spectra relative to the second harmonic of the exciting laser frequency. The scattering occurs into all solid angles, i.e., incoherent, and is very weak even at tens of atmospheres of gas pressure with very high intensity laser sources. The technique has no utility for combustion diagnosis.

Coherent anti-Stokes Raman scattering or spectroscopy (CARS) has received considerable attention over the past several years for combustion diagnosis based upon the pioneering investigations of Taran in France.[8] It is also schematically diagrammed in Fig. 3. In brief, two laser sources at ω_1 (pump) and ω_2 (Stokes) generate an intense coherent beam at $\omega_3 = 2\omega_1 - \omega_2$ (anti-Stokes or CARS) when the frequency difference $\omega_1 - \omega_2$ is tuned to a molecular vibrational resonance. The effect has no threshold per se and possesses two major advantages. The CARS signal produced is, first, many orders of magnitude stronger than conventional Raman scattering and, second, emerges as a coherent beam so that it can be completely collected. Since it is in the anti-Stokes region, it resides in a region free of fluorescent interferences for the most part. Because it is a signal generative process, it is easily multiplexed with broadband ω_2 Stokes sources and optical multichannel detectors. In addition, due to its dependence on $|\chi^{(3)}|^2$, it contains signal contributions from the nonresonant background and can be referenced to compensate for turbulent refractive effects. In fact, in certain concentration ranges, e.g., 0.5 to 30%, concentration measurements can be based on spectral signatures and not on absolute signal strengths, a significant experimental advantage. Among its disadvantages, relative to spontaneous Raman scattering, is its double-endedness, i.e., two opposed optical ports are required, its more complicated spectra, and its nominal restriction to measuring one species at a time. A major portion of this chapter will be devoted to CARS because of its very broad practical applicability. Advanced CARS techniques will also be examined which permit many constituents to be measured simultaneously.

From this review, two techniques emerge as being most attractive for temperature and major species concentration measurements in gas turbine combustors or simulations thereof, namely spontaneous Raman scattering and coherent anti-Stokes Raman spectroscopy (CARS). However a number of practical considerations need to be taken into account before the feasibility of application can be safely ascertained.

3. PRACTICAL CONSIDERATIONS FOR APPLICATION TO GAS TURBINE COMBUSTORS

Gas turbine combustion devices possess flames which, generally, from an instrumentation standpoint, differ markedly from flames often investigated in fundamental or laboratory devices. Many diagnostic approaches developed for use on relatively benign flames in ideal laboratory environments are often unsuited for practical application. Consequently, in evaluating any diagnostic approach, the ultimate environment of application must be factored into any decision concerning the feasibility of a given technique. The application of interest here is a gas turbine combustor. Like most practical combustors it contains flames which are highly luminous and particle laden when burning a hydrocarbon fuel. The high luminosity levels preclude the use of a continuous wave laser source for most diagnostic applications. However, use of a pulsed laser source engenders a variety of laser induced particulate effects which can mask detection of the sought-for signal. The soot particles produced will absorb and scatter the incident laser and emerging signal radiation. This, at the very least, will diminish species sensitivity and, at worst, preclude successful application of any optical approach. In addition, there are other effects common to both laboratory and practical devices, such as gas breakdown, which need to be considered. Here the more important of these various considerations will be reviewed to assist in assessing the feasibility of various diagnostic techniques. More detailed discussions are contained in Refs. 1 and 9.

3.1 Interferences

Interferences are radiations, either naturally occurring or laser induced, which interfere with or mask the detection of the signal radiation. These include background luminosity, laser-induced effects, laser-window interactions and spurious laser scattering. The latter can be a problem even when viewing spectrally-shifted signal radiation due to insufficient off-detection-band rejection. This can generally be overcome by increased spectral filtering and with suitable beam trapping geometries. Laser window interactions, i.e., Fresnel reflections and fluorescences, can also be avoided by proper experimental design.

3.1.1 Background Luminosity. Most laser diagnostic techniques are situated in the visible or ultraviolet portion of the spectrum, and consequently, concern is centered on flame source radiations in this spectral region. Here luminosity is comprised primarily of chemiluminescent emissions and the gray/blackbody continuum from soot. The former arise from electronic states excited during chemical reactions. Although these emissions are dominant in near stoichiometric, premixed flames, their contribution to the total luminous intensity in hydrocarbon-fueled diffusion flames is generally small compared to the soot continuum. Furthermore, chemiluminescent interferences can be avoided through proper laser wavelength selection.

For illustration purposes, Table 1 lists Raman signal/interference ratio calculations for background luminosity emitted by a dispersion of soot particles 40 nm in dia. at a concentration of 10^8 cm^{-3} and flame temperature of 2000 K. Such a dispersion, corresponding to a volume fraction of $3(10^{-9})$, would be typical of a lean to stoichiometric primary zone. The luminosity was calculated by merely summing the individual particle radiation contributions, i.e., optically thin, calculated from the Planck radiation formula accounting for small particle "radiation efficiencies." At 500 nm, the radiation energy density emitted by this dispersion is on the order of 10^2 nanowatts/(cm^3 sr nm). As can be seen from the Table, with proper laser selection and spatial resolution (which restricts the volume viewed), Raman scattering appears feasible, although potentially marginal in certain cases. At higher soot volume fractions corresponding to rich primary zones, background luminosity can be anticipated to be quite a severe problem.

Table 1

N_2 RAMAN SIGNAL/BACKGROUND LUMINOSITY INTERFERENCE RATIOS (FLAME: 2000 K, 1 ATM; PARTICULATES; 10^8 cm^{-3}, 40 nm)

Laser	Power (W)	Laser Wavelength (nm)	Raman Wavelength (nm)	Sampling Extent 1 cm	Sampling Extent 1 mm
2xNd	5(7)	532	607	11	1100
Dye	2(6)	590	684	.14	14
Ruby	6.7(7)	694	828	1.2	120
2xRuby	3.3(7)	347	378	3.8(3)	3.8(5)

3.1.2 Laser-induced Interferences. These involve laser interactions with both molecular constituents and particles. Of the former, laser-induced fuel fluorescences are likely to be the most severe particularly in fuel-rich primary zones. These are generally difficult to quantify a priori and can be minimized by using laser wavelengths in the visible region of the spectrum. The use of uv laser sources, which is attractive to minimize background luminosity and enhance Raman scattering signal strength, may not be a viable approach if fuel fluorescences are too severe. This is an area where further investigations in simulated gas turbine combustors would be desirable. In interactions with soot particles, the laser may elastically Mie scatter, Raman scatter, induce fluorescences and heat the soot particles to greater incandescence. Under high soot loadings, Mie scattering is a major cause of signal loss due to reduced laser and signal transmission. From an interference standpoint, the unshifted Mie scattering can generally be rejected with sufficient spectral filtering. Particulate Raman scattering is generally not a problem, nor is fluorescence with visible excitation.

Laser modulated soot incandescence is a very serious problem. It occurs when the already incandescent soot particles absorb the incident laser radiation, heat to temperatures far above the ambient flame temperature and thereby, due to the temperature dependence of the Planck function, emit greatly increased amounts of blackbody radiation. Figure 4 presents experimentally determined soot surface temperatures as a function of laser focal flux for values of interest in laser diagnostics.[10] The dashed curves are analytical predictions based upon heat conduction and vaporization heat loss models. Heat conduction is generally insufficient to restrain the particle surface temperature from reaching vaporization thresholds. Figure 5 presents Raman signal/interference ratio calculations for laser modulated soot incandescence for a pulsed laser in the visible spectral region at 532 nm. At soot loadings typical of practical flames, the signal to interference ratios are too low to permit successful Raman measurements.

Calculations for the other laser sources considered in Table 1 are similarly pessimistic. One intriguing option is the potential of uv lasers to overcome this problem. For a KrF excimer laser at 249 nm, the Raman scattering from N_2 would be a factor of twenty-two stronger. Furthermore, due to the trailoff of the Planck function in the

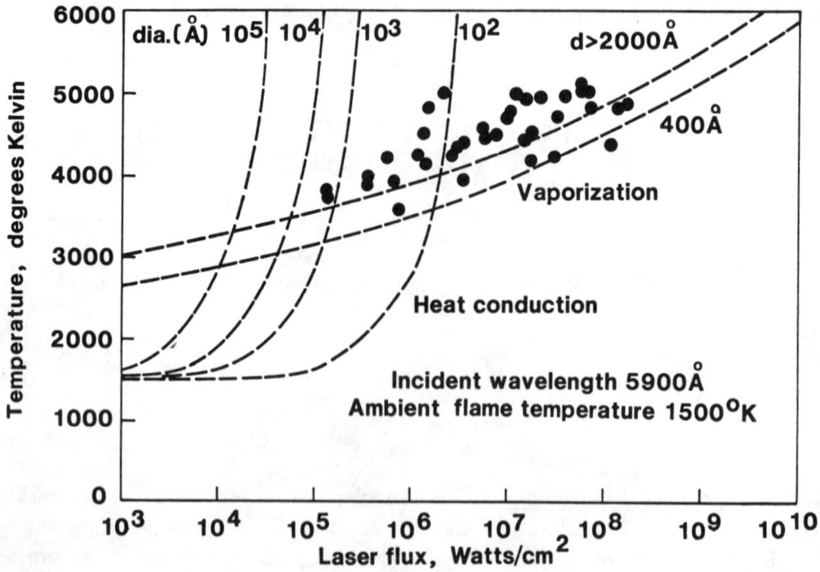

FIGURE 4. Laser irradiated soot surface temperatures. The dashed curves are analytical models for surface temperature based on either heat conduction or vaporization heat loss being predominant.

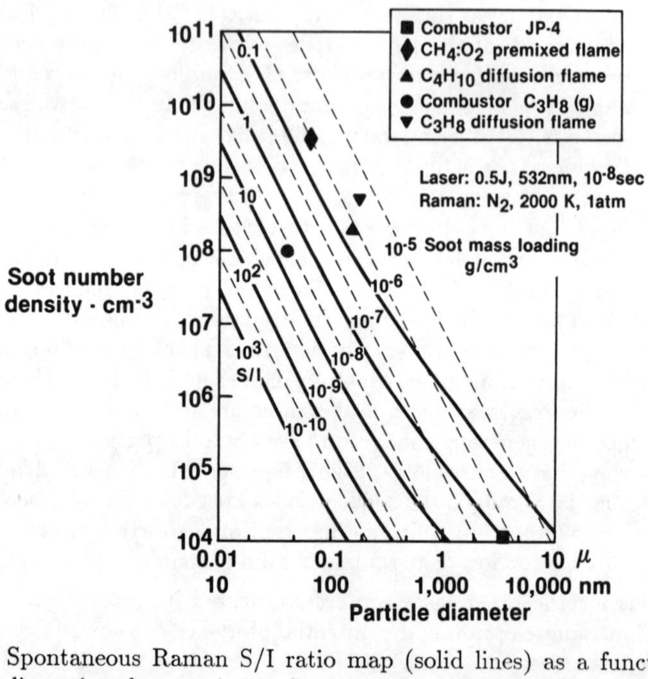

FIGURE 5. Spontaneous Raman S/I ratio map (solid lines) as a function of soot dispersion characteristics. Soot mass loadings are denoted by the dashed lines.

uv, the incandescence would diminish by a factor of four to five. This would increase the signal to interference ratio by nearly two orders of magnitude and possibly permit spontaneous Raman scattering to be used near the combustor exit plane, after some soot burnup has occurred through the secondary dilution zone. Again, the jeopardy introduced by laser-induced fluorescent interferences from fuel fragments requires investigation.

3.2 Perturbations

Since laser diagnostic techniques are nonintrusive, they are generally assumed to be nonperturbing. However, this is not necessarily the case unless reasonable care is exercised. In this section, potential perturbations will be briefly considered.

3.2.1 Medium Heating. In the absence of particulates, most combustion media are transparent to the incident laser radiation even when electronic resonances are being probed. Soot particulates, however, are highly absorbing over a wide spectral range and, for $\pi D/\lambda > 1$, have absorption cross sections approximately equal their geometric cross sectional areas. In sufficient abundance, they can absorb substantial fractions of the incident laser radiation. Of concern here is the potential thermal perturbation of the medium as heat is transferred from the soot to the gas. The medium could be perturbed if the characteristic time for heat to be transferred half the interparticle spacing is comparable to or less than the laser pulse length. Perturbation will only occur if the characteristic time for heat transfer from the particle to the gas is also less than the laser pulse length. For carbonaceous round particles 0.2μ in diameter, these times are comparable and approximately equal to $0.5(10^{-6})$ sec for a soot mass loading of 10^{-7} grams/cm^3. For smaller particles, the heat transfer time from the particles to the gas decreases slower than D^{-2} due to mean free path to particle diameter effects and reaches a value of $0.5(10^{-7})$ sec for 10 nm particles. For typical Q-switched solid state or excimer laser pulse lengths of 10^{-8} sec, there is no possibility of perturbing the medium due to particle absorption and subsequent heat transfer to the gas. Medium perturbations are potentially possible for flashlamp-pumped dye lasers with pulse lengths in the range of hundreds of nanoseconds to a microsecond for dispersions with particle diameters below 100 nm. Whether the medium will actually be perturbed depends on the amount of energy absorbed from the laser beam. This depends on the incident laser energy and soot mass loading and should be examined in each specific application.

3.2.2 Stimulated Raman Scattering. In spontaneous Raman scattering, only a very small fraction of the molecules, typically 10^{-12} or less at flame temperatures, change state. Thus, there is no concern about the molecular population distributions deviating from Boltzmann statistics. However, other processes can occur which can perturb the molecular population distributions. In the stimulated Raman gain process as seen earlier, a Stokes-shifted probe input wave will be amplified in the presence of an intense laser field. In the absence of the probe laser, a Stokes wave will grow from the spontaneously generated Raman photons, i.e., stimulated Raman scattering. For typical combustion gases and short, high intensity focal zones, calculations illustrate that stimulated Raman scattering is unlikely to perturb the populations being probed.

3.2.3 Optical Breakdown. With the application of very strong electric fields from high intensity laser radiation, the medium under examination may "break down," i.e., spark and become partially or highly ionized. When plasma formation occurs, the medium is substantially altered precluding diagnostic interrogation. Without going into the details of the physics,[11] clean air breakdown thresholds in the visible at atmospheric pressures are on the order of 10^{12} W/cm^2. In attempting to estimate breakdown thresholds for a practical combustion device, the effects of number density, composition and particulates must be accounted for. Particulates greater than a micron in diameter decrease clean air thresholds by about two orders of magnitude. Smaller particulates generally have no effect. Breakdown thresholds are inversely proportional to density and, thus, pressure. There should be no direct temperature dependence. In flame studies, however, breakdown thresholds have been found to be a factor two to three lower than expected perhaps due to higher ionization levels in flames than in air.[12] Hydrocarbon species also possess lower breakdown thresholds than air. In fuel-rich regions of hydrocarbon-fueled diffusion flames, breakdown is often encountered at laser intensities which are not problematical in other flame regions. In practical devices, the breakdown thresholds can be anticipated to be in the 10^{10} W/cm^2 range at atmospheric pressure and less at elevated pressures. Gas breakdown is a problem that limits usable laser intensities in intensity dependent processes such as CARS. In spontaneous Raman scattering, which is not intrinsically intensity dependent, softer, more diffuse focussing can be employed should breakdown be a problem.

3.2.4 Optical Stark Effects. The high electric fields in focussed laser radiation can become comparable to the internal fields in molecules and thus perturb the molecular energy levels leading to shifts and splittings in the transitions.[13] These result in alterations of the molecular spectrum and can contribute to inaccuracy in parameter measurements based on these spectra. Since the applied optical electric fields are not spatially uniform, different fields produce different shifts which manifest themselves as line broadening. Only purely rotational transitions split and do so at laser intensities about an order of magnitude lower than those for the spectral line shifts in pure rotational and rotational-vibrational Raman transitions to become pronounced. Pure vibrational transitions are even less sensitive and these are typically interrogated in both spontaneous and coherent Raman diagnostics. At the laser intensities required to stay below practical breakdown thresholds, these effects are generally unimportant. They can be important in clean flame and spectroscopic investigations.

3.3 Optical Access and Laser Transmission

To perform spatially-resolved laser measurements at a given point in a combustor, the laser beam(s) requires optical access into the region and must be transmitted with reasonable efficiency to the measurement region. In addition, the signal radiation must be extractable.

3.3.1 Optical Access. The primary combustor zone in modern jet engines is very difficult to access optically and, to the author's knowledge, no spectroscopically-based laser diagnostic has been employed for measurements inside actual gas turbine combustors. Rather measurements have been restricted to combustion tunnels, fitted with generous optical access, where primary zones are simulated to some degree or to engine exhausts where access is not a problem. Because of its double-ended nature, CARS will be difficult to implement in actual engines. With advances in long pulse length excimer lasers and high power, ultraviolet conducting fiber optics, spontaneous Raman scattering may be applied through borescope ports. Whether the $S/\bar{I}$ would be high enough for successful measurements, in view of laser modulated soot incandescence or fuel fragment fluorescences, remains to be ascertained. The suggestion here is that the optical engineering does not appear to be insurmountable.

Optical access also implies viewing the phenomenon under investigation through a window, often of limited aperture. The window must be capable of withstanding high combustion temperatures and transmitting the laser energy without suffering damage. Quartz and sapphire are commonly used to satisfy these requirements and are generally capable of transmitting fluxes without damage on the order of a few gigawatts/cm^2. The optical surface needs to be kept clean since dirt and soot buildup will decrease the damage thresholds and/or attenuate the laser beam(s) unacceptably. To minimize damage probability, the laser flux through the window should be kept as small as possible. One way to do this is to place lenses as close to windows as possible to maximize the beam diameter passing through the window. Particulate deposition can be minimized through suitable gas purging of the surfaces.

3.3.2 Transmission through the Medium. As the laser beam propagates through the combustor, particulate matter will attenuate the beam by scattering and absorption. As mentioned earlier, for soot particulates whose size is such that $\pi D/\lambda > 1$, the absorption cross sections are approximately geometric, and scattering and absorption become comparable. Thus the total extinction cross section is about twice the geometric cross sectional area and equals 10^{-8} cm^2 at $\pi D/\lambda = 6$ for carbonaceous particles. The total extinction can be calculated from the Beer-Lambert law knowing the soot number density and optical pathlength. Sensitivity will diminish as some power of the laser transmission. In Raman scattering, sensitivity will scale as the square of the transmission, i.e., laser propagation to the measurement point and signal transmission out. In CARS, the scaling is roughly to the fourth power. Since, with most of the spectroscopic techniques, signal level is at a premium, large beam attenuations are not tolerable. Unlike furnaces, most gas turbine combustors possess relatively short laser transmission distances transverse to the flow which is a compensating factor in this regard. Nevertheless, these effects are quantifiable either analytically or experimentally, e.g., with a helium-neon probe laser, and should be investigated before attempting diagnostic application.

In addition to energy delivery considerations, the laser beam needs to be delivered with minimum defocussing or steering to the measurement location. Clearly, if the medium is so turbulent that the integrity of beam delivery is compromised, diagnostic application will not be feasible. Again, the relatively short length of most gas turbine combustors is beneficial in this regard. Nevertheless, the severity of this problem can be investigated relatively inexpensively by conjuring up simple tests with a helium-neon laser to check for beam steering, defocussing or breakup.

4. COHERENT ANTI-STOKES RAMAN SPECTROSCOPY (CARS)

In the last section, it was seen, through calculation and actual experience, that spontaneous Raman scattering is generally incapable of probing the highly luminous, particle laden flames typical of gas turbine combustion media. To operate successfully in these environments, diagnostic approaches are required which provide stronger signal levels and/or possess alternative characteristics, such as coherence, which lead to sufficiently high S/I ratios. CARS is such a diagnostic approach.

Large increases in S/I relative to spontaneous Raman scattering are afforded by CARS due to both its high signal conversion efficiency and coherent, i.e., laserlike signal character. At atmospheric pressure and above, CARS signals are generally orders of magnitude more intense than spontaneous Raman scattering. In addition, the CARS radiation emerges in the form of a coherent beam permitting the entire signal to be captured. This is in contrast to an incoherent process where only a small fraction, $\leq 1\%$, of the total signal is collected even with fast low f-number optics. Furthermore, because the CARS signal is captured within a very small solid angle, interference collection is greatly minimized. With CARS, the interferences still exist; however one samples far less of them and they are not a problem. CARS detectivity is quite similar to spontaneous Raman scattering and it is most suited to thermometry and major species concentration measurements, i.e., typically $> 0.1\%$ at atmospheric pressure. In this section, CARS will be overviewed and its applications to gas turbine combustion will be surveyed.

4.1 Overview

4.1.1 Basic Principles.[1,8,9,14,15] CARS is produced, as illustrated in Fig. 6, when laser beams at frequencies ω_1 and ω_2 are "mixed" via any one of several geometrical phase-matching schemes to generate the coherent CARS signal at $\omega_3 = 2\omega_1 - \omega_2$. The laser interaction with the medium occurs through the third-order nonlinear susceptibility giving rise to an oscillating polarization at ω_3 and, thus, the signal radiation. Phase matching ensures that the CARS signal generated at location z will be phased with that generated at z + Δz so that constructive coherent signal growth occurs. ω_1 is termed the pump laser and ω_2, the Stokes, because it is downshifted in frequency by the vibrational/rotational Raman shift of the molecule being probed. Adjustment of the frequency difference $(\omega_1 - \omega_2)$ to a particular spectral region permits different molecular constituents to be examined. Single pulse measurements are possible using a broadband dye laser, i.e., $\Delta\omega_2 \sim 5$ nm, to generate the entire CARS signature by accessing simultaneously all the Raman resonances in a given molecular spectral band region. Broadband, multichannel detection then permits parameter measurements with a time resolution of the laser pulse lengths employed, typically 10^{-8} sec. The nonlinear susceptibility is both density and temperature dependent which provides the basis for diagnostics.

4.1.2 Advantages/Disadvantages. Several of the advantages of CARS have already been alluded to, namely its good signal strength and coherent character, which lead to high S/I ratios and make CARS broadly applicable to practical environments. In addition, these capabilities are possible employing commercially-available laser sources at laser pulse energies roughly an order of magnitude lower than those required for spontaneous Raman scattering. These lasers operate in the 10-30 Hz range leading to higher data rates than the 1 Hz typical of very energetic pulsed lasers.

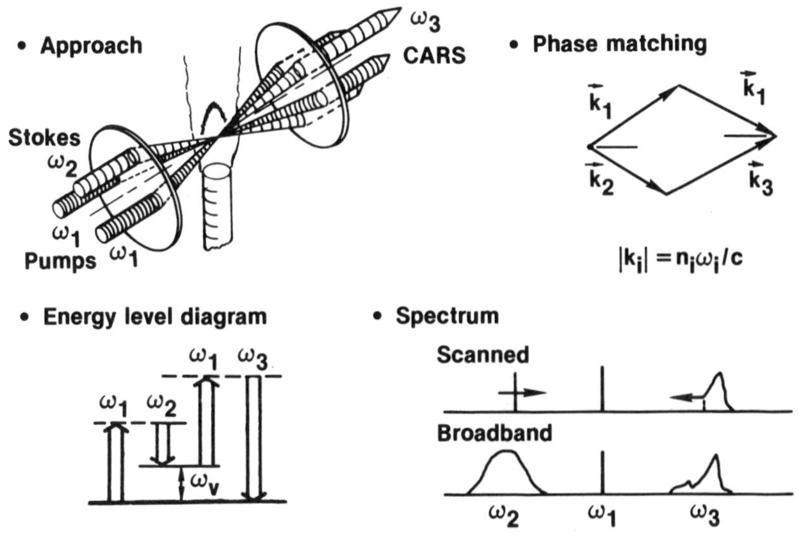

FIGURE 6. Coherent anti-Stokes Raman spectroscopy (CARS).

Relative to spontaneous Raman scattering, there are a number of disadvantages which one must bear in mind. For one, CARS is a double-ended technique requiring line-of-sight optical access. Even though high spatial precision is possible, phase matching dictates generation of the signal in the forward direction. Unlike incoherent techniques, CARS cannot be implemented with a single optical port in backscattering. Because of the forward nature of CARS generation, CARS has not been applied to the primary zone combustors in actual gas turbine engines. Second, CARS is generally restricted to monitoring a single constituent at a time. In Raman scattering, all constituents give rise simultaneously to the signal. Advanced, multi-color CARS approaches, which overcome this limitation, will be discussed in Section 5. Third, CARS spectroscopic modelling is considerably more complicated than spontaneous Raman. In the latter, spectral synthesis is easily accomplished by a simple summation of the transition magnitudes and the spectra are relatively insensitive to Raman transition linewidths or precise spectral location. CARS spectra exhibit both constructive and destructive interference effects and are quite sensitive to Raman transition linewidths and location. Computationally-intensive numerical programs are required to convolve over the laser profiles and nonlinear susceptibilities to synthesize the spectra.[1,16] Much progress has been made in this vein and some commercial software is emerging.

4.1.3 Phase Matching. For efficient signal generation, the incident laser beams must be properly pointed to phase the wave-mixing process. Phase matching ensures the constructive build-up of the coherent signal. The general phase-matching diagram is illustrated in the upper quadrant of Fig. 6 and requires that

$$2\vec{k}_1 = \vec{k}_2 + \vec{k}_3 \tag{5}$$

where k_i is the wave vector at frequency ω_i with absolute magnitude $\omega_i n_i/c$, where n_i is the refractive index at the appropriate frequency and c is the speed of light.

Because gases are nearly dispersionless, i.e., the refractive index is nearly invariant with frequency, the photon energy conservation condition $2\omega_1 = \omega_2 + \omega_3$ indicates that phase matching is satisfied when the input beams are aligned collinearly with each other. Collinear phase matching, although easy to implement, is seldom used in practical applications because of poor spatial discrimination particularly in the presence of temperature gradients.

It is thus desirable to avoid laser beam overlap and CARS generation in all regions except the desired measurement location. This can be accomplished using crossed-beam phase matching or BOXCARS as depicted in Fig. 7. The term BOXCARS derives from the general shape of the phase matching diagram. Note that crossed-beam phase matching needs not be restricted to a plane and can be "folded" into three dimensional configurations. In BOXCARS, the ω_1 pump frequency is split into two beams which are crossed at an angle which depends on the spatial resolution desired. Since the CARS signal scales as the square of the interaction length, one generally works at the minimum spatial resolution, i.e., maximum length, permissible. Depending on the ω_1 crossing angle and species being probed, the crossing angle for the ω_2 beam is specified and, as seen in Fig. 7, the beam may be introduced either in plane (planar) or out of plane (folded). The advantage of the latter is the complete separation of the signal beam from the incident wave-mixing beams. Cylindrical sample volumes a few hundred microns in diameter and a few millimeters long are typically employed.

There are also two-beam three-dimensional phase-matching approaches which are particularly attractive for practical application.[17] These take advantage of the annular output from the unstable resonators used in some high power lasers. In one approach, termed USED CARS (from unstable- resonator spatially enhanced detection), the ω_2 Stokes beam is aligned coaxially inside an annular ω_1 beam.[18] CARS generation occurs only in the beam overlap region of the laser foci and good spatial resolution can be obtained with low f number optics. Because of the common beam axis, these techniques are easy to align and are relatively insensitive to turbulence-induced flow distortions.

4.2 Parameter Measurement Approaches

4.2.1 Overview. Measurements of medium properties are performed from the shape of the CARS spectral signatures, i.e., the signal intensity variation with frequency, and/or the intensity of the CARS radiation. Temperature information derives from spectral shapes since these mimic the Boltzmann distribution of population over the

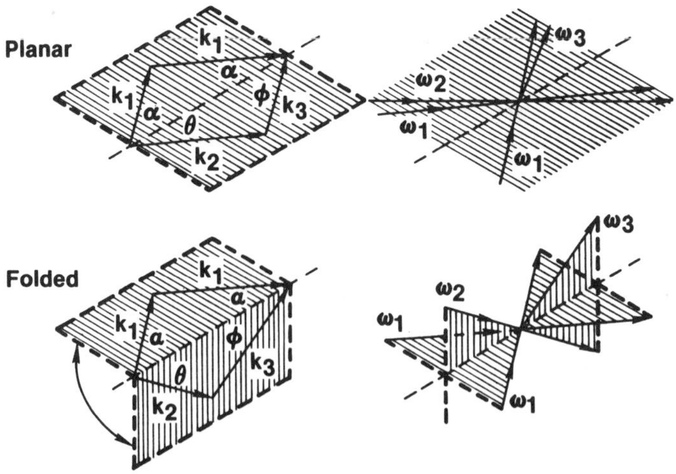

FIGURE 7. Planar and folded BOXCARS phase-matching diagrams and beam geometries.

various vibrational/rotational energy states. Concentration measurements are based on the strength of the CARS signal in general. Such measurements are not easy to perform and, in practical situations, may not be possible if extinction and refraction effects markedly affect signal levels. Fortunately, the shape of the CARS spectrum is also concentration sensitive in the range from approximately 0.5 to 30%, thus permitting fractional concentration measurements from spectral profiles. Absolute intensities are, therefore, unimportant per se as long as the signal levels are statistically significant.

4.2.2 <u>CARS Signal Level</u>. All of the above is most easily understood by considering the basic equations of CARS. The CARS signal intensity I_3 at frequency ω_3, for the case of monochromatic laser sources, is given by (mks units)

$$I_3 = \frac{{\omega_3}^2}{c^4 {\epsilon_o}^2} I_1^2 I_2 \left|\chi^{(3)}\right|^2 \ell^2 \tag{6}$$

where c is the speed of light; ϵ_o, permittivity of free space; and ℓ, the interaction length. The third order nonlinear susceptibility may be expressed as

$$\chi^{(3)} = \sum_j (\chi'_j + i\chi''_j) + \chi_{nr} \tag{7}$$

χ_{nr} is the nonresonant background susceptibility contributed by electron cloud distortion of all the molecules. The total nonresonant background is composition sensitive and is equal to the mole fraction weighted average of the nonresonant backgrounds of the various constituents present. The term in parentheses is the resonant susceptibility for the jth transition and is given by

$$\chi'_j + i\chi''_j = \frac{8\pi^2 \epsilon_o c^4}{\hbar \omega_2{}^4} N \Delta_j g_j \left(\frac{\partial \sigma}{\partial \Omega}\right)_j \frac{\omega_j}{\omega_j{}^2 - (\omega_1 - \omega_2)^2 - i\Gamma_j(\omega_1 - \omega_2)} \tag{8}$$

where N is the number density of the species being interrogated; Δ_j, the fractional population difference between the energy levels giving rise to the jth transition; g_j, the degeneracy; $(\partial\sigma/\partial\Omega)_j$, the Raman cross section; Γ_j, the Raman linewidth; and $\hbar$, Planck's constant divided by 2π. The real and imaginary components may be written

$$\chi'_j = K_j \frac{2\Delta\omega_j \Gamma_j}{4\Delta\omega_j{}^2 + \Gamma_j{}^2} \tag{9}$$

$$\chi''_j = K_j \frac{\Gamma_j{}^2}{4\Delta\omega_j{}^2 + \Gamma_j{}^2} \tag{10}$$

In the above, $\Delta\omega_j$ is the detuning and is defined as

$$\Delta\omega_j \equiv \omega_j - (\omega_1 - \omega_2) \tag{11}$$

K_j is the amplitude of the susceptibility and is given by

$$K_j = \frac{(4\pi)^2 \epsilon_o c^4 N \Delta_j}{\hbar \omega_2{}^4 \Gamma_j} \left(\frac{\partial \sigma}{\partial \Omega}\right)_j \tag{12}$$

In the above, the medium has been assumed dispersionless, i.e., all refractive indices are equal, and since attention is restricted to the gas phase, all refractive indices have been assumed equal to unity. The dependence on species density is clear. The temperature dependence enters through the population difference term Δ_j. Note that the real component of the resonant susceptibility displays dispersive behavior about line center ($\Delta\omega_j = 0$). That is, it goes from a negative value to a positive value as the detuning goes from negative to positive. The imaginary component possesses lineshape behavior. This is analogous to the behavior displayed by the real and imaginary parts of the refractive index, i.e., the linear susceptibility.

4.2.3 CARS Spectral Behavior. Due to the dependence of CARS spectra on $|\chi|^2$ and the presence of the nonresonant background contribution, CARS spectra are more complicated than spontaneous Raman scattering and exhibit both constructive and destructive interference behavior. These are readily illustrated by considering some simple examples. For the case of constructive interferences, consider two closely spaced Raman resonances and assume the nonresonant background is small. Then

$$\chi^{(3)} = \chi'_1 + i\chi''_1 + \chi'_2 + i\chi''_2 \tag{13}$$

hence

$$|\chi^{(3)}|^2 = (\chi_1'^2 + \chi_1''^2) + (\chi_2'^2 + \chi_2''^2) + 2(\chi'_1\chi'_2 + \chi''_1\chi''_2) \tag{14}$$

In addition to the "isolated" line contributions one would expect, the first two terms in parentheses, there are also the constructive interference cross terms, the third term in parentheses. The strength of these constructive interferences depends on the spacing between the transitions and the magnitude of the Raman linewidth.

Destructive interferences are apparent by considering a single transition in the presence of the nonresonant background, i.e.,

$$\chi^{(3)} = \chi' + i\chi'' + \chi_{nr} \tag{15}$$

thus

$$\left|\chi^{(3)}\right|^2 = \chi'^2 + \chi''^2 + 2\chi'\chi_{nr} + \chi_{nr}^2 \tag{16}$$

For a constituent in low concentration, χ', $\chi'' < \chi_{nr}$, hence

$$\left|\chi^{(3)}\right|^2 = \chi_{nr}^2 + 2\chi'\chi_{nr} \tag{17}$$

Far from resonance, the CARS spectrum is featureless and given by χ_{nr}^2. As resonance is approached for negative detuning, anti-resonance or destructive interference behavior is exhibited since the $2\chi'\ \chi_{nr}$ term goes negative and detracts from the χ_{nr}^2 contribution. When the detuning becomes positive on the other side of line center, the signal increases above the nonresonant background level. The CARS spectrum thus appears as a modulation of the nonresonant background. The modulation depth depends on the concentration of the species being probed and this forms the basis for fractional concentration measurements based on spectral shapes. When the fractional concentration becomes too low to produce a well-defined and measurable modulation, the species is no longer detectable unless the nonresonant background is suppressed. The most effective means of suppression is through polarization orientation of the wave mixing laser fields.[19] This leads to considerable signal loss and is seldom employed in practical applications unless the pressure levels, and hence the signals, are large.

Most species of interest in combustion possess tens if not hundreds of vibrational-rotational levels. In addition, the laser sources are generally not monochromatic (certainly not for broadband CARS), and convolution integrals must be performed over the laser lineshapes involved, accounting for the detunings of all of the transitions from the various laser frequency components involved. There are also partial coherence effects between the various contributions to a given CARS frequency component.[20] All of this dictates computer synthesis of CARS spectra and the various numerical codes employed are well documented in the literature.[16] Some software is available commercially as well.

4.2.4 Thermometry. Two examples of such CARS spectral calculations will be discussed here. In Fig. 8, the temperature variation of the N_2 CARS spectrum from 300 to 2300 K is displayed in 100 K increments. Nitrogen is the dominant constituent in airfed combustion and remains relatively unchanged through the course of the combustion process; thus, it is an ideal thermometric tracer. At low temperatures, a simple, narrow peak is seen consisting of unresolved Q-branch transitions between the vibrational v = 0 to 1 states. In a Q-branch transition, the rotational quantum number J does not change. As temperature increases, the band broadens as the rotational population distribution smears and shifts to higher rotational energies in accordance with Boltzmann statistics. At temperatures > 1200 K, the v = 1 to 2 band begins to appear as the first vibrationally-excited state becomes thermally populated. At even higher temperatures, v = 2 to 3 transitions emerge. Temperature measurements are made by ascertaining which theoretically-calculated spectrum best fits the experimentally-acquired CARS signature. With supercomputers, such theoretical spectra can be calculated as the data reduction proceeds; with smaller laboratory minicomputers, fitting is generally performed against a "library" of previously calculated spectra. Similar thermometric strategies pertain in the case of other molecules.

Time-averaged CARS measurements in N_2 possess accuracies on the order of ± 1% in carefully conducted laboratory experiments. Single pulse measurements are generally precise to about ± 3% due to CARS spectral distortion arising from amplitude irregularities in the broadband Stokes lasers.[21] Field measurements are probably accurate to about ± 3% at flame temperatures due to how often certain calibrations can be performed and other experimental factors.

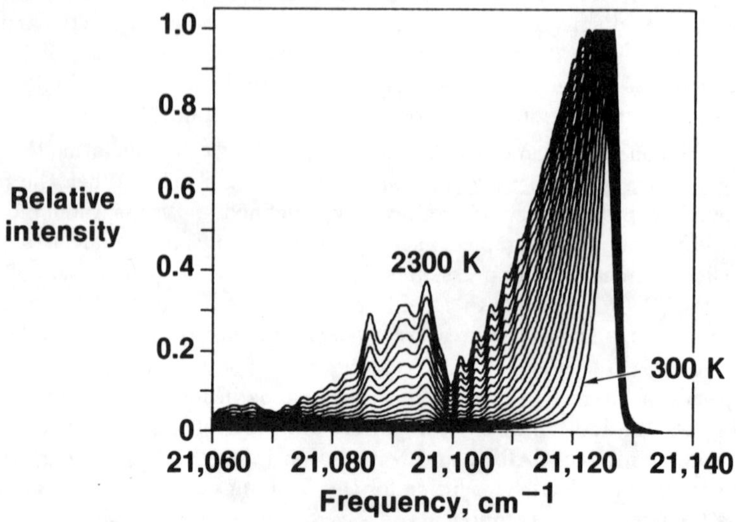

FIGURE 8. Temperature variation of CARS spectrum in 100 K increments from 300 K to 2300 K for a spectral resolution of 2 cm^{-1}.

4.2.5 Species Measurements. Figure 9 displays calculated CARS spectra of CO at 1800 K and a series of fractional concentration levels. As is apparent, the spectral shapes are quite concentration sensitive. As discussed earlier, at low concentrations (< 30%), these appear as modulated spectra due to the interference of the resonant susceptibility with the nonresonant electronic background. As long as the signal level is adequate, concentration measurements can be extracted from the spectral signature without regard to the absolute intensity level.

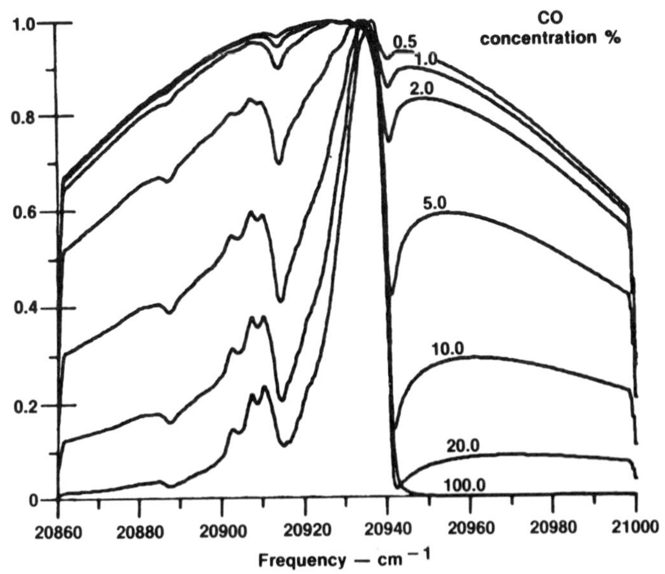

FIGURE 9. Concentration variation of CO CARS spectrum at 1800 K.

In the case of unmodulated spectra, concentration measurements must be made from the intensity of the CARS radiation. Spectra are unmodulated at high concentrations or when the nonresonant background contributions have been removed, e.g., by polarization orientation suppression. In these instances, measurements are based upon the absolute intensity of the spectrally-integrated signal. Since the CARS signal depends upon the product of $I_1^2 I_2$, this must be independently measured by a referencing system. The intensity product cannot be analytically estimated to sufficient accuracy due to temporal and spatial mode fluctuations in the laser beams. The referencing system contains a cell with a gas at known conditions whose susceptibility can be very accurately calculated. CARS is generated in the reference cell either in series with the measurement laser beams or in parallel by splitting off a small fraction of them. The reference and measurement signals are given by

$$I_{ref} = k_r \, I_1^2 \, I_2 \, \left|\chi_{ref}^{(3)}\right|^2 \tag{18}$$

$$I_{meas} = k_m \, I_1^2 \, I_2 \, \left|\chi_{meas}^{(3)}\right|^2 \tag{19}$$

where k is an apparatus constant accounting for interaction length, collection efficiencies, etc. Dividing the two then results in a measurement of the resonant susceptibility from which the species density can be calculated, thus

$$\left|\chi_{meas}^{(3)}\right| = \left(\frac{k_r I_{meas}}{k_m I_{ref}}\right)^{\frac{1}{2}} \left|\chi_{ref}^{(3)}\right| \tag{20}$$

The ratio of the apparatus constants is easily determined by calibration, e.g., using the same gas in the measurement and reference volumes.

Referencing is limited to those situations in which particulate extinction and turbulence do not markedly affect the measurement volume CARS signal level. For this reason, referencing is best performed after the measurement region rather than before. The reference signal should not change markedly from pulse to pulse for stable laser operation. Changes are indicative of medium perturbations; unfortunately, there is no way to correct for such effects since the measurement and reference signals are not affected in exactly the same fashion.

There are approaches to in-situ referencing where the reference is generated at the measurement location.[22] These are difficult to implement with broadband approaches due to signal level considerations since they involve polarization separation of the resonant CARS and background signals. These approaches use the nonresonant background explicitly as a $I_1^2 I_2$ reference. They are more complicated and more difficult to implement than concentration measurements based on spectral shapes which use the nonresonant background as an implicit reference. In any approach, relative inaccuracies are most likely in the $\pm$ 5-10% range due to the difficulty in measuring absolute intensities even with referencing or to uncertainties in the measurement volume nonresonant background susceptibility. Hydrocarbon species possess large nonresonant backgrounds and the total nonresonant background can vary markedly ($\sim$ 50%) as the mixing and burning proceed through the primary zone of a gas turbine combustor.

4.3 Experimental Methods

4.3.1 Laser Selection. There is no formal laser intensity threshold for generating CARS and CARS can be produced with cw laser sources albeit at very low signal levels. For time-resolved measurements, high intensity laser sources are required to generate statistically-significant signal levels well in excess of various sources of interference. For gas-phase combustion diagnostics, short duration ($\sim 10^{-8}$ sec), high intensity ($\geq 10^7$ W), lasers are required to satisfy both signal level and interference criteria. Although ruby lasers were first used as the pump laser in early CARS experiments, most diagnostic systems now use frequency-doubled neodymium:YAG lasers (2xNd:YAG) which are Q-switched to produce short, $\sim 10^{-8}$ sec, pulses with energies at 532 nm in the 200 to 400 mJ range. These lasers run at repetition rates of 10 to 30 pps far better than ruby lasers. Furthermore, the output at 532 nm is well suited to pumping quite efficiently the yellow/orange/ red laser dyes required for the Stokes laser. Systems using a high power excimer laser to pump two dye lasers for the pump and Stokes beams are also available, but have not been widely employed since their pulse energies tend to be modest in terms of single pulse measurement requirements.

4.3.2 Instrument Layout. In Fig. 10, a simplified schematic diagram of a CARS system is shown. For clarity, many of the optical details have been omitted. In essence, due to the line-of-sight optical requirements, CARS arrangements may be analyzed as consisting of a transmitter and receiver. The transmitter houses the laser sources required for the wave mixing as well as the necessary optical components to deliver the laser sources with the correct polarizations, beam geometry and focal parameters to the measurement location. These components include, for example, half-wave plates for polarization rotation, Galilean telescopes for beam expansion and subsequent focal volume control and beamsplitters/dichroics for achieving the proper phase matching geometry. In the systems using a 2xNd:YAG laser as the ω_1 pump source, the ω_2 Stokes laser is pumped by using a fraction (30 to 50%) of the pump laser as shown in Fig. 10. The dotted optical path shown represents the central optical axis for more complex phase-matching geometries such as BOXCARS or the actual laser beam axes for overlapped arrangements such as USED CARS.

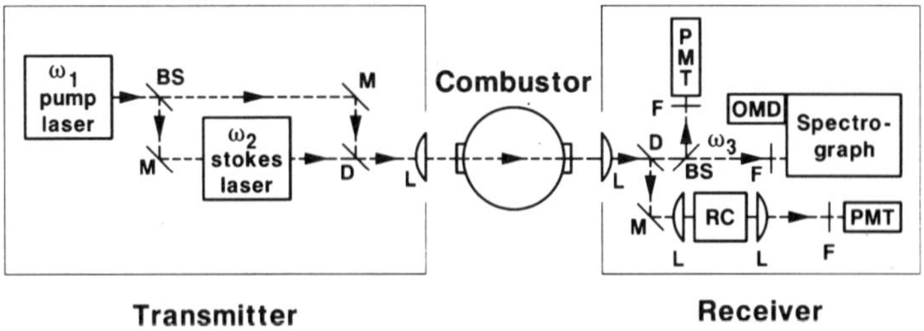

FIGURE 10. CARS experimental schematic.

The role of the receiver is basically to capture and record the CARS signal from the measurement volume and to generate and record the reference CARS radiation for signal normalization. There are many different approaches to this end and the one shown is only meant to be illustrative. The wave mixing beams are separated from the measurement volume CARS signal and sent to the reference cell to generate the normalizing CARS signature. The measurement CARS signal is focussed into a spectrograph where each pulse is dispersed and captured on an optical multichannel detector. Prior to this, a small fraction of the signal is split off and sent to a photomultiplier tube which monitors the spectrally- integrated signal. This signal is available as part of the data reduction and is also used to monitor the optical alignment of the system. Optical adjustments are made in the transmitter until the CARS signal is maximized.

Not shown in Fig. 10 is a means to traverse the measurement point through the combustor volume. A common approach is to translate the transmitter and receiver in tandem to move the measurement location.[18] Optical schemes also exist which employ a fixed transmitter and receiver with lightweight traversing frames which move the focussing optics and, thus, the measurement point while maintaining alignment with the transmitter and receiver. For measurement movement along the optical axis, adjustable, tandem Galilean telescopes can be used to effect large movement in the measurement volume with very small translation of one of the Galilean lenses.[23]

4.4 Gas Turbine Applications

As mentioned earlier, due to the line-of-sight optical requirement imposed by phase matching, CARS has not been applied to the primary zones of actual gas turbine combustors. Rather it has been used in combustion tunnels which contain burner cans or simulate primary zones. The only application to actual engines has been in the exhaust of afterburning engines[18] where optical access is, of course, not a problem.

4.4.1 Gas Turbine Simulations. Work in optically-ported combustion tunnels simulating gas turbine primary zones has been conducted by various research groups throughout the world and is summarized in Table 2. Early work concentrated on demonstrating the feasibility of CARS to turbulent, hydrocarbon-fueled diffusion flames. Later work has begun to employ CARS on a fairly routine basis to gather scalar data, generally temperature, in conjunction with laser velocimetry measurements. An example of work in this regard is displayed in Fig. 11 where CARS temperature and laser velocimeter measurements are compared against TEACH code predictions at two axial stations downstream of a bluff-body combustor.[27] From histograms of 1500 single pulse temperature measurements, average temperatures were obtained for comparisons with TEACH code calculations which were found to predict successfully several qualitative features of the mean temperature and axial velocity fields. From the bimodal shape of the temperature pdf's, phenomenological insight was obtained into the major transport mechanisms. The transport is intermittent and believed to be dominated by the large scale vortices shed from the bluff body.

Table 2

PRACTICAL UTILIZATION OF CARS IN SIMULATED GAS TURBINE PRIMARY ZONES

AERE, Harwell, U.K.[24]

Thermometry, BS360 combustor sector, natural gas fuel

National Aerospace Laboratory, Japan[25]

Thermometry @25Hz, triangular bluff centerbody combustor, premixed propane/air

ONERA, France[26]

Thermometry and O_2 concentration, research can combustor, kerosene

AFWAL, U.S.[27]

Thermometry @10Hz, axisymmetric bluff centerbody combustor, gaseous propane

UTRC, U.S.[28]

Thermometry, swirl burner and JT-12 can, Jet A fuel

4.4.2 Jet Engine Exhaust Measurements. One region in actual jet engines, which is easily accessed optically, is the exhaust region and this is the only region to date where CARS has been applied to real hardware.[18] The CARS instrument was assembled and installed on a traversing framework which permitted the CARS measurement point to be translated in two dimensions throughout the engine exhaust. A photograph of the instrument installed about the engine in an outdoor test stand is shown in Fig. 12. Because of the very high acoustic noise levels, ~ 150 dB, the instrument must be well shielded acoustically, hardened and capable of completely remote operation. The compact 1m x 2m x 0.7m transmitter houses the 2xNd:YAG laser, Stokes dye lasers, ancillary optics and stepper motors for remote alignment control. One feature

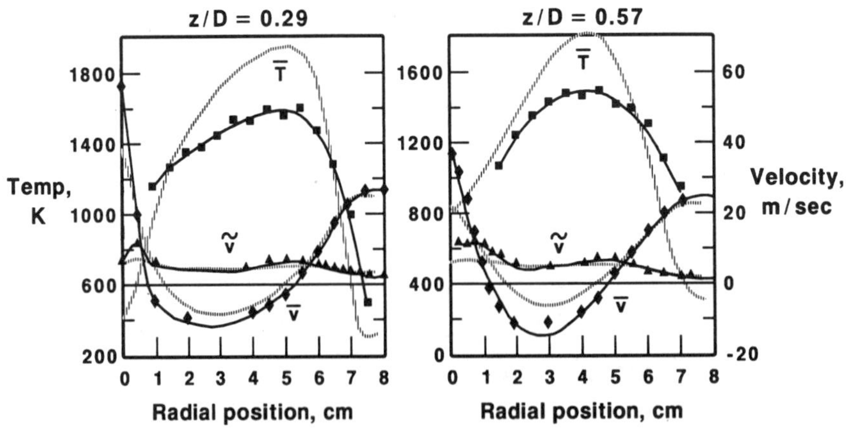

FIGURE 11. Comparisons of CARS temperature measurements and TEACH computer code calculations at two axial locations downstream of a bluff centerbody combustor.

is the chevron of five dye cells which can be translated sequentially into a common Stokes optical cavity to address the various exhaust constituents, N_2, O_2, CO, H_2O and unburned hydrocarbon species. Each dye cell is optically pumped in two separate spatial locations, one for the oscillator, the other to amplify the oscillator output which is folded back. The laser beams are focussed into the jet exhaust with a 1.1 m focal length lens in a USED CARS phase-matching configuration.

The 0.7m x 0.7m x 0.6m receiver separates the CARS signature from the incident wave mixing beams and focusses the CARS radiation into a fiber optic link for transmission to the remotely located spectrographic detection equipment in the control building. The residual laser beams are employed to generate a reference signal from a gas-filled cell for signal normalization. The instrument is capable of single pulse N_2 thermometry at a 20 Hz rate. All spectra are recorded on a single shot basis by a streaming tape drive. At low species concentrations typical of CO, O_2, the spectra are averaged in the computer to enhance the signal-to-noise ratio assuming fluctuations at the measurement location are small enough for averaging to be valid. Figure 13 displays a temperature histogram and time series at one location in the engine exhaust with the afterburner at full power. Time-averaged concentrations of O_2, CO and H_2O were also obtained.

4.4.3 Mobile CARS Instrument. Many, but not all, of the aforementioned CARS instruments were not easily transportable or represented dedicated test stand instruments. Recently construction of a relatively compact, mobile CARS instrument has been completed which can be easily moved from one application to another.[29] It is conceptually very similar to the instrument developed for jet engine measurements. It consists of a CARS transmitter on wheels approximately 2m long x 1.5m high x 1.5m wide with self-contained hydraulic jacks for height adjustment and leveling. The receiver is carriable, 0.6m square and 0.3m high and is typically installed on an inexpensive but rigid stand custom built for each application. The transmitter and receiver

FIGURE 12. CARS instrument mounted behind an augmented jet engine in an outdoor test stand.

contain identical periscope towers for beam exit and entry. Spanning the towers is an optical traversing framework containing the focussing and recollimating field lenses to move the measurement volume through the combustion field under study. The traversing framework is also envisioned to be specially designed for each application since each is likely to vary markedly. Optical fibers are again employed to pipe the CARS signatures to a remotely located, three cabinet instrumentation and control rack from which the instrument is operated. Stepper motors permit optimization of CARS signal generation and positioning of the optical fiber for maximum signal capture. A photograph

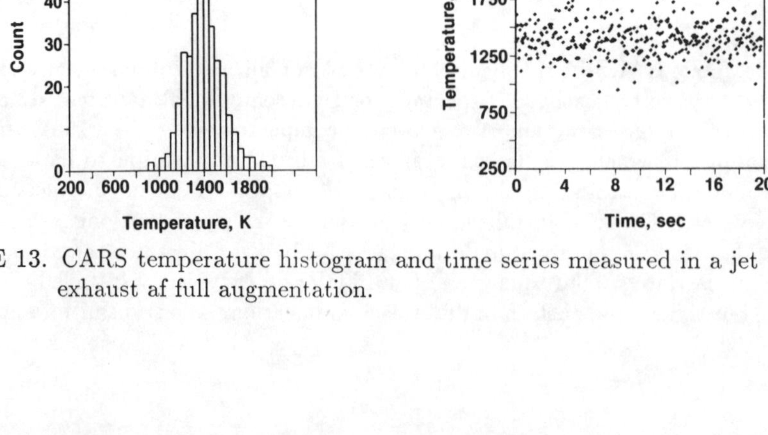

FIGURE 13. CARS temperature histogram and time series measured in a jet engine exhaust af full augmentation.

of the mobile instrument installed about a supersonic combustion tunnel is shown in Fig. 14. In this case as can be seen, the traversing optics were mounted on a beam bridging the test section. The instrument has been employed to demonstrate single pulse CARS thermometry in a supersonic combusting flow in its initial demonstration.

FIGURE 14. Mobile CARS instrument installed about a supersonic combustion test facility.

5. ADVANCED CARS TECHNIQUES

Despite the superiority of CARS vis-a-vis spontaneous Raman scattering for probing instrumentally hostile environments, CARS possesses a major disadvantage when compared to the latter, namely the inability to measure more than a single constituent at a time as normally implemented. Recently there has been a trend toward advanced multi-color techniques which overcome this limitation and which permit additional capabilities as well, e.g., simplified pure rotational CARS, conditional sampling.

5.1 Multi-Color CARS Techniques for Multiple Species Measurements

5.1.1 Multi-Color Wave Mixing. Figure 15 summarizes the various approaches conceived to date for simultaneous, multiple species CARS measurements.[30] Dual Stokes approaches are a straightforward extension of CARS. For each constituent to be measured, a separate Stokes laser is introduced. Three-dimensional phase matching (folded BOXCARS) is generally employed to produce spatially separable CARS signals. Due to the complexity of arranging the beams, generally just two Stokes lasers are used. One thus performs two separate, two-color wave mixing sequences to monitor two different constituents. The CARS signatures are located spectrally where they would normally occur, i.e., at the pump frequency plus the Raman shift,

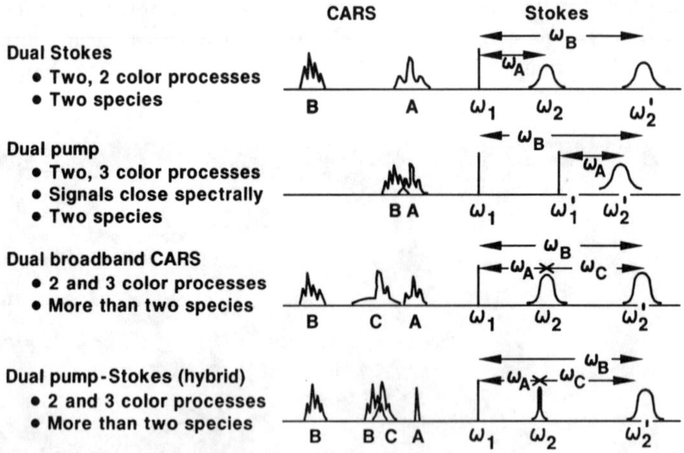

FIGURE 15. Summary of multi-color CARS techniques for simultaneous multiple species measurements.

$$\omega_3^{\mathrm{A}} = 2\omega_1 - \omega_2 = \omega_1 + \omega_{\mathrm{A}} \tag{21}$$

$$\omega_3^{\mathrm{B}} = 2\omega_1 - \omega_2' = \omega_1 + \omega_{\mathrm{B}} \tag{22}$$

In dual pump approaches, two narrowband pump lasers, ω_1 and ω_1', are used in conjunction with a single broadband Stokes laser to monitor two species via two separate, three-color wave mixing processes. Dual pump approaches are also termed "three-laser CARS" and are discussed in greater detail elsewhere in this volume. This is merely CARS in its most general form with frequency degeneracy removed, i.e., three different input waves. In frequency degenerate CARS, dual species capability is sacrificed. An interesting aspect of this approach is that the spectra from the two constituents reside in the same spectral vicinity simplifying optical mutichannel detection. The spectra occur at frequencies of

$$\omega_3^{\mathrm{B}} = \omega_1' + (\omega_1 - \omega_2') = \omega_1' + \omega_{\mathrm{B}} \tag{23}$$

$$\omega_3^{\mathrm{A}} = \omega_1 + (\omega_1' - \omega_2') = \omega_1 + \omega_{\mathrm{A}} \tag{24}$$

and are separated by

$$\omega_3^{\mathrm{B}} - \omega_3^{\mathrm{A}} = (\omega_1' - \omega_1) + (\omega_{\mathrm{B}} - \omega_{\mathrm{A}}) \tag{25}$$

By judicious selection of ω_1', spectral overlap can be avoided or minimized to a significant extent.

5.1.2 Dual Broadband CARS. In dual broadband CARS, two broadband Stokes lasers are used in conjunction with a pump laser and a combination of two- and three-color wave mixing processes simultaneously occurs assuming all are phase matched. The two, two-color sequences are normal two-color CARS processes as in the dual Stokes method described previously. However, in addition, there is the three-color wave mixing sequence for Raman resonances which correspond to the frequency difference between ω_2 - ω_2'. Since the frequency difference range spanned by two broadband sources is quite large, Raman resonances over a several hundred wavenumber range can be blanketed. The dual broadband CARS signature resides at the frequency

$$\omega_3 = \omega_1 + (\omega_2 - \omega_2') = \omega_1 + \omega_C \tag{26}$$

and is the same as would pertain in a two-color wave mixing process for an ω_C Raman resonance. In the same frequency region there is also a contribution from $\omega_1 - \omega_2'$ resonances, i.e.,

$$\omega_3 = \omega_2 + (\omega_1 - \omega_2') = \omega_2 + \omega_B \tag{27}$$

This contribution will be highly smeared spectrally since it is the broadband ω_2 scattering from the excited Raman coherences at $\omega_1 - \omega_2' = \omega_B$. Depending upon the relative concentrations of the species C and B, this background could be ignored or may have to be accounted for to perform accurate measurements.

The spectral positioning of the two broadband dye lasers clearly depends on the application and species of measurement interest. For hydrocarbon-fueled combustion, the spectral location of the various major species resonances is quite fortuitous. In the most common approach, the two Stokes sources are positioned to generate CARS from the major combustion products, namely CO_2 and H_2O. The H_2O Raman bandhead is situated at 3657 cm^{-1}, the major bands of CO_2 at 1285 and 1388 cm^{-1}. Centering the CO_2 Stokes source near 1326 cm^{-1} permits full coverage of the CO_2 band system. The frequency difference of 2331 cm^{-1} between the two broadband sources permits excitation of the N_2 Raman resonances from which the ω_1 pump beam scatters to generate the CARS signature. Despite the use of two broadband sources in the three-color wave mixing, the CARS spectral resolution is governed, as in two-color processes, by the spectral width of ω_1 and/or the resolution of the spectrograph employed. This is due to the Raman resonances being well defined by the specific molecular constituents and not by the manner in which the resonances are excited. Many other species possess Raman resonances in the broad frequency difference range spanned by the two broadband sources and are detectable if sufficiently abundant. These include CO (2143 cm^{-1}), N_2O (2224), HCN (2097) and NO (1876). Experimentally, if the laser dye DCM is employed for the H_2O laser source, very broad lasing bandwidths (FWHH) of 350 cm^{-1} occur. Coupled with a bandwidth of 150 cm^{-1} typical of the Rhodamine dyes used for CO_2, a spectral range of 500 cm^{-1} can be covered with less than a factor of four loss in peak signal.

Another spectral placement strategy is to position the low Raman shift dye laser near 1472 cm^{-1} to generate CARS from O_2 (1556 cm^{-1}) and CO_2 (1388) nominally at the half heights of the dye profile. With the second dye source centered at 3657 cm^{-1} (H_2O), one can generate dual broadband CARS from the CO resonances just slightly off the peak of the profile and from N_2 just above the H_2O Stokes source half height. Since N_2 is the dominant constituent in airfed combustion, this latter strategy is favorable from a dynamic range standpoint on the optical multichannel detector, i.e., the signal levels from the various constituents may not be too grossly disparate. Another strategy would be to place one dye laser near the CO peak (2143) and generate CARS from CO and N_2 simultaneously in a two-color wave mixing process. In conjunction with an H_2O Stokes source, dual broadband CARS would be generated from O_2 and CO_2. Depending on the specific situation to be examined, other strategies may be preferred.

It should be noted that the three-color dual broadband process will not be as intense as a normal two-color process due to the intensity difference between a pump laser and a Stokes source, i.e. $I_1^2 I_2$ versus $I_1 I_2 I_2'$. However, it will not be as weak as one might initially suspect due to the spectral integration aspect of dual broadband CARS, i.e., many frequency combinations drive each Raman resonance. For a resonance centered between two, equally-broad Stokes lasers, all of the energy in each laser is employed in the wave mixing. This is unlike normal two-color broadband CARS where only a thin spectral slice of the dye laser drives the Raman resonance. Due to the increase of the CARS signal with increasing pressure, dual broadband CARS should be capable of single pulse measurements of the major constituents at the elevated pressures typical of gas turbines, internal combustion engines and burning propellants. Furthermore, in dual broadband CARS, a beneficial spectral averaging may occur resulting in improved single pulse spectral quality.

For all wavemixing combinations to occur, each must be phase matched according to the diagrams displayed in Fig. 7. Using planar or folded BOXCARS or combinations thereof, there are several approaches for phase matching all processes.[31] USED CARS may even be employed. In this scheme for dual broadband CARS, the two Stokes sources are collinearly aligned and placed coaxially inside the annular ω_1 pump beam.[32] An example of simultaneous CARS generation from CO_2, N_2 and H_2O in the post reaction zone of a CH_4/air flame is displayed in Fig. 16. Since these signatures reside in their normal spectral locations relative to ω_1, they were sequentially recorded due to the high dispersion of the spectrograph employed. Optical approaches are being implemented to position all spectra onto a single optical multichannel detector at high dispersion.

5.1.3 Dual Pump-Stokes CARS. The last technique shown in Fig. 15 is a hybrid technique termed dual pump-Stokes due to the fact that the low Raman shift source, ω_2, can serve as both a pump and a Stokes source. It may be considered a specialized variation of dual broadband CARS. If the low frequency Raman resonance is narrow, the strength of the two-color CARS signal is greatly enhanced by contracting the ω_2 bandwidth. In so doing, this will sharpen the underlying, normally diffuse background spectrum in dual broadband CARS for $\omega_1 - \omega_2'$ resonances, Eq. (27), producing two well-defined spectral signatures in close proximity as in the dual pump approaches. This situation arises in examining multi-color strategies for H_2-air combustion.

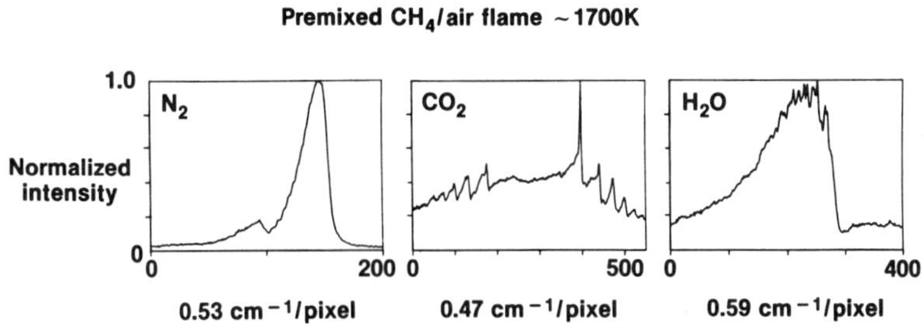

FIGURE 16. Simultaneous dual broadband CARS generation from the major constituents in the postflame region of a premixed methane-air flame.

5.2 Simplified Approach to Pure Rotational CARS

A variant of dual broadband CARS has led to a new and very simple approach to pure rotational CARS. Pure rotational CARS is of interest due to its thermometric accuracy at low temperatures, e.g., < 1000 K,[33] and the fact that its transitions in N_2 remain well separated even at high pressures. Thus collisional narrowing[19] is avoided and low spectral resolution will suffice. Despite these features, pure rotational CARS has not seen widespread diagnostic use since it is cumbersome to implement. To excite pure rotational resonances, the Stokes laser must be placed in close spectral proximity to the pump. However, dye laser physics precludes dye laser output spectrally close to the pumping source. Thus, with Nd:YAG based CARS systems, one must generate both the third and second harmonics simultaneously at considerable power loss to the latter. The third harmonic is then used to pump a dye laser spectrally adjacent to the pump to excite the rotational resonances. Because of the discreteness of the transitions and spiking in the dye laser amplitude profile, each single pulse spectrum must be referenced to account for the dye profile irregularities. This is done by wavemixing in some standard gas. Hence two spectra must be captured with each laser pulse and spectrally registered to one another, a complexity clearly desirable to avoid. The spectral integration aspect of dual broadband CARS may permit such amplitude irregularities to be averaged out and eliminate the need for referencing. In addition, dual broadband CARS for rotational resonances leads to a very simple experimental approach.

To perform dual broadband CARS for rotational Raman resonances in the 0 to 150 cm^{-1} range, the two Stokes sources have to be considerably overlapped. So overlapped in fact that one can simply employ a single broadband dye laser, Fig. 17. Different frequency combinations within the broadband laser excite the rotational Raman coherences from which the ω_1 pump scatters. Quite importantly, the broadband Stokes laser can be arbitrarily positioned in a spectral sense. Thus, third harmonic frequency generation from the Nd:YAG laser is not necessary. Since precise spectral positioning of the dye laser is not required, calibration is eliminated and spectral shifting of the Stokes laser does not affect measurement accuracy. One needs to know only the shape of the amplitude profile and not its precise spectral location. Thus the dye laser frequency can be selected on the basis of maximizing the output power or exciting a selected vibrational mode in addition.

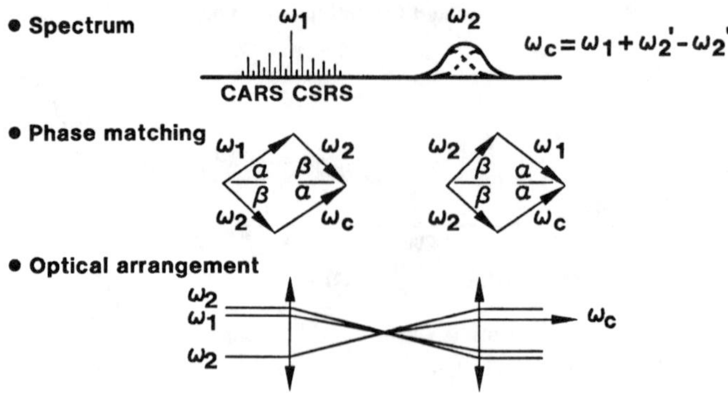

FIGURE 17. Simplified approach to pure rotational CARS based upon dual broadband CARS concept.

Another advantage of this new approach is the very good discrimination against the pump laser which can be obtained. This is achieved utilizing crossed-beam phase matching, Fig. 17, either folded or planar. Since the pure rotational CARS is situated in the green region of the spectrum relative to the 532 nm pump, discrimination is expedited further by placing the solitary Stokes laser at wavelengths beyond 600 nm. Whether the diagnostic potential of this new simplified approach to pure rotational CARS is realized vis-a-vis the elimination of spectral referencing requires further investigation. With this simplified approach, pure rotational CARS and vibrational CARS can be performed simultaneously by appropriately positioning the Stokes laser and phase matching the vibrational CARS process.[34]

5.3 Conditional Sampling

In studies of hydrocarbon-fueled turbulent combustion, it may be desirable to conditionally sample each temperature measurement for the presence of fuel and/or soot particulates. Such conditionally-sampled temperature pdf's should expedite insight into the fundamental sooting behavior of practical flames and may also aid in distinguishing between transport and chemical processes. Thus N_2 CARS thermometry must be augmented by additional species measurements and multi-color approaches become necessary.[35]

5.3.1 Soot Detection. As discussed earlier, soot particles scatter laser radiation quite efficiently and soot presence could be interrogated for by monitoring the resultant Mie scattering. In confined combustion vessels, it might be very difficult to discriminate the Mie signal from spuriously-scattered laser light. If this turns out to be the situation experimentally, then an alternate approach to soot detection would be necessary. A spectroscopic approach to soot detection has been formulated as outlined in Fig. 18. At the laser intensities used to generate CARS (10^9 to 10^{11} W/cm^2), the highly absorbing soot particles are laser heated into the 5000-6000 K range (Fig. 4) and vaporize. Major vaporization products include C_2 and C_3 and the presence of soot can be indirectly inferred by detecting the resultant C_2.

The C_2 can be detected either incoherently or coherently. Anti-Stokes shifted laser-induced fluorescence excited by the broadband Stokes ω_2 laser can be monitored to detect the C_2 incoherently. The C_2 Swan electronic system has many absorption bands throughout the spectral region occupied by appropriately positioned Stokes lasers. A Stokes dye laser positioned for N_2 CARS is particularly effective in this regard. Coherent detection is possible through conventional Raman-resonant CARS in the 1618 cm^{-1} C_2 vibrational mode or through electronically-enhanced mixing in the C_2 when the anti-Stokes signal coincides with a C_2 Swan band component in the blue region of the spectrum. This latter, purely electronically-enhanced, signal has been found to be weaker than the Raman-resonant signal, which is probably strengthened by some electronic enhancement as well.

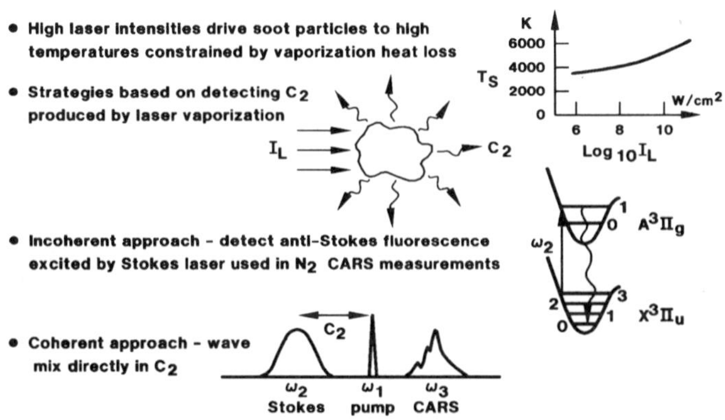

FIGURE 18. Strategies for spectroscopic detection of soot particles.

5.3.2 Fuel Detection. Most hydrocarbon fuels have very strong Raman cross sections in several spectral regions and all fuel detection approaches considered involve wave mixing in characteristic hydrocarbon vibrational band regions. Raman-active frequencies of most hydrocarbons exist in regions of both high Raman frequency, 2900-3200 cm^{-1} due to C-H stretch, and low frequency, 900-1700 cm^{-1}, from skeletal carbon stretches. Most olefinic species possess a strong Raman active mode near 1630 cm^{-1}, very near the C_2 resonance. Since the simpler hydrocarbons are major pyrolysis products of more complex hydrocarbons, their presence in a detectable abundance certainly signals the presence of fuel rich regions.

5.3.3 Integrated Strategies. If incoherent detection of the soot, i.e., C_2, is employed, then dual Stokes or dual pump strategies can be utilized to monitor N_2 for thermometry and the fuel in two, separate two- color or three-color wavemixing processes respectively. For Raman-resonant coherent detection of the C_2 (soot), dual Stokes or dual broadband CARS can be employed. The key here is to position the low Raman frequency Stokes source spectrally in such a manner that the fuel and C_2 are detected simultaneously by the Stokes source in the two-color wave mixing.

6.0 SUMMARY

Laser diagnostic approaches possess a number of attributes which make them quite attractive for probing the hostile combustion environments typical of gas turbines. Unfortunately, optical access to the primary zone of an actual gas turbine engine is quite a challenge which explains the lack of application to real engines other than to the exhaust regions. This situation may change in the future with advances in lasers and fiber optics. Practical application considerations screen the laser techniques which can operate successfully in the high interference environments typical of gas turbine combustion. For major species and temperature measurements, the focus here, CARS is the method of choice and has been demonstrated to be practically applicable to a broad range of device operating conditions permitting thermometry and species measurements in the 10 to 30 Hz data rate range. Advanced CARS techniques are now under development to remove the inability of CARS to measure more than one constituent at a time.

7.0 REFERENCES

1. A. C. Eckbreth: Laser Diagnostics for Combustion Temperature and Species. Abacus Press, Tunbridge Wells, Kent, U.K. (1987).

2. D. A. Long: Raman Spectroscopy. McGraw-Hill, New York, NY (1977).

3. N. Bloembergen: Nonlinear Optics. Benjamin, New York, NY (1967).

4. J. F. Driscoll, R. W. Schefer and R. W. Dibble: 477-485, Proceedings of the Nineteenth Symposium (International) on Combustion. The Combustion Institute, Pittsburgh, PA (1983).

5. M. C. Drake, M. Lapp, C. M. Penney, S. Warshaw and B. W. Gerhold: 1521-1531, Proceedings of the Eighteenth Symposium (International) on Combustion. The Combustion Institute, Pittsburgh, PA (1981).

6. W. J. Jones and B. P. Stoicheff: Phys. Rev. Letts. 13, 657 (1964).

7. P. Esherick and A. Owyoung: 130-187 in R. J. H. Clark and R. E. Hestor, Eds.: Advances in Infrared and Raman Spectroscopy - Volume 9. Heyden and Son, London, UK (1982).

8. S. A. J. Druet and J. P. E. Taran: Prog. Quant. Elec. 7, 1 (1981).

9. A. C. Eckbreth, J. F. Verdieck and P. A. Bonczyk: Prog. Energy Combust. Sci. 5, 253 (1979).

10. A. C. Eckbreth: J. Appl. Phys. 48, 4473 (1977).

11. D. C. Smith and R. G. Meyerand, Jr.: Chapter 11 in G. Bekefi, Ed.: Principles of Laser Plasmas. Wiley-Interscience, New York, NY (1974).

12. G. L. Switzer, C. G. Meyers and W. B. Roh: AIAA J. 16, 766 (1978).

13. L. A. Rahn, R. L. Farrow, M. L. Koszykowski and P. L. Mattern: Phys. Rev. Lett. 45, 620 (1980).

14. J. W. Nibler and G. V. Knighten: 253-299 in A. Weber, Ed.: Raman Spectroscopy of Gases and Liquids. Springer-Verlag, Berlin, FRG (1979).

15. R. J. Hall and A. C. Eckbreth: 213-309 in J. F. Ready and R. K. Erf, Eds.: Laser Applications, Vol. 5. Academic Press, New York, NY (1984).

16. R. L. Farrow and L. A. Rahn: J.O.S.A. B 2, 903 (1985).

17. K. A. Marko and L. Rimai: Opt. Lett. 4, 211 (1979).

18. A. C. Eckbreth, G. M. Dobbs, J. H. Stufflebeam and P. A. Tellex: Appl. Opt. 23, 1328 (1984).

19. A. C. Eckbreth and R. J. Hall: Combust. Sci. Tech. 25, 175 (1981).

20. R. E. Teets: Opt. Lett. 9, 226 (1984).

21. M. Pealat, P. Bouchardy, M. Lefebvre and J. P. Taran: Appl. Opt. 24, 1012 (1985).

22. R. L. Farrow, R. P. Lucht, G. L. Clark and R. E. Palmer: Appl. Opt. 24, 2241 (1985).

23. A. C. Eckbreth and J. H. Stufflebeam: Appl. Opt. 24, 1405 (1985).

24. D. A. Greenhalgh, F. M. Porter and W. A. England: Combust. Flame 49, 171 (1983).

25. S. Fujii, M. Gomi, K. Eguchi, S. Yamaguchi and Y. Jin: Combust. Sci. Tech. 36, 211 (1984).

26. R. Bedue, P. Gastebois, R. Bailly, M. Pealat and J. P. Taran: Combust. Flame 57, 141 (1984).

27. G. L. Switzer, L. P. Goss, D. D. Trump, C. M. Reeves, J. S. Stutrud, R. P. Bradley and W. M. Roquemore: AIAA J. 24, 1155 (1986).

28. A. C. Eckbreth: Combust. Flame 39, 133 (1980).

29. T. J. Anderson, G. M. Dobbs and A. C. Eckbreth: Appl. Opt. 25, 4076 (1986).

30. A. C. Eckbreth: SPIE 742, 34 (1987).

31. A. C. Eckbreth and T. J. Anderson: Appl. Opt. 24, 2731 (1985).

32. A. C. Eckbreth and T. J. Anderson: Appl. Opt. 25, 1534 (1986).

33. D. V. Murphy and R. K. Chang: Opt. Lett. 6, 233 (1981).

34. M. Alden, P-E. Bengtsson and H. Edner: Appl. Opt. 25, 4493 (1986).

35. A. C. Eckbreth, T. J. Anderson and G. M. Dobbs: Proceedings Twenty-first Symposium (International) on Combustion. The Combustion Institute, Pittsburgh, PA (1987).

RAMAN PROCESSES AND THEIR APPLICATION

ALFRED LEIPERTZ

RUHR-UNIVERSITÄT BOCHUM
INSTITUT FÜR THERMO- UND FLUIDDYNAMIK
D-4630 BOCHUM 1, FED. REP. GERMANY

1. INTRODUCTION

The necessity for non-perturbing, remote temperature and concentration measurements in combustion research forces further development of laser techniques for diagnostic purposes. High local and temporal resolution can be achieved only by laser induced fluorescence (LIF) and by laser scattering techniques (Mie, Rayleigh, Raman). LIF suffers from an only limited species applicability and from quenching problems (1,2). Mie scattering is very well suited for a large variety of flow velocity measurements (3) and also for selected different applications in flames and engines (2). Rayleigh scattering is so far restricted to density/temperature measurements in clean particle free (non-sooting) flames (2,4). The most general applicability is given for the Raman scattering processes. This is explained in some detail here in connection with fleasibility studies. Selected applications related as well to combustion modeling as to more industrial fields are presented. Recent developments in some aspects of the nonlinear CARS technique are covered in this volume in connection with applications in propulsion systems by A.C. Eckbreth and with engines by R.P. Lucht.

2. RAMAN PROCESSES

On irradiating a gas sample with laser frequencies ν_1 and ν_2, besides other optical processes an interaction between laser photons and gas molecules takes place in form of Raman scattering (Fig. 1). Basic information on these processes and their technical usability can be obtained from a variety of text and research books, see e.g. Refs. 5-9. Spontaneous linear Raman scattering (LRS) appears in every direction with frequencies $\nu_{1/2}\pm\nu_M$, ν_M being the rotational or rotational-vibrational resonance frequency of the molecule under investigation. This frequency shift is due to energy conservation during the interaction (Fig. 2) leading to Stokes $(\nu_{1/2}-\nu_M)$ and anti-Stokes $(\nu_{1/2}+\nu_M)$ spectral lines.

Besides LRS, simultaneously various nonlinear Raman processes occur in distinct directions relative to the irradiated laser beams (10). These directions are governed by phase matching conditions (momentum conservation during the processes). A few of them are indicated in Figure 1. For Stimulated Raman Gain Spectroscopy (SRG) and Inverse Raman Spectroscopy (IRS) these conditions are fulfilled in any case.

D. F. G. Durão et al. (eds.), Instrumentation for Combustion and Flow in Engines, 107–122.

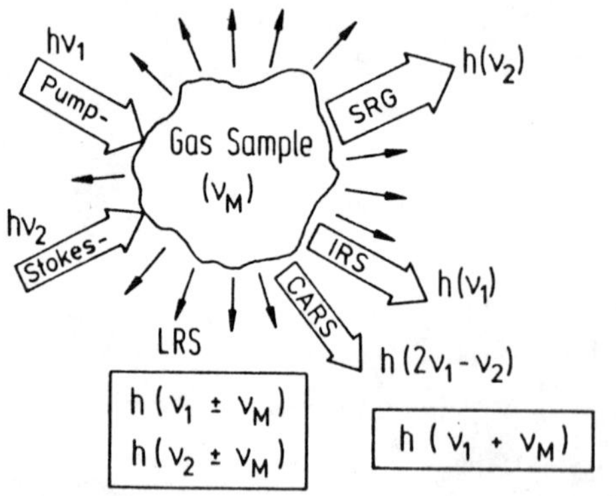

FIGURE 1. Schematic of different laser Raman processes.

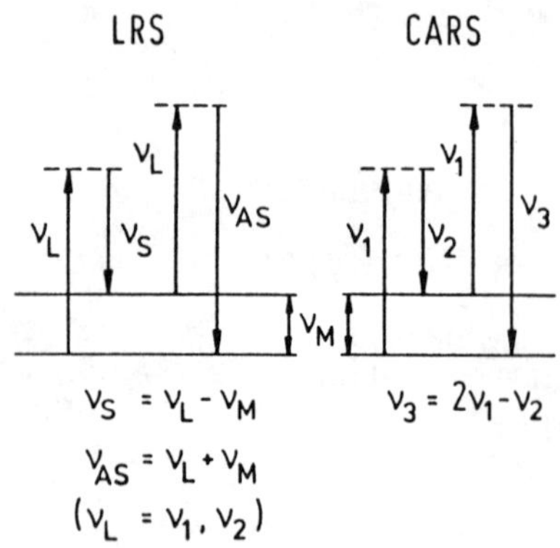

FIGURE 2. Energy level diagramm for spontaneous Raman scattering (LRS) and for CARS.

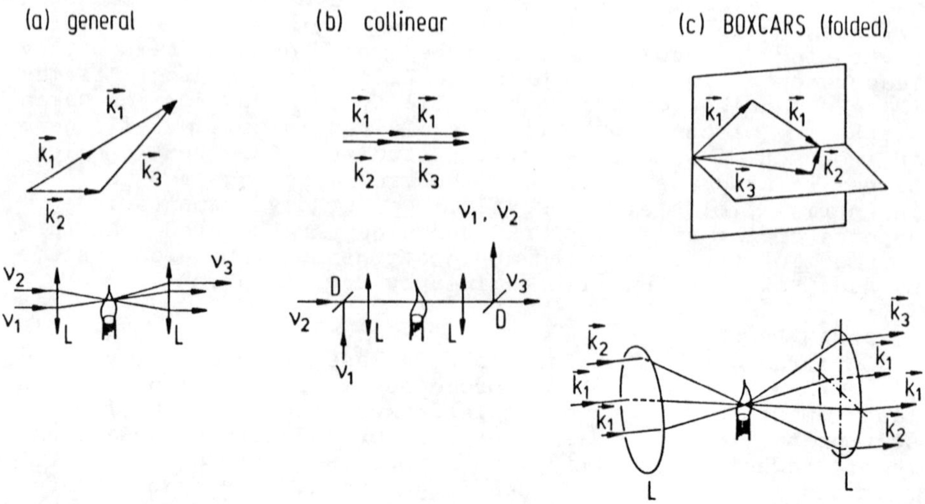

FIGURE 3. Schematic of different approaches for meeting the phase matching requirements for CARS generation. Shown are the wave vectors and the beam direction for the two laser beams and for the CARS beam. The BOXCARS configuration can also be mounted in a plane (14).

Due to problems with separating a relatively small signal from a large laser signal with same frequency, both processes have only poor technical significance. IRS is however very suitable for the measurement of fundamental physical parameters, e.g., the Raman line widths (11), and has recently also been used for velocity measurements in wind tunnels (12). The only nonlinear process with high technical performance so far is coherent anti-Stokes Raman scattering (CARS). It appears in the form of a laserlike beam with the difference frequency $2\nu_1-\nu_2$ (see Fig. 2). Selecting the laser frequencies appropriately ($\nu_1-\nu_2=\nu_M$), its intensity is by the molecule resonantly enhanced to a signal strength which is several orders of magnitude larger than the weak LRS signal. The CARS process can thus also be used successfully for measurements in hostile and industrial environments.

The phase matching condition requires the wave vectors $\vec{k}_3+\vec{k}_2=\vec{k}_1+\vec{k}_1$. Using $k_i=2\pi n_i\nu_i/c$ and $\nu_M<<\nu_1,\nu_2,\nu_3$ with refractive index n, $k_3+k_2=2k_1+(4\pi/c)\nu_M\,\partial n/\partial\nu$. Thus, for normal dispersion $\partial n/\partial\nu>0$, an angle arrangement between the beam vectors must be met (Fig. 3a) leading to the distinct direction of the CARS beam (13). For gases, however, $\partial n/\partial\nu\approx 0$. Thus, a simplified collinear arrangement can be used (Fig 3b). Higher local resolution can be achieved by a BOXCARS configuration (Fig. 3c) (14).

3. LINEAR (SPONTANEOUS) RAMAN SCATTERING

3.1. Temperature and concentration from LRS spectra

Figure 4 shows a schematic figure of the oxygen LRS spectrum for three different temperatures using a ruby laser for generation. Pure rotational and vibrational parts are generated simultaneously and can thus be collected within the time period of a single pulse event when a pulsed laser is used, or can be collected by scanning a monochromator over the spectral range of interest when using a CW laser. Simultaneously also the LRS spectra of other present gas species are generated and also those spectra for other exciting frequencies, e.g., ν_1 and ν_2 in Fig. 1. Thus, LRS has a multiple species capability.

Different possible approaches for obtaining temperature information from the spectra can be realized in Fig. 4 (5,8). Using the vibrational parts, the spectral shape of both branches (Stokes -S, anti-Stokes -AS) is strongly temperature dependent, as is the intensity ratio of both integrated branches (S/AS). Temperature can also be extracted from the intensity ratio of different bands within the branches, e.g., ground state band (0-1) and first excited band (1-2). Using the pure rotational wings which consist of a large number of single rotational lines as indicated for one of the temperature contours, again the spectral shape shows a strong temperature dependence: it is broadening, and the maximum position shifts further away from the generating wavelength with increasing temperature. The intensity distribution between the lines is varying with temperature which can be used for temperature evaluation from the ratio of single line intensi-

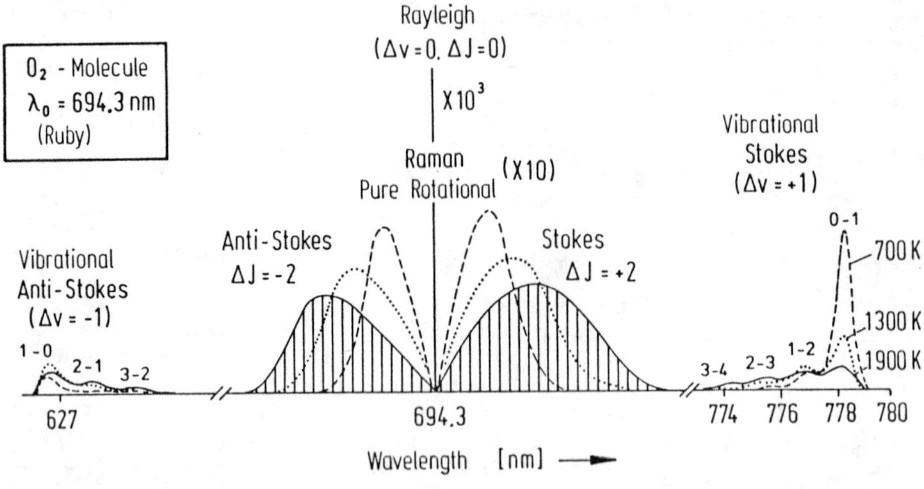

FIGURE 4. Schematic of the oxygen LRS spectrum for three different temperatures. The spectrum is generated by the irradiation of a (pulsed) ruby laser.

ties or, more accurate, by using all the line intensities for a Boltzmann plot.

Concentration information can be obtained from the integrated line intensities of the vibrational branches or of single rotational lines. As absolute measurements are difficult to perform, calibration at room temperature will simplify the procedure (15).

3.2. Feasibility of LRS measurements

On basis of the investigations of Goulard (16,17), the feasibility of LRS measurements will be studied for different laser systems. Starting from the basic equation for the LRS intensity

$$E_{LRS} = E_L \, (d\sigma/d\Omega)_L \, N \, l \, \Omega \, e, \tag{1}$$

an useful feasibility index

$$X \equiv N \, (d\sigma/d\Omega)_o \, l \, \Omega \, e \tag{2}$$

has been developed and applied to several applications which are characterized by the space (length l) and time resolution required as indicated in Fig. 5 (A - F) (16). For the investigation of fluid dynamic processes, suitable measurement resolution is given by the Kolmogorov length and time scales (see e.g., (18)), which also are basis of the additionally indicated combustion applications (G,H,I) as adopted from Ref. 17. In Eq.(1,2), E_{LRS} is the energy of the Raman scat-

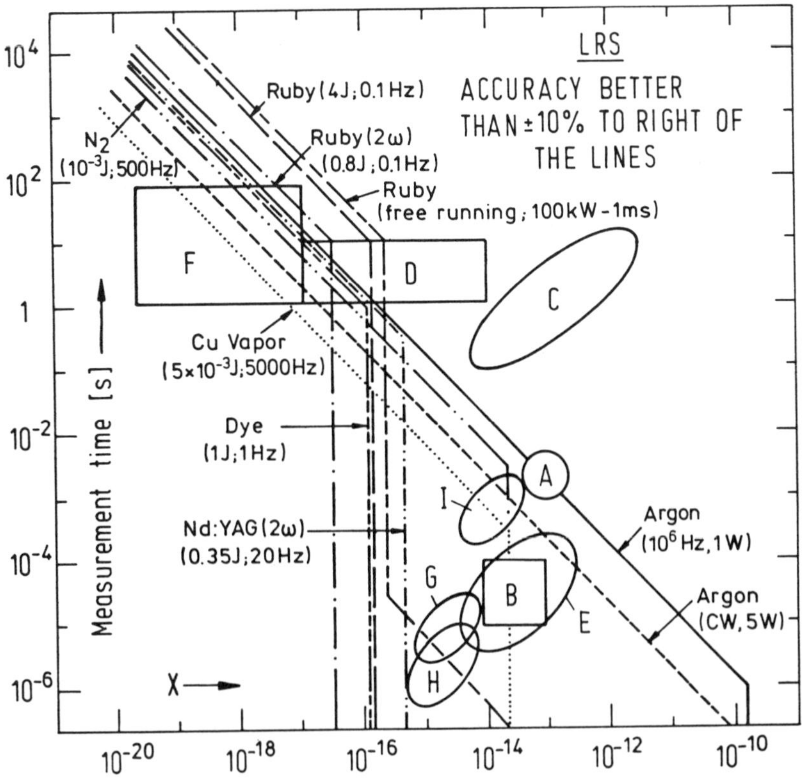

FIGURE 5. Feasibility of spontaneous (linear) Raman measurements. Application key: A - Compressor inlet; B - Compressor blade passage; C - Laboratory temperature and concentrations; D - Remote atmospheric constituents and temperature; E - Small scale turbulence in laboratory flow; F - Remote pollution sources; G - Gas turbine combustor; H - Afterburner combustor; I - Fires.

tered photons received by the sensor; E_L is the energy of the irradiated photons; l is the length of the gas sample in the direction of the laser beam; N is the number density of the gas molecules investigated, which is dependent on pressure and gas temperature; e is the optical efficiency of the system; Ω is the solid angle of the collection optics; $(d\sigma/d\Omega)_{L,o}$ is the differential Raman scattering cross section of the molecules for laser frequency ν_L and for a reference frequency ν_o, respectively $((d\sigma/d\Omega)_L/(d\sigma/d\Omega)_o=(\nu_L/\nu_o)^4\equiv 1/q^4)$. In order to meet a desired measurement accuracy, an appropriate number of Raman photons has to reach the detector. This number is governed by photon statistics (Poisson statistics) and is $N_{LRS}=100/\eta$ for an inaccuracy of ±10% and $N_{LRS}=10^4/\eta$ for ±1% (η: quantum efficiency of the detector for the Raman frequency). For the comparison of different laser-detector systems, from Eq. (1) a laser and detector dependent index can be derived - similar to the application index in Eq.(2) -,for example for an inaccuracy of ±10%

$$X_{min} = 100\ q^4 / \ E_L\ \eta. \qquad (3)$$

Figure 5 shows these values for several different laser systems indicated, and it also indicates how the laser-detector capabilities match against the feasibility index regions of various applications. From this comparison it appears, that a measurement inaccuracy of ±10% can be obtained for nearly all applications when pulsed high power lasers are employed, even ±1% when using the frequency doubled ruby laser (for this accuracy the lines have to be shifted to the right by two orders of magnitude). If the time history of the measured values is needed, this can be realized using cw or quasi-cw laser operation and is possible only for a few of the applications. For high frequency processes time history can be observed by the free running ruby laser during its single pulse operation time of about 1 ms (19). Emphasis must, however, be laid on the fact, that these considerations are only related to signal strength. Problems arising from signal interferences from particle scattering (e.g., soot) or from species fluorescence are not considered here. In general these influences lead to a small signal-to-interference-ratio and thus to poorer accuracy. These problems are treated in more detail by A.C. Eckbreth in this volume. They restrict the application of the LRS technique to relative clean environments.

3.3. LRS applications

Even when a few of the LRS applications are directed to more technical fields (e.g., turbojet engine (20), IC engine (21), coal furnace (22)), the majority of those measurements are related to fundamental research, system developments or combustion modelling (see e.g., (15,19,23-37)). Premixed flames (23-27) have been investigated as well as non-sooting diffusion flames (28-34). Using the photon-correlation technique, structures in flows and flames can be detected easily (35). For the simultaneous collection of all major combustion species, polychromators have been used (30,31). In those configurations in connection with a LDA system the simultaneous measurement of vector and scalar quantities is possible (32-34). Forming the laser beam to a light sheet and using a two-dimensional detector, simultaneously a two-dimensional field information can be obtained, e.g., of the fuel gas concentration (2,36). In a recent preliminary approach LRS measurements are used for determining the fuel/air-ratio before injection into an IC engine (37).

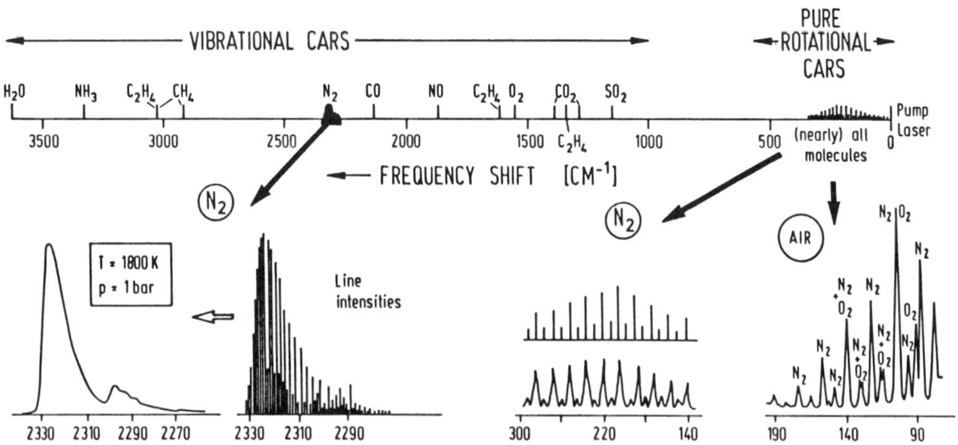

FIGURE 6. Schematic of the CARS spectra (vibrational CARS and pure rotational CARS).

4. COHERENT ANTI-STOKES RAMAN SCATTERING

4.1. Temperature and concentration from CARS spectra

Figure 6 shows a schematic of the CARS spectra generated by appropriately selecting the Stokes laser due to the molecule of interest. As the CARS frequency is identical to the difference frequency of both irradiated laser beams (Fig. 1), in general just a single gas species can be investigated by using vibrational CARS. This is mainly due to the fact, that the radiation of the Stokes dye laser is spectroscopically not broad enough to cover all the spectral range of interest for detecting several different molecules simultaneously. Recent approches to overcome these limitations are discussed by A.C. Eckbreth and R.P. Lucht in this volume. This is different for pure rotational CARS. Except the spherical top molecules (e.g., CH_4) every gas molecule gives contibutions in this spectra range. Thus all present molecules can be investigated simultaneously using the same Stokes dye laser. Problems may arise from linie overlap between different molecules, see e.g. the air spectrum in Fig. 6, which in general is not the case for vibrational CARS.

Temperature information can be obtained from the vibrational CARS spectra, see Fig. 7, by a weighted fit of the theoretical and experimental spectra (38). As the shape of the spectra is governed by the convolution of the single line intensities with the spectral resolution of the system, all quantities influencing this shape have to be considered carefully: e.g., laser bandwidths (39) with cross-coherence effects (40), collisional narrowing at high pressures (41) and the scaling of the nonresonant contributions (42). Besides that, difficulties may occur from Stark shift (43,44) and

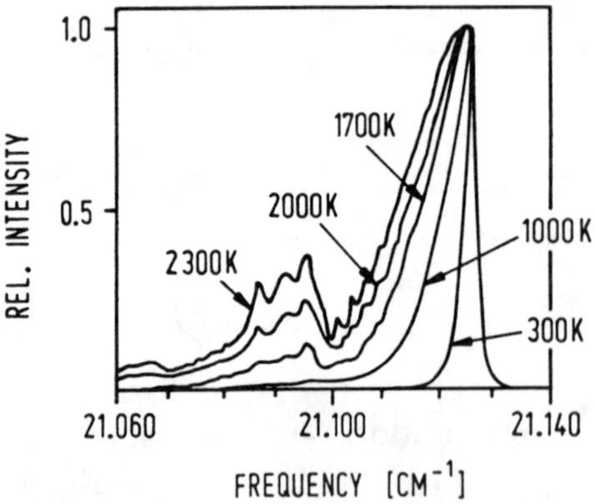

FIGURE 7. Schematic of the temperature dependence of the vibrational CARS spectrum of N_2.

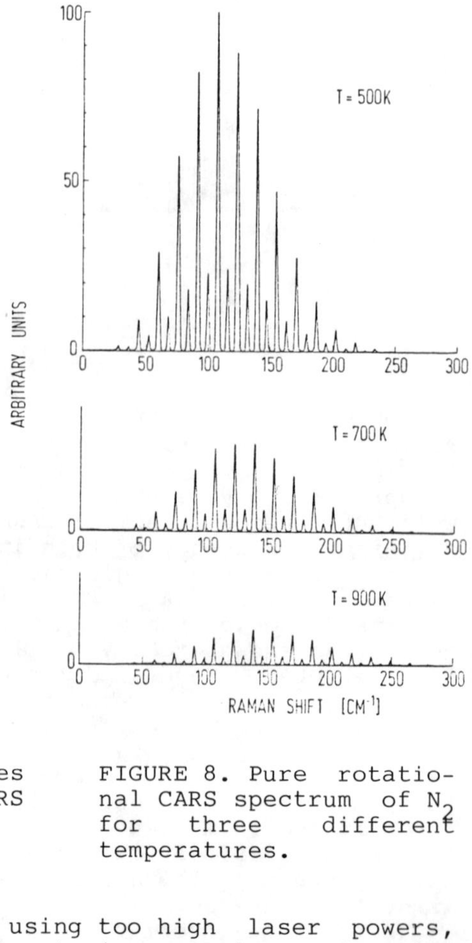

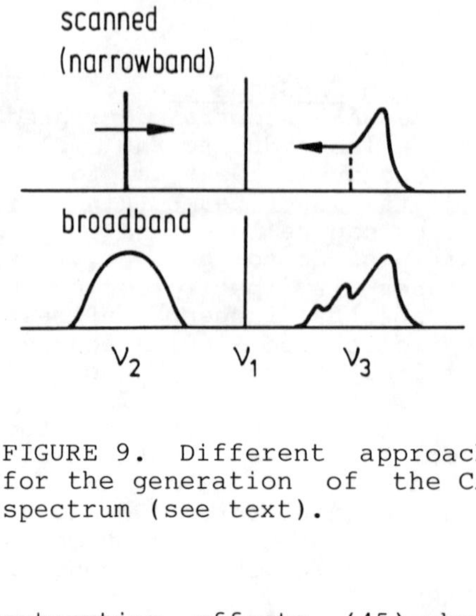

FIGURE 9. Different approaches for the generation of the CARS spectrum (see text).

FIGURE 8. Pure rotational CARS spectrum of N_2 for three different temperatures.

saturation effects (45) when using too high laser powers, leading in summary alltogether to inaccuracies and noise problems. The evaluation procedure is time consuming. In complicated cases a single fit may take up to a few hours. Quick-fitting techniques can sometimes be used to fasten the evaluation process (46). This situation is somewhat less complicated when using the pure rotational CARS spectrum. Due to an improved line separation (see e.g. N_2 in Fig. 6) a simplified modeling of the spectra is possible with faster evaluation procedures (47,48). The temperature dependence of the pure rotational CARS spectrum of N_2 is shown in Fig. 8. Pure rotational CARS gives extrem high accuracies for T<1000 K (see e.g., (47) for T<300 K), suffers however - although possible (49) - at flame temperatures from low signal intensities. Problems with line overlap from different molecules may be overcome by using the Fourier analysis technique (48,50).

Density or concentration determination from CARS spectra is more difficult to perform than thermometry because of the nonlinear nature of CARS generation. In general, again the integrated line intensities give information on the number density of the gas molecules considered. This, however, is not as simple done for vibrational CARS as for pure rotational CARS. Different approaches have been used which cannot be discussed here in detail, see e.g., (44,51-53).

CARS generation is possible in two different ways, see Fig. 9. Using a narrowband Stokes dye laser, the CARS spectrum is generated consecutively by scanning the Stokes laser over the spectral range of the molecule resonances. This procedure has the advantage of providing a higher spectral density of the laser radiation leading to higher CARS intensities. It takes, however, an extanded time period, over which the measurement information is accumulated in the CARS signals. Due to the nonlinear nature of the signal strengths, see Eq. (4), this procedure is not appropriate in situations with fluctuating concentrations and temperatures as is nearly every combustion process. Here, a broadband Stokes laser has to be employed generating simultaneously the entire CARS spectrum of interest within the instant of the pulse length. This is the only appropriate operating mode for combustion reseach.

4.2. Feasibility of CARS measurements

Similar feasibility considerations as for LRS measurements can be done for CARS measurements (8,54). The CARS intensity can be expressed by the dependence

$$P_{CARS} \sim \eta \, P_p^{\,2} \, P_s \, \{\Delta N(v,J) L (d\sigma/d\Omega)/\Gamma_J\}^2 \qquad (4)$$

Here, P_p and P_s is the power of the pump and Stokes laser, respectively; $\Delta N(v,J)$ is the population difference of the molecular vibrational (v) and rotational (J) energy levels involved in the CARS process with the energetic distance $h\nu_M$; Γ_J is the Raman linewidth; and L is the interaction length of pump and Stokes laser giving the local resolution in the measurement. Again, a separation can be done between quantities which are related to the laser/detector system to be used (P_p, P_s, η) and those which are dependent on the particular application case (local resolution L and gas quantities ΔN, $(d\sigma/d\Omega)$ and Γ_J which are function of pressure, concentration and temperature). Using a few simplifications (8,54), again a feasibility index Y for the laser/detector system and for the measurement applications can be defined. This is for an inaccuracy of ± 1 % indicated in Fig. 10 in

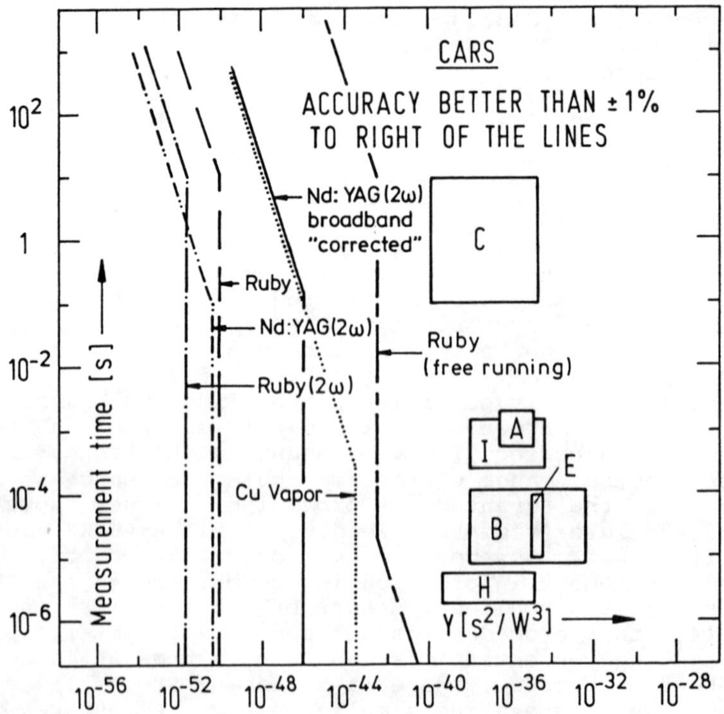

FIGURE 10. Feasibility of coherent anti-Stokes Raman scattering measurements. Application key identical to Fig. 5.

form of different laser/detector lines and application regions as defined by the alphabetic key in Fig. 5. Only high power lasers are considered, because cw CARS, which in general is possible (55), has no technical capability due to the low cw laser power.

It is obvious, that all CARS systems considered are capable of measuring to a high accuracy. Even the "corrected" Nd:YAG system can easily obtain a ± 1 % inaccuracy. This correction has been done taking into account a folded BOXCARS configuration, a broadband CARS generation and an energy limitation of 10 mJ for the pump laser in order to avoid Stark shifting and saturation effects. Again, however, emphasis must be laid on the fact that only signal strengsth considerations have been done neglecting all the above mentioned problems and their influences on accuracy.

4.3. CARS applications

The high signal strength allows the utilization of the CARS technique for concentration and temperature measurements also in hostile and industrial environments. It has been

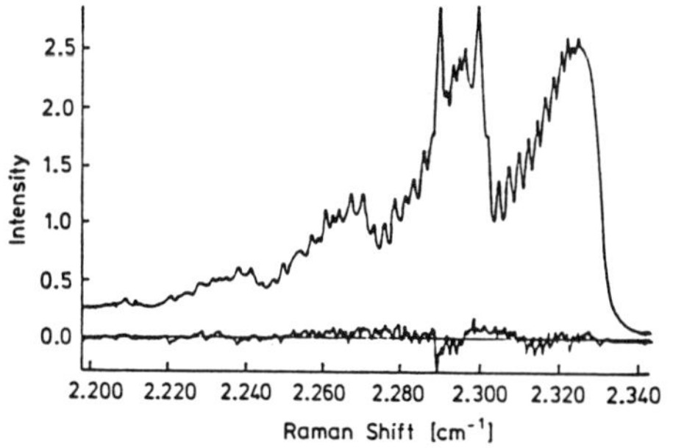

FIGURE 11. N_2 CARS spectrum measured inside of the coiled tungsten filament of a tungsten lamp. Evaluated temperature: 3467 K (82).

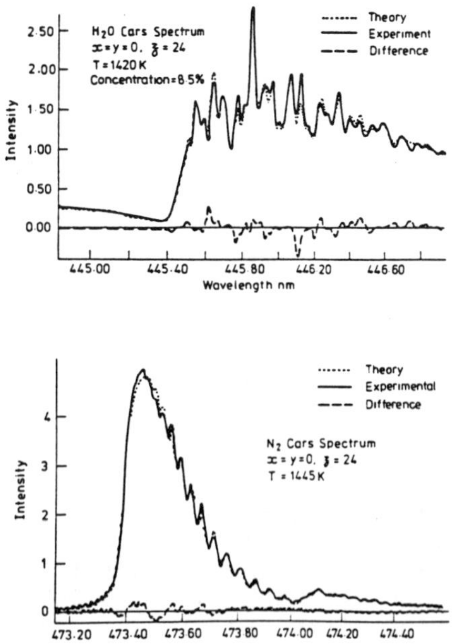

FIGURE 12. Average CARS spectra of water vapor and nitrogen measured at the same point in the post-flame zone of a 30 kW oil spray furnace (53).

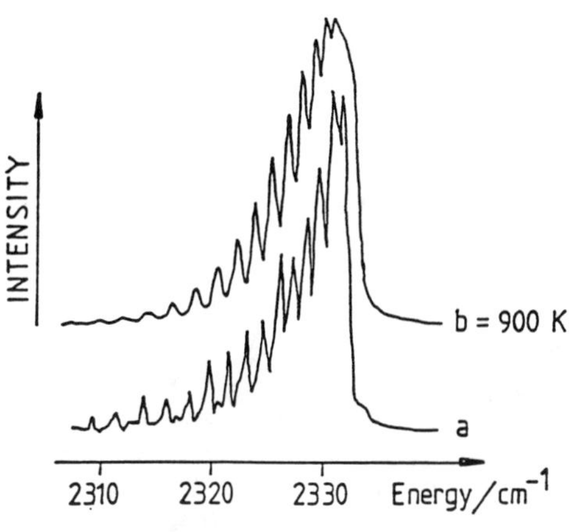

FIGURE 13. N_2 CARS spectrum measured in a full-scale industrial coal furnace (a) and (b) calculated spectrum for 900 K (79).

employed to the investigation of turbulent, sooting and

propellant flames (56-65) as well as for measurements in technical combustion systems (51,53,66-81). These technical applications range from the exhaust regions of afterburning jet engines (66), over gas and liquid fired simulation burners (51,67,68), IC engines (69-77), combustors and furnaces (53,78,79) to chemical reactors (80) and simulated coal-fired MHD exhausts (81). Applications in propulsion systems are considered in more detail by A.C. Eckbreth in this volume and IC engine measurements by R.P. Lucht. Here, as well gas as petrol, but also diesel (71) fired engines have been investigated. The highest temperature measured so far was taken inside a coiled tungsten lamp with 3467 K (82). Figure 11 shows the N_2 CARS spectrum and the deviation plot between theoretical and experimental shape for that temperature value. Measurements outside the filament were found to be in good agreement with a two-dimensional fluid model. The experimental error was claimed to be in the order of ± 20 K.

CARS spectra measured in furnaces are shown in Figs. 12 and 13. Figure 12 shows the average CARS spectra of water vapor and of nitrogen measured at the same point in the post-flame zone of a 30 kW oil (Kerosene) spray furnace (53). In this furnace instantaneous temperature p.d.f.'s have been taken at various radial and axial positions within the reaction zone and in the post-flame regions as well indicating a high standard deviation due to large fluctuations at several positions. The agreement between water vapor and nitrogen temperatures in Fig. 12 is excellent and the calculated water concentration of 8.5 % seems reasonable. Figure 13 shows the N_2 CARS spectrum taken in a full-scale (10 m * 10 m) 150 MW industrial coal furnace (79). The beam transmission through the furnace was found to be about 5 %. Still CARS measurements were possible but only by using the narrowband scanning technique giving too low temperature results due to the nonlinear nature of the CARS intensity, Eq.(4). This application case shows the border-line which can be reached so far by the CARS technique.

Thermometry mostly has been done with the nitrogen molecule, but also other molecules have been used for CARS measurements, see e.g., Refs. (83,52,84,85,86,87) for O_2, CO, CO_2, H_2O, N_2O and SO_2, respectively. For more complex molecules the calculation of the CARS spectra is difficult. Thus, temperature information from the fit procedure is not possible to obtain. A first approach for measuring the fuel/air ratio between iso-octane and nitrogen was performed by calibrating the iso-octane signal by generating experimentally a computer library for the spectra at different temperatures and concentrations (76).

5. CONCLUSIONS

Raman processes are very well suitable for concentration (c) and temperature (T) measurements in combustion systems achieving high temporal (ns - µs) and spatial resolution (10^{-3} mm^3; typical < 1 mm^3). They are applicable to all gas molecules. For a typical combustion system with air, the performance of LRS and CARS systems in terms of accuracy and detection limit is given in the following table:

	LRS		CARS	
	Single Puls	CW(1 sec)	Broadband (Single Puls)	Narrowband (Scanning)
Inaccuracy $T(N_2)$	5 - 10 %	< 1 % (Room) 25-50K (Flame)	5-10 % (Room) 50-100K (Flame)	3-5 % (Room) 30-50K (Flame)
Inaccuracy $c(N_2)$	10 - 20 %	< 5 %	10 - 20 %	
Detection Limit	< 100 ppm	1ppm	>1%	100ppm

The accuracy for the single-puls measurements can be improved according to signal statistics by averaging over a large number of pulses. Those averages must be calculated, however, for the evaluated concentration and temperature values, but not for the signal intensities. Due to the nonlinear dependence of the Raman processes on temperature, signal accumulation will give erroneous values. Whenever possible, the application of the more simple LRS technique is recommended.

REFERENCES

1. Eckbreth AC, Bonczyk PA, Verdieck JF: Prog. Energy Combust. Sci. 5 (1979) 253-322.
2. Leipertz A: Quantitative two- and three-dimensional measurement techniques (in this volume).
3. Durst F, Melling A, Whitelaw JH: Principle and Practice of Laser-Doppler Anemometry. London-New York: Academic Press, 1976. See also the chapters written by F. Durst and D. Durao in this volume.
4. Arcoumanis C: The laser Rayleigh scattering technique (in this volume).
5. Lapp M, Penney CM (Eds.): Laser Raman Gas Diagnostics. New York: Plenum Press, 1974.
6. Long DA: Raman Spectroscopy. London: Mc Graw-Hill, 1977.
7. Weber A (Ed.): Raman Spectroscopy of Gases and Liquids. Berlin: Springer Verlag, 1979.
8. Leipertz A: Temperaturbestimmung in Gasen mittels linearer und nichtlinearer Raman-Prozesse. Habilitationsschrift, Ruhr-Universität Bochum, 1984.
9. Eckbreth AC: Laser Diagnostics for Combustion Temperature and Species. Cambridge (Mass.): Abacus Press, 1987.
10. Levenson MD: Introduction to Nonlinear Laser Spectroscopy. New York: Academic Press, 1982.
11. See e.g., Rosasco GJ, Lempert W, Hurst WS, Flin A: Chem. Phys. Lett. 97 (1983) 435-440; Rahn LA, Palmer RE: J. Opt. Soc. Am. B 3 (1986) 1164-1169.
12. Exton RJ, Hillard ME: Appl. Opt. 25 (1986) 14-21; AIAA paper 87-1531 (1987).
13. Tolles WM, Nibler JW, Mc Donald JR, Harvey AB: Appl. Spectrosc. 31 (1977) 253-270.

14. Eckbreth AC: Appl. Phys. Lett. 32 (1978) 431-432; Shirley JA, Hall RJ, Eckbreth AC: Opt. Lett. 5 (1980) 380-382.
15. Leipertz A, Haumann J, Fiebig M: Chem. Eng. Technol. 10 (1987) 190-203.
16. Goulard R: J.Quant. Spectrosc. Radiat. Transfer 14 (1974) 969-974.
17. Goulard R, Mellor AM, Bilger RW: Combust. Sci. Technol. 14 (1976) 195-219.
18. Tennekes H, Lumley JL: A First Course in Turbulence. Cambridge (Mass.), London: MIT Press, 1973.
19. Pealat M, Bailly R, Taran JP: Opt. Commun. 22 (1977) 91-94.
20. Roquemore WM, Yaney PP: NBS Spec. Publ. 561 (1979) 973-1025.
21. Smith JR: AIAA paper 80-1359 (1980); SAE paper 800137 (1980).
22. Harvey SM: Ontario Hydro Research Div. Rep. 79-51-K (1979).
23. Bechtel JH, Blint RJ Dasch CJ, Weinberger DA: Combust. Flame 42 (1981) 197-213.
24. Bechtel JH, Blint RJ: Appl. Phys. Lett. 37 (1980) 576-578.
25. Setchell RD, Miller JA: Combust. Flame 33 (1978) 23-32.
26. Drake MC, Hastie JW: Combust. Flame 40 (1981) 201-211.
27. Stephenson DA: Proc. 17th Symp. (Int.) on Combustion, pp. 393-399. Pittsburgh: The Combustion Institute, 1979.
28. Aeschliman DP, Cummings JC, Hill RA: J. Quant. Spectrosc. Radiat. Transfer 21 (1979) 293-307.
29. Setchell RE: AIAA J. 18 (1980) 181-187.
30. Drake MC, Lapp M, Penney CM, Warshaw S, Gerhold BW: Proc. 18th Symp. (Int.) on Combustion, pp. 1521-1531. Pittsburgh: The Combustion Institute, 1981.
31. Dibble RW, Kollmann W, Schefer RW: Combust. Flame 55 (1984) 307-321.
32. Drake MC, Pitz RW, Lapp M: AIAA J. 24 (1986) 905-917.
33. Johnston SC, Dibble RW, Schefer RW, Ashurst WT, Kollmann W: AIAA J. 24 (1986) 918-937.
34. Dibble RW, Hartmann V, Schefer RW, Kollmann W: Exp. Fluids 5 (1987) 103-113.
35. Birch AD, Brown DR, Dodson MG, Thomas JR: J. Phys. D 8 (1975) L 167-170; J. Fluid Mech. 88 (1978) 431-449.
36. Long MB, Levin PS, Fourguette DC: Opt. Lett. 10 (1985) 267-269.
37. Heinze T, Daams HJ: SAE paper 870098 (1987).
38. See e.g., Snelling DR, Smallwood GJ, Sawchuk RA, Parameswaran T: Appl. Opt. 26 (1987) 99-110.
39. Yuratich MA: Mol. Phys. 38 (1979) 625-655.
40. Teets RE: Opt. Lett. 9 (1984) 226-228; Greenhalgh DA, Hall RJ: Opt. Commun. 57 (1986) 125-131.
41. Hall RJ: Opt. Eng. 22 (1983) 322-329; Koszykowski ML, Farrow RL, Palmer RE: Opt. Lett. 10 (1985) 478-479.

42. Rosasco GJ, Hurst WS: Phys. Rev. A 32 (1985) 281-299; Lundeen T, Hou SY, Nibler JW: J. Chem. Phys. 79 (1983) 6301-6306.
43. Rahn LA, Farrow RL, Koszykowski ML, Mattern PL: Phys. Rev. Lett. 45 (1980) 620-623; Farrow RL, Rahn LA: Phys. Rev. Lett. 48 (1982) 395-398.
44. Farrow RL, Lucht RP, Clark GL, Palmer RE: Appl. Opt. 24 (1985) 2241-2251.
45. Weil H, Schreiber PW: Appl. Opt. 21 (1982) 941-948; Gierulski A, Noda M, Yamamoto T, Marowsky G, Slenczka A: Opt. Lett. 12 (1987) 608-610.
46. Eckbreth AC, Dobbs GM, Stufflebeam JH, Tellex PA: Appl. Opt. 23 (1984) 1328-1339; Pealat M, Bouchardy P: ONERA Rep. RF 30/7131 PY (1982).
47. Murphy DV, Chang RK.: Opt. Lett. 6 (1981) 233-235.
48. Lasser T, Magens E, Leipertz A: Opt. Lett. 10 (1985) 535-537.
49. Zheng JB, Snow JB, Murphy DV, Leipertz A, Chang RK, Farrow RL: Opt. Lett. 9 (1984) 341-343.
50. Leipertz A, Magens E, Lasser T: AIAA paper 85-1569 (1985).
51. Attal B, Pealat M, Taran JP: AIAA paper 80-0282 (1980); Goss LP, Switzer GL, Schreiber PW: J. Energy 7 (1983) 389-395.
52. Eckbreth AC, Hall RJ: Combust. Sci. Technol. 25 (1981) 175-192.
53. Greenhalgh DA, Porter FM: Proc. Combust. Res. Conf., Harwell, 1986, pp. 20-33.
54. Leipertz A: in Optical Methods in Dynamics of Fluids and Solids (Ed. M. Pichal), pp. 21-27. Berlin: Springer Verlag, 1985.
55. Hirth A, Vollrath K: Opt. Commun. 18 (1976) 213-214.
56. Eckbreth AC, Hall RJ: Combust. Flame 36 (1979) 87-98.
57. Switzer GL, Roquemore WM, Bradley RB, Schreiber PW, Roh WB: Appl. Opt. 18 (1979) 2343-2345.
58. Harris LE, Mc Ilwain M: Combust. Flame 48 (1982) 97-100.
59. Fujii S, Gomi M, Jin Y: Combust. Flame 48 (1982) 233-240.
60. Greenhalgh DA, Porter FM, England WA: Combust. Flame 49 (1983) 171-183.
61. Goss LP, Mc Donald BG, Trump DD, Switzer GL: AIAA paper 83-1480 (1983).
62. Fujii S, Gomi M, Eguchi K, Yamaguchi S, Jin Y: Combust. Sci. Technol. 36 (1984) 211-226.
63. Hall RJ, Boedecker LR: Appl. Opt. 23 (1984) 1340-1346.
64. Antcliff RR, Jarrett Jr.O, Rogers RC: J. Propulsion 1 (1985) 205-209.
65. Roquemore WM, Bradley RP, Jackson TA, Kizirnis SW, Goss LP, Switzer GL, Trump DD, Sarka B, Ballal DR, Lightman AJ, Yaney PP, Chen TH: Development of laser dignostics for combustion research, 1986 Spring Techn. Meeting-CSS/Combust. Inst., May 1986.
66. Eckbreth AC, Dobbs GM, Stufflebeam JH, Tellex PA: AIAA paper 83-1294 (1983)
67. Eckbreth AC: Combust. Flame 39 (1980) 133-147.

68. Bedue R, Gastebois P, Bailly R, Pealat M, Taran JP: Combust. Flame 57 (1984) 141-153.
69. Stenhouse IA, Williams DR, Cole JB, Swords MD: Appl. Opt. 18 (1979) 3819-3825.
70. Klick D, Marko KA, Rimai L: Appl. Opt. 20 (1981) 1178-1181.
71. Kajiyama K, Sajiki K, Kataoka H, Maeda S, Hirose C: SAE paper 821036 (1982).
72. Alessandretti GC, Violino P: J. Phys. D Appl. Phys. 16 (1983) 1583-1594.
73. Lucht RP, Green RM, Palmer RE, Teets RE, Ferguson CR: SAND Report 86-8749 (1986).
74. Marie JJ, Cottereau MJ: SAE paper 870458 (1987).
75. Lucht RP, Maris MA: SAE paper 870459 (1987).
76. Greenhalgh DA, Williams DR, Baker CA: Proc. ISATA Conf., Florence May 1987.
77. Brüggemann D, Hassel E, Dittie G: VDI Berichte 645 (1987) 603-614.
78. Ferrario A, Malvicini C: in Analytical Laser Spectroscopy (Eds. S. Martellucci, A.N. Chester), pp. 89-105. New York: Plenum Publ., 1985.
79. Alden M, Wallin S: Appl. Opt. 24 (1985) 3434-3437.
80. England WA, Milne JM, Jenney SN, Greenhalgh DA: AERE Report R10945 (1983).
81. Beiting EJ: Appl. Opt. 25 (1986) 1684-1692.
82. Greenhalgh DA et al: Chem. Phys. Lett. 129 (1986) 191-196; 132 (1986) 283-290; 133 (1987) 458-464.
83. Shirley JA, Hall RJ, Eckbreth AC: NASA Rep. 159280 (1980).
84. Hall RJ, Stufflebeam JH: Appl. Opt. 23 (1984) 4319-4327; Papineau N, Pealat M: Appl. Opt. 24 (1985) 3002-3009.
85. Hall RJ, Shirley JA: Appl. Spectrosc. 37 (1983) 196-202.
86. Hall RJ, Stufflebeam JH in Ref. (84).
87. Alden M: private communication, July 1987.

QUANTITATIVE TWO- AND THREE-DIMENSIONAL MEASUREMENT TECHNIQUES

ALFRED LEIPERTZ

RUHR-UNIVERSITÄT BOCHUM
INSTITUT FÜR THERMO- UND FLUIDDYNAMIK
D-4630 BOCHUM 1, FED. REP. GERMANY

1. INTRODUCTION

Real turbulent flow and combustion processes are three-dimensional in nature revealing large- and small-scale structures in the flow and combustion field. Those coherent structures play an important role in the development of turbulence and also for the interaction of flowfield and chemistry in combustion (1,2). Their existence can easily be seen, e.g., from Schlieren pictures (3) providing however only qualitative information. For modelling purposes and thus for a better understanding of the phenomena, quantitative analysis is necessary. Using point-measurement techniques with quantitative potentials, two- and three-dimensional structures cannot be resolved easily. Extending these techniques, however, by forming a light sheet and detecting the signal intensity within this sheet by a two-dimensional photo-electrical detector, the required informations can be obtained for a large number of fundamental applications, but also for a few more technical situations. Well-established quantitatively working laser techniques providing high spatial and temporal resolution are laser-induced fluorescence (LIF) and the laser scattering techniques (Mie, Rayleigh, Raman). These will be treated here in some detail as they are mostly employed to two-dimensional (2-d) planar field measurements. More recent approaches for three-dimensional (3-d) imaging (4-6) are described in the last chapter of this paper.

2. PRINCIPLE OF THE TECHNIQUES

2.1. Measurement techniques

On irradiating a laser beam into a gas sample of molecules and particles (dust, soot, etc.), different interaction processes between matter and the laser photons may occur. This is shown schematically and very much simplified in Fig. 1. Here it is assumed that the molecules may have an initial energy state due to vibrations with frequency ν_1 and rotations with frequency ν_2. Besides light absorption processes by the molecules or the particles, which will not be considered here because of its lack of spatial resolution, the different forms of interaction can be recognized in every direction and distinguished, e.g., by the emitted frequency.

The different techniques are indicated in Table 1 in con-

*Figures 9, 10, 11, 14, 17, 18, 20 and 22 in this article were submitted in colour, but have been reproduced in black and white.

D. F. G. Durão et al. (eds.), Instrumentation for Combustion and Flow in Engines, 123–140.

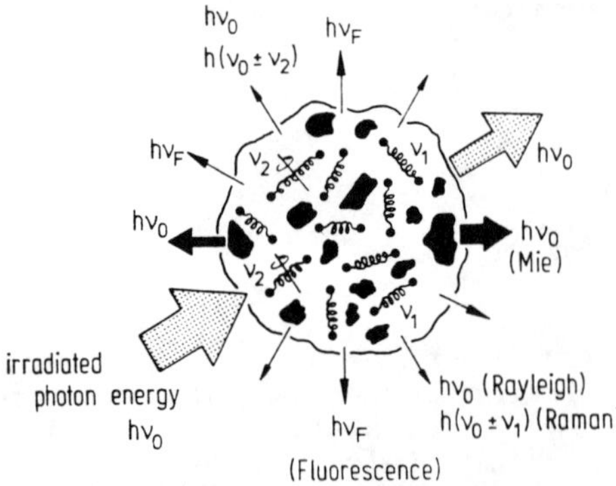

FIGURE 1. Simplified schematic of the interaction of laser photons with matter.

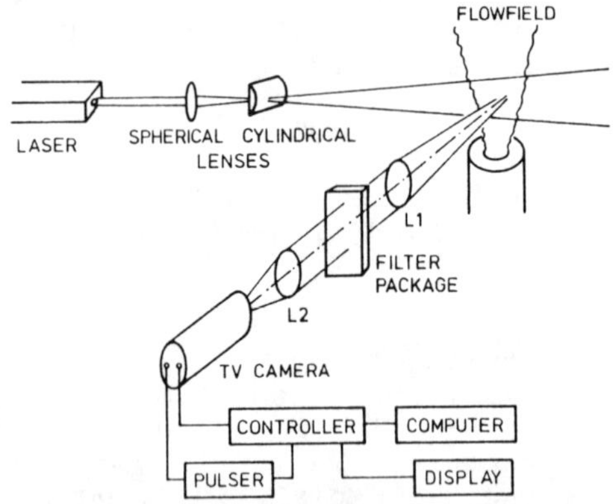

FIGURE 2. Schematic of the general experimental setup of the 2-d laser probe.

Technique	Particle	Frequency	$d\sigma/d\Omega$ (cm^2/sr)
Mie	ϕ 10 µm	ν_0	10^{-7}
Mie	ϕ 0.1 µm	ν_0	10^{-13}
Fluorescence	Molecule	$\nu_F \geq \nu_0$	$10^{-19} - 10^{-24}$
Rayleigh	Molecule	ν_0	10^{-27}
Raman	Molecule	$\nu_0 \pm \nu_R$	$10^{-30} - 10^{-31}$

Table 1: Comparison of different laser techniques suitable for 2-d/3-d imaging.

nection with their appearance frequency from Fig. 1 and by the magnitude of the scattering cross section. For comparable setup configurations of the laser probe, the collected signal intensity can in principle be given for all different techniques considered by

$$I_s = k\, I_0\, l\, \Omega\, (d\sigma/d\Omega)\, n_i . \tag{1}$$

Here, I_0 is the irradiated laser intensity; l is the length of the laser beam which is observed by the collection optics with solid angle Ω; n_i is the number density of the scattering particles or molecules investigated; k is a calibration constant and $(d\sigma/d\Omega)$ the differential scattering cross section. Thus, for the same experimental condition the scattering cross section gives direct information on the mangnitude of the collectable signal intensity. Roughly, a distinction between particle (Mie) and molecular techniques can be done. Mie and also Rayleigh scattering are elastically working processes and appear thus frequency-unshifted. Fluorescence and Raman processes appear frequency-shifted due to inelastic interactions between molecules and photons. These different techniques will be discussed in more detail in the next chapters in connection with possible applications in 2-d and 3-d flow and combustion diagnostics.

2.2. <u>General setup considerations</u>

A schematic of the general setup configuration is shown in Fig. 2. The laser beam is formed to a light sheet by using appropriate spherical and cylindrical lenses. The light sheet is typically 50-200 μm thick when it passes the flow or combustion field of interest. The scattered light or fluorescence intensity is collected perpendicular to the sheet using a lens combination and filter package suitable for the employed laser technique and a 2-d photo-electrical detector (e.g., a TV camera as indicated in Fig. 2 or a diode array or CCD camera) which is controlled by a laboratory computer which simultaneously stores the data and allows the evaluation procedures. The selection of these single probe components is of high importance but cannot be treated here (see e.g., Refs. (7,8)). Hanson (8) also presented a detailed overview of this technique for combustion diagnostics.

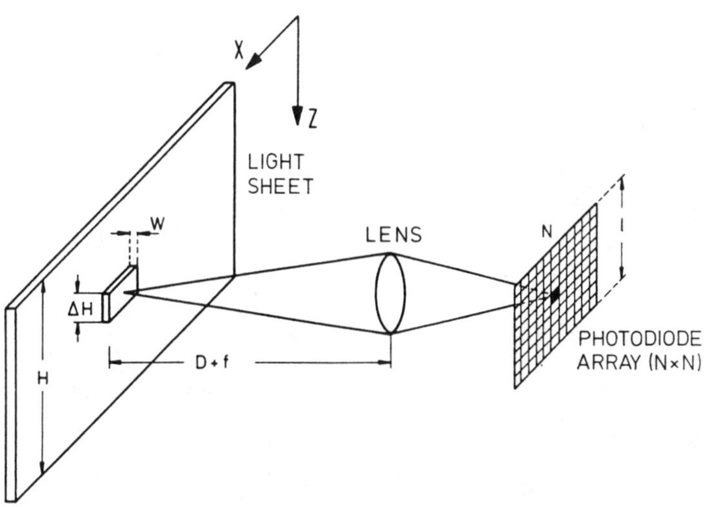

FIGURE 3. Local resolution and sensitivity of the 2-d probe.

Between probe sensitivity (accuracy) and achievable local resolution a compromise must be found (9). Figure 3 shows the schematic for imaging parts of the light sheet onto the surface of a quadratic NxN photodiode array with length l. The magnification of the optical system is M and thus the achievable local resolution $\Delta H = l/NM$. The number of detectable signal photons N_P per pixel of the diode array can be calculated from Eq.(1) which can be developed to give

$$N_P = k\ E_0\ (1/N^2)\ (d/f)^2\ n_i\ f_i\ (d\sigma/d\Omega)\ M/(M+1)^2 \qquad (2)$$

Here, E_0 is the irradiated laser energy (I_0); d is the diameter and f the focal length of the lens being used; and f_i is the population factor of a particular energy level of the specific molecule of number density n_i. The cut-off between sensitivity and local resolution becomes clear: Higher local resolution means smaller ΔH and thus larger magnification M for a given detector. This however leads to a smaller N_P for a particular n_i and thus to lower sensitivity. Using LIF, concentration values in the lower ppm range can be detected with a spatial resolution of 0.1-0.2 mm^3 (10).

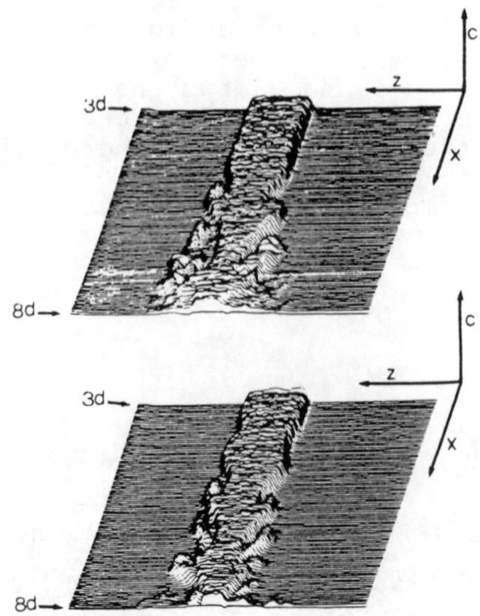

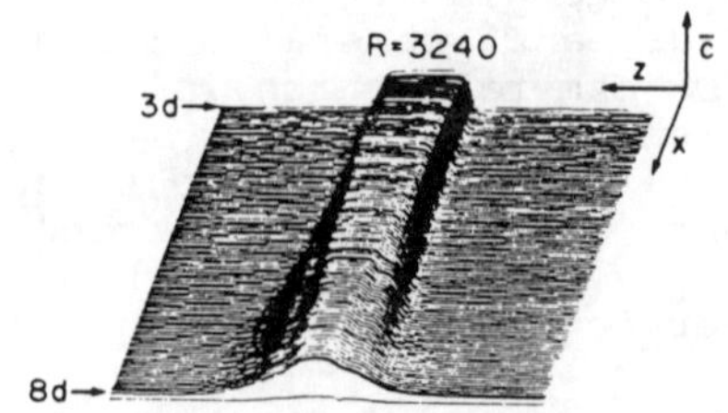

FIGURE 5. Average concentration of an axisymmetric jet calculated from 1000 instantaneous measurements of Fig. 4 (12).

FIGURE 4. Instantaneous nozzle fluid concentration for an axisymmetric jet at Re-number of 3240 (12).

3. MIE SCATTERING

Particle Mie scattering is the strongest process considered (Tab. 1). It is well known from being used in laser Doppler velocimetry (11). Its intensity is function of the particles diameter (d_P), the number of the particles contributing to the signal (n_P) and of the direction of observation relative to the laser direction (Θ): $I=f(d_P,n_P,\Theta)$. For an experimental situation with constant collection angle (Θ=const) and when seeding appropriately a flow component with particles of constant diameter (d_P=const), the collected intensity is a direct measure of the number of particles and thus of the concentration of the seeded gas component: $I=f(n_P)=kc_i$.

This has been used for imaging the gas concentration in a nozzle flow by Long et al (12) using sugar particles of constant diameter, see Fig. 4. The existence of large cohe-

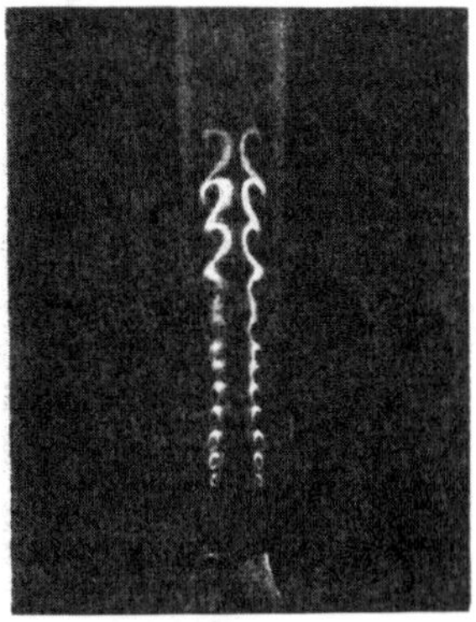

FIGURE 6. Mie imaging in a propane-air jet flame (13).

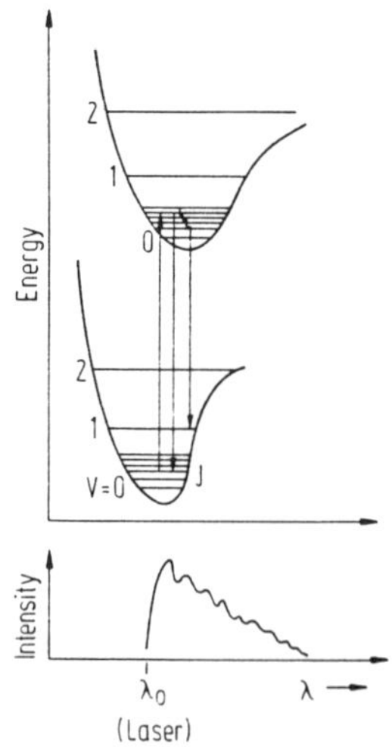

FIGURE 7. Schematic of the laser-induced fluorescence (LIF) process.

rent flow structures is evident. The instantaneous character of such measurements with time resolution in the order of micro-seconds and below is of high importance for flow investigations and combustion research as well. This can be seen in comparison to Fig. 4 from Fig. 5 where the average of 1.000 instantaneous measurements is shown. Clearly can be seen that the coherence structure in the flowfield has disappeared. This structure, however, indicates in reality those positions in combustion where the instantaneous concentration and temperature requirements may be met for the chemical reaction and where thus combustion takes place.

A different seeding technique has been used by Chen and Roquemore (13) for flame investigations. Figure 6 shows the visualization of a propane-air jet diffusion flame which gives new insights in the combustion processes by discovering structures and the zone of the reactive interface. This could be done by seeding the fuel gas with $TiCl_4$. By the water formed within the combustion process, $TiCl_4$ is reduced to give submicron particles of TiO_2 which serve as Mie scatterers indicating the place where water is formed. The same technique has been used by Bracco and his group at Princeton University (14) for observing the flame fronts inside IC engines. They found, that the flame front becomes more and more wrinkled with increasing engine speed which is in good agreement with the length scale considerations of the flamelet model (15).

4. LASER-INDUCED FLUORESCENCE

4.1. Basics of the fluorescence technique

Among the molecular processes, LIF gives the highest signal intensities. Fluorescence is a photon-molecule interaction process by which the molecule is excited to a higher electronic energy level (Fig. 7). In general it appears 10^{-5}-10^{-10} s after this excitation with an intensity which is given by

$$I_F = c\ SV(p,T)\ B_{12}\ n_i\ f_i(T)\ I_0. \tag{3}$$

Here, c is a calibration constant; SV(p,T) is the Stern-Vollmer factor with $SV=A_{21}/(A_{21}+Q)$; B_{12} and A_{21} are the Einstein coefficients for the energetic transition; and Q is the quenching rate. An electronically excited molecule does not in any case loose energy by radiation processes alone but also by collisions with other molecules or by chemical reactions, in general called "quenching effects". Thus, the collected signal intensity is in some parts smaller than expected theoretically for radiation processes alone, expressed by the Stern-Vollmer factor SV. Due to its pressure and temperature dependence, SV cannot easily be calculated for all flame and combustion situations which in general makes the application of LIF difficult.

Assuming a Boltzmann population distribution in the different molecular energy levels and in particular in the ground state to be excited by the laser photons irradiated, the number of molecules in this energy level follows from $n_i f_i$ with

$$f_i(T) = \frac{2\ (J+1)}{Q_E\ Q_V\ Q_R} \exp\left(-\frac{hc}{k}\ \frac{T_E(n) + G(v) + F(J)}{T}\right) \tag{4}$$

The population factor f_i is dependent on the quantum numbers of the energy level J for the rotational, v for the vibrational and n for the electronic state and on the temperature (Q_E, Q_V and Q_R are the partition distribution functions for the electronic, the vibrational and the rotational levels, respectively; h, k and c are the Planck constant, the Boltzmann constant and the velocity of light, respectively).

From Eqs.(3,4) follows, that concentration measurements are only possible by LIF, if $SV*f_i$ can be considered to be constant all over the area of interest

$$I_F = K\ n_i \tag{5a}$$

(for $SV*f_i$=const), and that temperature determination is possible from the signal of two different pump levels ($I_F(1)$, $I_F(2)$), if the total gas pressure remains constant over the field of interest leading only to a temperature dependence of the Stern-Vollmer factor.

$$I_F(1)/I_F(2) = (SV\ f_i)_1/(SV\ f_i)_2 = F(T) \tag{5b}$$

A more general problem for LIF measurements is contained in the requirement for exciting the gas molecule electronically.

Commercially available laser sources do not have enough energy to excite simple molecules by a single photon event. So it is not easy or even impossible to investigate such molecules as N_2, CO_2, H_2O etc. Using excimer lasers, recently a simple gas component like O_2 has been tracked (16) and also CO by using two-photon excitation processes (17). The variety of species used in 2-d LIF imaging is given in table 2 in connection with an indication for what purposes in combustion research they have been used for.

Species	Application
CH, C_2, O, H	Reaction Zone
CO	All Combustion Zones
O_2	All Combustion Zones, Temperature
C_2H_2, NO_2	Pre-reaction Zone (Fuel Gas)
OH	Post-reaction Zone (Burned Gas), Temperature
NO	Flowfield/Flame Visualization, Temperature

Table 2: Planar LIF: Species used for different combustion applications.

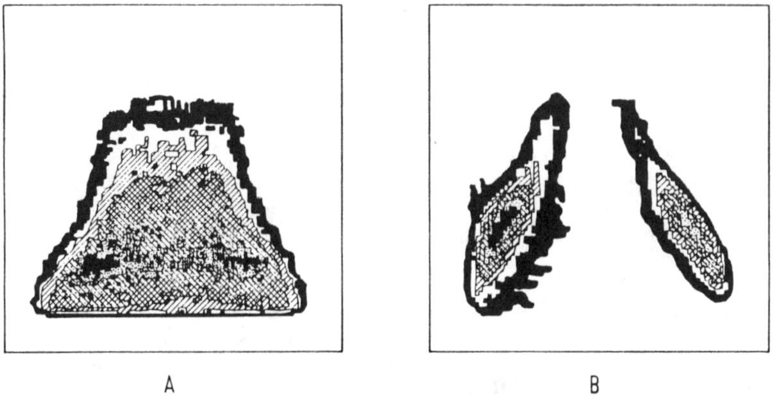

FIGURE 8. Instantaneous OH-concentration in a laminar, premixed CH_4-air flame for (a) ϕ=1.0 and (b) ϕ=1.15 (18).

4.2. Concentration imaging

Most of the 2-d LIF imaging work has been concerned with concentration or mole fraction measurements. Figure 8 shows OH-measurements in a flat flame burner for different equivalence ratios giving the digital pictures of the instantaneous OH-concentration in a laminar, premixed methane-air flame (18). In Fig. 8a the stoichiometric flame is pictured, in Fig. 8b the flame with an equivalence ratio of ϕ=1.15. The maximum OH-concentration is 1400 ppm. For simplicity only a few different distinction steps are used. In reality, the dynamic range of those measurements is about 200 steps, providing a real detection limit of 7 ppm. Spatial resolution is 0.9mm * 0.9mm * 0.2mm, temporal resolution due to the laser pulse length 8 ns. In Fig. 8b one can realize that for under-stoichiometric conditions the reaction zone moves into the mixing region with ambient air.

Figure 9 shows a sequence of four single-shot measurements of the instantaneous NO-distribution in the same burner running again a laminar, stoichiometric methane-air flame (10,18). The different flame structures at different instants can be seen. Here, NO-molecules were seeded to the fuel gas in order to allow the visualization of the flame structures. The creation and propagation of large-scale structures within the flame were found to be clearly evident also from frame sequences taken in an unstable fuel-rich methane-air flame at a repetition rate of 200 frames/second tracking the O_2-molecule (16). By those measurements a periodic behaviour of the flame was observed completing a cycle in normally 75 ms (about 13 Hz), a value which also was found in similar flames by Schlieren techiques (2,3) or by using laser Rayleigh correlation techniques for flame thermometry (19). Figure 10 shows one of these O_2-distributions at an instant. An inner conical region of strong signal is observed which is due to unburned air in the reactants. A signal from the secondary flame zone is seen further downstream which is due to the entrainment, heating and combustion of O_2 from the room air. Oxygen measurements were possible recently by tuning an excimer laser (ArF) to 193 nm. The asymmetric variation of the signal is due to strong attenuation of the exciting laser beam through the flame by hot O_2.

Species LIF imaging has also been done successfully using the CO-molecule (17) for quantitative concentration measurements and the CH- (20) and the C_2-molecule (20) for imaging the flame front/reaction zone. The NO_2-molecule has been used for tracking the non-reacted gases (21) seeding the air with a low level of NO_2 being decomposed in the flame zone. Also for marking the unburned fuel regions, C_2H_2 has been used recently (22).

Already very close to industrial applications are the measurements of the OH-distribution in spray flames (23), which are typical for oil and patrol combustions, and the simultaneous imaging of vapor and liquid distributions in evaporating sprays (24). Figure 11 shows the instantaneous OH-concentration in a heptane-air spray flame near the tip of the primary reaction zone (23). The spatial resolution is 0.12mm * 0.12mm * 0.2mm. By setting appropriate filters in the collection optics, simultaneously some parts of the Mie scattered light from the heptane-droplets can be collected indicating by white spots the position of the droplets in the combustion field. The maximum LIF intensity from the OH-molecules corresponds to 1,000 ppm. An example for simultaneously imaging vapor and liquid distribution is shown in Fig. 12 (24) using organic exciplexes. After laser excitation, the fluorescence colour is different for the liquid and for the vapor phase. Both contributions can be separated by using appropriate absorbing glass filters.

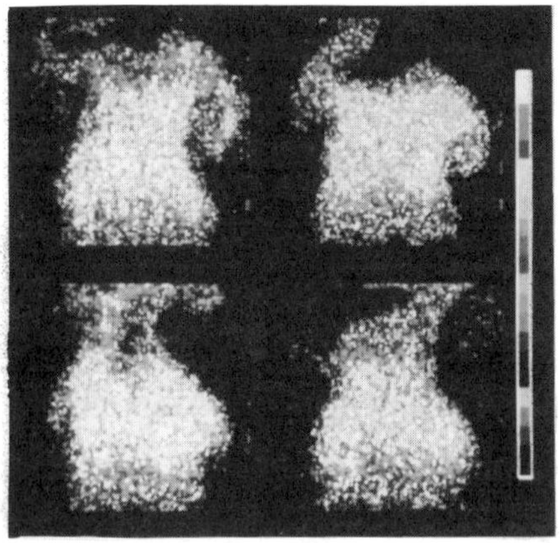

FIGURE 9. Sequence of 4 instantaneous NO-concentrations in a NO-seeded (2000 ppm) CH_4 air flame (10,18).

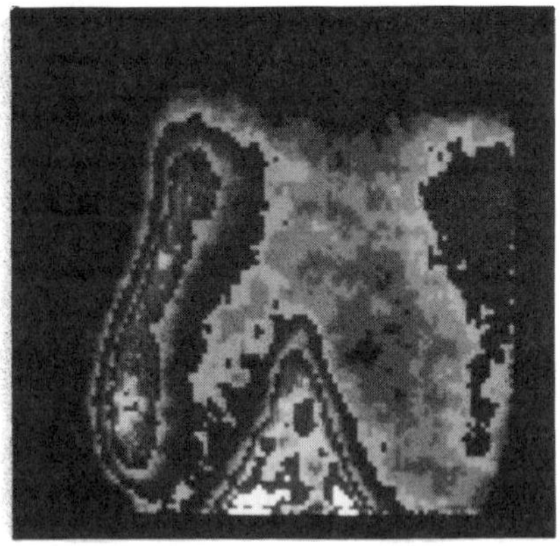

FIGURE 10. Instantaneous oxygen concentration in an unstable fuel-rich CH_4-air flame (16).

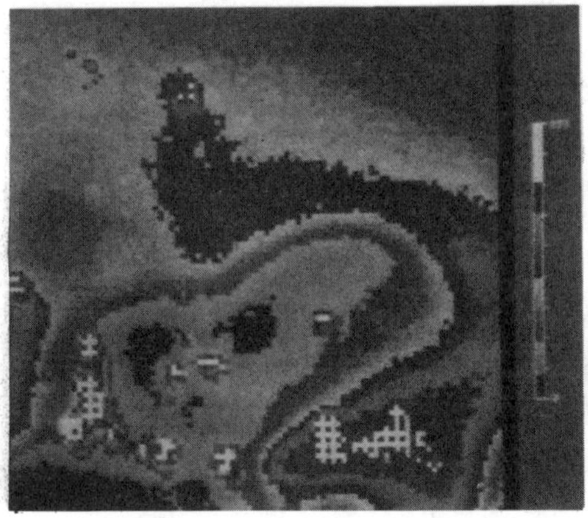

FIGURE 11. Instantaneous OH-concentration and position of the droplets (white points) in a heptane-air spray flame (23).

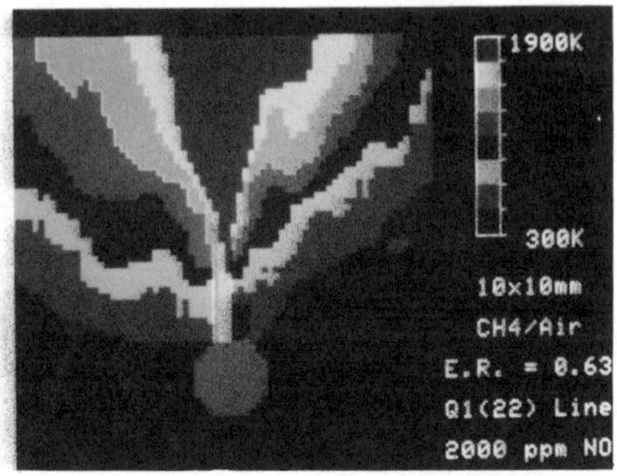

FIGURE 14. Temperature distribution in a rod-stabilized CH_4-air flame by NO-LIF (25).

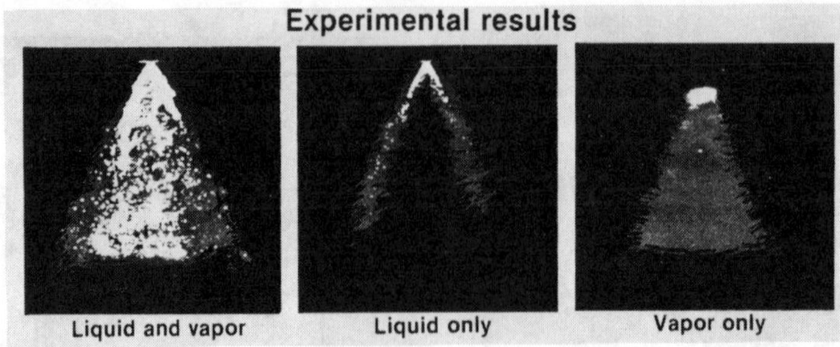

FIGURE 12. Laser-induced exciplex fluorescence from conical fuel spray. Commercial nozzle operating with cetane doped with naphthalene/tetramethyl phenylene diamine (TMPD) (24).

4.3. <u>Temperature imaging</u>

Temperature determination from LIF signals is in general possible by using Eq.(5b) assuming constant pressure. Planar 2-d temperature imaging has been done successfully so far with the NO-molecule (25) and with the O_2-molecule (26) as well. Using the NO-molecule two approches are possible (Fig. 13). In the more general case in accordance to Eq.(5b) one has to use the signal of two initial states, e.g., the $Q_1(35)$ and $R_1(16)$ lines for NO. Seeding the flame with NO-molecules and thus providing constant NO concentration in the combustion field, the procedure can be more simple. Then, also the signal from a single fluorescence line may be used for temperature evaluation e.g., the $R_1(16)$ line or the

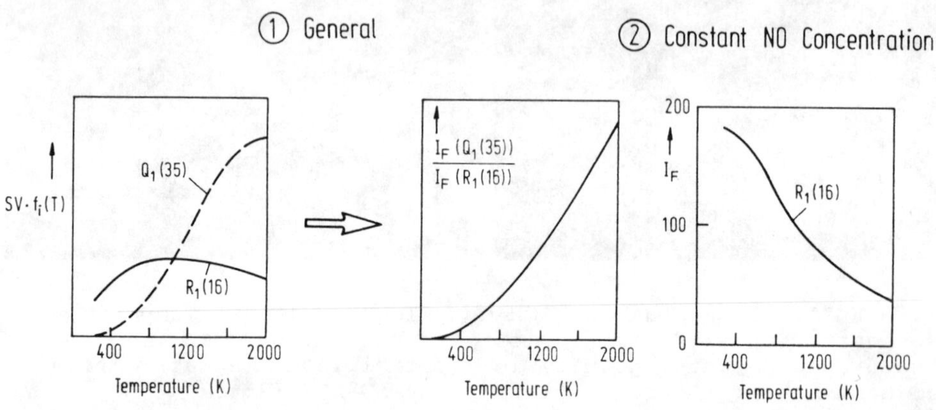

FIGURE 13. LIF temperature measurements using e.g., the NO-molecule (general case and for constant NO concentration).

$Q_1(22)$ line which has been selected for temperature imaging in a rod-stabilized methane-air flame (Fig. 14). The uncertainty in these investigations has been assumed to be 20-100 K in different regions.

5. RAYLEIGH SCATTERING

The frequency-unshifted Rayleigh signal contains contributions of all gas species present in the sample. Its intensity is given from Eq.(1) by

$$I_{Ray} = k_1 \, I_0 \, n \, (d\sigma/d\Omega)_{eff} \tag{6}$$

with total number density

$$n = p \, A/R \, T \tag{7}$$

and effective scattering cross section

$$(d\sigma/d\Omega)_{eff} = \Sigma \, c_i \, (d\sigma/d\Omega)_i . \tag{8}$$

Here, p is the total gas pressure; A is the Avogadro number; R is the gas constant and c_i are the concentrations of the gas species present with the individual scattering cross sections $(d\sigma/d\Omega)_i$. The utilization of Rayleigh scattering for 2-d or 3-d imaging follows from these equations.

Concentration measurements are possible by tracking the magnitude of the effective scattering cross section, but only in two-component gas systems under isothermal conditions and with constant gas pressure, if the two individual cross sections are known or can be measured independently.

$$I_{Ray} = K_1 \, (c_1 \, (d\sigma/d\Omega)_1 + c_2 \, (d\sigma/d\Omega)_2) \tag{9a}$$

$$1 = c_1 + c_2 \tag{9b}$$

Figure 15 gives as an example for such measurements in basic fluid dynamic research the instantaneous freon concentration distribution in a turbulent jet issuing into coflowing nitrogen (27). As has been seen already in the Mie and LIF images, large-scale coherence structures are clearly evident in the flowfield. Due to the low signal intensity obtainable by Rayleigh scattering (Tab. 1), the 2-d light field has to be generated more efficiently. This can be achieved e.g., by using a multipass-cell being formed by a pair of cylindrical reflectors (Fig. 16) which gives an increase in the laser energy in the light sheet by a factor of 10-30 relative to the simple lens configuration in Fig 2.

Temperature measurements are possible through density measurements, if the flame to be investigated can be modelled in such a way that $(d\sigma/d\Omega)_{eff}$ and pressure remain constant all over the combustion field.

$$I_{Ray} = K_2 \, (d\sigma/d\Omega)_{eff}/T \tag{10}$$

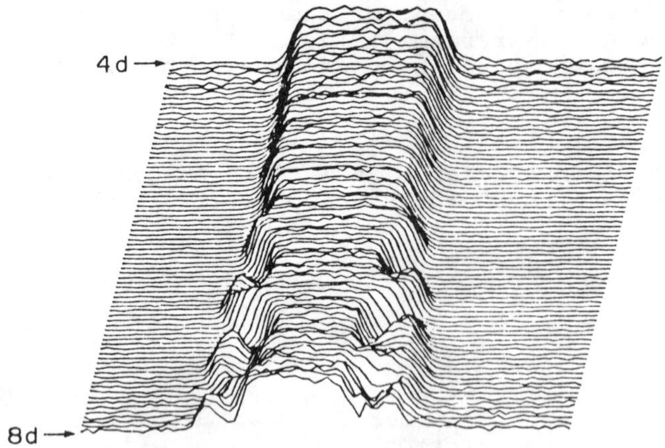

FIGURE 15. Instantaneous freon concentration distribution in a coflowing turbulent jet obtained by Rayleigh scattering (27). Spatial resolution: 50 μm * 150 μm * 130 μm.

This requirement seems to be difficult to meet due to the large number of different species in a flame. In reality, however, nearly each hydrocarbon combustion can be modelled in such a way (28,29). Figure 17 shows the instantaneous 2-d temperature field between 20 and 30 nozzle diameters downstream from the burner exit measured by Rayleigh scattering in a turbulent, non-premixed methane-hydrogen diffusion flame (30). Each indicated step represents a temperature range of 143 K between 300 K and 2000 K. The shape of the flame front with its large temperature gradients gives again a clear evidence of large-scale structures appearing in the combustion field which also can be realized in some way in the unburned gas. The importance of the instantaneous character of such measurements with a time resolution in the order of micro-seconds can be seen in comparison to Fig. 17 from Fig. 18 where the average of 800 individual instantaneous temperature field measurements is shown. Again as already realized from the flowflied studies in Figs. 4 and 5, the large-scale structure is washed out by averaging the instants. Of importance in particular for combustion studies is the fact, that additionally the maximum temperature obtained in the combustion field has decreased from ∿ 2000 K for the instantaneous events to now ∿ 1800 K in the averaged data.

There is one severe problem in the utilization of the laser Rayleigh techniques which strongly restricts its application to clean environments and thus to basic investigations only. As being unshifted in frequency, interferences may occur from the much stronger particle (Mie) scattering process or from laser reflections within the probe hiding the Rayleigh signal in part or even totally. The separation of the different contributions is not easy or even impossible preventing the application of Rayleigh scattering, e.g., in dusty flows or sooting flames.

Using two 2-d detectors simultaneously for imaging the same flow or combustion field within the same instant, different quantities can be obtained at the same time. This has

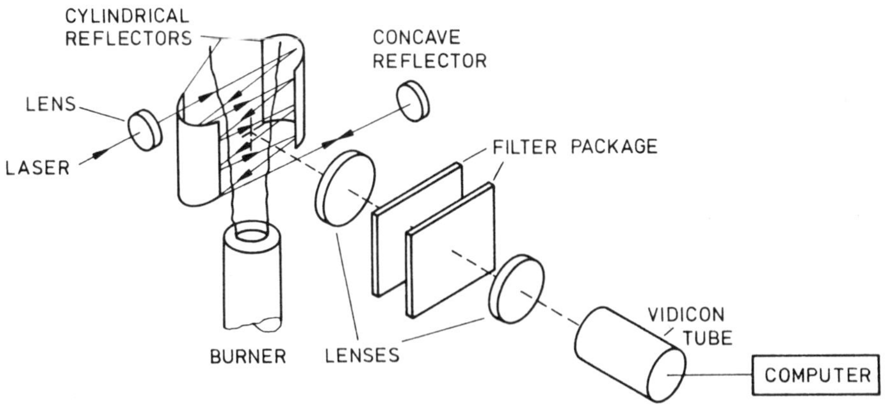

FIGURE 16. Schematic of a modified setup configuration for Rayleigh and Raman work using a multipass-cell.

the potential for extending tremendeously the information contents of those measurements. It has been done successfully e.g., for detecting simultaneously the flame front by the C_2 LIF signal and the 2-d temperature field from Rayleigh measurements (31).

6. RAMAN SCATTERING

Raman scattering is the technique with the most extended application range. Usable for all gas molecules, it in general can be utilized for temperature and concentration measurements as well (32). It suffers, however, from its weakness (Tab. 1), which did not allow so far planar temperature measurements.

Figure 19 shows schematically the shape of the vibrational Raman Q-branch of oxygen for three different temperatures using a ruby laser for excitation. The intensity ratio I_2/I_1 gives an indication of gas temperature and the integrated Q-branch intensity a measure for oxygen concentration. Using an interference filter of FWHM=4.25 nm and of peak position at 775 nm, for the temperature range of 700-1900 K nearly temperature independent concentration measurements are possible.

Planar 2-d concentration mapping was first reported in a high pressure gas cell (33). Using a cw laser source, the averaged concentration field in the flowfield of axisymmetric jets at atmospheric pressure has been determined by Webber at al. (34). Information on turbulent mixing mechanism requires time resolution in the order of micro-seconds. Those measure-

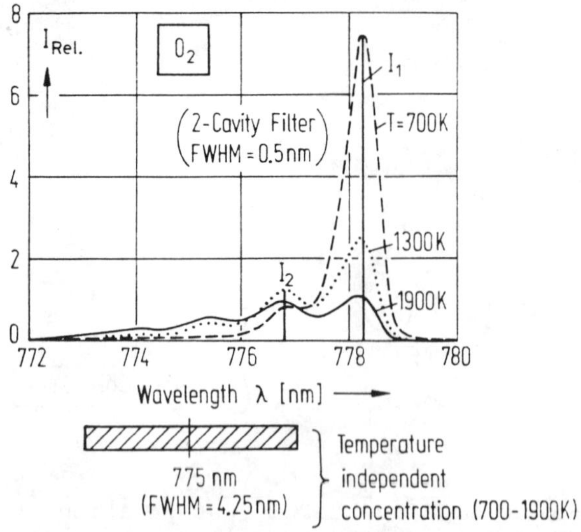

FIGURE 19. Schematic for using Raman scattering for temperature determination and for nearly temperature independent concentration measurements.

ments first were done in co-operation between Yale University and Sandia National Laboratories (35) and recently also in flames (36). Figure 20 shows the methane concentration distribution for three different instants of 1.8 μsec obtained from a turbulent nonpremixed CH_4-H_2-air diffusion flame. Large-scale coherence structures again are evident. In this investigation (36) simultaneously also the temperature distribution has been measured for the same instants collecting the 2-d Rayleigh scattering distribution by a second 2-d detector. In the nonpremixed (Fig. 20) and in a premixed flame as well, in general the same structures were found for the c- and for the T-fields: low concentration for high temperature and vice versa. Sometimes, however, this tendency was different in some flame regions, where low CH_4 concentration was related to low temperature. Those situations indicate entrainment of room air into the flame body.

7. THREE-DIMENSIONAL IMAGING

For many types of events two dimensions are fine. Some phenomena, however, and in particular turbulent flow and combustion processes require a three-dimensional visualization. Recently those approaches have been performed successfully using LIF, Mie or Rayleigh scattering (4-6,37,38). There are three essentially different approaches how to realize 3-d imaging: (1) by irradiating two light sheets of different colour simultaneously and thus providing the third dimension by the distance between both sheets (4); (2) by stepping the investigation object through the illumination plane of a single light sheet and recording the successive slices of the flow which only can be done very well defined in clearly defined stable flow situations, e.g., in a photoacoustically perturbed jet (37); and (3) by scanning

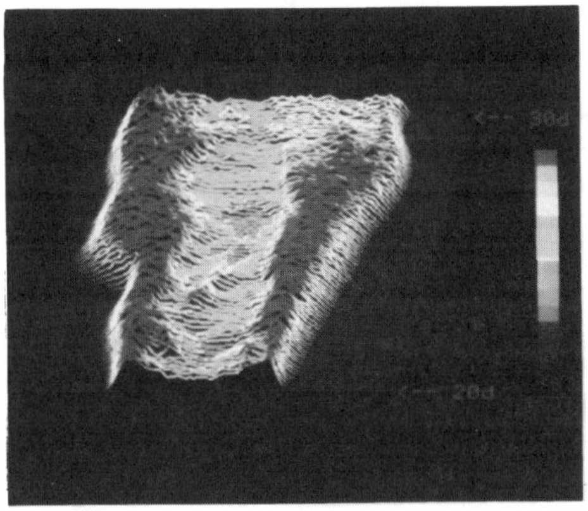

FIGURE 17. Instantaneous temperature field measurement in a turbulent, non-premixed CH_4-H_2 diffusion flame using Rayleigh scattering (30).

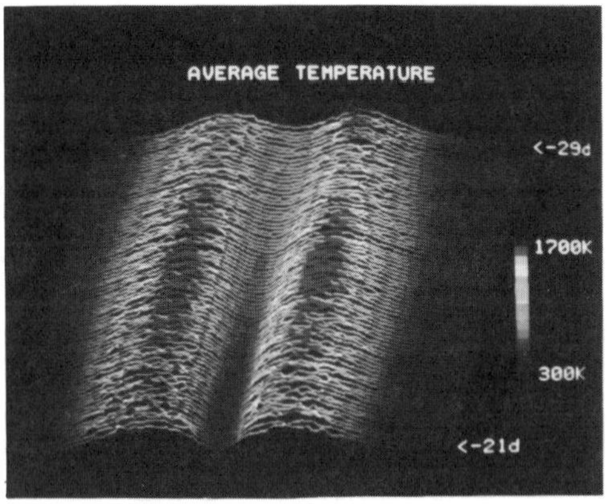

FIGURE 18. Average temperature field calculated from 800 invidual instantaneous temperature field measurements of Fig. 17 (30).

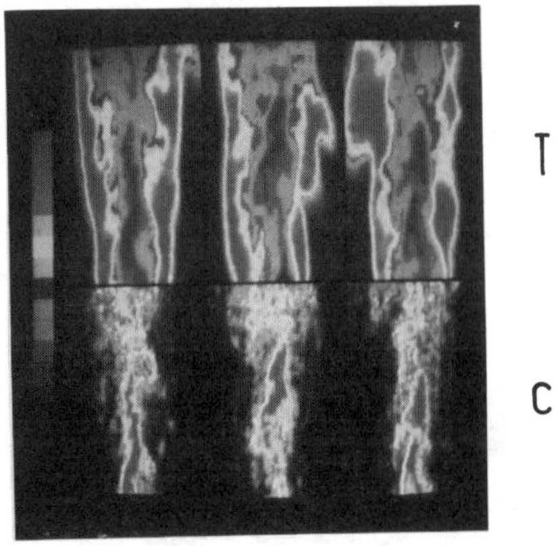

FIGURE 20. Three pairs of instantaneous 2-d concentration and temperature distributions obtainded by Rayleigh (T) and Raman (c) measurements from a turbulent non-premixed CH_4-H_2 diffusion flame (36).

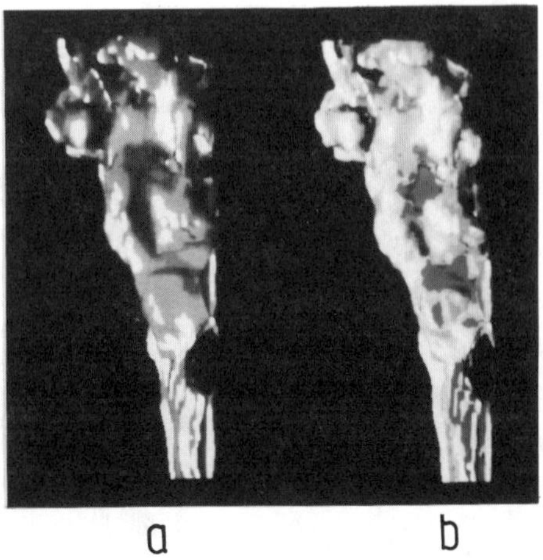

FIGURE 22. Three-dimensional constant concentration surface in the transitional region of a gas jet (a) and (b) representation of the calculated magnitude of the concentration gradient vector (6).

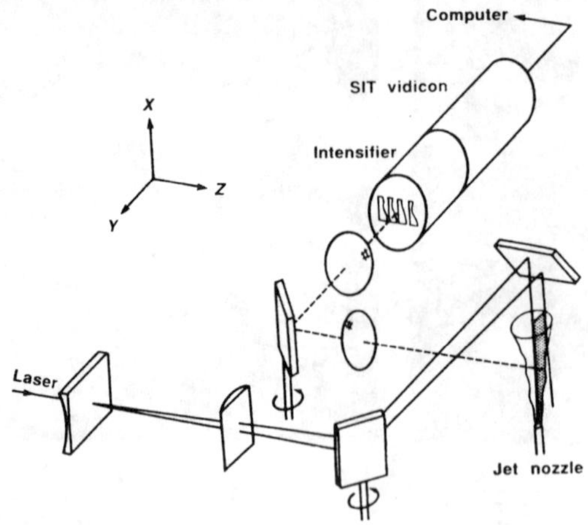

FIGURE 21. Schematic of the experimental arrangement used for the 3-d imaging of concentration in a gas jet (6).

rapidly the laser sheet through the flow or flame to be investigated so that a series of parallel planes within the volume is illuminated and recorded within a short time period (∿ micro-seconds) (6,38). Here, a second rotating mirror was used to sweep several consecutive images onto different rectangular regions of the detector (Fig. 21).

Figure 22 shows an example for the results obtained in a gas jet (6). In Fig. 22a the constant concentration surface in the transitional region of the jet is displayed. As many of those surfaces are stored in the computer, the concentration gradient in each direction can be calculated. Fig. 22b shows the same surface colour-coded by the magnitude of the calculated concentration gradient vector decreasing from bright yellow to dark blue. In combustion, even Fig. 22b alone provides inside in the process, e.g., the concentration contour giving the surface where chemical reaction takes place (flame sheet) and the value of the gradient indicating the amount of heat release. In a methane-air flame which was seeded with biacetyl vapor as a fluorescent marker, 3-d concentration imaging was done very recently (38).

8. CONCLUSIONS

Extending quantitative point measurement techniques (Mie, Rayleigh, Raman, LIF) to 2-d and 3-d resolving techniques by using laser sheets and 2-d detectors, instantaneous field measurements are possible providing as well qualitative visualization of the coherent structures of the flow or combustion field as quantitative results with high temporal (< micro-seconds) and spatial (< mm^3) resolution. From those quantitative measurements all statistically important data are obtainable concerning the field distribution as well as temporal and local correlations in the field. In combustion this can be done for concentration mapping of major species (Raman) and of minor species (LIF) as well as for temperature imaging. Furthermore, LIF also can be used in selected applications for the 2-d planar imaging of pressure (39) and velocity (40) which has not been covered here in this paper.

REFERENCES

1. Jimenez J: The Role of Coherent Structures in Modelling Turbulence and Mixing. Berlin-Heidelberg: Springer, 1981.
2. Peters N, Williams FA: in Ref. (1), pp. 364-393.
3. See e.g., Knoche KF, Janicka J, Oebels R: VDI-Berichte 423 (1981) 142-147.
4. Yip B, Long MB: Opt. Lett. 11 (1986) 64-66.
5. Kychakoff G, Paul PH, van Cruyningen I, Hanson RK: in press (1987).
6. Yip B, Lam JK, Winter M, Long MB: Science 235 (1987) 1209-1211.
7. Chang RK, Long MB: in Light Scattering in Solids II (M. Cardona, G. Guntherodt, Eds.) pp. 179-205. Heidelberg: Springer, 1982.
8. Hanson RK: Proc. 21st Symposium (Int.) on Combustion. Pittsburgh: The Combustion Institute (in press 1987).
9. Kychakoff G, Howe RD, Hanson RK: Appl. Opt. 23 (1984) 704-712.
10. See e.g., Kychakoff G, Knapp K, Howe RD, Hanson RK: AIAA J. 22 (1984) 153-154.
11. See e.g., Durst F, Melling A, Whitelaw JH: Principles and Practice of Laser-Doppler Anemometry. London-New York: Academic Press, 1976
12. Long MB, Chang RK, Chu BT: AIAA J. 19 (1981) 1151-1157.
13. Chen L-D, Roquemore WM: Combust. Flame 66 (1986) 81-86.
14. See e.g., Zur Loye AO, Bracco FV: SAE paper 870454 (1987).
15. Peters N: Proc. 21st Symposium (Int.) on Combustion. Pittsburgh: The Combustion Institute (in press 1987)
16. Lee MP, Paul PH, Hanson RK: Opt. Lett. 11 (1986) 7-9.
17. Haumann J, Seitzman JM, Hanson RK: Opt. Lett. 11 (1986) 776-778; Appl. Opt. 26 (1987) 2892-2899; see also the paper by the same authors in this Volume.
18. Kychakoff G, Howe RD, Hanson RK, Mc Donald JC: Appl. Opt. 21 (1982) 3225-3227; Kychakoff G, Knapp K, Howe RD, Hanson RK: Paper 82-60, Westen State Sect./The Combustion Institute, Oct. 1982 (Livermore, CA)
19. Haumann J, Leipertz A: Opt. Lett. 9 (1984) 487-489.
20. Allen MG, Howe RD, Hanson RK: Opt. Lett. 11 (1986) 126-128.
21. Cattolica RJ: Proc. 21st Symposium (Int.) on Combustion. Pittsburgh: The Combustion Institute (in press 1987)
22. Allen MG, Hanson RK: Proc. 21st Symposium (Int.) on Combustion. Pittsburgh: The Combustion Institute (in press 1987)
23. Allen MG, Hanson RK: Opt. Eng. 25 (1986) 1309-1311.
24. Melton AL, Verdieck JF: Combust. Sci. Technol. 42 (1985) 217-222

25. Seitzman JM, Kychakoff G, Hanson RK: Opt. Lett. 10 (1985) 439-441.
26. Lee MP, Paul PH, Hanson RK: Opt. Lett. 12 (1987) 75-77.
27. Ecoda MC, Long MB:AIAA J. 21 (1983) 81-84.
28. Dibble RW, Hollenbach RE: Proc. 18th Symposium (Int.) on Combustion, pp. 1489-1499. Pittsburgh: The Combustion Institute, 1981.
29. Namer I, Schefer RW: Exp. Fluids 3 (1985) 1-9.
30. Fourguette DC, Zurn RM, Long MB: Combust. Sci. Technol. 44 (1986) 307-317.
31. Dibble RW, Long MB, Masri A: Progr. Astr. Aeron., Vol. 105 (1986) 99-109.
32. See e.g., Leipertz A: Habilitationsschrift, Ruhr-Universität Bochum 1984. See also the chapter "Raman processes and their application" in this Volume.
33. Hartley DL: in Laser Raman Gas Diagnostics (M. Lapp, C.M. Penney, Eds.) pp. 311-314. New York: Plenum Press, 1974.
34. Webber BF, Long MB, Chang RK: Appl. Phys. Lett. 35 (1979) 119-121.
35. Long MB, Fourguette DC, Escoda MC, Layne CB: Opt. Lett. 8 (1983) 244-246.
36. Long MB, Levin PS, Fourguette DC: Opt. Lett. 10 (1985) 267-269.
37. Yip B, Fourguette DC, Long MB: Appl. Opt. 25 (1986) 3919-3923.
38. Yip B, Schmitt RL, Long MB: to be published in Opt. Lett. 12 (1988)
39. Hiller B, Cohen LM, Hanson RK: AIAA paper 86-0161 (1986).
40. See e.g., Mc Daniel JC, Hiller B, Hanson RK: Opt. Lett. 8 (1983) 51-53; Hiller B, Mc Daniel JC, Rea Jr. EC, Hanson RK: Opt. Lett. 8 (1983) 474-476; Hiller B, Hanson RK: Opt. Lett. 10 (1985) 206-208.

QUANTITATIVE TWO-DIMENSIONAL IMAGING OF CO IN COMBUSTION GASES USING LIF

Juergen Haumann, Jerry M. Seitzman*, Ronald K. Hanson*
ASEA BROWN BOVERI
CORPORATE RESEARCH
CH 5405 Baden / Switzerland
* Stanford University
High Temperature Gasdynamics Laboratory
Department of Mechanical Engineering
Stanford, CA 94305, USA

1. INTRODUCTION

Two- and three- dimensional imaging of species concentration distributions is a powerful diagnostic tool in reacting and nonreacting flows[1]. Planar laser-induced fluorescence(PLIF) is especially suited for mapping the concentration of radicals and stable species, such as OH, CH, C_2, O_2 and NO. Recently PLIF was extended to the measurements in two-phase flows, like spray flames, increasing substantially the possibilities of its application in practical combustion systems. Several minor and major species (e.g., N_2, CO and O), however, have transitions in the VUV, which are inaccessible using linear excitation. For this reason, two-photon excitation techniques were developed for single-point detection of O[3,4] H[5] and OH[6,7] in flames, as well as for the one dimensional imaging of O[8] and CO[9].

In this paper, a technique for the two-dimensional imaging of CO-concentration in combustion gases is demonstrated using two-photon-excited PLIF, which is the first use of a nonlinear process for 2-D imaging. The extension of PLIF is important owing to the significance of this molecule as a combustion intermediate and pollutant specie. Some aspects concerning the potential of this nonlinear LIF technique for making quantitative measurements will be discussed, too. The most important of these are the modeling of the temperature and laser-intensity dependencies of the detected fluorescence and the photochemical effects induced by the intense laser pulses that are required for the multiphoton excitation.

The illumination technique of our CO-probe is based on the combination of a nonlinear absorption scheme[10,11], in which two photons at 230.1 nm excite transitions of the $B^1\Sigma^+ \leftarrow X^1\Sigma^+$ system, and the use of an ultraviolet multipass cell for producing the laser sheet. The subsequent visible fluorescence ($B^1\Sigma^+ \rightarrow A^1\Pi$) is imaged onto an intensified two-dimensional photodiode array.

2. THEORY

A quantitative four-level model for the nonlinear pumping and detection scheme illustrated in Fig.1 is used for the rate equation analysis. Although several rotational levels of the ground ($X^1\Sigma^+$) and excited ($B^1\Sigma^+$) electronical states in the so called Hopfield-Birge system of CO are coupled by the laser, they are represented only by single rovibronic levels this model. An excited molecule can relax to the rovibronic levels of the ground state and the first excited state ($A^1\Pi$), in the model regarded as a third single level. In our experiment the broadband fluorescence of the $B^1\Sigma^+ \rightarrow A^1\Pi$ Angstrom system within the spectral range of 451nm

*Figures 5, 6, 7 and 10 in this article were submitted in colour, but have been reproduced in black and white.

D. F. G. Durão et al. (eds.), Instrumentation for Combustion and Flow in Engines, 141–150.

(v''=0) to 725nm (v''=6) is collected by the detector. The spectral seperation between the laser radiation and the fluorescence allows for easy discrimination of the elastic scattering. A complication of the analysis arises as the energy of a third 230.1nm photon (5.4eV) is sufficient to ionize the CO in a 2+1 photon process (the fourth model level).

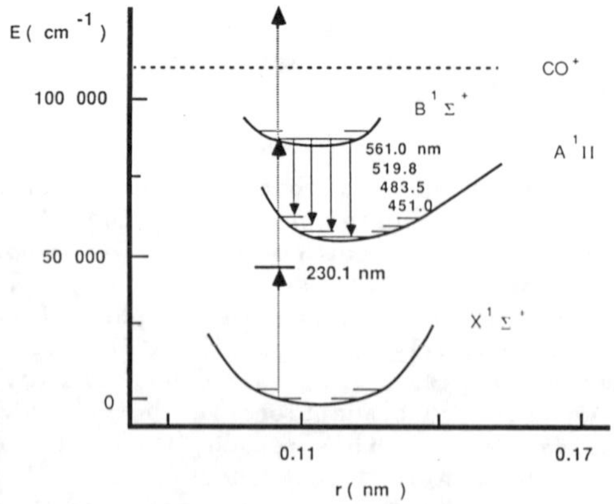

FIGURE 1. Energy level diagram of CO showing states involved in the excitation and detection process for 2-photon LIF of CO at 230nm.

Assuming linearly polarized light and a laser intensity well below the saturation value, a solution for the described system is obtained, which express the fluorescence emission rate, when the laser couples only a single rovibronic level in the lower and upper electronic states. For the case in which the laser excites multiple rotational lines, the total fluorescence rate is found by summing over the different transitions.

With some further assumptions, which are described elsewhere in detail[11], the rate of collected photons R_c [photons/s] can be modeled by

$$R_c(\nu_o) = \eta\cdot(\Omega/4\pi)\cdot V_c\cdot N_{co}\cdot\alpha(T,\nu_o)\cdot\frac{(I/hc\nu_o)^2\cdot A}{(A+Q+\sigma_i\cdot\phi_i\cdot I/hc\nu_o)} \qquad (1),$$

where η is the optical efficiency of the detection system; Ω is the collection solid angle; V_c is the collection volume; N_{co} is the number density of CO; $\alpha(T,\nu_o)$ is an effective two photon absorption coefficient[11] and is a function of temperature T and excitation frequency ν_o; I is the laser intensity; A is an effective Einstein A coefficient for broadband collection; Q is the electronic quench rate and $\sigma_i\phi_i(\nu_o)$ is the photoionization cross section.

From this equation it is evident that although the two photon stimulated absorption rate is proportional to I, the measured fluorescence rate will in general be proportional to I^n, where $1<n<2$.

If the photoionization rate dominates the quenching and fluorescence rates, the measured fluorescence becomes linearly proportional to the laser intensity. In addition, the temperature dependence of the Stern-Vollmer factor, $A/(A+Q+\sigma_i\phi_i I/hc\nu_o)$ is negligible and the signal's temperature dependence is given by that of $\alpha(T,\nu_o)$ and N_{co} alone. In this case, the usual complexities introduced by the specie and temperature dependence of the quenching Q are absent. Measurements by Loge et al. /10/ as well as by our own have shown at low-pressure and room-temperature that n indeed approaches 1 for moderate intensities.

This simplifies the conversion of the detected fluorescence images to quantitative CO concentration fields significantly.

A similar nonlinear excitation technique, based on the use of two lasers, has recently been demonstrated for quenching-independent flame measurements of atomic hydrogen[13]. With the assumption of a dominating photoionization rate the total number of photons collected, N_p, is

$$N_p(v_o) = \eta \cdot (\Omega/4\pi) \cdot V_c \cdot (E/hcv_o) \cdot N_{co} \cdot \alpha(T,v_o) \cdot \frac{A}{\sigma_i \cdot \phi_i(v_o)} \quad (2),$$

where E [J/cm^2] is the total energy fluence of the laser pulse. Thus measurements of the fluorescence from V_c yields the average number of CO molecules within the collection volume.

Theoretical two photon absorption coefficients and excitation spectra for the CO Q-branch of the $B^1\Sigma^+ \leftarrow X^1\Sigma^+$ system were calculated on basis of the eqs. (1) and (2) for a temperature of 300K and 700K and compared with measured spectra. The experimental data points in Fig. 2 have been taken at atmospheric conditions and represent an average of approximately 100 laser shots per point. The good agreement between the experimental and predicted spectra indicates that the spectral and temperature dependence of the two-photon excitation process is accurately modeled by eq.(2).

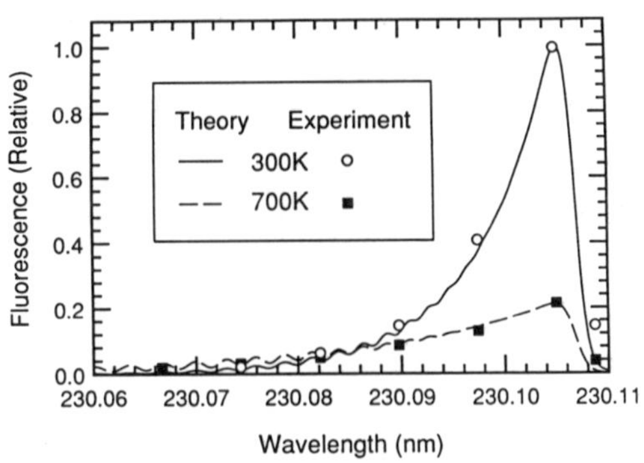

FIGURE 2. Two-photon excitation spectra for the Q-branch of the CO $B^1\Sigma^+(v'=0) \rightarrow X^1\Sigma^+(v''=0)$ transition from 230.06 to 230.11 nm at 300K and 700K. The open circles and filled squares are experimental points, and the solid and dashed curves are theoretical predictions (Eq.(2)).

Additionally, Fig.3 shows the theoretical temperature dependence of the detected fluorescence for specific laser frequencies, based on a computer calculation of eq.(2). As no absolute measurements of the CO two-photon absorption cross-section were available, the calculated emission spectra are presented on a relative scale. With the other parameters (e.g. CO mole fraction) fixed, 230.10nm excitation produces a strong fluorescence signal with a significant temperature dependence, while excitation at 230.08nm is less sensitive to temperature variations. It is clear from these results that a determination of the optimum excitation frequency depends on a tradeoff between the magnitude of signal (or the SNR) and the signal's temperature dependence. Thus, extraction of CO concentration values from the PLIF signals can require knowledge of the temperature field, but the simultaneous two-dimensional mapping of species concentration and temperature in flames has been demonstrated already by Long et. al.[14] using Raman and Rayleigh scattering and two imaging detectors.

A more detailed description of the used quantitative four-level model of CO can be found elsewhere[12].

3. EXPERIMENTAL SETUP

The schematic of the probe, used for the 2-d imaging of the two-photon LIF of CO, is illustrated in Fig.4. The frequency-doubled output of a Quanta-Ray DCR-1A Nd-YAG laser pumped a pulsed dye laser (PDL 1, Kiton Red dye). The dye-laser output was frequency-doubled and then mixed with the residual 1064nm YAG pulse in KD*P crystals to obtain the output at 230.1nm (0.5cm^{-1} bandwidth). The laser pulse energy was measured with a Scientech model 38-0101 calorimeter. The maximum pulse energy at 230.1nm was 3.4mJ. For most measurements, however, the crystals were tuned to achieve optimum spatial beam quality yielding a pulse energy of 2mJ.

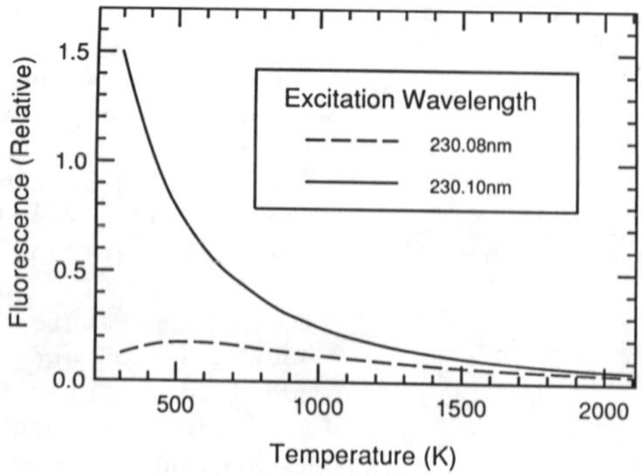

FIGURE 3. Calculated temperature dependence of the two-photon-excited CO fluorescence for fixed CO mole fraction and for excitation at 230.08 and 230.10 nm.

Converting the laser beam into a sheet by using the simple lens arrangements of previous PLIF applications yielded insufficient fluorescence signals for sheets larger than 0.5cm. Therefore a multipass cell was designed for producing a high-intensity laser sheet. The cell consisted of two 10-cm-diameter cylindrical mirrors, coated for high reflectivity (96-98%) at 230nm. Each mirror's radius was 15cm, resulting in a cell length of 30 cm. A single 50cm focal-length cylindrical lens was used to focus the beam into a small sheet between the two mirrors, as the intensity reaching the cell was highly dependent on the number of surfaces (e.g. prisms and lenses) in the beam path. The mirrors were aligned so as to have the beam walk down and up the mirrors, refocusing at the center of the cell on each pass.

A gain of 15 to 20 in sheet intensity was thus obtained compared with an equivalent sized sheet formed using lenses alone. Loose focusing and overlapping passes in the cell produced a uniform laser sheet over a maximum of 4.5cm x 4.5cm area in the present experiment. Both mirrors were mounted to allow three degrees of freedom, two translation and one rotation, with the second mirror having one additional translational degree of freedom (Fig.4). With this design, cell alignment was relatively simple. In all the combustion experiments, the effect of beam steering on sheet focussing proved to be negligible, even for the largest flame zone (taller than the multipass cell) studied.

The resulting right-angle fluorescence was collected using a standard camera lens (Pentax 50mm f/1.2), imaged onto an ITT dual-microchannel-plate intensifier tube, and detected using a Reticon photodiode array (100 x 100) camera. The glass camera lens served as an edge filter, blocking elastically scattered laser light. In some experiments a 486nm

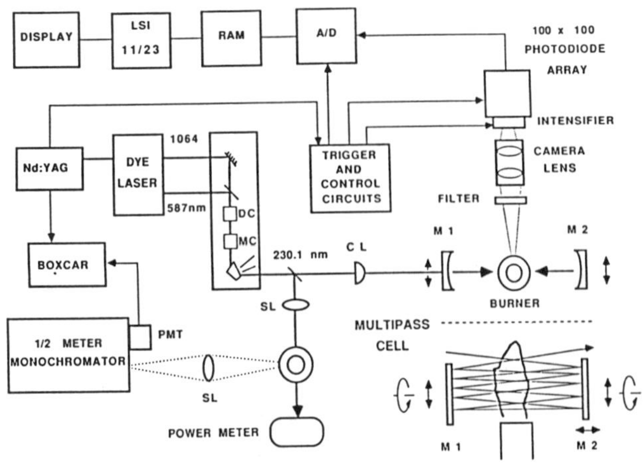

FIGURE 4. Schematic of the experimental apparatus used for the single-point measurements and the two-dimensional imaging experiments: DC, doubling crystal; MC, mixing crystal; SL, sperical lens; CL, cylindrical lens; M1, M2, cylindrical mirrors. The translational and rotational degrees of freedom for the mirrors in the multipass cell are indicated by the arrows.

bandpass filter (FWHM=10nm) was used to isolate a single CO vibrational band, the (0,1) band at 483.5nm. Time gating (1 μs) the intensifier discriminated against the flame emission. For the experiments, described here, the spatial volume imaged by an individual pixel was approximately $0.06mm^{-3}$. Other details of the camera system and the data-acquisition system are thoroughly described elsewhere[15].

In the single point measurements, the laser beam was focused with a 20cm focal length CaF_2 lens. The fluorescence signal was subsequently collected with a 5cm focal length quartz lens, spectrally filtered by a Jarrel-Ash Ebert-type monochromator, and detected with a RCA 1P28A PMT. The output of the PMT was recorded using an EG&G PAR model 162 gated boxcar integrator.

As explained above, conversion of the LIF images to CO mole fraction profiles required the knowledge of the corresponding temperature profiles. Point measurements of temperature were made with a butt-welded PT/PT 10% Rh thermocouple (Omega Corp.). The output of the thermocouple (50 μm diameter wire with a 125 μm bead) was corrected for radiation losses. Mole fraction profiles of CO were also determined by a suction quartz probe measuring the CO absorption of the withdrawn gas samples with a Beckman model 865 infrared analyzer.

4. RESULTS

Qualitative and quantitative imaging of CO were performed at atmospheric pressure in a variety of cold CO jets, CO-air and methane-air flames. The raw data of each laser pulse were corrected by the appropriate image processing for fixed pattern noise, pixel gain variations, background radiation and laser sheet nonuniformities. The corrections for variations in laser sheet intensity were established by using the fluorescence recorded after each series of experiments from laminar CO jets (as sources of uniform CO concentration).

Fig 5. shows the single shot image of the CO fluorescence of a laminar CO diffusion flame. The exit of a 6mm dia fuel tube was 1.2cm below the bottom of this 4cm x 4cm image. The image exhibits a low-frequency oscillation resulting from a Helmholtz instability. The peak CO signal occurs just above the center of the tube, where the CO concentration is maximum, and the signal decays slowly as the temperature increases. As a function of radial distance from the center line, the signal decays rapidly, the result of the sharp temperature increase and CO mole fraction decrease across the reaction zone.

As an example for a hydrocarbon flame Fig.6 shows two images acquired in an unsteady, premixed, fuel-rich methane-air flame. The flame was produced by a rectangular flat flame burner. At this equivalence ratio, both the primary and secondary reaction zones are detatched from the burner surface. The 3.6cm x 3.6cm area imaged here is just above the very thin primary zone, which converts a fraction of the fuel to CO. Fig.6 (left) is

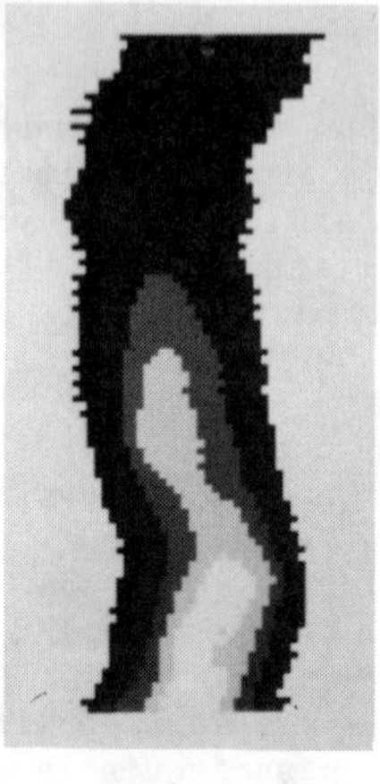

FIGURE 5. CO-distribution in an atmospheric-pressure, laminar CO diffusion flame

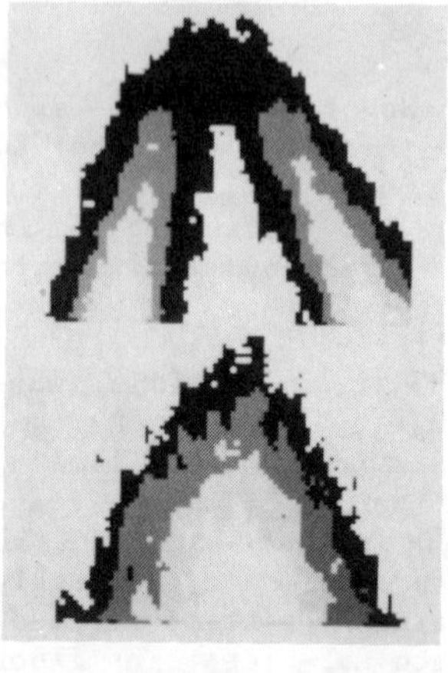

FIGURE 6. Images taken above the detached, primary reaction zone of an unsteady, premixed methane-air flame. Top, ungated flame emission, no laser illumination, bottom, CO flurescence.

the ungated flame emission without laser illumination. Thus it provides an indication of the flame structure. The location of the secondary diffusion flame is evident. Fig.6 (right) shows the corresponding single-shot CO distribution measured using the 230.1nm excitation and a 1.2 μs intensifier gate. The grey scale is a linear mapping of the intensity, with black representing the lowest intensity and white the highest. It is evident, that the maximum CO concentration occurs just above the primary reaction zone and that little CO survives the secondary zone.

In addition, the theoretical model was used to convert CO fluorescence images, measured in atmosperic-pressure jets and flames, to CO concentration profiles. As the simplest quantitative example, Fig.7 represents the measured single-shot CO concentration distributions of cold, nonreacting CO jets at atmospheric conditions. The 4cm x 4cm CO images were recorded 1.5cm above the orifice of a 6mm diam cylindrical tube at four different flow conditions. The theoretical model described was used to convert the fluorescence signals to CO mole fractions, assuming a constant gas temperature in the entire flow field (separate normalizations were required since different voltages were applied to the intensifier and the laser intensities varied for the different conditions).

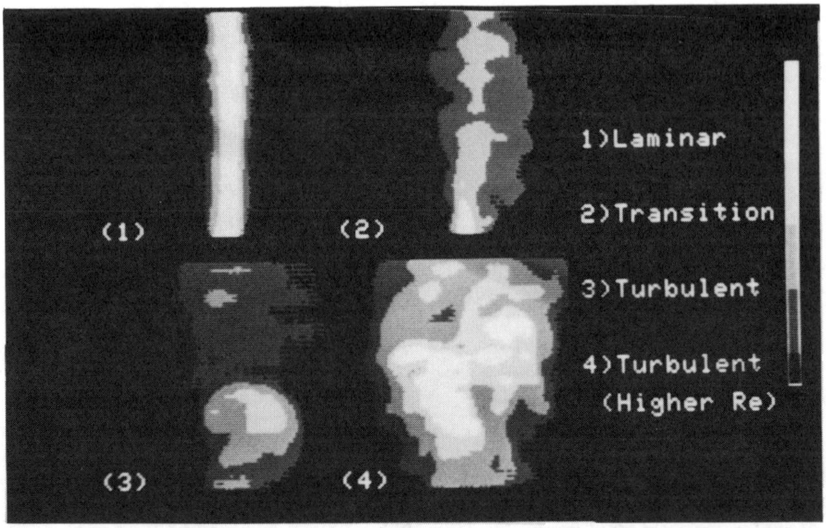

FIGURE 7. Single shot images (4 x 4 cm) of the CO concentrations fields in a nonreacting CO jet at four different flow conditions

A laminar jet at a relative low exhaust velocity can be seen at condition (1). By increasing the velocity a transition from a laminar to a turbulent flow occurs with a break-up of the jet (condition (2)). In the turbulent jet of condition (3) single vortices can be distinguished with a marked intermittency in the CO concentration. A further increase in the Re-number (4) illustrates the scaling of the flow structures in dependence of the Re-number. These results demonstrate, that the small and large scale structures in flows can be visualized quantitatively very well.

Fig.8 represents the 2.5 x 3cm fluorescence image of a laminar CO-air diffusion flame at the same experimental conditions as in Fig.5, but at a lower Re-number. The 8-bit data are presented with a six-level gray scale in this example. Conversion of the measured fluorescence images to CO mole fraction profiles is possible through eq. (2), but this requires knowledge of the corresponding temperature distribution. To verify our model, the centerline (1-d) temperature distribution of the diffusion flame was measured with a thermocouple. The corresponding fluorescence profile (Fig.8) was converted to a mole fraction distribution using the 230.1nm result of eq. (2), plotted in Fig.3. As shown in Fig.9, the result of the optical probe compares well to the profile determined by the suction probe. Since the fluorescence was measured on a relative scale, the fluorescence profile was normalized to the suction probe profile by match-

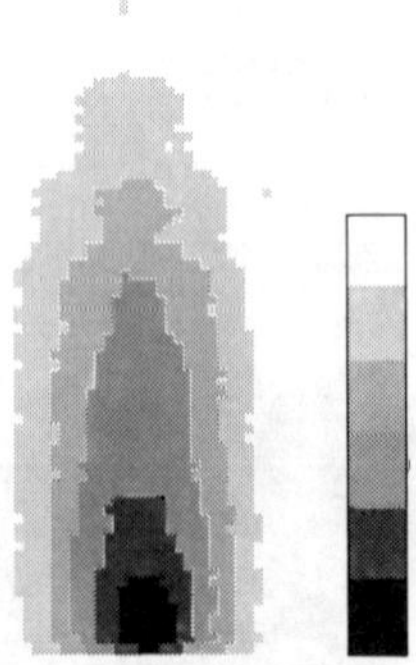

FIGURE 8. Image (2.5 x 3 cm) of CO fluorescence (230.10 nm excitation) measured in a laminar, atmospheric-pressure, CO-air diffusion flame, stabilized above the exit of a 6 mm diameter fuel tube. The original digitization of 256 levels has been reduced to 6 levels in this presentation.

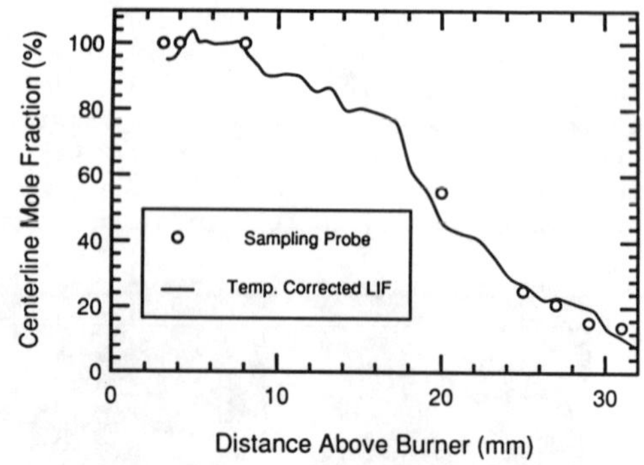

FIGURE 9. Comparison of LIF and probe measurements of the CO mole fraction profile along the centerline of the CO-air diffusion flame depicted in Fig. 8. The two-photon LIF signals of Fig.8 were converted to mole fraction measurements using a measured temperature profile and the model results shown in Fig. 3

ing the two for the 100% CO reading. The lowest CO mole fraction measured in this example was approximately 14% (5.9 x 10^{17} cm^{-3} at 1900K). For the experimental conditions used in the above experiment, however, we estimate the single shot CO detection limit to be 1200ppmV (4.9 x 10^{15} cm^{-3}) at 1900K and 15ppmV (4.0 x 10^{14} cm) at 300K and a detection limit in earlier single point measurements[9] of below 0.2ppmV have been achieved without optimizing the optical setup.

The results described show that eq. (2) accurately models the CO fluorescence for this experiment (e.g., it is shown that photoionization is dominant and collisional quenching is not an important decay process). Thus it is possible to make quantitative field measurements using two-photon PLIF of CO.

Significant optical interferences from other species were not observed in the experiments described above, although a problem arises from the photochemical effects produced by the intense laser radiation required for the nonlinear excitation.

Very small interferences from the C_2 Swan band emission of laser-produced C_2 ($d^3\pi_g \rightarrow a^3\pi_u$) [e.g. 474nm (1,0); 513nm (1,1); 517nm (0,0); 564nm (0,1)] was found in the images of the cold jet and to an even lesser extent in the CO-air diffusion flame. In the hydrocarbon flames , however, the C_2 signals were greater.As an example, Fig.10 presents three images (2.2 x 2.7cm) acquired in a fuel rich premixed methane-air flame, without the 486nm bandpass filter. The ungated flame emission (Fig.10a) shows the flame structure and indicates the location of the reaction zone. Figs.10b and 10c are gated images of the fluorescence recorded with the laser tuned onto the CO 2-photon resonance (230.1nm) and off-resonance (230.5nm), respectively. In the off-resonant case (no CO fluorescence), the C_2 emission is restricted to a narrow region near the reaction zone. For the resonant excitation, fluorescence is excited in a much larger region, including both C_2 emission from the reaction zone and CO fluorescence from the postflame gases.

FIGURE 10. Three (non-simultaneous) images acquired in an atmospheric-pressure, fuel-rich, premixed methane-air flame: (a) 100 μsec exposure of flame emission; (b) 1 μsec image of the fluorescence (CO + C_2) produced by laser excitation at 230.10 nm; (c) 1 μsec image of the C_2 fluorescence (from photodissociation of C_2H_2) excited by a 230.50 nm laser pulse. For (b) and (c), all camera system parameters (e.g. intensifier gain) and the nominal pulse energies were fixed.

Two mechanisms for the C_2-formation were found. In CO-air mixtures C_2 is most[9] likely created by a resonant UV photodissoziation of CO and is highly dependent on the laser wavelenght, decreasing by at least a factor of 20, when the laser is tuned off-resonance. In hydrocarbon flames the laser photodissoziation of higher hydrocarbons[17] (e.g. C_2H_2 and the C_2H fragment) is the most probable C_2 production mechanism, as the production of C_2 in these flames is more efficient than in the CO flames, is not sensitive to changes in the laser wavelength and occurs only near the narrow primary reaction zone. In sooting flames a third C_2 creation mechanism is possible by the photodecomposition of soot particles, since C_2 interferences have been previously observed in CARS studies[16] at such flame conditions. In addition, it should be noted, that the photodissociation of acetylene could perturb the local CO concentration by altering the flame chemistry and is therefore a potential problem in some regions of combustion systems.

The C_2-emission itself could be sufficiently surpressed by gating the intensifier ($\lesssim$1μs) and spectral filtering of the CO-signal by isolating a single unperturbed CO band (483.5nm) with a narrow bandpass filter. However, even without filtering measurements were often possible due to the clear spatial separation of the C_2 -emission (in the primary reaction zone) and the CO-signal (in the postreaction zone).

5. CONCLUSION

It was demonstrated, that the described CO detection technique based on two-photon-excited PLIF is a valuable tool for carrying out quantitative measurements of CO concentration fields in combustion processes.

A quantitative four-level model for the nonlinear pumping scheme used is presented and experimentally verified by comparison of predicted and measured excitation spectra.

Qualitative and quantitative imaging experiments were performed at atmospheric pressure in a variety of cold CO jets, CO-air and methane-air flames. The quantitative model described was used to convert the fluorescence images to CO mole fraction distributions. These profiles agree well with absolute profiles obtained with a standard sampling probe.

These results indicate, that the photoionization of the CO molecule by a 2+1 photon process is the dominating loss mechanism in the $B^1\Sigma^+$ state. Thus it is demonstrated, that the usual complexities of LIF measurements in reacting flows introduced by the species and temperature dependence of the collisional quenching are absent. This simplifies the conversion of the detected fluorescence images to quantitative CO concentration profiles significantly.

The photochemical effects (e.g. C_2 emission) produced by the intense 230nm radiation were also studied. Finally, the presence of two distinct C_2 creation mechanisms (photofragmentation of CO and higher order hydrocarbons) became apparent during the imaging experiments because of their spatial separation within flames. Thus the advantage of imaging techniques is illustrated in cases, where the measurement of time-resolved spatial structures is important.

6. REFERENCES

1. R.K. Hanson, "Combustion Diagnostics: Planar Flowfield Imaging", in Twenty-First Symposium (International) on Combustion (The Combustion Institute, Pittsburgh, Pa., 1987).
2. M.G. Allen, R.D. Howe., R.K. Hanson. Opt. Lett. 10, 267 (1985)
3. M. Alden, H. Edner, P. Grafstrom, and S. Svanberg, Opt. Commun. 42, 244 (1982).
4. A.W. Miziolek and M.A. DeWilde, Opt. Lett. 9, 390 (1984)
5. R.P. Lucht, J.T. Salmon, G.B. King, D.W. Sweeney, and N.M. Laurendeau, Opt. Lett. 8, 365 (1983).
6. D.R. Crosley and G.P. Smith, J. Chem. Phys. 79, 4764 (1983)
7. J.E.M. Goldsmith and N.M. Laurendeau, Appl. Opt. 25, 276 (1986).
8. M. Alden, H.M. Hertz, S. Svanberg, and S. Wallin, Appl. Opt. 23, 3255 (1984)
9. M. Alden, S. Wallin, and W. Wendt, Appl. Phys. B 33, 205 (1984)
10. G.W. Loge, J.J. Tiee, and F.B. Wampler, J. Chem. Phys. 79, 196 (1983).
11. J. Haumann, J.M. Seitzman, and R.K. Hanson, Opt. Lett. 12 (1986)
12. J.M. Seitzman, J. Haumann, R.K. Hanson, Appl. Opt. 26, 2892 (1987)
13. J.T. Salmon, N.M. Laurendeau, Opt. Lett. 11, 419 (1986)
14. M.B. Long, P.S. Levin, D.C. Fourguette, Opt. Lett. 10, 267 (1985)
15. G. Kychakoff, R.D. Howe, and R.K. Hanson, Appl. Opt. 23, 704 (1984)
16. D.A. Greenhalgh, Appl. Opt. 22, 1128 (1983)
17. B.B. Craig, W.L. Faust, and L.S. Goldberg, J. Chem. Phys. 76, 5014 (1982)

LASER VELOCIMETRY FOR COMBUSTION

D.F.G. Durão, J.M.C. Mendes-Lopes

Instituto Superior Técnico
Mechanical Engineering Department
Av. Rovisco Pais
1096 Lisbon Codex
Portugal

1 - INTRODUCTION

It is well known that Laser-Doppler velocimetry (LDV) has become the most suitable and the most used technique to make measurements of velocity in flows, especially under difficult conditions.

For many years, those measurements relied on the use of mechanical measuring probes, such as total-pressure and static-pressure probes, and hot-wire and hot-film anemometers. However, total and static-pressure probes present relatively poor spacial resolution and very poor temporal resolution, and so they have only been (and will continue to be) used to make measurements of mean velocity. Hot-wire and hot-film anemometers allow intantaneous velocity to be measured, and hence mean values, rms velocities and velocity correlations can be calculated. However, the application of these techniques has been substancially limited to constant property flows of low temperature, low velocities and low turbulence intensity, to regions outside recirculating zones, and to cases where the disturbance caused by the introduction of the probe did not alter the main characteristics of the flow field.

Laser-Doppler velocimetry can overcome all of those limitations. Indeed, LDV is an optical technique, and hence it does not disturb the flow. Also, it is not the velocity of the flow that is measured, but the velocity of particles suspended in the flow. Therefore, measurements do not depend on the properties of the fluid and, consequently do not depend on its temperature and

D. F. G. Durão et al. (eds.), Instrumentation for Combustion and Flow in Engines, 151–177.

composition. A number of particles that are suitable for the use of this technique exist in many flows. If its number is too low, the flow can be seeded with appropriate particles to allow the measurements to be made. LDV provides means of measuring instantaneous velocity and with high data rate, so that mean and rms velocities and velocity correlations can be calculated.

The points mentioned above make it clear that LDV is capable of coping with flows under difficult conditions, and is therefore the most suitable technique to make measurement in reactive flows. Measurements are possible not only in laboratory burners but also in a wide range of practical combustors, including gas turbine engines, internal combustion engines and industrial furnaces.

This paper reviews some typical optical arrangements, control volume and light collection characteristics, photodetectors and signal processors. Factors that can influence the accuracy of the measurements are mentioned and the sources of error are described and discussed.

2 - OPTICAL ARRANGEMENTS

The principle upon which the LDV technique is based was layed down more than one century ago. When a particle of velocity U is lighted by a light of frequency ν_0, the detected light frequency is ν_p, slightly different from ν_0. A fixed observer looking at this moving lighted particle receives a frequency ν_D slightly different from ν_p. The frequency ν_D is related to ν_0 by a function of the velocity U of the particle [1]:

$$\nu_D = \nu_0 \frac{1 - \dfrac{U\, r_i\, n}{c}}{1 - \dfrac{U\, r_d\, n}{c}} \simeq \nu_0 \left[1 - \frac{U\,(r_i - r_d)\, n}{c}\right] \tag{1}$$

where n is the refractive index of the medium where the particle travels, c is the speed of the light, and r_i and r_d are the unitary vectors along the directions of the light source and of the observer, respectively.

Usual photodetectors are not able to detect frequencies of the order of magnitude of ν_D and ν_0, which are about 10^{14} Hz. However, if two waves interfere on the photodetector, the constant term ν_0 is eliminated and the resulting frequency can be read. There are three ways to eliminate ν_0 that correspond to three optical set-ups [2]: the two-scattered beam or one beam anemometer, the reference beam anemometer, and the dual-beam or fringe anemometer.

On the two-scattered beam set-up (Fig. 1) a single focussed beam is directed into the flow (direction r_i) and light scattered by a particle in two directions (r_{d1} and r_{d2}) is collected symetrically about the axis of the system. In this set-up the frequency measured ν_A is the difference between the two frequencies ν_{D1} (r_{d1}) and ν_{D2} (r_{d2}):

$$\nu_A = \nu_{D_1} (r_{d_1}) - \nu_{D_2} (r_{d_2}) = \frac{\nu_0 \, U \, (r_{d_1} - r_{d_2}) \, n}{c} =$$

$$= \frac{U \, (r_{d_1} - r_{d_2}) \, n}{\lambda_0} = \frac{2 \, U}{\lambda_0} \sin \theta_A \tag{2}$$

where λ_0 is the wavelenght of the incident light and θ_A is the angle between the incident light and the collecting beams. Note that in air $n \simeq 1$. Hence, thereafter n will be taken as 1.

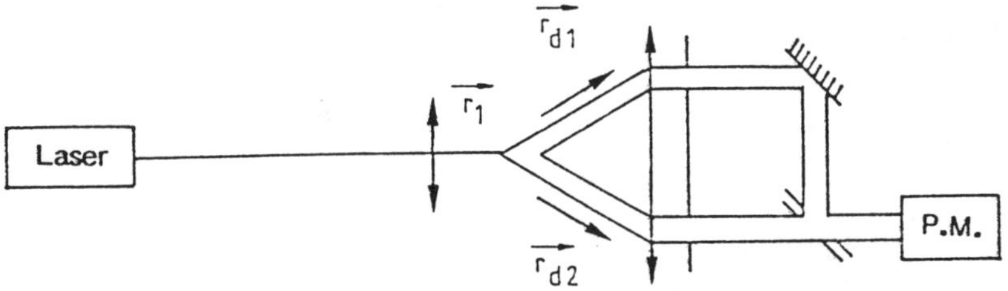

Fig. 1 - One beam or two-scattered beam set-up

On the reference beam set-up (Fig. 2), two beams of different intensities of ratio approximately 1:9 are used. The lower intensity beam, the reference beam of frequency ν_0, is directed onto the detector, in the direction r_d. The light scattered by the particles (also in the direction r_d and with frequency ν_D) passing through the incident beam (in the direction r_i) interferes with the light from the reference beam. The frequency measured ν_B is the difference between the frequencies ν_D and ν_0:

$$\nu_B = \nu_D - \nu_0 = \nu_0 \frac{U\,(r_i - r_d)\,n}{c} =$$

$$= \frac{U\,(r_i - r_d)\,n}{\lambda_0} = \frac{2\,U}{\lambda_0} \sin \theta_B \tag{3}$$

where θ_B is the semi-angle between the incident and the reference beam.

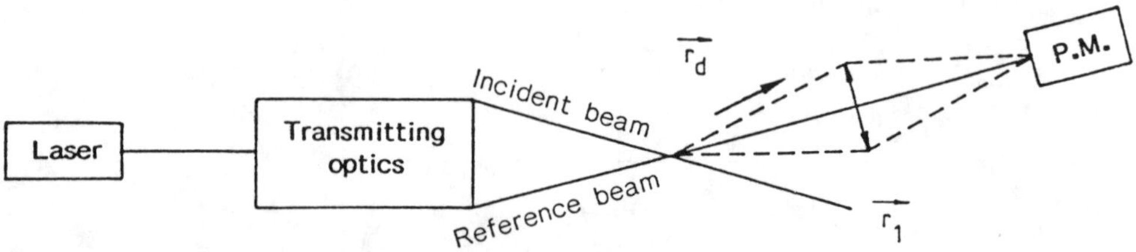

Fig. 2 - Reference beam set-up

On the dual-beam set-up (Fig. 3), two laser beams in the directions r_{i1} and r_{i2} of equal intensity are used. They intersect at the measuring volume. The frequency measured, in the direction r_d, is the difference between the frequencies ν_{D1} (r_{i1}) and ν_{D2} (r_{i2}):

$$\nu_C = \nu_{D_1}\,(r_{i_1}) - \nu_{D_2}\,(r_{i_2}) = \frac{\nu_0\, U\,(r_{i_2} - r_{i_1})\,n}{c} =$$

$$= \frac{U\,(r_{i_2} - r_{i_1})}{\lambda_0} = \frac{2\,U}{\lambda_0} \sin \theta_C \tag{4}$$

where θ_c is the semi-angle between the two incident beams.

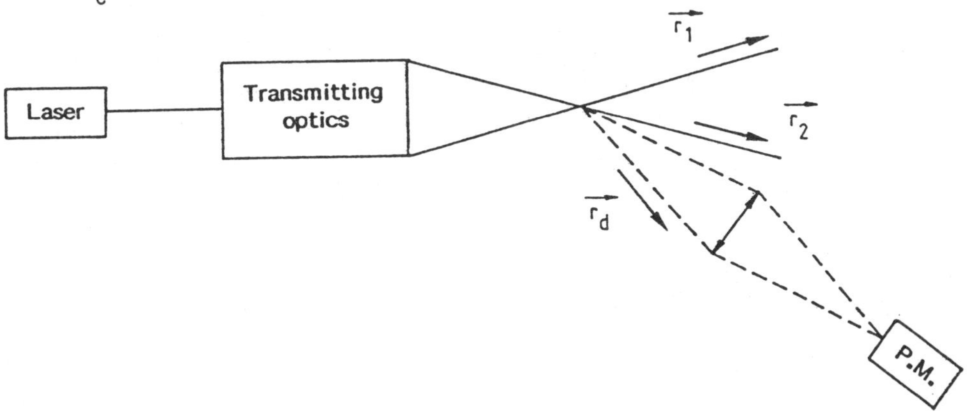

Fig. 3 - Dual-beam or fringe set-up

It can be seen from equations (2-4) that in the dual-beam set-up the measured frequency is independent of the direction r_d of the light collection, which is not the case of the other two set-ups. This is a big advantage. This set-up is by this way the easiest to build operational instruments. The scattered light may be collected in a large solid angle (i.e. in various r_d) without creating an instrumental broadening of ν_c, and the receiving optics can be placed anywhere, where it is more convenient or where the intensity of the scattered light is highest.

In the dual-beam set-up, the two coherent laser beams are crossed and focussed at the measuring (or control) volume. This produces and interference fringe pattern composed of successive light and dark equidistant planes, parallel to the interior bissectrix of the crossing laser beams. In the plane defined by the two laser beams, the light and dark planes appear as light and dark fringes, and so this set-up is also known as the fringe set-up. When a particle passes through, a periodic intensity variation in the light scattered exists. The frequency of this intensity variation is directly proportional to the velocity of the particle in the direction perpendicular to the fringes.

As described above, the signal received when a particle crosses the fringe pattern does not carry information about its direction of motion. This ambiguity in the determination of the direction of motion needs to be removed.

The accepted solution is to make the fringe pattern move at a constant and known velocity along the direction of the measured velocity component. When the absolute value of this velocity is higher than highest of the absolute values of the flow velocities, the ambiguity is removed. There are three types of devices for shifting the frequency of the laser beam [3]: mechanical, electro-optic and acousto-optic.

The rotating diffraction grating is the most used mechanical device (Fig. 4). Two first order beams are scattered by the rotating grade and there frequencies are shifted by Doppler effect. The shifted frequency ν' is given by:

$$\nu' = \frac{U_{grating}}{d_{grating}}$$

where $d_{grating}$ is the groove spacing of the grating and $U_{grating}$ is the linear velocity of the grating. Hence, two laser beams are obtained with frequencies of $\nu_0 + \nu'$ and $\nu_0 - \nu'$. The values of ν' may go up to 10 MHz. The maximum efficiency of the transmitted light is about 70%.

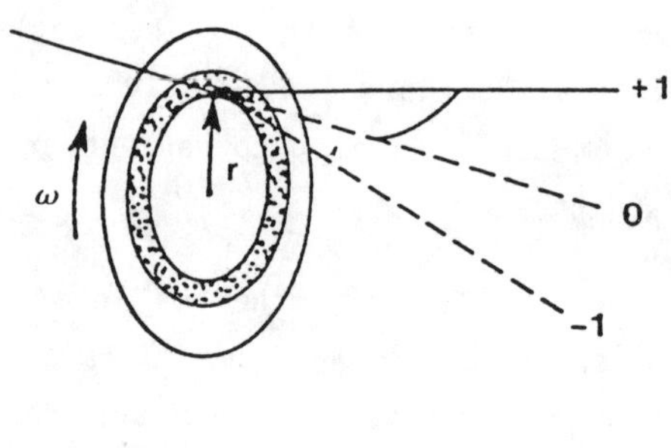

Fig. 4 - Rotating diffraction grating

Electro-optic devices are no longer widely used. They are based in Pockels electro-optic effect in crystals.

The most commonly used frequency shifting device is the Bragg cell (Fig. 5). This is an acousto-optical device where acoustic sinusoidal waves of frequency ν', issued from a piezoelectric crystal, propagate through a liquid.

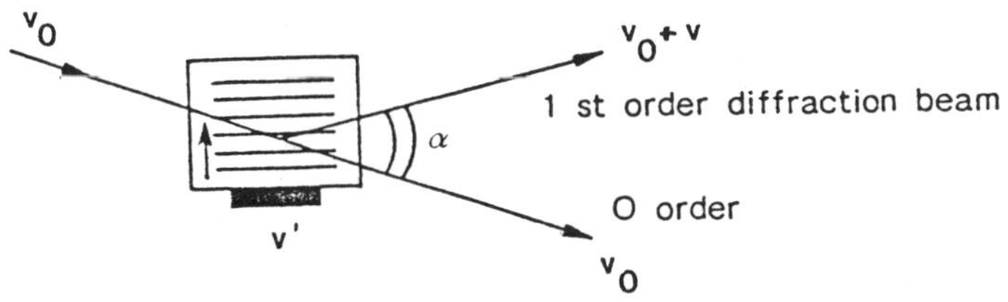

Fig. 5 - Bragg cell

The incident laser beam of optical frequency ν_0 is scattered by Brillouin effect, and the frequency of the first order scattered beam is $\nu_0 + \nu'$. The current values of ν' vary between 40 and 200 MHz. When even the lower values obtained of ν' are too high for a given application, two Bragg cells are used, shifting the frequency ν_0 by ν' and ν''. The frequency shifting of the fringe pattern is thus $\nu' - \nu''$, and this value can be set to be a few MHz. The efficiency of the transmited beams is superior to 85%.

3 - CONTROL VOLUME AND LIGHT COLLECTION CHARACTERISTICS

As mentioned in paragraph 2, in the dual-beam or fringe set-up, the two coherent laser beams are crossed and focussed at the control volume. The beam diameter d_b at the focal point, when it is focussed by a lens of focal length f, is given by [4]:

$$d_b = \frac{4}{\pi} \frac{\lambda f}{D_b} \tag{5}$$

where D_b is the diameter of the unfocussed beam. The control volume defined by the crossing of the two beams at their focal points is an ellipsoid of boundary (Fig. 6) [5, 6]:

$$x^2 \cos^2 \theta + y^2 + z^2 \sin \theta = \left(\frac{2}{\pi} \frac{\lambda f}{D_b}\right)^2 \tag{6}$$

the dimensions of which are:

$$\Delta x = \frac{4}{\pi} \frac{\lambda f}{D_b} \frac{1}{\cos\theta}$$

$$\Delta y = \frac{4}{\pi} \frac{\lambda f}{D_b} \qquad (7)$$

$$\Delta z = \frac{4}{\pi} \frac{\lambda f}{D_b} \frac{1}{\sin\theta}$$

where θ is the semi-angle between the two intersecting laser beams.

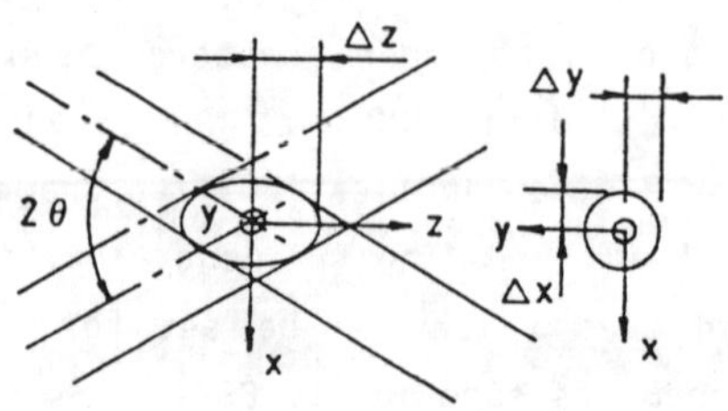

Fig. 6 - Control volume

The distance Δx_f between consecutive light fringes is given by [7]:

$$\Delta x_f = \frac{\lambda}{2 \sin\theta} \qquad (8)$$

and the number of fringes can be calculated by:

$$N_{fr} = \frac{\Delta x}{\Delta x_f} = \frac{8}{\pi} \frac{f}{D_b} \tan\theta \qquad (9)$$

Interference considerations, together with the known Gaussian light distribution and control volume dimensions allow the intensity distribution within the control volume to be written as [8]:

$$I_f(x, y, z) = 2\, I_{b\,max} \{\exp [-(\frac{a^2 x^2 + y^2 + b^2 z^2}{\frac{4}{\ell}\sigma^2})]\}$$

$$[1 + \cos (2 \frac{2 \sin \theta}{\lambda} x)] \qquad (10)$$

where $a = \cos \theta$, $b = \sin \theta$ and $\sigma = \frac{\lambda f}{\pi D_b}$. This equation assumes an equal light intensity distribution in each of the two crossing beams. $I_{b\,max}$ is the maximum intensity of each beam before focussing and is half of the maximum intensity, I_{max}, of the beam emerging from the laser. The beam diameter D_b is related to the $e^{-\ell}$ intensity location, i.e.:

$$\frac{I_b}{I_{b_{max}}} = e^{-\ell} \qquad (11)$$

The equation (10) for I_f can be rewritten in terms of the $1/e^2$ point as:

$$I_f(x, y\, z) = I_{f_{max}} \{\exp [-(\frac{a^2 x^2 + y^2 + b^2 z^2}{2\sigma^2})]\}$$

$$[1 + \cos (2 \frac{2 \sin \theta}{\lambda} x)] \qquad (12)$$

where $I_{f\,max}$ is half the maximum light intensity in the control volume and is equal to the sum of the maximum light intensity in each of the laser beams.

When a particle crosses the control volume with an orthogonal velocity component U, the scattered light has an overall Gaussian envelope, or pedestal, due to the $I_{b\,max}$ exp (...) part of equations (10, 11), and a periodic intensity variation due to the [1 + cos (...)]. The frequency ν of this variation can readily be related to U by:

$$\nu = \frac{U}{\Delta x_f} \tag{13}$$

so that

$$U = \nu \frac{\lambda}{2 \sin \theta} \tag{14}$$

The scattered light is collected by the light collecting system. This system transmits then the scattered light, in form of photons, to the cathode of the photodetector. The maximum voltage amplitude of the photodetector output signal depends on the number of electrons that reach the anode. The maximum number of photons scattered by a particle of diameter d_p, when located in the centre of the control volume, is

$$N_{ph_{sc_{max}}} = \frac{I_{max}}{I_{mean}} \frac{P_L \lambda}{h c} \frac{d_p^2}{\Delta x^2} \cdot Q_{sc} \tag{15}$$

where I_{max} and I_{mean} are the maximum (in the centre) and mean light intensities in the control volume, P_L is the laser power, h is the Planck's constant, and Q_{sc} is the light scattering efficiency. For two crossing beams of equal intensity and for a very small particle, $I_{max}/I_{mean} = 4$.

The intensity of scattered light is a complex function of the scattering angle, beam-angle, particle size and shape, particle material, and ratio of fringe spacing to particle diameter (Fig. 7).

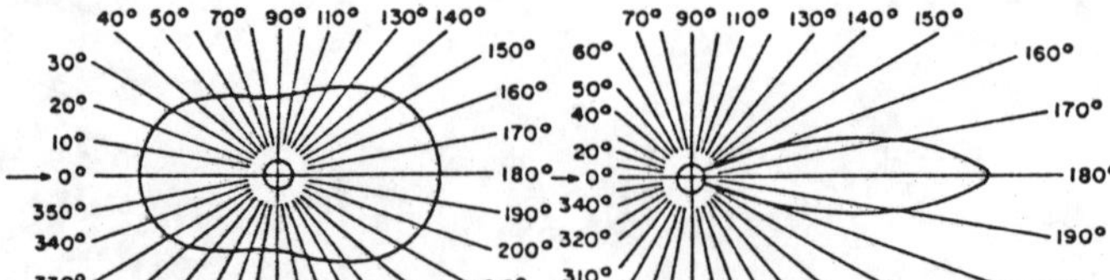

The intensity of scattered light depends on the particle size.

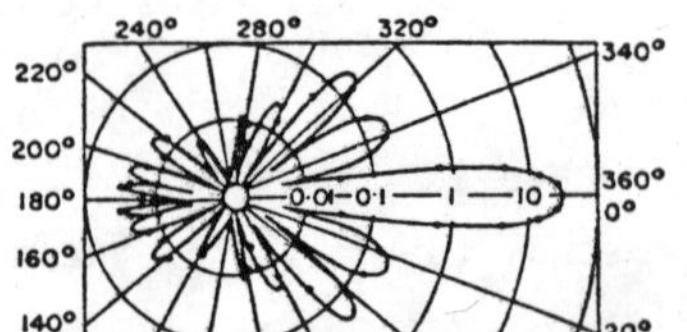

Light intensity variation as function of the direction of observation.

Fig. 7

Exact values for the scattering efficiency can be determined only for spherical and cylindrical particles. Accurate evaluation of the total scattering light, taking into account the light intensity variation in the space occupied by the particle and the shape of the particle, is very difficult. A first approximation of the number of the scattered photons can be calculated assuming an integration of the light over a fraction of the plane fringe pattern equal to the dimension of the particle (as explained in Fig. 8). The resulting equation for the maximum number of photons is similar to equation (15), with I_{max} replaced by I'_{max}, i.e.:

$$N'_{ph_{sc_{max}}} = \frac{I'_{max}}{I_{mean}} \cdot \frac{P_L \lambda}{h\ c} \frac{d_p^2}{\Delta x^2} Q_{sc} \qquad (16)$$

where I'_{max} is given by:

$$I'_{max} = \frac{2}{d_p} \int_{x = x_1}^{x = \frac{\pi}{2}} I_{max} \sin x\ dx \qquad (17)$$

$$\text{and } x_1 = \pi \frac{\Delta x_f - d_p}{\Delta x_f} \qquad (18)$$

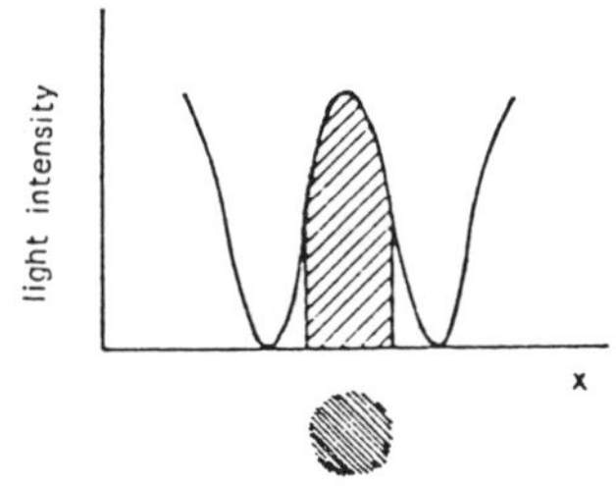

Fig. 8

The signal strength S is defined as the difference between the maximum and minimum value of the instantaneous light intensity. It is given by:

$$S = I_{sc_{max}} - I_{sc_{min}} \qquad (19)$$

where $I_{sc_{max}}$ and $I_{sc_{min}}$ are the intensity of the light scattered by the particle when located at (x, y, z) equal to (0, 0, 0) — centre of the control volume — and equal to $(\pm \lambda/(4 \sin \theta), 0, 0)$, respectively.

The signal quality or signal visibility η is defined as the ratio of the Doppler signal amplitude to the pedestal amplitude. It is given by:

$$\eta = \frac{I_{sc_{max}} - I_{sc_{min}}}{I_{sc_{max}} + I_{sc_{min}}} = \frac{N'_{ph_{sc_{max}}} - N'_{ph_{sc_{min}}}}{N'_{ph_{sc_{max}}} + N'_{ph_{sc_{min}}}} \tag{20}$$

4 - PHOTODETECTORS

The light scattered by the particles moving through the control volume shows intensity variations and has frequencies which contain the required velocity information. Opto-electronic devices are used to process this type of signals. The optical signal is converted into an electrical signal that is then analysed.

The most common of the photodetectors used in LDV is, by far, the photomultiplier. The photomultiplier consists of a photocathode, an electron-optical input system, a dynode chain and an anode (Fig. 9).

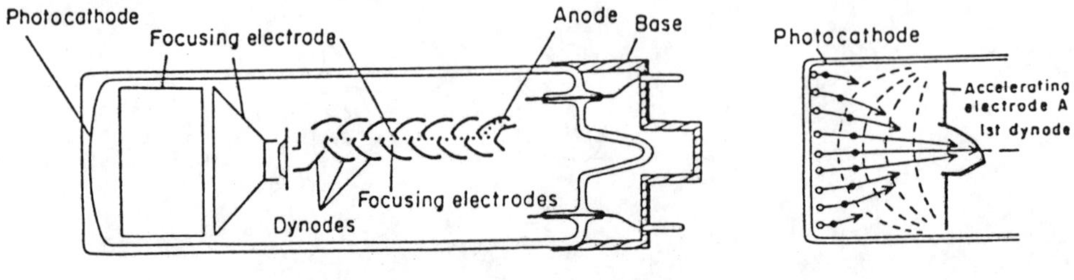

Fig. 9 - Photomultiplier

The photocathode is made up of a light sensitive film that emits electrons when excited by light. The electron-optical input system focuses the electrons on the first dynode. The dynode chain is a secondary emission multiplier, made up in several stages. The anode is the end of the dynode chain and collects the electrons that originate from the cathode and are "multiplied" by each dynode stage to yield a final gain of $G = \delta^k$ where k is the number of dynode stages and δ is the ratio of secondary electrons leaving a dynode to the number of electrons falling on to it.

The maximum number of electrons leaving the cathode $Ne_{c,max}$ is related with the maximum number of photons scattered $N_{ph_{sc_{max}}}$ by the collecting efficiency of the collecting optical system η_c and by the quantum efficiency of the cathode η_q, i.e.

$$Ne_{c,max} = N_{ph_{sc_{max}}} \eta_c \eta_q \tag{21}$$

The maximum current in the anode is given by:

$$I_{a_{max}} = Ne_{c,max}\ e\ G \tag{22}$$

and the maximum amplitude of the photomultiplier output signal is given by:

$$V_{max} = I_{a_{max}} \Omega$$

where e is the electrical charge of an electron, and Ω is the anode resistance.

Values that may characterize a typical anemometer and its calculate performance are given in table 1.

TABLE 1

Characteristics and performance of a typical anemometer			
P_L	$= 5$ mW $= 5 \times 10^{-10}$ J/s	$N_{ph_{max}}$	$= 6.4 \times 10^{16}$ photons/s
λ	$= 632.8 \times 10^{-9}$ m	$N_{ph_{sc_{max}}}$	$= 1.3 \times 10^{11}$ photons/s
h	$= 6.62 \times 10^{-34}$ J/s	$N'_{ph_{sc_{max}}}$	$= 1.1 \times 10^{11}$ photons/s
c	$= 3 \times 10^{8}$ m/s	$Ne_{c_{max}}$	$= 5.5 \times 10^{8}$ electrons/s
e	$= 1.6 \times 10^{-19}$ coulombs	$Ne_{c_{min}}$	$= 10^{8}$ electrons/s
G	$= 0.90 \times 10^{12}$	$I_{a_{max}}$	$= 1.3 \times 10^{3}$ amperes
Ω	$= 50$ ohms	V_{max}	$= 66$ mV
f	$= 200$ mm	V_{min}	$= 12$ mV
θ	$= 9.0^{o}$	ν	$= 0.69$
d_p	$= 1\ \mu$m	$Ne_{c_{max}}$	HAS BEEN CALCULATED BASED UPON $N'_{ph_{sc_{max}}}$
Q_{sc}	$= 50\%$		
η_c	$= 5\%$		
η_q	$= 10\%$		
Δx	$= 0.20$ mm		

5 - SIGNAL PROCESSING

To study the electrical signal from the photodetector different types of instrumentation can be used: spectrum analysers, filter banks, frequency trackers, frequency counters, transient recorders, and photon correlators.

Several interacting technical factors are relevant to the choice of the most appropriate instrument. These technical factors are, in general, the primary factors in the choice of the most suitable instrumentation. Considerations of cost and convenience apply only when deciding between instruments of the same type from different manufacturers. The technical factors can be grouped into those related to the nature of the Doppler signal and those related to the necessities of the experimenter. It should be pointed out that these two sets of factors are not necessarily compatible in all respects. Broadly speaking, the factors related to the signal to be processed are the type of signal (continuous, burst or photon resolved —— Fig. 10), the

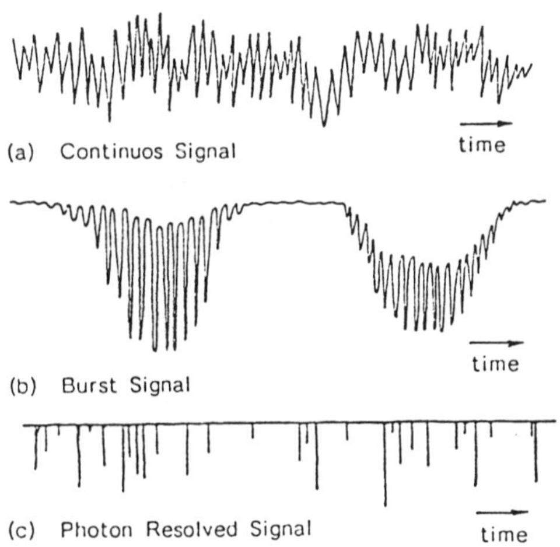

Fig. 10 - Type of signal

signal-to-noise ratio SNR, the frequency range and the maximum frequency expected. The experimenter specified factors are the data rate, the type of output (analog or digital), and the number of velocity components to be measured simultaneously (one, two, or three).

In a spectrum analyser the Doppler frequency range of interest is scanned by a narrow bandpass filter. When a signal at the input contains a frequency component which matches the centre frequency of this narrow filter within a bandwidth tolerance of the filter, a pulse is produced during the short period when the Doppler signal frequency and the continuoustly changing filter frequency are matched (Fig. 11). The amplitude of the pulse is proportional to

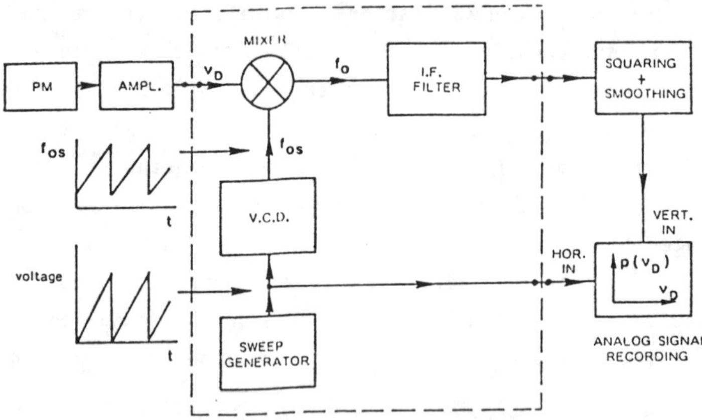

Fig. 11 - Block diagram of a spectrum analyser

the Doppler signal amplitude. An integration of many such pulses yields the Doppler spectrum of the photodetector signal (Fig. 12) which is equivalent to

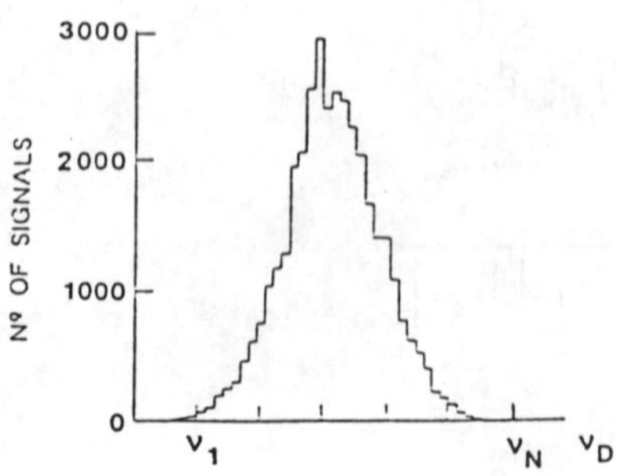

Fig. 12 - Doppler spectrum of the photodetector signal

the probability density function of the Doppler signal. Its main advantages are its capability to cope with low SNR and with high frequencies (usually higher than 100 MHz). Also, the time averaged display of all signal frequencies occuring over its frequency range is useful for diagnostic purposes. Its main disadvantages are the impossibility to obtain real time demodulation of the signal, the impossibility to perform correlations between frequency measurements from the same particle in multi-component velocity set-ups, the distortion of the Doppler spectrum when time-varying particle concentration is recorded by a single slow sweep. Also, only a low proportion of the available signals is utilized because of the narrow filter bandwidth in comparison with the frequency range scanned.

Similarly to the spectrum analyser, the filter bank provides as output the time-averaged Doppler frequency spectrum. However, this is not obtained by sequencial examination of narrow frequency bands across the desired frequency range using a single narrow bandpass filter. This is obtained by simultaneous surveillance of the entire frequency range using a bank of parallel overlapping filters (Fig. 13). This allows a much higher utilization of the

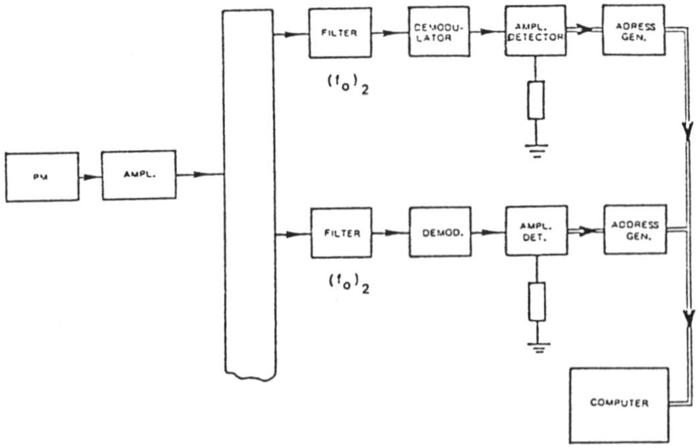

Fig. 13 - Block diagram of a filter bank

available signals. The main advantages of the filter bank are its capability to process low SNR signals, it is applicable in highly turbulent flows, 100% signal detection probability (for signals exceeding the detection threshold), and, unlike the spectrum analyser, the Doppler spectrum will not be distorted by a slowly varying particle concentration (because the entire frequency range is surveyed simultaneously). Its main disadvantages are the impossibility to obtain real time demodulation of the signal (unless a very high data rate is available), and the impossibility to perform correlations of velocity measurements in multi-component velocity set-ups. Also, its feasibility for complete Doppler signal processing above about 10 MHz has not yet been demonstrated.

The frequency tracker attempts to follow the frequency of the input signal and provides a real-time output proportional to the input frequency. The voltage is constantly revised and updated with a feedback loop (Fig. 14).

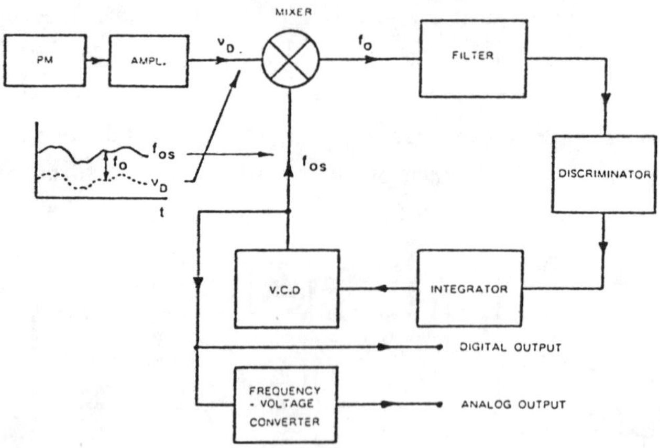

Fig. 14 - Block diagram of a frequency tracker

Under ideal conditions the time variation of the measured velocity component can be obtained. By suitably processing the output, statistics of the time averaged velocity distribution may be determined. The main advantages of the frequency tracker are the capability to use 100% of the signal (in the absence of extended periods of drop-out), the possibility to obtain real-time demodulation (allowing the determination of mean velocity, turbulence intensity and turbulence energy spectra), and its capability to cope with moderate SNR. Its main disadvantages are a relatively low higher processable Doppler frequency (below 20 MHz), the necessity of a nearly continuous signal (implying a high particle concentration), and limitations on the maximum turbulence intensity measurable (15% without frequency shifting).

The frequency counter determines the Doppler frequencies from individual signals burst and passes the frequency information in digital form to a computer for further processing. The frequency is determined from high-pass filtered, ideally symetrical signal burst about the zero level, by counting pulses from a stable, high frequency clock, during a fixed number of zero crossings of the Doppler signal (Fig. 15). There are several possibilities of occuring errors (Fig. 16) and the counter suitable for LDV has to be able to

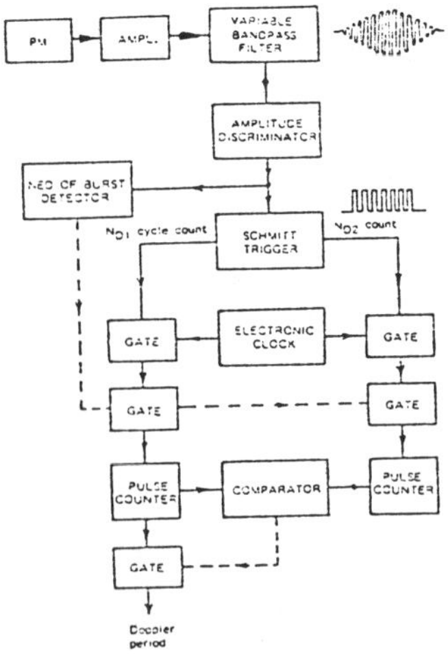

Fig. 15 - Block diagram of a frequency counter

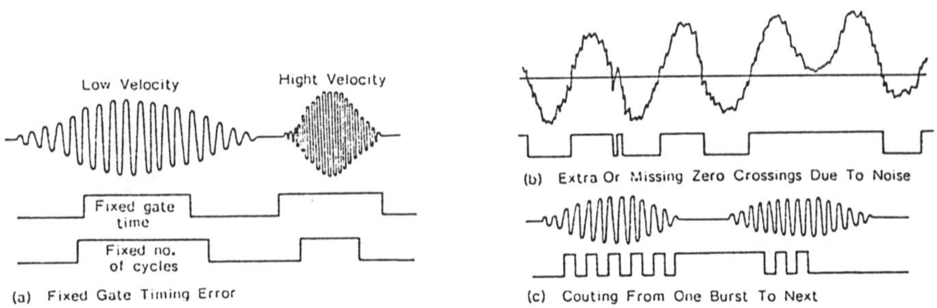

Fig. 16 - Possible counter errors

cope with them. For these counters the gate is controlled by the zero crossings of the signal and, usually, a fixed number of cycles are used. Several checking devices are incorporated to validate each measurement. The counter has many important advantages that has make it the most versatile signal analyser for LDV: it functions well with low concentration of scattering particles since it operates on individual signal burst, it can measure frequencies up to 100 MHz, it can measure in highly turbulent flows, and is appropriate for simultaneous 2-component velocity measurements. However, it needs a high SNR, interfacing with a minicomputer (or other peripherical) is essential, and, if not suitably processed, the results are subjected to bias.

A transient recorder captures a transient event and after a rapid analog-digital conversion stores the digital version of the signal. This signal can then be processed. This signal is usually transferred to a computer. The transient recorder operates well with any SNR and turbulence intensity, and measurements are possible with low particle concentrations. However, data processing is slow, the maximum frequency is limited (15 MHz) and the analysis of the signal requires the use of a computer.

The photon correlator operates with photon resolved signals to determine the autocorrelation function of the photon arrival rate at the photodetector (Fig. 17). This autocorrelation function is theoretically the Fourier transform of the Doppler spectrum (Fig. 18). It thus contains the desired

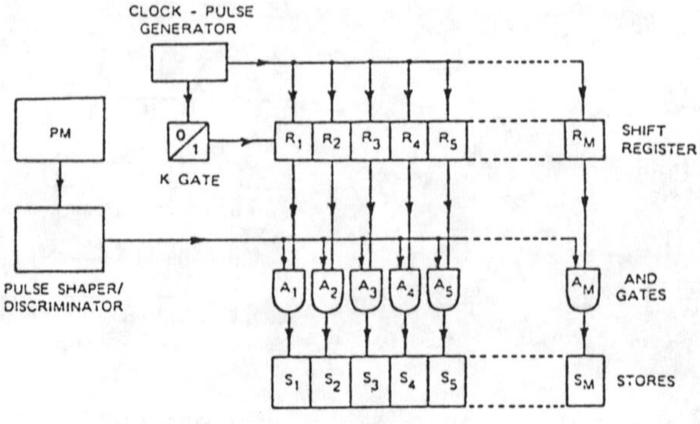

Fig. 17 - Block diagram of a photon correlator

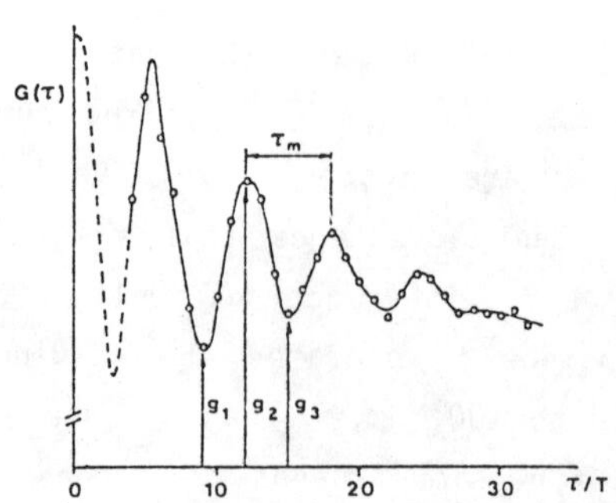

Fig. 18 - Autocorrelation function of Doppler signal

information on the statistics of a turbulent flow. Photon correlation is an extremely sensitive technique and, hence, low laser power and little or no seeding can be used. Since there is no need to filter the signal or trigger the instrument, 100% utilization of the available information is possible. It can measure highly turbulent flows. However, it requires extensive computation, and is very sensitive to background flare.

7 - FACTORS INFLUENCING ACCURACY

Many factors can influence the measurements accuracy. Those can be related to the optics or optical effects, to the seeding particles, to the instrumentation, and to the statistical nature and analysis of the measurements.

To make accurate measurement one has to use high quality optics. These have to be set up appropriately. One small problem that usually is not accounted for but that may become important in some applications, such as investigation in large flow sections like industrial furnaces, is the crossing of the beams out of their focal point. This happens because when a single lens is used to focus two input parallel beams, the axes of the beams intersect at the lens focal point and this point does not, generally, correspond to the focal point of each Gaussian beam [3]. The fringe spacing is no longer constant (Fig. 19). Optimum beam intersection can be achieved adding another

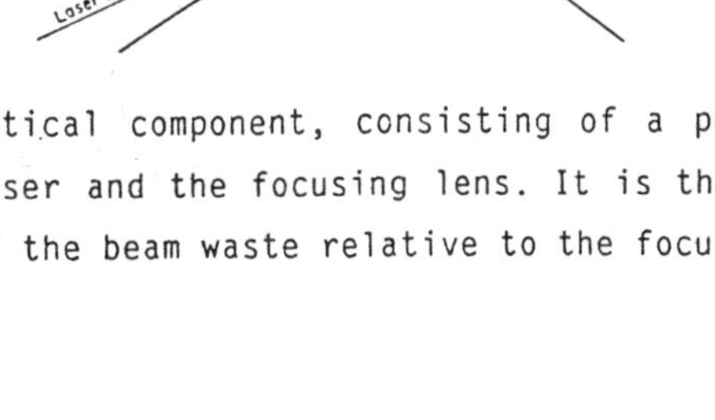

Fig. 19 - Improperly focused laser beams

optical component, consisting of a positive and negative lens, between the laser and the focusing lens. It is then possible to vary freely the position of the beam waste relative to the focusing lens.

The necessity for scattering particles to obtain signals in any laser velocimetry experiment demands the motion of the particles to be representative of the fluid motion. This is just one of the characteristics required (others being good light scattering, chemical inactivity, etc), but it is one that may affect significantly the accuracy of the measurements. The problem of fixing an upper limit on particle diameter to follow the flow within a prescribed tolerance has been approach analytically for spherical particles (see [9]). Table 2 shows the maximum diameter of typical particles used in LDV that can follow the flow within 1% for frequencies up to 1kHz and 10 kHz.

TABLE 2

Particle	Fluid	Density ratio	Viscosity ($kg\ m^{-1}\ s^{-1}$)	Diameter 1 kHz	Diameter 10 kHz
silicone oil	air (NTP)	900	1.8×10^{-5}	2.6	0.8
TiO_2	air (NTP)	3.5×10^3	1.8×10^{-5}	1.3	0.4
MgO	Methane-air flame (1800 K)	1.8×10^4	5.9×10^{-5}	2.6	0.8
TiO_2	oxygen plasma (2800 K)	3.0×10^4	1.1×10^{-4}	3.2	0.8
PVC	water	1.54	1.0×10^{-3}	16	5.0

The size of the particles are also related to the characteristics of the control volume. Particles of diameter equal to the fringe spacing give visibilities around 0.2. Particles of diameter larger than twice the fringe spacing give visibilities worse than 0.1.

Another problem that may arise due to the set-up of the optical system is poor spacial resolution and velocity gradient broadening. When the intersection angle of the two beams is very small (which may be a necessity due to restricted optical access or due to measurements made at long distances) the control volume is alongated (Fig. 20) and the spacial

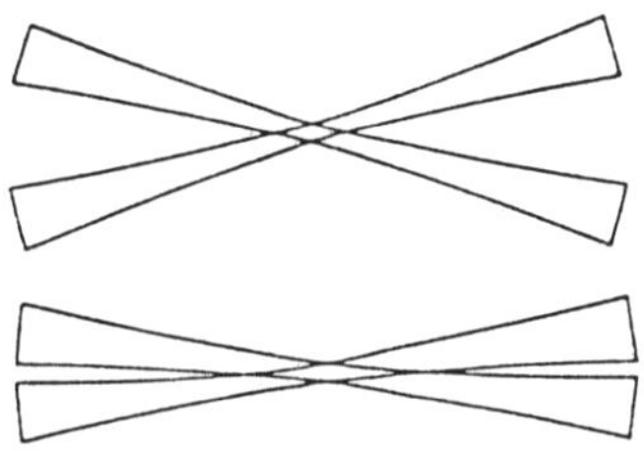

Fig. 20 - Influence of the intersection angle on the dimensions of the control volume

resolution is reduced accordingly. Alongation of the control volume also leads to inaccuracy when the velocity gradient $\partial u/\partial y$ is large (where u is the velocity component perpendicular to the direction y of the fringes).

In highly turbulent flows, in turbulent flows with mean velocity near zero, in recirculating zones, in transient situations where the velocity direction may change, it is necessary to provide means to resolve the ambiguity in the direction of the velocity component measured. As mentioned in paragraph 2, this is done by shifting the frequency of one laser beam relative to the other. The result is the motion of the fringe pattern at constant velocity. When choosing the frequency shift to be used, care must be taken to relate it to the flow velocities (and hence measured frequencies) expected. The sum of the frequency shift and the frequency of the maximum velocity in the direction of the shift must be smaller than the maximum frequency that the photomultiplier and remaining instrumentation can handle. The difference between the frequency shift and the frequency of the maximum velocity in the opposite direction of the shift must be positive (Fig. 21).

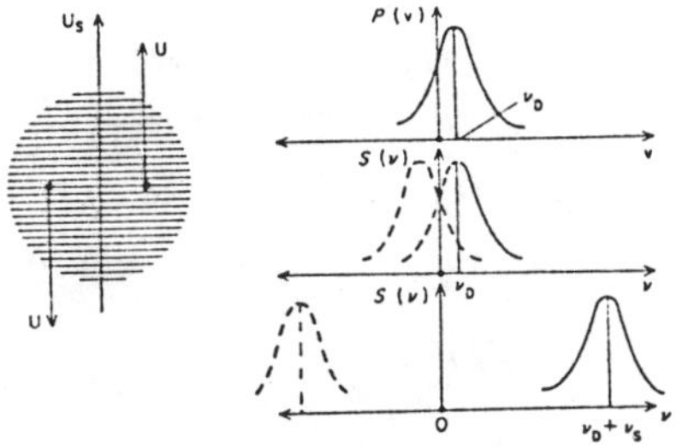

Fig. 21 - Frequency shifting

Another problem typical of the combusting flows and related to the scattering particles is the inherent inhomogeneous seeding due to the thermal expansion of the hot products. The thermal expansion modifies the volumetric particle concentration since a volume of cold gas containing a number of particles increases to a larger volume of hot gas but the number of particles remains the same. A temperature-velocity correlation near a corrugated flame front (Fig. 22) can bias the particle velocity information in favour of those

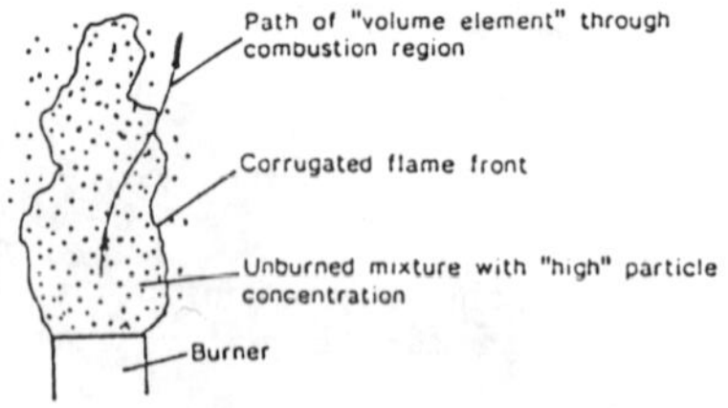

Fig. 22 – Inhomogeneous seeding near a corrugated flame front

velocities at the lower temperatures because the cold and hot regions contribute to the signal output proportionally to their periods of retention in the measuring volume and to their respective particle concentration.

Instrumentation inappropriate to the flow being studied and/or inappropriate analysis of the signal obtained, can give rise to various degrees of inaccuracy. The frequency response characteristic of photomultipliers is an important parameter in the selection of the suitable photomultiplier for a given maximum expected frequency of the scattered light. Also, the frequency of the clock of a counter, for instance, has to be sufficiently high to avoid inaccuracies in the measurements of the photomultiplier output.

A very important factor that can influence the accuracy is the statistical bias. Some sources of statistical bias may be considered [10]. The velocity bias is a consequence of the arrival rate of particles to the control volume not being statistically independent of the flow velocity. It is more likely that more particles of high velocity arrive at the control volume than particles of low velocity. Hence, a simple average cannot be done. Filter bias can also occur. This is due to the signal of a particle not being taken by the instrumentation because it falls off the range set in the processor, or because the filter efficiency is low for that given signal frequency and hence it fails to exceed the processor's threshold. Angle bias can exist when the trajectory of a particle is such that it forms a small angle with the fringes. In this case the particle does not cross enough fringes for its signal to be accepted by, for example, a counter.

8 - CONCLUDING REMARKS

Laser-Doppler velocimetry can be used for quantitative measurements of the mean and rms velocity components in many flow situations including those with combustion, high turbulence and recirculating zones. The appropriate adjustment of the optical and signal analysis equipment to the flow under consideration is essential to obtain accurate velocity measurements.

REFERENCES

[1] CLARKE, W.B.
"Fluid velocity measurement by Doppler shift of scattered light"
Massachussetts Institute of Technology AD-68421 (1968)

[2] BOUTIER, A.
"Velocimètre compact pour measures dans des écoulements très turbulents"
N.T. 237 ONERA (1974)

[3] DURST, F., MELLING, A. & WHITELAW, J.H.
"Principles and practice of laser-Doppler anemometry"
Academic Press, London (1981)

[4] KLEIN, M.V.
"Optics"
Wiley International Edition (1970)

[5] BRAYTON, D.B.
"A simple laser Doppler velocimeter with self-aligning optics"
Proc. Electro-Optics System Desing Conference, New York, p. 168 (1969)

[6] KREID, D.K.
"Laser-Doppler velocimeter measurements in non-uniform flow: error estimates"
Appl. Optics, 13, p. 1872 (1974)

[7] MILLER, C.M.
"A geometric analysis of the probe volume in a dual beam laser Doppler velocimeter"
Proc. Conference on Electro-Optic Systems in Flow Measurements, University of Southampton, paper 2 (1972)

[8] DURAO, D.F.G.
"The application of laser anemometry to free jets and flames with and without recirculation"
Ph.D. Thesis, Imperial College of Science and Technology, London (1976)

[9] MELLING, A.
"Scattering particles for laser anemometry in air: selection criteria and their realization"
Imperial College of Science and Technology, ET/TN/B/7 (1971)

[10] EDWARDS, R.V.
"Report of the special panel on statistical particle bias problems in laser anemometry"
J. Fluids Engineering, 109, pp. 89-93 (1987)

SOME PROBLEMS AND SOLUTIONS IN THE APPLICATION OF LASER VELOCIMETRY TO CONTINUOUS COMBUSTION

J. LABBE and A. JEROT

Office National d'Etudes et de Recherches Aérospatiales
BP 72 - 92322 Châtillon Cedex (France)

The use of LV for measurement in combustion is presently one of the best way to determine the flow field and to localise the recirculation zone.

Nevertheless, some cautions have to be applied in order to obtain a valuable accuracy in despite of the hostile environment.

Some practical problems encountered in such experiments are overviewed and some solutions are proposed in order to overcome these difficulties.

1. PROBLEM ANALYSIS

1.1. Location of the measuring volume (MV)

The most simple problem, but perhaps one of the most important is the precise location of the MV.

Though the accuracy of the 3D traversing systems are typically 10 µm, the expansion of the set-up is much larger and can reach several centimeters.

- The first solution is to fix the frame of the test rig on the same baseground than the velocimeter and to insert several mechanical bellows to dump the effect of the expansion.

  Unfortunately, this obvious solution cannot be chosen always because the facilities were previously designed without taking into account the needs of the laser velocimetry.

  In a general manner, it is benefit to take into account the needs of the laser velocimetry at the beginning of each new design of test rig.

- The second solution is to use a little auxiliary laser (an He/Ne for example), to fix it on the velocimeter frame and to weld a little plate drilled with a thin pinhole on the experimental set-up.

  By means of a photodetector fixed on an unmovable support, it is possible , time to time to check the alignment of the laser beam through the hole and eventually to take into account the apart distance.

  This solution is generally chosen when the set-up is able to reach a threshold of temperature and to become practically steady after this time.

- The third possibility which, by example, is applied on turbomachines consists in equip the set-up with one or several detectors fixed on steady support and to continuously record its displacement.

D. F. G. Durão et al. (eds.), Instrumentation for Combustion and Flow in Engines, 179–190.

The corresponding data can be directly introduced into the loop which monitores the traversing systems so that the values displayed on the screen became perfectly right in despite of expansion and mechanical stress.

1.2. Thermal effects and hostile environment

An other difficulty which has to be surmounted is the protection of the optical components against the effects of temperature and acoustic noise; often generated in combustion phenomena (about 110 db).

In order to avoid a disalignment of the optics, it is judicious to put a cover which stops the thermal radiations and the low frequency vibrations.

It can be noticed concerning this particular operation that there is a contradiction between the bright coating suitable to decrease the thermal radiation and the security rules which recommends the use of black coating.

A good precaution to maintain the cleaness of the Brewster's windows of the laser is to inject continuously a little flow of dry and filtered nitrogen to the exit zone of the laser. Nevertheless, the gaseous pressure had to be fixed at a very low level because a high pressure could modify the local index of refraction, involving a lot of instabilities.

1.3. Flame effects

The following point concerns the visible radiation of the flame which disturbs the emitted light and obliges to place an optical filter on the photomultipliers which selects a narrow pass band centered on the laser wavelengths.

In order to maintain a good signal/noise ratio in such a condition, it is advised to adopt the configuration in "forward scattered light" as soon as the geometry is technically possible.

The magnitude of the signal collected being about 100 time the magnitude of the signal collected in "back scattered light".

For the same reasons, it is necessary to use a laser of high optical power (≥ 8 W) and to be sure that all the optical components are rightly coated for the corresponding wavelengths.

The quality of this signal/noise ratio is more so important as in combustion flows, the turbulence is generally very high in the boundaries of the recirculation zone.

Sometimes, intermittence phenomena can be observed due to the fact that a laser velocimeter gives a local and instantaneous determination of the velocity.

In such a case which often appears close to the instable boundaries of mixed jets, the measuring volume is at a given time located inside the first flow and records the velocity of this flow, but at the next time without moving, it can be located now inside the second flow and records the velocity of the last one.

From a statistical point of view, it seems that there is a large turbulence, but if the shape of the velocity histograms are examined in details, several peaks can be observed.

Each peak corresponds to the mean velocity of each flow and consequently it is recommended to determine the velocities on the graphs instead of using an automatic computation which would mask the physical phenomena. To a certain extent, it would be possible by the improper way to compute a mean velocity which does not exist because it is an average of several velocities.

About this matter, it is noticeable that the use of a laser two focus velocimeter in combustion would be hazardous, the limit of turbulence which is acceptable being less than 30%.

1.4. Seeding

Finally, the last problem to solve is the seeding of the flow. It is clear that in laser velocimetry, the velocity which is measured is not actually the velocity of the flow but the velocity of particles carried along with the flow.

Theoretically, the best particles for this seeding would be light, spherical, .5 micrometers of diameter and able to resist the flame temperature which can reach 2100 K.

Practically, the choice is very poor and the particles which are more fluently selected for combustion experiments are the alumina and the zirconium dioxyde.

Titanium dioxyde can be used if the temperature does not reach 1500 K.

The mean diameter of these particles is very difficult to be estimated "in-situ", due to their random shape which prevents the use of Mie's theory.

Furthermore, these particles are subject to agglomeration by dampness and electrostatic effects, so that the direct optical measurements of sampling collected on solid plates are strongly biased by the impact phenomena.

2. EXPERIMENTAL SET-UP

The LDV fluently used in the Energetic Department of ONERA in order to analyse flows in combustion experiments was previously designed by the Physical Department for use in large wind tunnels.

It was the reason why it had to be deeply modified to be adapted to measurements in continuous combustion flows.

This velocimeter is a bidimensional system using two wavelengths corresponding to green and blue lines of a 15 W argon ion laser.

The optical system is placed on a heavy mounting frame, fixed itself on a concrete block injected in the ground. The weight of the complete velocimeter is about 350 kg which gives a good stability.

The traversing system is able to travel 200 mm along three perpendicular directions.

On an optical bench movable along xy (fig. 1) are located:

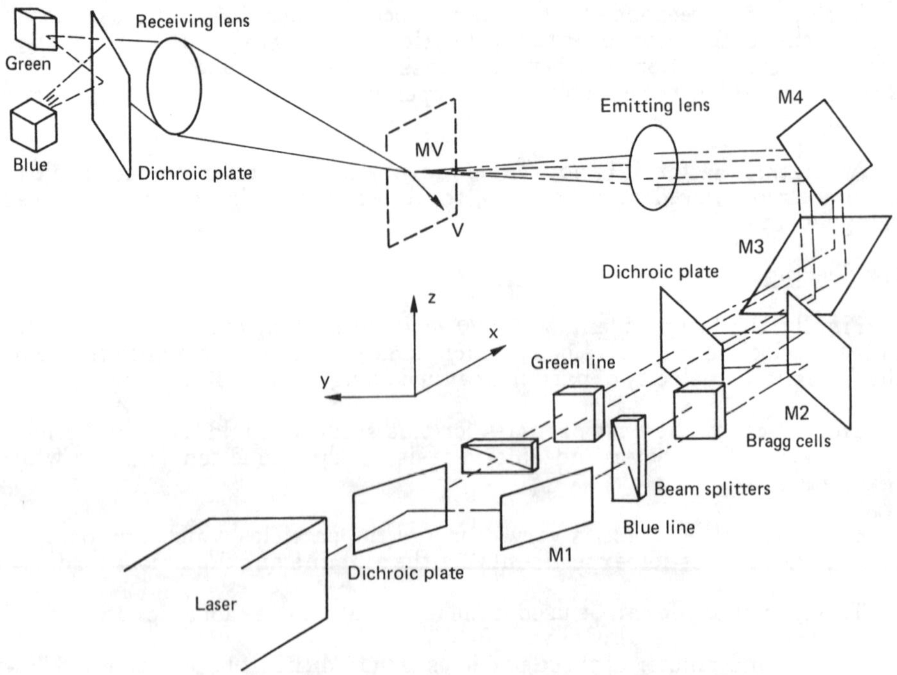

Fig. 1 - Optical arrangement (forward scattered light).

- the first dichroic plate which separates the green and the blue line,
- the afocal telescope which adjusts the beam waist into the measuring volume,
- the beam splitters and the 4 Bragg cells (one on each beam),
- the mirrors and the second dichroic plate which forms finally four parallel beams vertically directed with 30 mm apart.

On a vertical column, high of which being adjusted by use of auxiliary elements, a large plate can be vertically moved (fig. 2).

This plate bears the receiving optics and a strong tubular beam which generally overlaps the experimental set-up.

The emitting optics weight of which does not exceed some hundred grams are brung at the tip of the tube so that the incidents beams are directed toward the main frame by means of auxiliary mirrors.

This reverse configuratin working in forward scattered light was adopted to avoid bending and vibration of the tube involved by a heavy optical system.

This influence of the weight is particularly sensitive with use of a Cassegrain telescope, focal length of which can be adjusted from 0.75 to 1.5 m in order to increase the distance between the set-up and the telescope and to minimize the thermal effects on this element.

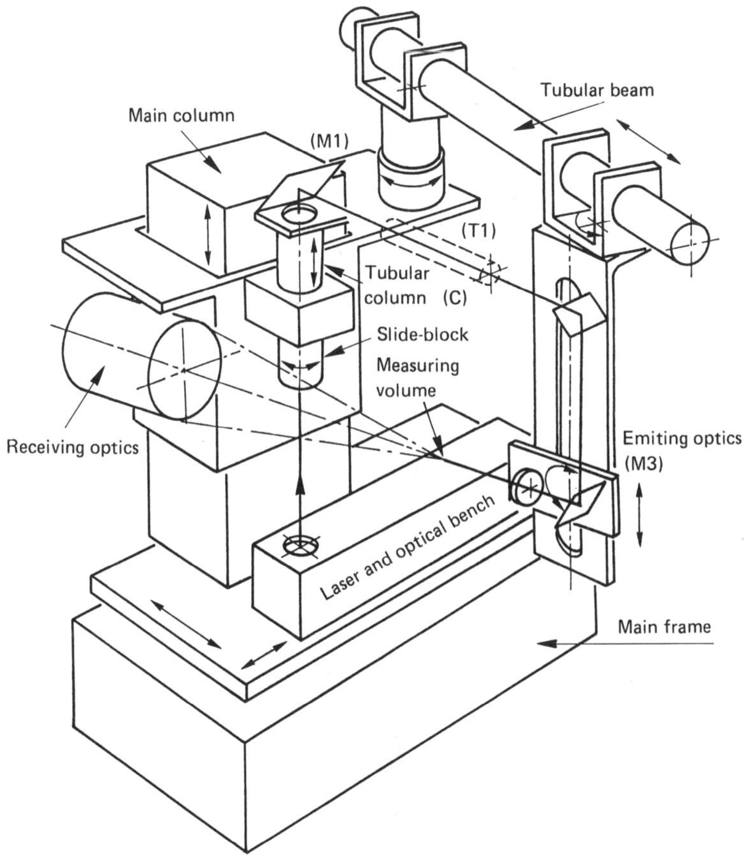

Fig. 2 - LDV used for measurements in combustion flows.

3. APPLICATIONS

The combustion studies using laser velocimetry are summarized on the figure 3.

- The first experiment concerns the mixing of two parallels flows in a rectangular vein. The lower is a mixing of air and methane, the upper is a flow of preheated air. It is a typical experiment where several peaks clearly appear in the velocity histograms.

- The following application of LDV was the study of a combustion flame of air and methane stabilised by a step (fig. 4). The flow lines plotted from data obtained by LDV can be compared to those computed from a theoretical model.

  The shape of the recirculation zones, with (fig. 5) or without combustion (fig. 6) is very similar.

- The third experiment was the determination of the flow field inside an experimental model of a combustion chamber close to one designed by SNECMA Company.

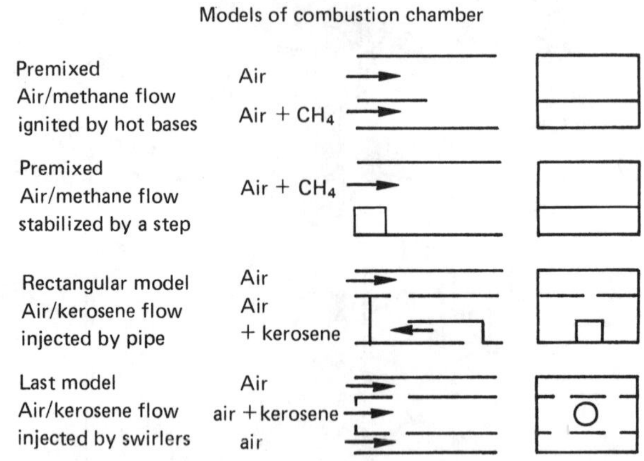

Fig. 3 - Models of combustion chamber using LDV.

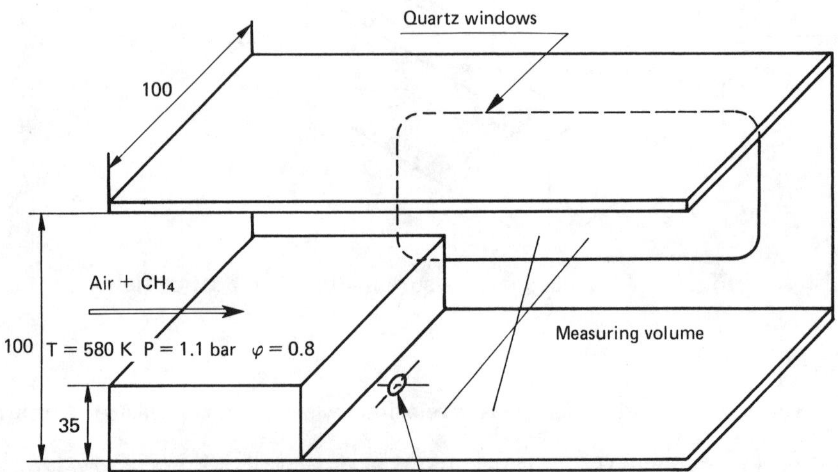

Fig. 4 - Combustion flame stabilized by a step.

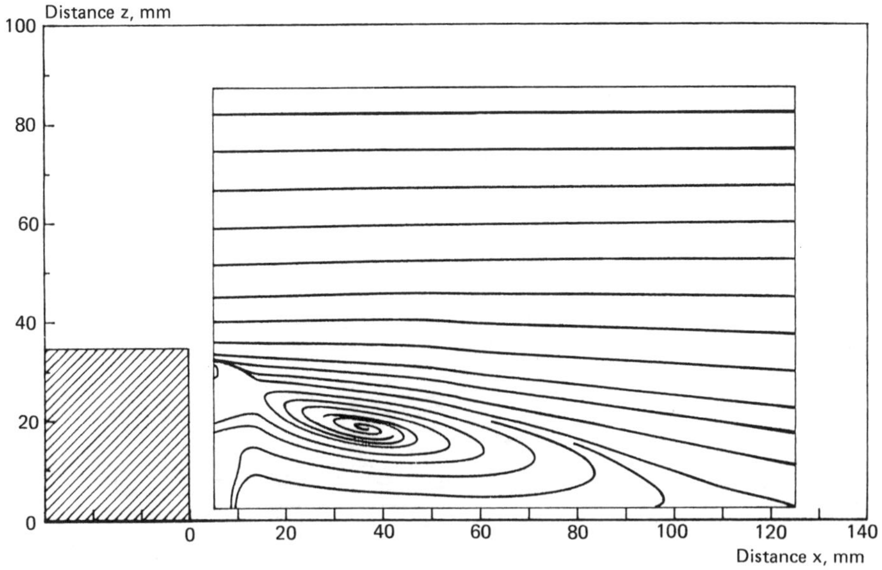

Fig. 5.a - Trajectories with combustion (Experimental).

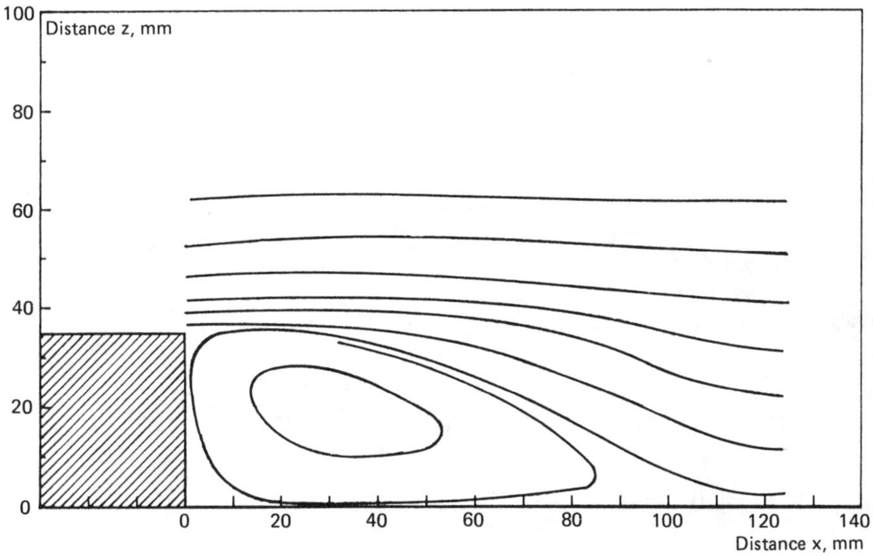

Fig. 5.b - Trajectories with combustion (Computed).

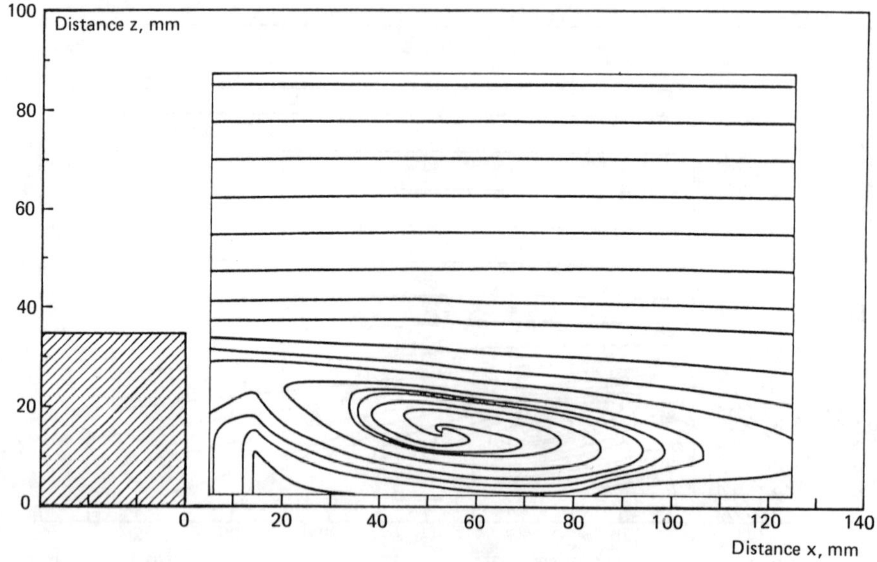

Fig. 6.a - Trajectories without combustion (Experimental).

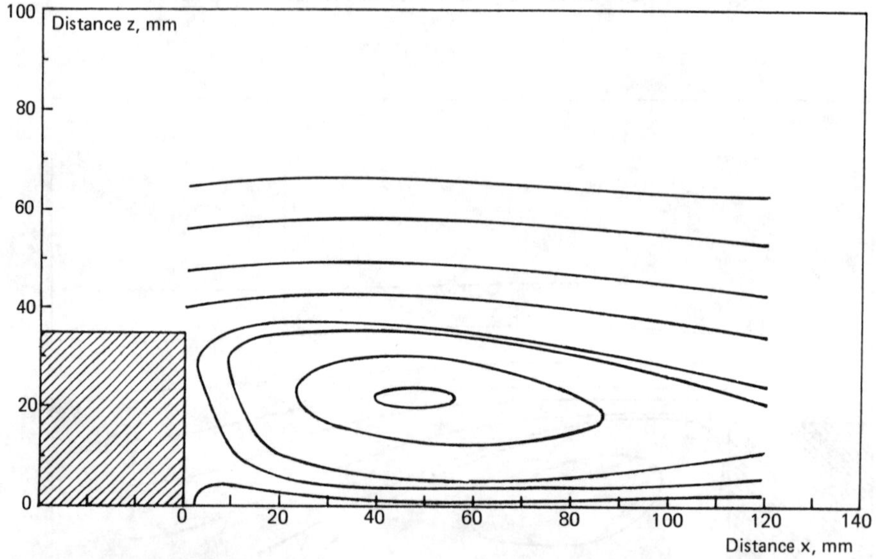

Fig. 6.b - Trajectories without combustion (Computed).

The geometry of the set-up was fitted with the mathematical model using rectangular meshes. It was supplied by a mixing of air and kerosene through several injection pipes (fig. 7).

The influence of the combustion on the flowfield and the corresponding displacement of the recirculation zone clearly appears on figures 8 and 9.

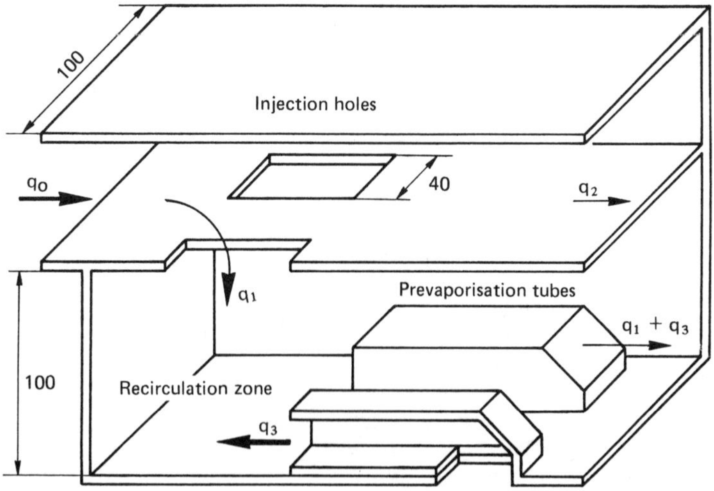

Fig. 7 - Schematic view of the combustion chamber model equipped with injection pipes.

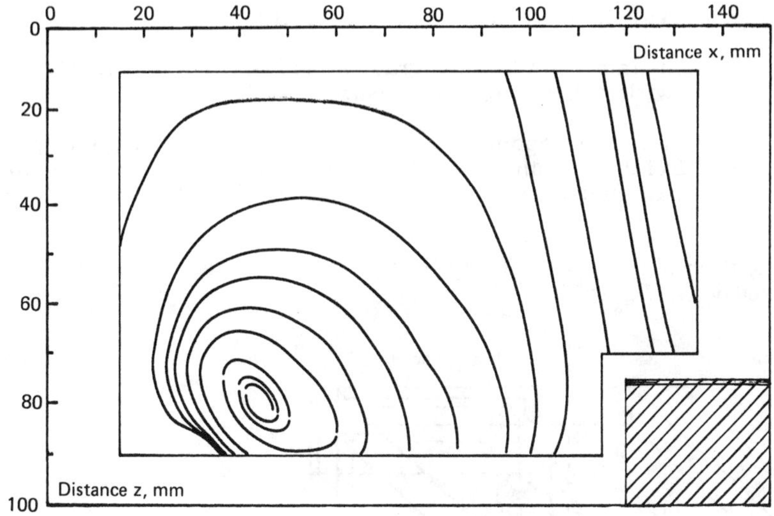

Fig. 8 - Trajectories with combustion.

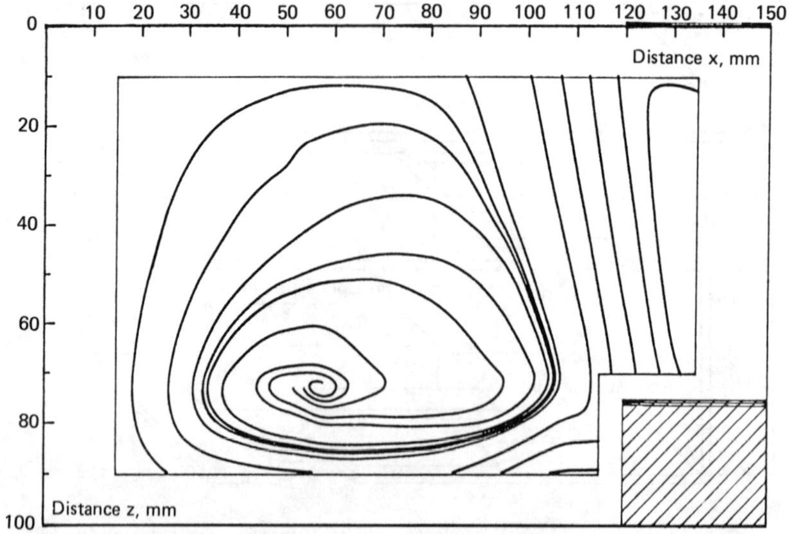

Fig. 9 - Trajectories without combustion.

In the last study involving combustion flows, the injection pipes of the previous experiment were replaced by a lot of swirlers (fig. 10).

Nevertheless, the important thermal gradient and the high turbulence close to the exit of the swirlers, LDV pure the measurement off.

The figure 11 shows the mapping of the field velocity in a vertical plane passing through the *CL* of the median swirler, until figure 12 gives the same mapping in a plane passing through the *CL* of an injection hole and located just between two swirlers.

The evolution of the axial velocity fluctuations along the *CL* of a swirler are presented on the figure 13.

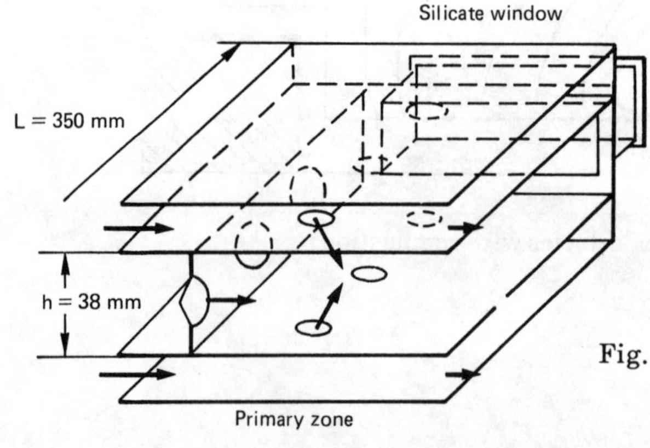

Fig. 10 - Schematic view of the experimental model of combustion chamber.

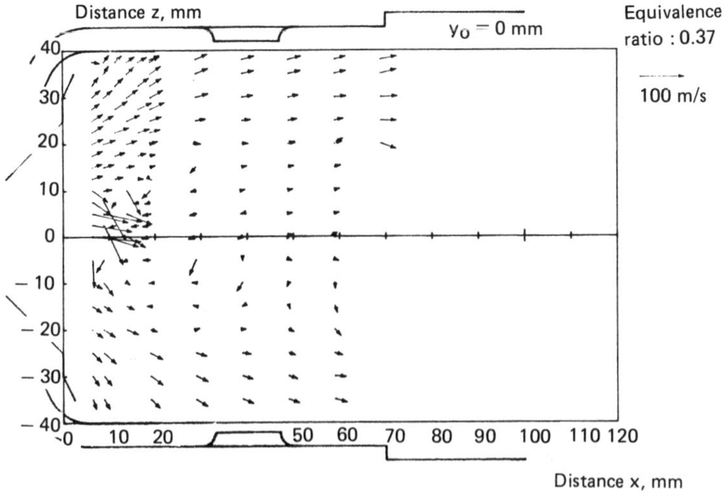

Fig. 11 - Experimental model of combustion chamber equipped with swirlers (Mapping in the *CL* of the median swirler).

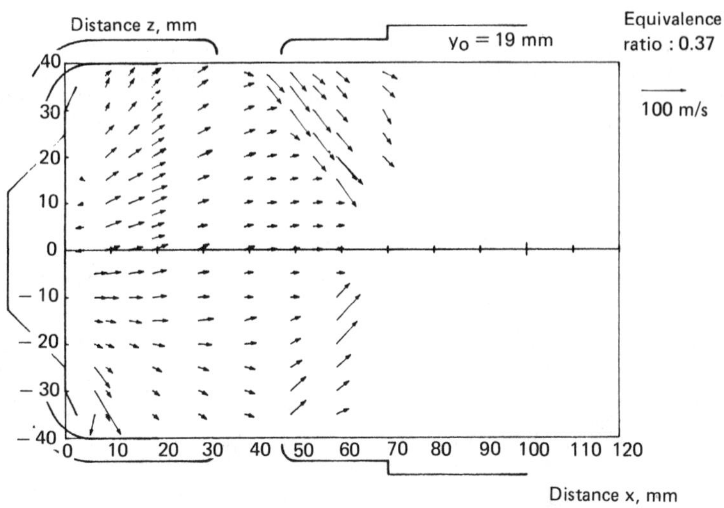

Fig. 12 - Experimental model of combustion chamber (Mapping in the *CL* of a primary injection hole).

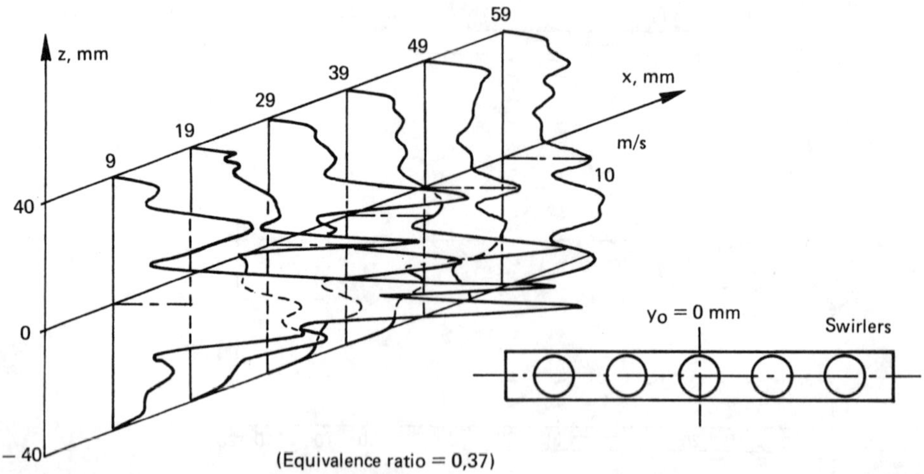

Fig. 13 - Evolution of the axial fluctuations in a combustion chamber model.

4. CONCLUSION

These applications of LDV in a very hostile environment clearly demonstrates that if a minimum of precautions are taken, and if it is possible to get a laser velocimeter equipped with good optics, this method of measurement is very well-suited to analyse the flows in continuous combustions.

THE SIMULTANEOUS MEASUREMENT OF VELOCITY AND SCALAR PROPERTIES

M.V. HEITOR
Instituto Superior Técnico
Department of Mechanical Engineering
Av. Rovisco pais
1096 Lisboa Codex
PORTUGAL

ABSTRACT

The techniques which can be used to measure simultaneously velocity and scalar properties are reviewed in this paper. The aim of the search is to elucidate the importance of such joint measurements to improve understanding of turbulent combustion.

It is concluded that although optical techniques for scalar measurements may offer exciting possibilities, their combination with laser velocimetry may be affected by the seeding particles and subsequent joint statistical analysis limited by the random characteristics of the Doppler signals. The combination of LDV with digitally-compensated thermocouples is particularly analysed and, on the basis of the results of Heitor et al (1987), is shown to provide useful information to the study of turbulent combustion.

1. INTRODUCTION

The direct measurement of the fluxes of turbulent momentum, heat transfer and mass transfer has prompted many experimental investigations in simple flow geometries which, in the recent past, have had the advantage of instruments capable of useful frequency resolution. The measurements have, therefore, been extended to include spectral information and joint probability density distributions of vector and scalar properties, their correlations and conditionally-sampled values as discussed, for example, by Libby et al (1982) and Bilger (1986). This paper presents a brief summary of the combination of avanced diagnostic techniques, which have been used to measure turbulent scalar fluxes.

There are a number of additional incentives for making such measurements. For example, the correlation between velocity and temperature fluctuations is closely related to the turbulent transport of the products of combustion, which has been shown in premixed flames to occur in the "countergradient" sense by recent analytical work of Bray et al (1981) and Masuya and Libby (1981). Measurements in Bunsen-like premixed

D. F. G. Durão et al. (eds.), Instrumentation for Combustion and Flow in Engines, 191–214.

flames, such as those of Moss (1980), Shepherd and Moss (1981), Yanagi and Mimura (1981) and Tanaka and Yanagi (1983), confirm these trends but detailed comparisons with experiments are required for flows with recirculation, such as those of Heitor et al (1987). Frequency resolved values of velocity and temperature or density will also allow conclusions to be drawn as to whether Favre-, or Reynolds-averaged data is measured and will enable an advance towards turbulence closure, as shown, for example, by Starner and Bilger (1987) in non-premixed flames. Also, in premixed turbulent flames the moments of velocity are made up of separate contributions from the reactants and products and this occurs whenever the rate of reaction is fast compared to that of turbulent mixing, as described by Libby and Bray (1981). The simultaneous measurement of velocity and temperature can lead by means of conditional averaging to comparison of the velocity characteristics in the reactants separately from those in the products as illustrated, for example, by Shepherd, Moss and Bray (1982) and Heitor et al (1985). Finally, information on the relative phase between local temperature and velocity will be useful in providing a fuller description than is currently available of a range of practical flows as, for example, in afterburners of gas-turbine engines which are subject to violent combustion-induced oscillations.

The following section presents a brief summary of combination of techniques which have been used to measure, simultaneously, velocity and scalar properties. It is intended as a background to the formulation and assessment of the choices of instrumentation for future works. An example of the use of combined techniques which have improved knowledge of turbulent combustion is summarized in section 3 on the basis of the work of Heitor et al (1987). The final section summarizes the main conclusions of the present review and provides some suggestions for future research.

2. THE COMBINATION OF TECHNIQUES FOR VELOCITY AND SCALAR MEASUREMENTS

Detailed measurements of relevant correlations have been widely reported for simple non-reacting flows and are summarised in table 1. Probe techniques (table 1-A) based on hot-wires with different levels of overheat, hot-films and resistance thermometers as well as non-intrusive optical techniques (tables 1-B and C) have been applied in air flows with small temperature gradients and in isothermal flows involving the mixing of the gases of widely different molecular weight. Optical techniques have also been used to provide scalar information in water flows with a tracer, but they are usually expensive to assemble and to use, although very desirable in recirculating flows.

Joint measurements in turbulent reacting flows are fewer in number and have greater experimental uncertainty than those in non-reacting flows, as may be seen from table 2. Since hot-wire anemometry is unsuited to most combusting flows, instantaneous measurement of velocity is possible only by laser-Doppler velocimetry. Simultaneous scalar measurements can then be affected by the particles used as light scatterers and subsequent joint statistical analysis further limited by the random characteristics of the Doppler signals. These aspects of the combination of experimental techniques have only been partly reported in the literature and are analysed in the subsections below.

2.1 The combination of LDV with fine-wire thermocouples

Precise measurement of the time resolved characteristics of no scalar has been so well developed as that of velocity. A relatively simple method which can be used accurately in a range of flows is to measure temperature with a fine-wire thermocouple. The limitations of this technique are briefly analysed in the following paragraphs and its suitability assessed in relation to alternative optical methods, such as those discussed by Lapp (1980), Bilger (1986), Dibble and Hollenbach (1981), Eckbreth (1981), Hall and Eckbreth (1983), Greenhalgh (1983) and Penner, Wang and Bahadori (1984).

A thermocouple probe, although inexpensive and relatively easy to fabricate and use, can cause local and global flow field perturbations as discussed by Bilger (1977). Probe configurations and orientations can, however, be improved to minimize these effects except in some local regions of recirculating flows, as observed by LaRue, Samuelsen and Seiler (1984) in a confined swirling flame. The probes can also be designed to make conductive heat losses from the thermocouple beads negligibly small, and the work of Ballantyne, Boon and Moss (1976) and Heitor et al (1985) have suggested that losses from the thermocouple by radiation (as well as catalytic effects) may affect the mean temperature but are unimportant for the time-resolved measurement.

The thermal inertia of the thermocouple gives rise to first-order damping of the frequency response, for which compensation must be made. There is an extensive body of literature, for example Ballantyne, Boon and Moss (1976), Ballantyne and Moss (1977), Yule, Taylor and Chigier (1978), Yoshida and Tsuji (1979), Lenz and Guenther (1980), Yoshida and Guenther (1980), Lockwood and Moneib (1980 and 1981), Yanagi and Mimura (1981), Tanaka and Yanagi (1983), Sivasegaram and Whitelaw (1983), LaRue, Samuelsen and Seiler (1984), Nina and Pita (1985), Cambray et al (1985) and Katsuki et al (1987), describing how such compensation can be effected, once the time constant of the thermocouple has been found. Since this constant depends on the heat transfer coefficient, its value is a function of the local velocity, temperature and composition. It is stressed that although we are considering turbulent flows, the local Reynolds number of wire (based on a wire diameter of below 0.040 mm) is small and hence heat transfer is determined by the local conditions. Thus Strahle and Muthukrishnan (1976), Yule, Taylor and Chigier (1978), Lockwood and Moneib (1980) and LaRue, Samuelsen and Seiler (1984) have obtained an average value for the time constant at every point at which temperature measurement is required. However, the large temporal changes of temperature and velocity imply that the response of the thermocouple is non-linear and the use of an average time constant leads to some error. The effect is analogous to that which occurs in the context of hot-wire anemometry, as shown by Comte-Bellot (1976). The technique of compensation for thermal inertia is, however, well established for use in midly heated, non-reacting air flows, as discussed by Paranthoen et al (1983) and Verriopoulos (1983), but not in reacting flows. Chandran et al (1984-a) have obtained a close agreement between the time resolved temperature measurement from a compensated thermocouple (d_w = 0.025 mm) and that obtained simultaneously by laser Rayleigh scattering in an open premixed methane-air flame. Although similar agreement with other optical techniques have also been reported for the mean temperature by Roquemore and Yaney (1979), Chauveou et al (1981), Farrow, Mattern and Rahn (1982) and Reuss (1983), further research of this kind is required to establish the accuracy of time-resolved thermocouple measurements.

Thermocouple probes have been combined with laser velocimetry to measure turbulent scalar fluxes and table 2-A presents a summary of the flows studied and conclusions obtained. Errors may arise from different sources and their quantification was an objective of the investigation of Heitor et al (1985). Specifically, the main conclusions are:

1. Such measurements are possible provided that:
   (a) the time constant of the thermocouple is a function of the instantaneous temperature and velocity. This requires numerical compensation techniques and adequate sampling rates of the temperature signal.
   (b) the thermocouple wires used are no larger than 0.040 mm in diameter. This entails some loss of the highest spectral components of temperature fluctuations as compared with a 0.015 mm wire but improves thermocouple lifetime and accuracy of positioning.
   (c) the seeding density is low (for example, no greater than 60 particles per second in the premixed flame of Heitor et al, 1987) and that each thermocouple is used for a limited period of time (no longer than 15 minutes).
   (d) the separation of the anemometer probe volume from the thermocouple junction is as small as possible (about 1 mm) in zones of high spatial gradients.
2. The probability density function of measurements conditioned by velocity are close to Favre-weighted.
3. The conditional velocity statistics are insensitive to the details of the compensation procedure and the conditioning levels. In contrast, the correlation $\widetilde{u''c''}$ is sensitive to these details and is underestimated.

This technique can provide qualitative information which is useful to the study of turbulent premixed flames. The tolerances in the measurements require further experimental investigation, possibly using laser-based techniques which are currently under development.

2.2 The combination of optical techniques

Tables 2-B to 2-E summarise works which have combined laser velocimetry with an optical technique, as an alternative to thermocouple probes, to obtain simultaneous measurements of a vector and a scalar. The main conclusions are described in the following paragraphs.
The intensity of light scattered from a small optical measuring volume containing a high concentration of seeding particles ("marker nephelometry" after Becker, 1977) can be interpreted in terms of the instantaneous gas density and linked to temperature with an equation of state and the assumption of constant pressure, as shown by Kennedy and Kent (1979 and 1981). However, the need of dense particle concentrations may affect the flame, impair the quality of the simultaneous Doppler signal and decrease the spatial resolution of the scalar measurement if an enlarged control volume is used, as reported by Moss (1980, 1981). Table 2-B presents a summary of the references which have combined this technique with laser velocimetry to measure turbulent scalar fluxes

either in premixed or non-premixed flames. For the latter, additional problems may arise due to differential effects between marker particles and fuel atoms which have been partly quantified by Starner and Bilger (1981-b). Comparisons with Raman scattering and probe techniques, reported by Drake, Bilger and Starner (1982) for hydrogen non-premixed jet flames, have shown that the Mie scattering technique largely overestimates the mixture fraction fluctuations, although a good correspondence was obtained for the mean values.

Driscoll, Schefer and Dibble (1982) reported measurements in a non-premixed flame with a combination of laser velocimetry and Rayleigh scattering (table 2-C); the main problem arises from the need to separate Mie from Rayleigh scattered light which, in turn, restricts the data-colletion rate and the simultaneity of the measurements. Further problems inherents to the Rayleigh scattering technique have been discussed, for example, by Rajan, Smith and Rambach (1984) and Namer and Schefer (1985).

Warshaw et al (1980), Dibble and Schefer (1983), Dibble, Kollman and Schefer (1984) and Dibble et al (1984) made use of a combination of laser velocimetry and Raman scattering to measure correlations between velocity and temperature, density or mixture fraction in hydrogen jet non-premixed flames (table 2-D). Again, the technique involves the separation of two types of scattered light which, on this occasion, can be theoretically achieved by frequency, rather than by amplitude discrimination. The weakness of the Raman signal requires, however, a time delay between the two measurements to filter the Mie scattered light and, in addition, implies the use of pulsed lasers, limiting the sampling frequency to about 10 Hz and consequently the population size for joint statistics. Other aspects of the establishment of the accuracy of Raman measurements have been discussed by, for example, Drake, Pitz and Lapp (1984). Bias errors in both velocity and scalar averages have been analysed by Dibble et al (1984) and can be an additional significant problem.

Some of the weaker aspects of Raman spectroscopy could be improved with the use of CARS (coherent anti-Stokes Raman spectroscospy), as discussed by Farrow, Mattern and Rhan (1982), and some attempts to combine this technique with laser velocimetry have been reported, although in very early stages of development (table 2-E). The use of pulsed lasers entails high cost and general complexity in data handling and processing, and therefore limits the applicability of the technique to very difficult combustion environments not measurable by other methods, such as those described by Eckbreth (1980).

3. AN INVESTIGATION OF TURBULENT PREMIXED FLAMES BY COMBINING LDV WITH DIGITALLY COMPENSATED THERMOCOUPLES

Although optical techniques for scalar measurements may offer exciting possibilities, they are, as yet, not widely available and their combination with LDV may not bring further insight in the study of the structure of simple laboratory flames.

This section summarizes the results of Heitor et al (1987) obtained in disc-stabilized turbulent premixed flames by combining LDV with fine-wire thermocouples. The main objective is to elucidate the importance of such combination of techniques, which are available in most research laboratories.

The premixed flames were stabilized on a disk baffle located on the axis, and at the exit, of a pipe: the experimental conditions are summarized in table 2-A together with the main characteristics of the experimental technique (see Heitor, 1985 for details). On the basis of the work of Williams (1984), the structure of the flames studied is unlikely to be determined by chemical kinetics but thin-flame models, such as those of Bray, Libby and Moss (1984 and 1985) are also unlikely to represent the flames.

Figure 1 shows vectors of mean velocity and turbulent heat flux superimposed on isotherms: note that the subscript 0 refers to the approaching reactant stream. These are highly curved and, in the terminology of Masuya & Libby (1981) and Bray et al (1985), reveal a non-planar flame oblique to the oncoming reactants. The reaction region occurs outside the locus of the mean separation streamline (Ψ=0) and is curved along its length. This curvature imposes mean velocity effects on the turbulence field and the result is a highly strained flame different from those for which detailed velocity and temperature measurements have been reported.

The turbulent heat fluxes tend to be restricted to the shear layer with a large component directed along the isotherms rather than normal to these, as would be expected from gradient-transport models of the kind used in non-reacting flows.

The radial heat flux is always positive - in other words, a heat flux directed away from the centreline. This is the expected direction in a baffle-stabilized flame since the ignition of the reactants (which are characterized by $c''<0$ where $C=\tilde{C}+c''$ is the reaction progress variable and the $\sim$ means a favre average) must be associated with movement towards the recirculation zone ($v''<0$, hence net correlation $\widetilde{v''c''}>0$). The flame is then established by the heat transfer between the hot products and the cold reactants. Note also that the sign of the radial heat flux is in qualitative agreement with gradient-transport models ($\widetilde{v''c''}= -\nu_c.\partial\tilde{C}/\partial r$, where ν_c is a scalar turbulent thermal diffusivity).

The axial flux, $\widetilde{u''c''}$, is strongly influenced by the velocity field in addition to the initial dependence on the mean scalar distribution. For $X/D<1$, the peak absolute values of the axial flux are larger (by at least a factor of two) than the values of the radial flux, even though the axial temperature gradients are negligible as compared to those in the radial direction. The observation is not unusual, however, in the sense that similar ratios are found in heated, non-reacting flows such as curved boundary layers (e.g. Gibson and Verriopoulos, 1984; Dakos, Verriopoulos and Gibson, 1984), round jets (e.g. Chevray and Tutu, 1978), co-axial jets (e.g. Johnson and Bennett, 1983) as well as near the burner tip of the premixed flame of Tanaka and Yanagi (1983). It should be noted that it is the spatial gradients of the heat flux components, and not the components themselves, that give rise to heat transport. The axial gradients of $\widetilde{u''c''}$ are substantially smaller than the radial gradients of $\widetilde{v''c''}$ and hence the turbulent transport of heat is principally in the radial and not in the axial direction.

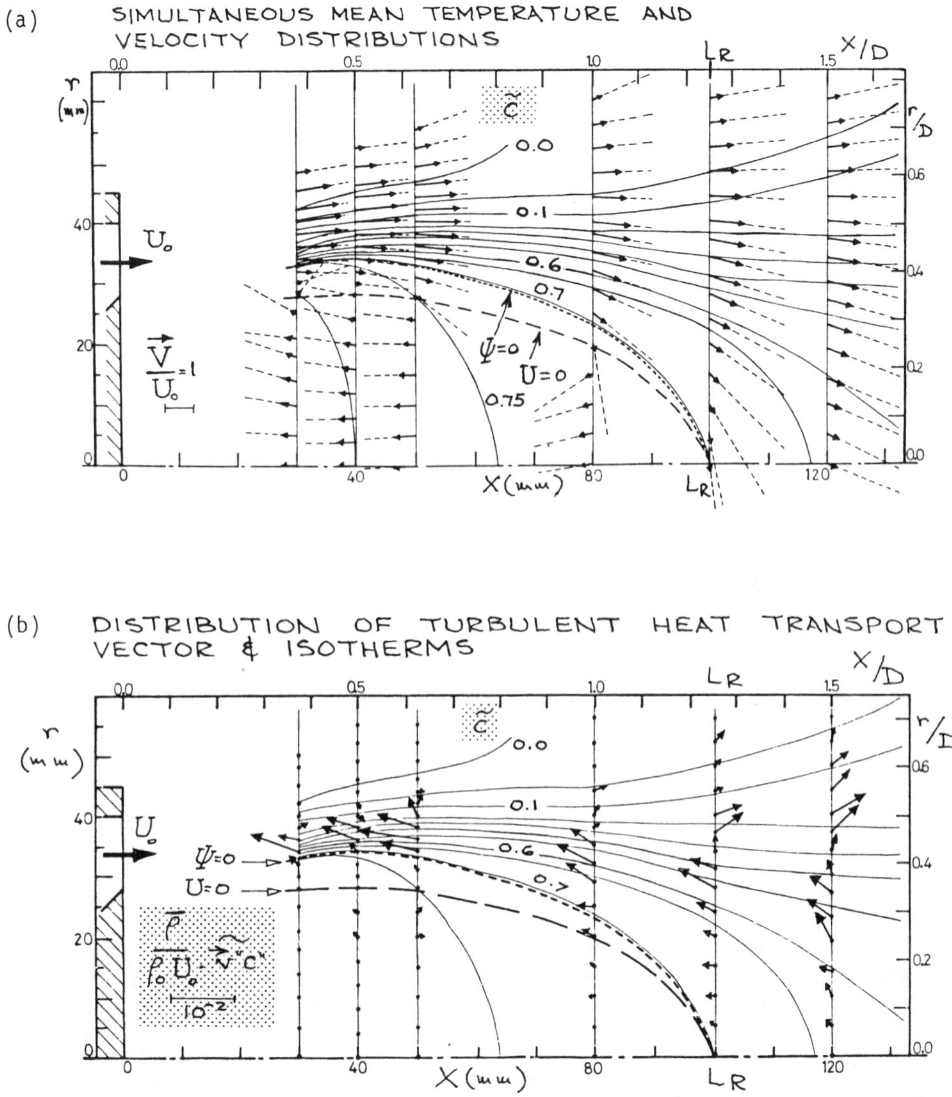

Figure 1. Simultaneous velocity and temperature characteristics of the disc-stabilized, turbulent, premixed flame of Heitor et al (1987). Vectors of (a) mean velocity and (b) turbulent heat transfer rate superposed in isotherms and separation steamline, locus = 0. The scale for the magnitude of the vectors is given in the lower left-hand corner of each figure.

As the flow develops downstream, and particularly for X/D>1 the sign of $\widetilde{u''c''}$ reverse from negative (along the reaction zone, where $\widetilde{u''v''}<0$) to positive (in the outer flame, where $\widetilde{u''v''}>0$) about the location where the intensity of temperature fluctuations is maximum (i.e. around $\widetilde{u''v''} = 0$). In addition, the direction of the axial flux is opposite to that of the respective mean scalar gradients, except around the zone where $\widetilde{u''c''} \simeq 0$. Similar behaviour has been observed in other turbulent premixed flames (e.g. Moss, 1980; Shepherd and Moss, 1981; Shepherd, Moss and Bray, 1982; and Tanaka and Yanagi, 1983) and has also been predicted analytically (e.g. Bray et al, 1981 and Masuya and Libby, 1981). The resulting turbulent heat flux vectors do not intercept the isotherms normally over the whole length of the measurements, but indicate, instead, a strong diffusion flux along the flame in both the reaction zone for counter-flowing heat flux and the outer flame for co-flowing heat flux, in a similar way to that reported by Tanaka and Yanagi (1983). In general, the results show that the net turbulent scalar flux occurs along directions in which the mean scalar gradient is not large and, therefore, calculations using conventional gradient transport hypothesis would be inaccurate for any system of co-ordinates. If the present results are transformed into streamline co-ordinates (see Heitor, 1985 for details) the net turbulent scalar transport occurs mainly along the streamlines as a consequence of the highly oblique flame.
Provided that chemical reaction occurs in "thin flames" (e.g. Libby, 1985), then the turbulent heat flux can be represented as

$$\widetilde{u''c''} = \widetilde{C}\,(1-\widetilde{C})\,(\overline{U}_{ip} - \overline{U}_{ir})$$

This means that the sign, and to a lesser extent the magnitude (because the above formula is an approximation) of the heat flux in the flames studied is given by the difference in the product and reactant mean velocities.

For these purposes, velocity measurements conditioned by temperature were obtained along several radial and axial profiles and figure 2 shows sample results in streamline co-ordinates (therefore, $\widetilde{U}_s=0$) along the line r/D=0.46, which is within the inside edge of the annular jet, close to the radius of maximum axial velocity. The results have shown that in the plane of maximum width of the recirculation bubble (X/D=0.63) the mean axial velocity of the products is smaller than that of the reactants, with mean differences around 0.15 U_o (i.e. 2 m/s; U_o is the annular bulk velocity) and maxima about 0.27 U_o = 4.8 m/s. However, downstream of X/D = 1.25 the mean velocity profiles are flatter than those along the extent of the recirculation zone and products are observed in outer flame zones (i.e., for r/D>0.44) with axial velocities higher than those of the reactants, although the differences do not exceeded 1 m/s (i.e. 0.05 U_o). The products are deflected outwards relative to the reactants by up to 1.7° in the plane of maximum width of the recirculation bubble, but by up to 5° in upstream locations (X = 40 mm = 0.50 D) and by 3.5° far downstream, at X/D = 2.50. The behaviour, although different from that previously reported in the flames of Shepherd and Moss (1983), Shepherd, Moss and Bray (1982), Cheng, Talbot and Robben (1984) and Cheng (1983), suggests that the relative magnitude of the axial mean pressure gradient across the shear layer is important

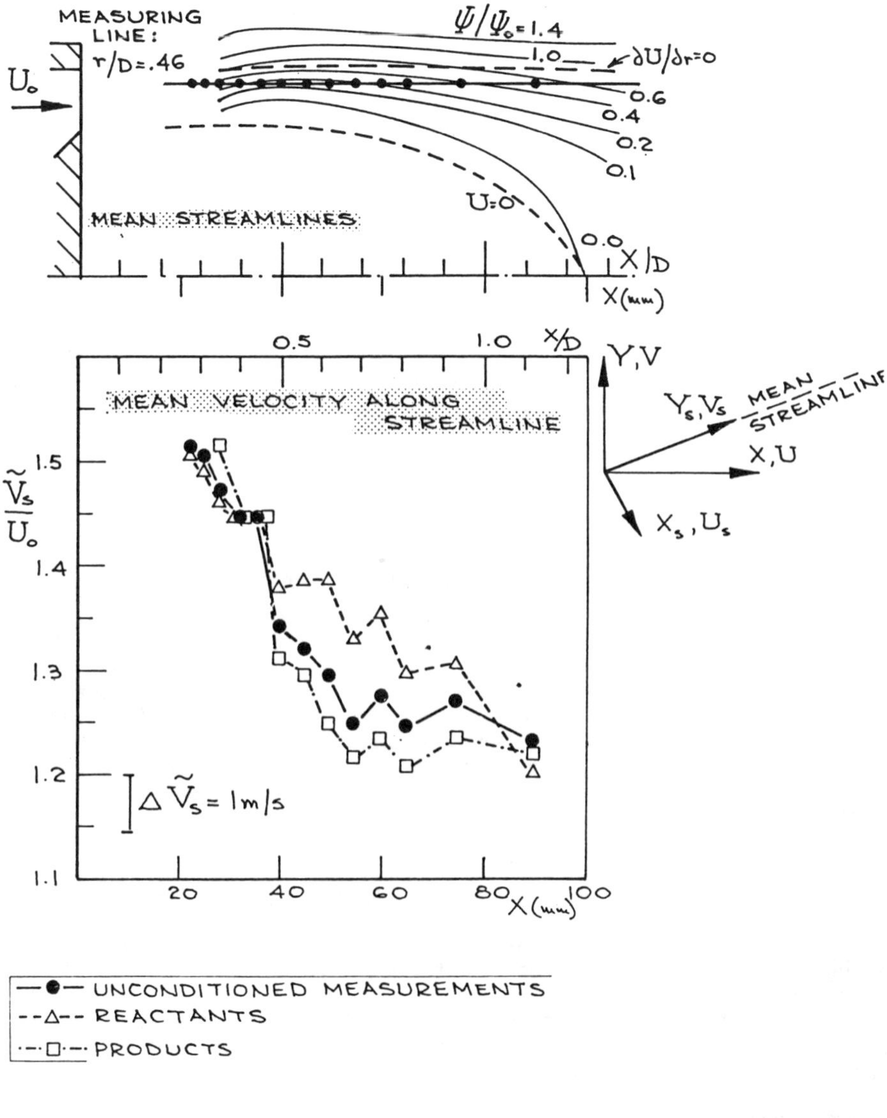

Figure 2. Unconditioned and conditioned mean velocity characteristics in streamline coordinates along the axial profile at r/D=0.46 in the disc-stabilized flame of Heitor et al (1987). The conditioning levels are considered in such way that products are for C>0.7 and reactants for C<0.15.

in the axial-momentum balance. In the present recirculating flow the low-density hot products exist mainly in zones dominated by adverse pressure gradients and, consequently, are preferentially decelerated relative to high-density cold reactants. However, the effect of the radial positive mean pressure gradient does not prevail in the momentum balance, since it has not preferentially deflected the light products towards the centre of the flame.

Further analysis based on examination of the terms in the momentum equations is provided by Heitor et al (1987), but it is still important to note that the differences in the conditioned mean velocities measured here are higher than those in the V-shaped flames of Cheng (1983) (i.e. up to 1.5 m/s = 0.27 U_o) and in the bunsen-like flames of Shepherd and Moss (1981) (up to 0.35 m/s = 0.13 U_o) and are similar to those in the downstream flame zones of the confined flow of Shepherd, Moss and Bray (1982) (up to 5.2 m/s = 0.28 U_o). Although comparisons are difficult to establish, the numbers are consistent with the expected magnitude of the mean pressure gradient in each flow.

Based on the preceding discussion of the axial and radial conditioned velocities, the reasons for the large non-gradient heat flux in the axial direction $\widetilde{u''c''}$, and the much smaller gradient transport in the radial direction $\widetilde{v''c''}$, can be explained. The direction of the axial flux is towards the baffle for r/D<0.5 because there $\overline{U}_R > \overline{U}_P$ for reasons stated above. For r/D > 0.5 (and for x/D > 1.25) the direction changes because there $\overline{U}_R < \overline{U}_P$, but the magnitude of the axial component is larger than that in the radial direction in both the zones because $|\overline{U}_P - \overline{U}_R|>|\overline{V}_P - \overline{V}_R|$. The radial flux is in the gradient sense only by virtue of the fact that $\overline{V}_P > \overline{V}_R$ everywhere. There does not appear to be any fundamental aerodynamic restriction that prevents $\overline{V}_R$ becoming larger than $\overline{V}_P$ except that, in that case, $\widetilde{v''c''}$ would represent counter-gradient heat transfer into the recirculation zone and it is not clear how a stable flame could exist. The conditions under which $\overline{V}_R$ would become larger than $\overline{V}_P$ are, presumably, values of equivalence ratio near unity when the density difference between products and reactants is largest, giving rise, therefore, to the maximum preferential acceleration between the two states.

The implications of these results for the calculation of turbulent premixed flames is that the magnitude and direction of heat flux cannot be represented by gradient transport models. It is necessary to solve the appropriate transport equations and thus represent the pressure gradient as an important source term.

4. CONCLUSIONS

Improved knowledge of turbulent combustion and the associated development of current theories and modelling techniques depends upon the efficient use and combination of advanced diagnostic techniques to allow the measurement of joint probability density distributions of vector and scalar properties, their correlations and conditionally-sampled values. This paper reviews the techniques which have been used to measure these quantities in premixed and non-premixed flames and is intended to help on the formulation and assessment of the choices of instrumentation for future works. In particular, the major interest in non-premixed flames is on measurements of mixture fraction and density and their correlations with velocity components, while in premixed flames attention is on reaction progress variable and its correlation with velocity.

In general, non-intrusive optical techniques are desirable to eliminate probe interference effects and, in some circumstances, may improve the tolerance in the measurement of scalar fluctuactions and can provide additional information on time and length scales of the scalar field, which are required because of their importance to some models of turbulent combustion. However, the combination of these techniques with laser velocimetry can be affected by the particles used as light scatterers and subsequent joint statistical analysis further limited by the random characteristics of the Doppler signal. Digitally compensated thermocouples may then be prefered to be combined with laser velocimetry and this paper reports a successful combination of these techniques in a laboratory turbulent flame with premixed reactants.

It is shown that in a disc-stabilized premixed flame the moments of velocity are made up principally of the separate contributions from the reactants and products. The existence of heat transfer in the flame thus depends on the difference between the mean velocities of the reactant and product gases, which is, as expected, large in the shear layer where the temperature fluctuations are large. The direction of the flux is not, however, aligned with the direction of the gradient of mean temperature because burned and unburned gases are differentially accelerated by the mean pressure gradient.

ACKNOWLEDGEMENTS

The author is pleased to acknowledge many useful discussions with colleagues in the Fluids Section of the Imperial College and, in particular, the unvaluable assistance of Dr. A.M.K. P. Taylor and of Prof. J.H. Whitelaw.

REFERENCES

R.A. ANTONIA, A.J. CHAMBERS, M. SOKOLOV and C.W. VAN ATTA (1981). On the similarity between velocity and temperature fields within a turbulent spot. Proc. 3rd Symp. on Turb. Shear Flows, Univ. California, Davis, pp. 10.1-10.6

A. BALLANTYNE, D.J. BOON and J.B. MOSS (1976). Measurements of fluctuating temperature an open diffusion flames employing fine wire thermocouples. A.A.S.U. Memorandum 76/3.

A. BALLANTYNE and J.B. MOSS (1977). Fine wire thermocouple measurements of fluctuating temperature. Comb. Sci. Tech. 17, pp. 63-72.

H.A. BECKER (1977). Mixing, concentration fluctuations and nephelometry. In "Studies in Convection" (B.E. Launder, Ed.) Vol. 2, pp. 45-139.

R.W. BILGER(1977) . probe measurements in turbulent combustion. Experimental diagnostics in gas phase combustion systems. Progress in Aeronautics and Astronautics, Vol. 53 (Eds. B.T. Zinn, et al), pp. 49-69.

R.W. BILGER(1986). The use of advanced laser diagnostics for the study of turbulent reacting flows. 9th Australian Fluid Mech. Conference. Auckland, December 8-12.

C. BORREGO, and D. OLIVARI (1980). A method for the measurement of mixing properties in a flow. AGARD-CP-281, paper 23.

K.N.C. BRAY, P.A. LIBBY, G. MASUYA and J.B. MOSS (1981). Turbulence production in premixed turbulent flames. Comb. Sci. Tech., 25, pp. 127-140.

K.N.C. BRAY, P.A. LIBBY and J.B. MOSS (1984-a). Flamelet crossing frequencies and mean reaction rates in premixed turbulent combustion. Comb. Sci. and Tech.

K.N.C. BRAY, P.A. LIBBY and J.B. MOSS (1984-b). Scalar length scale variations in premixed turbulent flames. Presented at the 20th Symposium (Intl.) on Combustion.

K.N.C. BRAY, P.A. LIBBY and J.B. MOSS (1985). Unified modelling approach for premixed turbulent combustion - PART I: General formulation. Comb. and Flame, 61, pp. 87-102.

P. CAMBRAY, M. VACHON, T. MACIASZEK and J.C. BELLET (1985). Instantaneous compensation of fine wire thermocouple inertia in a buoyant turbulent diffusion flame. ASME-HTD, 45, pp.293-298.

S.B.S. CHANDRAN, N.M. KOMERATH, W.M. GRISSOM, W.C. STRAHLE and J.I. JAGODA (1984a). Time resolved thermometry by simultaneous thermocouple and Rayleigh scattering measurements in a turbulent flame. Comb. Sci. and Tech.

S.B.S. CHANDRAN, N.M. KOMERATH and W.C. STRAHLE (1984). Scalar-velocity correlations in a turbulent premixed flame. 20th Symposium (Intl.) on Combustion

M.S. CHANDRASEKHARA, and B.R. RAMAPRIAN (1983). Measurements in two-dimensional jets and plumes. Proc. 4th Symp. on Turb. Shear Flows, Karlsruhe, pp. 7.7-7.12.

Y. CHAUVEAU, P. CAMBRAY, E. GENGEMBRE, M. CHAMPION and J.C. BELLET (1981). Characterisation of a turbulent combustion zone in a tubular reactor. Presented at the 8th Intl. Colloquium on Gas Dynamics of Explosions and Reactive Systems. Minsk, Aug/1981.

R.K. CHENG(1983). Conditional sampling of turbulence intensities and Reynolds stresses in a premixed turbulent flame. Lawrence Berkeley Lab. Report LBL-16938. Comb. Sci. and Tech.

R.K.CHENG, L. TALBOT and F. ROBBEN (1984). Conditional velocity statistics in premixed CH_4-air, C_2H_4-air turbulent flames. Presented at the 20th Symposium (Intl.) on Combustion.

R. CHEVRAY, and N.K. TUTU (1978). Intermittency and preferential transport of heat in a round jet. J.F.M., 88, pp. 133-160.

G. COMTE-BELLOT (1976). Hot-wire anemometry. Annual review of fluid mechanics, Annual Review Inc.,Palo Alto, 1976, pp.209-231.

T. DAKOS, C.A. VERRIOPOULOS and M.M. GIBSON (1984). Turbulent flow with heat transfer in plane and curved wall jets. J.F.M., 145, pp. 339-360.

I. DEKEYSER, and B.E. LAUNDER (1983). A comparison of triple-moment temperature-velocity correlations in the asysmmetric heated jet with alternative closure models. Proc. 4th Symp. on Turb. Shear Flows, Karlsruhe, pp. 14.1-14.8.

R.W. DIBBLE, V. HARTMAN. R.W. SCHEFER and W. KOLLMANN (1984). Conditional sampling of velocity and scalars in turbulent flames using simultaneous LDV-Raman scattering. Sandia Report DAND84-8860. Also, ASME 105th Winter Annual Meeting, Dec. 9-13, New Orleans, Louisianna.

R.W. DIBBLE, and R.E. HOLLENBACH (1981). Laser- Rayleigh thermometry in turbulent flames. 18th Symp. (Intl.) on Combustion, pp. 1489-1499.

R.W. DIBBLE, W. KOLLMAN and R.W. SCHEFER (1984). Conserved scalar fluxes measured in a turbulent non-premixed flame by combined laser-Doppler velocimetry and laser Raman scattering. Comb. and Flame, 55, pp. 307-321.

R.W. DIBBLE, G.D. RAMBACH, R.E. HOLLENBACH and J.T. RINGLAN (1981). Simultaneous measurement of velocity and temperature in flames using LDV and CW laser Rayleigh thermometry. Proc. 7th Biennial Symposium on Turbulence, Rolla-Missouri, pp. 64-72.

R.W. DIBBLE, and R.W. SCHEFER (1983). Simultaneous measurement of velocity and scalars in a turbulent non-premixed flame by combined laser-Doppler velocimetry and laser Raman scattering. 4th Symp. on Turb. Shear Flows, pp. 10.13-10.18.

M.C. DRAKE, R.W. BILGER and S.H. STARNER (1982). Raman measurements and conserved scalar modelling in turbulent diffusion flames. 19th Sympmosium (Intl.) on Combustion, The Combustion Institute, 1982, pp.459-467.

M.C. DRAKE, R.W. PITZ and M. LAPP (1984). Laser measurements on non-premixed hydrogen-air flames for assessment of turbulent combustion models. Paper: AIAA-84-0544. AIAA 22nd Aerospace Science Meeting, Jan. 9-12, 1984/Reno, Nevada.

J.F. DRISCOLL, R.W. SCHEFER and R.W. DIBBLE (1982). Mass fluxes 'u' and 'v' measured in a turbulent non-premixed flame. 19th Symposium (Intl.) on Combustion, The Combustion Institute, pp. 477-485.

F. DURST, and F. SCHMITT (1984). Joint laser-Doppler-laser induced fluorescence measurements in a turbulent jet. 2nd Intl. Symp. on Appl. of Laser Anemometry to Fluid Mechanics. Lisbon, July-1984. Paper 8.5.

A.C. ECKBRETH (1980). CARS thermometry in practical combustors. Comb. and Flame, 39, pp. 133-147.

G. FABRIS(1979). Turbulent temperature and thermal flux characteristics in the wake of a cylinder. In Turbulent Shear Flows I (ed. F. Durst et al), pp. 55-70. Springer-Verlag.

R.L. FARROW, P.L. MATTERN and L.A. RAHN (1982). Comparison between CARS and corrected thermocouples temperature measurements in a diffusion flame. Applied Optics, 21, pp. 3119-3125.

S. FUJII, M. GOMI, K. EGUCHI, S. YAMAGUCHI and Y. JIN (1984). Time-resolved LDV and CARS measurements in a premixed reacting flow. Comb. Sci. and Tech., 36, pp. 211-226.

M.M. GIBSON and C. VERRIOPOULOS (1984). Turbulent boundary layer on a mildly curved convex surface. Part 2: Temperature field measurements. Exp. in Fluids, 2, pp. 73-80.

L.P. GOSS, D.D. TRUMP and W.M. ROQUEMORE (1984). An investigation of temperature and velocity correlations in tirbulent flames. Presented at ASME 105th Winter Annual Meeting, New Orleans, Dec.-1984.

D.A. GREENHALGH (1983). Gas phase temperature and concentration diagnostics with lasers. AERE R-10764. Also, paper C66/83; Inst. Mech. Eng. Conference - Oxford, 1983.

R. GUNTHER, and V. WITTMER (1981). The turbulent reaction field in a concentric diffusion flame. 18th Symposium (Intl.) on Combustion, The Combustion Institute, pp. 961-967.

A.K. GUPTA, D.G. LILLEY and N. SYRED (1984). Swirl flows. Abacus Press, Energy and Engineering Science Series.

M.V. HEITOR(1985). Experiments in turbulent reacting flows. Ph. D. Thesis, University of London.

M.V. HEITOR, A.M.K.P. TAYLOR and J.H. WHITELAW (1985). Simultaneous velocity and temperature measurements in a premixed flame. Exp. in Fluids, 3. Also, Proc. ASME 105th Annual Winter Meeting, New Orleans, Dec. 9-13 (1984).

M.V. HEITOR, A.M.K.P. TAYLOR and J.H. WHITELAW (1987). The interaction of turbulence and pressure gradients in a baffle-stabilised premixed flame. J. Fluid Mechanics, 181, pp. 387 - 413.
B.V. JOHNSON, and J.C. BENNETT (1983). Mass and momentum turbulent transport experiments with confined co-axial jets. Proc. of 4th Symposium on Turbulent Shear Flows, pp. 18.4-18.19 - Karlsruhe, 1983.
M. KATSUKI, Y. MIZUTANI and Y. MATSUMOTO (1987). An improved thermocouple technique for measurement of fluctuating temperatures in flames. Comb. and Flame, 67, pp. 27-36.
I.M. KENNEDY and J.H. KENT (1979). Measurements of a conserved scalar in turbulent jets diffusion flames. 17th Symposium (Intl.) on Combustion. The Combustion Institute, pp. 279-287.
I.M. KENNEDY and J.H. KENT (1981). Scalar measurements in a co-flowing turbulent diffusion flame. Comb. Sci. and Tech., 25, pp. 109-119.
M. LAPP(1980). Raman-Scattering measurements of combustion properties. In: Laser Probes for Combustion Chemistry (ed. D.P. Crosley) pp. 207-229, Amer. Chem. Soc.
J.C. LARUE, G.S. SAMUELSEN and E.T. SEILER (1984). Momentum and heat-fluw in a swirl-stabilised combustor. 20th Symposium (Intl.) on Combustion. The Combustion Institute.
J.C. LaRUE, and P.A. LIBBY (1977). Measurements in a turbulent boundary layer with slot injection of helium. Phys. Fluids, 20, pp. 192-202.
J.C. LaRUE, P.A. LIBBY and D.V.R. SESHADRI (1981). Further results on a thermel mixing layer downstream of a turbulence grid. Proc. 3rd Symp. on Turb. Shear Flows, Univ. California - Davis, pp. 15.1-19.6.
J.C. LaRUE, G.S. SAMUELSEN and E.T. SEILER (1984). Momentum and heat-flux in a swirl-stabilised combustor. 20th Symposium (Intl.) on Combustion. The Combustion Institute.
W. LENZ and R. GUENTHER (1980). measurements of fluctuating temperature in a free-jet diffusion flame. Comb. and Flame, 37, pp.63-70.
P.A. LIBBY(1985). Theory of normal premixed turbulent flames revisited Prog. Energy Combust. Sci., 11, pp. 83-96.
P.A. LIBBY and K.N.C. BRAY (1981). Counter-gradient diffusion in premixed turbulent flames. AIAA J., 19, pp. 205-213.
P.A. LIBBY, N. CHIGIER and J.E. LaRUE (1982). Conditional sampling in turbulent combustion. Prog. Energy Combust. Sci., 8, pp. 203-221.
F.C. LOCKWOOD and H. MONEIB (1980). Fluctuating temperature measurements in a heated round free jet. Comb. Sci. Tech., 22, pp. 63-81.
F.C. LOCKWOOD and MONEIB (1981). A new on-line pulsing technique for response measurements of thermocouple wires. Comb. Sci.Tech., 26, pp. 177-181.
G. MASUYA and P.A. LIBBY (1981). Non-gradient theory for oblique turbulent flames with premixed reactants. AIAA J., 19,pp. 1590-1599.
A. MELLING(1975). A technique for simultaneous velocity and concentration measurements. Proc. LDA Symp., Copenhagen, pp. 505-509.
J.B. MOSS(1980). Simultaneous measurements of concentration and velocity in an open turbulent flame. Comb. Sci. and Tech., 22, pp. 119-129.
J.B. MOSS(1981). Light scatter turbulence measurements in thin-flame premixed combustion. Paper AIAA-84-1198.
I. NAMER and R.W. SCHEFER (1985). Error estimates for Rayleigh scattering density and temperature measurements in premixed flames. Exp. in Fluids, 3, pp. 1-9.
N.N.R NINA and G. P. PITA (1985). Measurements of fluctuating gas temperatures usingcompensated fine wire thermocouples. AGARD: 65th PEP Symp., pp. 32.1-32.10; Bergen, Norway - May 1985.

F.K. OWEN(1977). Measurements of instantaneous velocity and concentration in complex mixing flows. AIAA Progress in Astronautics and Aeronautics, Vol. 53, (ed. B.T. Zinn et al), pp. 293-303.
P. PARANTHOEN, J.C. LECORDIER, C. PETIT and P. GAJAN (1983). Survey and recent developments of frequency response studies of cold wires and fine wire thermocouples in turbulent heated flow. 8th Australasian Fluid Mech. Conf., Univ. newcastle, N.S.W., pp. 6B.1-6B.4.
S.P. PENNER, C. P WANG and M.Y. BAHADORI (1984). Laser diagnostics applied to combustion systems. presented at the 20th Symposium (Intl.) on Combustion, invited lecture.
W.M. PITTS, B.J. McCAFFREY and T. KASHIWAGI (1983). A new diagnostic technique for simultaneous, time-resolved measurements of concentration and velocity in simple turbulent flow systems. Proc. 4th Symp. on Turb. Shear Flows, Karlsruhe, pp. 15.22-15.27.
S. RAJAN, J.R. SMITH and G.D. RAMBACH (1984). Internal structure of a turbulent premixed flame using Rayleigh scattering. Comb. and Flame, 57, pp. 95-107.
D.L. REUSS (1983). Temperature measurements in a radially symmetric flame using holographic interferometry. Comb. and Flame, 49, pp. 207-219.
W.M. ROQUEMORE and P.P. YANEY (1979). Comparison of thermocouple, gas sampling and Raman measured temperatures in an afterburning turbojet engine plume. Proc. 10th Materials Res. Symp. on Characterisation of High Temperature Vapors and Gases, pp. 973-1025, NBS, Maryland, Sept. 1978.
R.W. SCHEFER and R.W. DIBBLE (1985). Simultaneous measurements of velocity and density in a turbulent non-premixed flame. AIAA J., 23, pp. 1070-1078.
E.T. SEILER(1984). Compensated fine-wire thermocouple measurements in a complex reacting flow. M.Sc. Thesis, Univ. California, Irvine.
I.G. SHEPHERD and J.B. MOSS (1981). Measurements of conditioned velocities in a turbulent premixed flame. AIAA J., 20, pp. 566-569.
I.G. SHEPHERD and J.B. MOSS (1983). Characteristic scales for density fluctuations in a turbulent premixed flame. Comb. Sci. and Tech., 33, pp. 231-243.
I.G. SHEPHERD, J.B. MOSS and K.N.C. BRAY (1982). Turbulent transport in a confined premixed flame. 19th Symposium (Intl.) on Combustion. The Combustion Institute, pp. 423-431.
S. SIVASEGARAM and J.H. WHITELAW (1983). Temperature characteristics of turbulent, premixed flames stabilised on a step. Expts. in Fluids, 1, pp. 161-165.
K.R. SREENIVASAN, S. TAVOULARIS and S. CORRSIN (1981). Turbulent transport in passively heated homogeneous and inhomogeneous flows. Proc. 3rd Symp. on Turb. Shear Flows, Univ. California-Davis, pp. 15.28-15.33.
R.A. STANFORD and P.A. LIBBY (1974). Further applications of hot-wire anemometry to turbulence measurements in helium-air mixtures. Phys. Fluids,17, pp. 1353-1361.
S.H. STARNER(1983). Joint measurements of radial velocity and scalars in a turbulent diffusion flame. Comb. Sci. and Tech., 30, pp. 145-169.
S.H. STARNER(1985). Conditional sampling in a turbulent diffusion flame. Comb. Sci. and Tech., 42, pp. 283-299.
S.H. STARNER, and R.W. BILGER (1980). LDA measurements in a turbulent diffusion flame with axial pressure gradient. Comb. Sci. and Tech., 21, pp. 259-276.
S.H. STARNER, and R.W. BILGER (1981a). Measurement of scalar-velocity correlations in a turbulent diffusion flame. 18th Symposium (Intl.) on Combustion, The Combustion Institute, pp. 921-930.

S.H. STARNER and R.W. BILGER (1981b). Differential diffusion effects on measurements in turbulent diffusion flames by the Mie scattering technique. Proc. 8th ICOGER, Minsk, USSR - Aug., 1981, pp. 81-104.
S.H. STARNER and R.W. BILGER (1987). Some budgets of turbulent stresses in round jets and diffusion flames. Combust. Sci. and Tech., 53, pp. 377-398.
W.C. STRAHLE and M. MUTHUKRISHNAN (1976). Thermocouple time constant measurement by cross power spectra. AIAA J., 14, pp. 1642-1644.
T. TAKAGI, T. OKAMOTO and A. OHTANI (1983). Simultaneous measurement of velocity and concentration in two-component gas flows by LDV and Laser Rayleigh scattering. Proc. ASME-JSME Joint Conf., 1983, pp. 317-321.
H. TANAKA, and T. YANAGI (1983). Cross-correlation of velocity and temperature in a premixed turbulent flame. Comb. and Flame, 51, pp. 183-191.
J.P. VALLET, J.P. MAYE and J. NERAULT (1981). Measurements of velocity-temperature correlations of different orders in hot turbulent flows. Proc. 7th Biennial Symp. on Turbulence, Rolls-Missouri, pp. 165-171.
C.A. VERRIOPOULOS (1983). Effects of convex surface curvature on heat transfer in turbulent flow. Ph. D. Thesis, University of London.
S. WARSHAW, M. LAPP, C.M. PENNEY and M.C. DRAKE (1980). Temperature-velocity correlation measurements for turbulent diffusion flames from vibrational Raman-scattering data. In: "Laser Probes for Combustion Chemistry" paper 19, pp. 239-246, (ed. D.R. Crosley), Amer. Chemical Soc.
J. WAY and P.A. LIBBY (1970). Hot-wire probes for measuring velocity and concentration in He-Air mixtures. AIAA J., 8, pp. 976-980.
J. WAY and P.A. LIBBY (1971). Applications of hot-wire anemometry and digital techniques to measurements in a turbulent helium jet. AIAA J., 9, pp. 1567-1573.
F.A. WILLIAMS (1984). Some aspects of the interaction between turbulent flows and combustion processes. Eastern States Section, The Combustion Institute, Dec., 1984.
V. WITTMER and R. GUNTHER (1980). Correlation measurement of velocity and temperature fluctuations in a free jet diffusion flame. AGARD CP 281, paper 10.
T. YANAGI and Y. MIMURA (1981). Velocity-temperature correlation in a premixed flame. 18th Symposium (Intl.) on Combustion. The Combustion Institute, pp. 1031-1039.
A. YOSHIDA and R. GUENTHER (1980). Experimental investigation of thermal structure of turbulent premixed flames. Comb. and Flame, 38, pp. 249-258.
A. YOSHIDA and H. TSUJI (1979). Measurements of fluctuating temperature and velocity in a turbulent premixed flame. 17th Symposium (Intl.) on combustion. The Combustion Institute, pp. 945-956.
A.J. YULE, D.S. TAYLOR and N.A. CHIGIER (1978). On-line compensation and processing of thermocouple signals for temperature measurement in turbulent flames. AIAA paper 78-30.

MEASUREMENT OF SCALAR FLUX IN TURBULENT FLOWS

TABLE 1 - TURBULENT NON-REACTING FLOWS

A: Combination of probes techniques.

REFERENCE	EXPERIMENT	TECHNIQUE	MEAS.
Way & Libby (1970,1971)	Helium jet mixing into air	Hot-film and hot-wire at 90°	U,C_{He}
Stanford & Libby (1974)	Porous pipe with Helium injection	Two hot-films and hot wire at 60°	U,V,C_{He}
LaRue & Libby (1977)	Boundary layer with slot injection of helium	Two hot-films and hot wire at 45°	U,V,C_{He}
Chevray & Tutu (1978)	Round heated air jet	Two hot-wires; Cross-wire and hot-wire at 90°	U,V,T
Fabris (1979)	Far wake of heated cylinder	Four-wire probe: cross-wire, yawed sensor and temperature-sensor	U,V,W,T
Vallet et al (1981)	Forced convection between two hot plates	Two-parallel-hot-wire probe at a range of rotational positions	U_i,T
Antonia et al (1981)	Turbulent spot in heated laminar boundary layer	Cross-wire and temperature sensor; hot-wire and cold-wire (parallel)	V,T
LaRue et al (1981)	Thermal mixing layer downstream of a heated turbulence grid	Cross-wire and cold-wire at 90°	U,V,T
Sreenivasan et al (1981)	Three homogeneous and one inhomogeneous heated shear flows	Cross-wire and temperature sensor	U_i,T
Dekeyser & Launder (1983)	Heated asymmetric two-dimensional turbulent jet	Cross-wire and cold-wire at 90°	U_i,T
Dakos et al (1984)	Heated wall jet on a convex wall	Cross-wire and cold-wire; parallel twin-wire probe	U_i,T

B: Combination of probe with optical techniques.

REFERENCE	EXPERIMENT	TECHNIQUE	MEAS.
Chandrasekhara & Ramaprian (1983)	Plane asymptotic non-buoyant jets and buoyant plumes	Two-channel LDV and resistance thermometer	U_i,T
Pitts et al (1983)	Axisymmetric air jet in a secondary gas (CH_4 or C_3H_8) co-flowing	Laser Rayleigh scattering and hot wire or hot-film	U_i,C_g

C: Combination of optical techniques.

REFERENCE	EXPERIMENT	TECHNIQUE	MEAS.
Melling (1975)	Axisymmetric free air jet (seeded with oil smoke)	LDV and Marker Nephelometry (Mie scattering)	U,f
Borrego & Olivari (1980)	Axisymmetric free air jet (seeded with oil smoke); Axisymmetric gas-smoke jet in co-flowing stream	LDV and Marker Nephelometry	U,f
Owen (1977-b)	Mixing zone downstream confined co-axial liquid jets (with fluorescent dye)	LDV and LIF (laser induced fluorescence)	V,C_d
Johnson & Bennett (1983)	Mixing zone dowstream confined coaxial liquid jets (with fluorescent dye)	LDV and LIF	U_i,C_d
Durst & Schmitt (1984)	Round liquid jet (with fluor. dye)	LDV and LIF	U,C_d
Takazi et al (1984)	Axisymmetric jet fo CO_2 surrounded by uniform stream of N_2	LDV and laser Rayleigh scattering	U,C_{CO_2}

MEASUREMENT OF SCALAR FLUX IN TURBULENT FLOWS

TABLE 2 - TURBULENT REACTING FLOWS

A: Combination of LDV with a thermocouple probe (continued).

REFERENCE	MEASUREMENTS	TECHNIQUES	COMMENTS
Wittmer & Gunther (1980) Gunther & Wittmer (1981)	Free jet diffusion flame: $D = 8$ mm; $Re = 24\times10^3$; 37×10^3. $\overline{U}, \overline{u'^2}, \overline{v'^2}, \overline{u'v'}$ $\overline{T}, \overline{t'^2}$ R_{UT}	LDV: 0.15x2.0 (mm); M_g0; Frequency tracker. Thermocouple: Pt/PT:10% Rd; $d_W = 50$ μm; electrically compensated; uncoated.	Error estimated for R_{UT} = 10% Results do not agree with gradient transport: $R_{UT} > 0$ for $\frac{\partial \overline{T}}{\partial x} > 0$. Not explained. Maximum correlations outside reaction zone: $Re = 37000$ - $R_{UT} = 0.25$; $\theta = 64\%$.
Yanagi & Mimura (1981) Tanaka & Yanagi (1983)	Open burner premixed flame: $D = 8$, 10 mm; $Re = 3.2\times10^3$ CH_4/Air: $\phi = 0.6$. $\overline{U}, \overline{V}, \overline{u'^2}, \overline{v'^2}$ $\overline{T}, \overline{t'^2}$ R_{UT}, R_{VT}	LDV: 0.2x2.0 (mm); M_g0 - $d_N < 3$ μm; Frequency tracker. Thermocouples: Pt/Pt:13% Rd; $d_W = 50$ μm ($\tau \simeq 25$ ms); electrically compensated; uncoated. ($f_S < 5$ kHz).	Thermocouple may be placed separately, just downstream of c.v. of LDA. Deposition of M_g0 particles on thermcouple surface affects mainly $\overline{u't'}$ and θ, but not R_{UT}. Thermocouple was cleaned every 3 meas. Error estimated for R_{UT}, R_{VT}: 10%. Maximum values measured: Reaction zone - $R_{UT} = -0.2$; $R_{VT} = 0.4$; $\theta = 30\%$ Outer zone - $R_{UT} = 0.7$; $R_{VT} = 0.6$; $\theta < 7\%$. Results suggest "countergradient transport of heat" in reaction and mixing zones.

TABLE 2

A: Combination of LDV with a thermocouple probe

REFERENCE	MEASUREMENTS	TECHNIQUES	COMMENTS
LaRue et al (1984) Seiler (1984) (also table 1.8.1)	Swirl-stabilised combustor: D = 80 mm; U_{ref} = 15 m/s. C_3H_8 $\overline{U},\overline{W},\overline{u'^2},\overline{w'^2}$ $\overline{T},\overline{t'^2}$ R_{UT},R_{WT} PDF(T)	LDV: two-channel; d = 0.36 mm; V = 0.022 mm^3 at 20°; Alumina - d_N = 1 μm; Counter processors. Thermocouple: Pt/Pt: 10% Rd; d_W {25 μm; τ = 5-25 ms; 50 μm; τ = 8-30 ms; electrically compensated; uncoated.	$\overline{T}$ and $\overline{t'^2}$ insensitive to time constant (τ), but flux varies up to 50% for 10% variation in τ. Probe perturbation leads up to 25% errors. The measurements show: Near-normal PDF(T) with $\sqrt{\overline{t'^2}}$ max = 250^oC; maximum correlations about 0.3; and that the axial heat flux downstream of the recirculation zone is in the non-gradient sense.
Chandran et al (1984-b)	Open burner premixed flame: D = 25 mm; Re = $10x10^3$ C_3H_8/Air: ϕ = 0.8 $\overline{U},\overline{V},\overline{u'^2},\overline{v'^2}$ $\overline{T},\overline{t'^2}$ R_{UT},R_{UP}	LDV: two-channel; 0.1x0.1 (mm); Al_2O_3, d_N = 1 μm; two counter processors. Thermocouple: Pt/Pt:10% Rd; d_W = 25 μm, coated. Microphone pitot probe: OD = 1.6 mm.	Time resolved temperature measurement validated by simultaneous Rayleigh scattering measurement (Chandran et al, 1984-a) Probe interference effects lead to separation between measuring points up to 3 mm. In general, errors are not quantified. "Counter-gradient" diffusion of heat is observed at the end of the cold inner cone, where maximum fluctuations were found. Pressure-velocity correlation is weak in this zone.

TABLE 2

A: Combination of LDV with a thermocouple probe (concluded).

REFERENCE	MEASUREMENTS	TECHNIQUES	COMMENTS
Heitor (1985) Heitor et al (1986) Heitor et al (1987)	Bluff- body stabilised turbulent premixed flame Unconfined disc - d = 40,56mm BR = 25; 50% Re = 43.4×10^3 to 46.8×10^3 U_0 = 11 to 18 m/s T_0 = 300 k; P = 1 atm Methane/air: Φ = 0.6 to 0.8 $\tilde{U}$, $\tilde{V}$, $\tilde{W}$, $\widetilde{u''^2}$, $\widetilde{v''^2}$, $\widetilde{w''^2}$, $\widetilde{u''w''}$, $\widetilde{u''v''}$ $\tilde{T}$, $\bar{T}$, $\overline{t'^2}$, $\widetilde{t''^2}$ $\widetilde{u''t''}$, $\widetilde{v''t''}$, $\widetilde{w''t''}$ Conditioned velocities	LDV: 0.26 X 5.19mm Al_2O_3 - $d_N < 1\,\mu m$ Frequency Counter Thermocouple: dw = 15; 40 µm digitally compensated $\tau = \tau$ (U,T)	Dependence of time constant on velocity and temperature taken into account. Seeding density up to 60 part./sec: thermocouples can be used for 15 minutes. Separation of the LDV control volume from the thermocouple up to 1.5 mm does not affect conclusions. The results are close to density-weighted averages. The interaction between gradients of mean pressure and the large density fluctuations which occur in the flame represents the largest source term in the transport equations for the turbulent heat fluxes. Non-gradient diffusion was found.

TABLE 2

B: Combination of LDV with marker nephelometry (Mie scattering) - (continued).

REFERENCE	MEASUREMENTS	TECHNIQUES	COMMENTS
Moss (1980) (also Bray et al, 1981)	Open burner premixed flame. $D = 50$ mm; $Re = 13\times10^3$ LPG/Air. $\overline{U},\overline{u'^2}$; conditioned values $\overline{I},\overline{i'^2}$ R_{UI}, R_{Uc} PDF(U), PDF(U,I)	LDV: 0.2x1.2 (mm); {Titanium dioxide - $d_N = 0.7$ {Alumina, $d_N = 1.4$ µm; Counter processor. Mie scatter light collected with separate optics.	Reduction in scattering properties of T_iO_2 on passing through a flame used to enhance bimodal nature of scalar pdf's, but the quality of LDV signal is reduced (errors not quantified). Turbulence levels reach a peak in the reaction zone. Since both R_{UC} and $\partial C/\partial x$ are positive, the turbulent flux appears in the "counter-gradient" sense. Mean and fluctuating velocities for products are higher than those for reactants.
Shepherd & Moss (1981)	Open burner premixed flame. $D = 50$ mm; $Re = 9300$; $U_o = 2.8$ m/s C_3H_8/Air - $\Phi = 1.6$ $\overline{U},\overline{u'^2}$; conditioned values $\tilde{C}$ R_{Ue} PDF(U)	LDV: 0.2x1.2 (mm); {Alumina, $d_N \approx 1$ µm; {Titanium dioxide, $d_N \approx 1$ µm; Frequency tracker. Mie scatter light collected with same PM: common focal point ($\Delta x = 0$).	T_iO_2 adopted as seed material to improve discrimination of scalar states. Tracker allowed heavy seeding levels (errors not quantified). Results showed "countergradient diffusion" and an increase in the turbulence levels through the flame. Conditioned measurements revealed: $\overline{U}_P-\overline{U}_R = 0.25$ m/s; $\sqrt{\overline{u_P'^2}} - \sqrt{\overline{u_R'^2}} = 0.1$ m/s.
Starner & Bilger (1981-a,b)	H_2 diffusion flame in co-flowing stream with pressure gradient. $D = 7.6$ mm; $Re = 11.2\times10^3$ $\overline{U},\overline{u'^2}$ } $\overline{\rho},\overline{\rho'^2},\overline{f},\overline{f'^2}$ } weighted and unweighted $\overline{\rho'u'}$, $\overline{f'u'}$ }	LDV: 0.2x1.4 (mm); Alumina, $d_N = 0.3$ µm; Counter processor. Light scatter: 1.0x1.2 (mm). Separation between laser probes: $\Delta x = 1.4$ mm ($f_S = 8$ kHz).	Seeding number density ≈ 100 - 200/mm^3. Shot noise in LDV signal less than 5%. Diffusive error on f range to 10% on the mean and 25% on variance and flux. Error in density are half of these. It was found that the term $\overline{u''}.\partial\overline{P}/\partial x$ is important in the balance of t.k.e.

TABLE 2

B: Combination of LDV with marker nephelometry (Mie scattering) - (concluded).

REFERENCE	MEASUREMENTS	TECHNIQUES	COMMENTS
Shepherd et al (1982) Shepherd & Moss (1983)	Rectangular duct: D_{eq} = 0.038 m L_D = 8.2 D_{eq} h = 12.5 mm; BR = 16.4% Re = 38 x 10^3 Re_h = 12 x 10^3 U_a = 15.5 m/s U_o = 18.5 m/s T_o = 300 K P_o = 1 atm Propane/Air; ϕ = 1 to 1.2	Velocity: LDV, 0.2 x 1.2 (mm); Titanium dioxide particles (d_N = 0.2 μm); frequency tracker; Density: Mie scattering, 0.2 x 1.2 (mm). Measurements with same optical collection arrangement.	Bimodal nature of scalar PDF's suggests thin flame turbulent burning. Turbulent flux is in the "counter-gradient" sense and is accompanied by turbulence generation. Time duration for the scalar states is log normally distributed and confirms convoluted nature of burnt-unburnt interface.

C: Combination of LDV with laser Rayleigh scattering.

REFERENCE	MEASUREMENTS	TECHNIQUES	COMMENTS
Dibble et al (1981)	Jet diffusion flame. (Hydrogen: 24% Methanol D = 4.6 mm; Re = $7x10^3$	LDV: 0.24x0.70 (mm); Alumina, d_N = 0.3 μm; Counter processor. cw-LRS in separate laser system. ($f_S \approx 1\ min^{-1}$).	Interference of Mie scattering limits the technique. A manual system is described but measurements are not presented.
Driscoll et al (1982)	Jet diffusion flame. (Hydrogen: 22% Argon) D = 5.3 mm; Re = $24x10^3$ $\overline{U}, \overline{u'^2}$ $\overline{\rho}, \overline{\rho'^2}$, PDF($\rho$) $\overline{\rho' u'}, \overline{\rho' v'}$	LDV: 0.5x2.0 (mm); Alumina, d_N = 0.3 μm; Counter processor. cw-LRS in separate laser system. Common focal point. Time delay between meas.: 60 μs (f_S < 5 Hz).	Data acquisition system is improved to avoid Mie scattered light. Errors not quantified. Gradient modelling agrees with radial correlation data, but is seldom consistent with axial data.

TABLE 2

D: Combination of LDV with laser Raman scattering.

REFERENCE	MEASUREMENTS	TECHNIQUES	COMMENTS
Warshaw et al (1980)	H_2 jet diffusion flame in co-flowing air. D = 3 mm; Re = $3x10^3$ $\tilde{U}$, $\tilde{T}$	LDV: Alumina, d_N = 1 μm; Titanium, d_N = 0.25 μm; Counter processor. Spatial overlap between laser probes about 20%. Time delay between measurements: $\Delta t \approx 10$ μs. ($f_S \approx 2$ Hz).	Weakness of Raman scattered light and use of pulsed laser limit the technique. Frequency discrimination is not enough to filter noise from Mie scatter light. A time delay (Δt) between the two measurements is required. Calibration is difficult to achieve.
Dibble & Schefer (1983)	Jet diffusion flame. (Hydrogen: 22% Argon) D = 5.3 mm; Re = $24x10^3$ $\overline{U},\overline{u'^2},\tilde{U}$ $\overline{T},\overline{t'^2},\overline{\rho},\overline{\rho'^2},\overline{u'\rho'}$	LDV: 0.5x2.0 (mm); Alumina, d_N = 0.3 μm; Counter processor. RAMAN: 1.0x1.0 (mm). Common focal point. $\Delta t \approx 40$ μs. ($f_S < 1$ Hz).	As time delay (Δt) increases, the correlation $\overline{\rho'u'}$ decreases and reaches zero for $\Delta t \approx 1$ s. Technique is limited by calibration procedure which is difficult to be extended to hydrocarbon fuels. Results reveal $\tilde{U}$ smaller than $\overline{U}$ about 10%.
Dibble et al (1984) Schefer & Dibble (1985)	Jet diffusion flame. (Hydrogen: 27% Argon) D = 5.3 mm; Re = $19x10^3$ $\overline{U},\overline{V},\overline{u'^2},\overline{v'^2},\overline{u'v'}$ $\overline{f},\overline{f'^2},\overline{v'f'}$,PDF(u,v,f) (also non-reacting, with C_3H_8)	LDV: two channel; 0.25x1.0 (mm); Alumina, d_N = 1 μm; Counter processor. RAMAN: 1.0 x 1.0 (mm). Common focal point. Δt = 40 μs. ($f_S < 2H_Z$)	Technique used as Dibble and Schefer (1983). Bias due to unequal seed density in the mixing streams is investigated. Bias in LDV is propagated to the Raman measurements. Mean mixture fraction is the weighting factor for the error.

TABLE 2

E: Combination of LDV with CARS.

REFERENCE	MEASUREMENTS	TECHNIQUES	COMMENTS
Fujii et al (1983,1984)	Confined premixed bluff-body stabilized flame. (table 1.5). $\overline{U},\overline{u'^2},U,u''^2$ $\overline{T},\overline{t'^2}$ $\overline{u't'}$	LDV: 0.060x0.3 (mm); Alumina; Frequency tracker. CARS: 0.035x3.0 (mm) f = 25 Hz. Common focal point. Time delay between measurements (Δt) set to be less than 20 ms. ($f_S < 25$ Hz).	The signals are sampled independently of each other. The inevitable time lag varied in average between 2 and 10 ms. Maximum number of shots per measurement: 300. Results show that R_{UT} depends strongly on $\overline{t'^2}$. $\widetilde{U}/\overline{U}$(= 0.98 to 1.01) is weakly correlated with $\overline{t'^2}$ and not at all with $\overline{u'^2}$.
Goss et al (1984)	Open premixed and non-premixed C_3H_8 flame. $\overline{U},\overline{u'^2},V,\overline{v'^2}$ $\overline{T},\overline{T'^2}$ PDF(U),PDF(T) R_{UT},R_{VT}	LDV: Titanium, d_N = 0.2 μm; Counter processor. CARS: d ≈ 1.5 mm; f = 10 Hz. Common focal point. $\Delta t < 4$ μs. (f_S 0.4 Hz).	Low time delay between the two measurements results in a very low sampling frequency. For each point were obtained 200 samples, which required ≈ 12 min. Error in correlation coefficient was, at least, 30%.

FUTURE TEST RIGS

A. CADIOU

Office National d'Etudes et de Recherches Aérospatiales
BP 72 - 92322 Châtillon Cedex (France)

1. FUTURE OF COMBUSTION RESEARCH

The fierce competition between manufacturers makes mandatory an ever improving performance for future aircraft engines. That means that the aims of the engine manufacturers are to increase the thrust/weight ratio, the durability and the reliability and to decrease the specific fuel consumption and the production costs.

These aims for future aircraft engines mean that for future combustors we must have the following goals:

- The operating conditions

The increase in the compression rate and the extension of the flight area lead to the enlarging of the pressure, the temperature and the equivalence ratio field for the combustors.

The air pressure at the inlet of the combustor (at the outlet of the compressor) varies from 0.3 bar to 40 bar.

The upstream air temperature varies from 250 K to 1000 K and the fuel-air ratio from 5 to 40 10^{-3}.

Research on the injection system will allow to develop efficient combustors for such operating conditions.

- The volume

Nowadays, manufacturers know how to design combustors with very high combustion efficiencies even for idling conditions.

Therefore, the improvement of the thrust/weight ratio can be obtained only by reducing the volume of all the components of the motor and particularly of the combustor. This can be achieved by increasing the combustion intensity with a better injection system and with the use of numerical calculation and of new measurement techniques allowing us to improve knowledge of the combustion process inside a combustor.

- The increase of durability

For commercial engine, a durability of 5000 hours is expected. Such results require very accurate control of the temperature of the flame tube.

D. F. G. Durão et al. (eds.), Instrumentation for Combustion and Flow in Engines, 215–226.

The increase of the pressure and of the temperature downstream of the compressor and upstream of the turbine demands the improvement of the cooling systems and of the prediction techniques used to calculate the wall temperatures.

- <u>The control of the temperature profile in the outlet cross section of the combustor</u>

The combustor is located just upstream the first stage of the turbine and the optimization of the turbine's durability and of the motor's performances can be realized with a precise control of the temperature field in the outlet cross section of the combustor.

These different goals for the future combustor lead to the following research necessary for the improvement of combustors:

- Technological research in order to control new techniques before their application to industrial aircraft motors. The main areas of this research are the injection process which is very important because of new operating conditions and the cooling process which allows the increase of the combustion products temperature and of the combustor's durability without increasing too much the percentage of air mass flowrate used for cooling.

- Numerical research to produce calculation codes predicting the velocity, the combustion efficiency, the pollutants emissions, the temperature and the combustion stability.

- Development of measuring techniques: non-intrusive optical methods are developed such as the CARS, the Laser Velocimetry and the Infra Red Thermography.

This research in combustion led to 2 different types of experiments. The first type is used for technological research and requires testing of complete or of a module of industrial combustors with realistic operating conditions. These tests are done in order to validate a new injection or cooling process for the design of future aircraft motors.

The second type of experiments is the type used to validate the models in numerical codes, the turbulence and the combustion models. For these experiments, combustors tested have simplified shapes. The test conditions don't necessarily correspond to operating conditions of motors. Often the fuel burnt is not the same as in motors: for example, methane is often used because its kinetics is simplified compared to the kinetics of kerosene.

But if this type of experiment doesn't require very large test facilities, it does need very accurate measurements and thus, a very large instrumentation. This instrumentation uses spectroscopic techniques which provide spatially precise and time resolved measurements.

2. DESCRIPTION OF A NEW EXPERIMENTAL SET-UP

The experimental set-up that is described here corresponds to the first type of experiment given above.

The aim of this set-up is the testing of airblast atomizers for combustors of future motors which have very high operating conditions.

An example of these motors is the Unducted Fan which is a General Electric project and where SNECMA, the French manufacturer is responsible for the combustor design. The operating conditions for this combustor are a pressure of 40 bar, a temperature of 1000 K upstream of the combustor and an air mass flowrate of 35 kg/s. There are 17 injectors in the combustor and each injector must be tested with a maximum air mass flowrate of 4 kg/s.

With such conditions of pressure and temperature there is no problem of combustion efficiency. Kerosene burns very quickly and may burn too near the injector.

The high temperatures can cause damage to the injectors.

The existing calculation codes are not sufficient to predict this. That explains why experiments are necessary.

2.1. Test facilities existing in France

To test these injectors it is necessary to reproduce the operating conditions of the combustor.

We have listed the main air supplies for combustion experiments existing in France.

There are 2 different possibilities for having pressurized air: air delivered directly by compressors or air coming from a high pressurized storage. Generally, the compressors are used when high air mass flowrates are necessary. The high pressure air storage is used for high pressure and low air mass flowrate experiments or short duration tests.

The different companies in France that have such air supplies are the following:

• SNECMA (Société Nationale d'Etudes et de Construction de Moteurs Aéronautiques) has a set of several rig facilities feeded by compressors delivering up to 40 kg/s with a pressure of 6 bar and 3 kg/s with a pressure of 9 bar.

• CEPr (Centre d'Essais des Propulseurs). It is a national company where the French manufacturers test their engines. Their compressors deliver 28 kg/s with a pressure equal to 22 bar.

• ONERA (Office National d'Etudes et de Recherches Aérospatiales). It is a national research center for aeronautics and aerospace. The air supplies existing at ONERA are an H.P. air storage located in Modane (Southeast of France), the tanks capacity is 28 m^3 with a pressure of 400 bar. There are also compressors and an H.P. air storage in Palaiseau near Paris. Compressors deliver 10 kg/s at 12 bar. The volume of the air storage is 24 m^3 with a pressure of 250 bar.

In this list, only appear the air pressure and flowrate or volume possibilities. The temperature of the air delivered is not given. For our experiments the temperature of the air is very high, 1000 K, and in this list the only air supply able to reproduce the test conditions is in Modane. Therefore, the test facility in Modane is described below.

Figure 1 represents a diagram of this facility. The air provided by compressors (pressure 9 bar) is compressed to 400 bar with a piston engine compressor and is stored in 60 tanks. Then, this air goes through a heater that has been previously heated and then goes to the test rig.

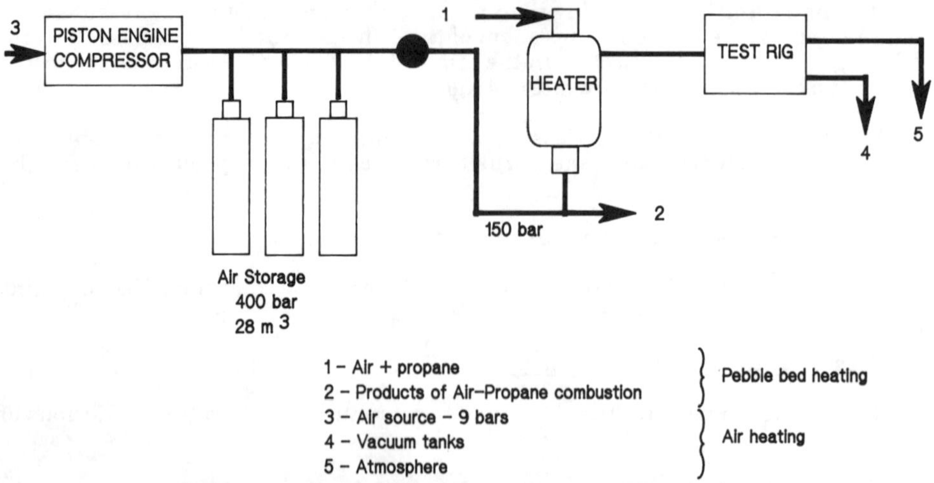

Fig. 1 — Diagram of the test facility of ONERA (Modane).

There is a second circuit for the air to heat the heater. The heat source is a propane-air combustion. This heater is an interesting technique that is very useful for providing very hot air.

Two schematic cross sections of the heater are represented on figure 2.

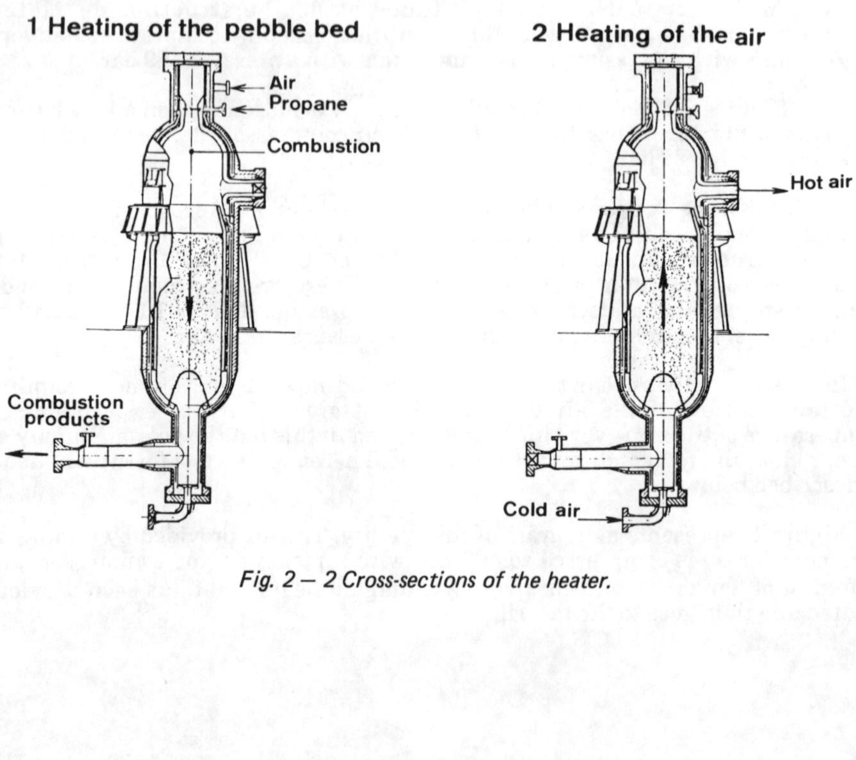

Fig. 2 — 2 Cross-sections of the heater.

The heater is a large cylindrical tank with a diameter equal to 2 meters and a height of 11 meters. This tank is filled with alumina balls (2 cm in diameter).

During the previous heating of the pebble bed, air and propane are injected into the upper part of the heater. There is a burner that produces hot gases with the air-propane combustion. The hot gases cross the pebble bed and heat the alumina balls.

When the temperature profile in the heater is suitable, the combustion of air and propane is stopped and the cold air goes through the hot pebble bed and is heated by it.

This test facility was developed during the seventies for aerodynamic studies on hypersonic propulsion. It is possible to reproduce the flight conditions for Mach numbers up to 7.

Now with the future hypersonic planes, this test facility might be used for supersonic combustion studies.

But the problem with this facility for the testing of injectors is its working cost. This test facility is disproportionate compared to our needs. For these reasons this air supply solution has not been selected.

2.2. Solutions to reproduce the test conditions

Other solutions have been considered to produce air with a pressure of 40 bar and a temperature of 1000 K in using the test facilities of ONERA at Palaiseau. These test facilities are the following:

- *compressors* which deliver an air mass flowrate sufficient for the test of one injector but with too low pressure and temperature,

- *high pressure air storage* which can deliver the right air mass flowrate 4 kg/s with a pressure of 40 bar during 30 minutes but with a temperature lower than the atmospheric temperature.

With these 2 elements, three solutions have been considered to produce 4 kg/s of 40 bar -1000 K air, 2 solutions using the air from the storage and one solution using the air from the compressors.

2.2.1. Solution 1

The first solution to produce 4 kg/s of 40 bar -1000 K air uses the air from the H.P. air storage (Figure 3).

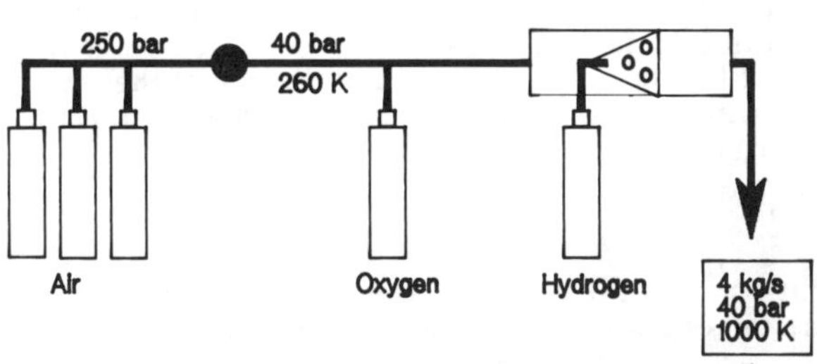

Fig. 3 – Diagram of the test facility . Solution 1.

The principle is based on combustion of hydrogen with the oxygen in the air and in order to have the same percentage of oxygen as in the air, oxygen is injected. The mass flowrates of hydrogen and oxygen injected are calculated so that the temperature is equal to 1000 K and the percentage of oxygen is appropriate.

The hot air produced with this method has H20 molecules instead of nitrogen molecules.

In our particular case, vitiation would be equal to 12%; it is the ratio of H20 molecules and H20 + N2 molecules.

The second drawback of this solution is that the duration of test is limited to fifteen minutes by the storage of oxygen. Its important advantage is its very low cost and the short time necessary to set it up.

With such facilities it is possible to have a high temperature level, impossible to obtain with a heater. This enable us to do experiments for supersonic combustion which will be a future area of combustion research. Such a facility has already been used at ONERA.

2.2.2. Solution 2

The second solution also uses the air from the storage. The increase of the air temperature is obtained with a heater. The hot gases of the heater are produced by a kerosene-air combustion. The air for this combustion is delivered by the compressors (Figure 4).

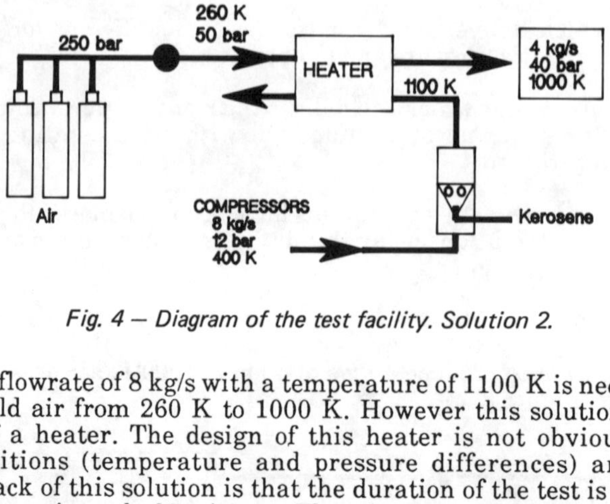

Fig. 4 – Diagram of the test facility. Solution 2.

An air mass flowrate of 8 kg/s with a temperature of 1100 K is necessary to heat the 4 kg/s of cold air from 260 K to 1000 K. However this solution requires the manufacture of a heater. The design of this heater is not obvious because its operating conditions (temperature and pressure differences) are very high. Another drawback of this solution is that the duration of the test is limited by the air storage to approximately 30 minutes. The advantage is that the air produced is not vitiated with water.

2.2.3. Solution 3

The third solution uses the air from the compressors available at Palaiseau. These compressors can deliver 10 kg/s of air with a pressure equal to 12 bar.

4 kg/s of this air is compressed in a new compressor to 40 bar. The 6 kg/s of remaining air after a combustion with kerosene are used in a turbine that makes the compressor work.

The possibility of such a solution is shown on an enthalpy-entropy diagram (Figurc 5):

a) is the air at the atmospheric conditions,

b) is the outlet of the Palaiseau compressors, pressure 12 bar - temperature only 400 K because there is cooling between the different stages of compression for technological reasons. An isentropic compression up to 40 bar from *b* would lead to *d* with a temperature of only 550 K.

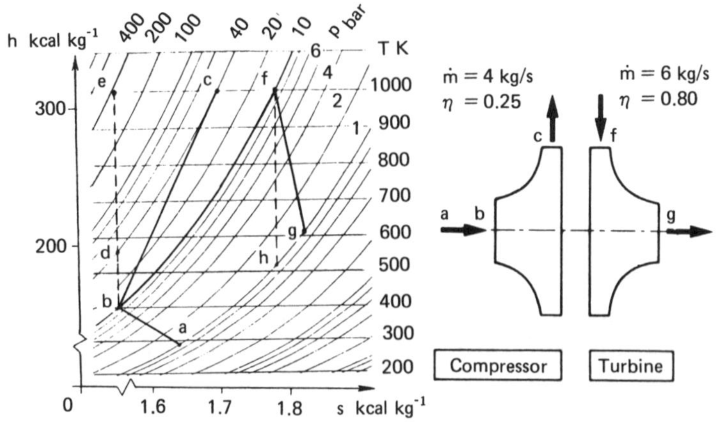

Fig. 5 – Enthalpy-entropy diagram for air.

As we need 1000 K the compressor has to compress from *b* to *c* and this leads to a compressor having an enthalpic efficiency of 0.25. This is very low compared to usual values of 0.80.

The other part of the air, the 6 kg/s are heated by combustion; this leads to *f* for a temperature of 1000 K for the combustion products. Across the turbine the decrease of pressure from 12 bar to the atmospheric pressure leads to *g* with an enthalpic efficiciency of 0.80 for the turbine. This value is realistic.

So, if we compare the variations of enthalpy in the compressor and in the turbine with the different mass flowrates we can conclude that this solution is feasible.

For technological considerations there would be 2 stages of compressor and turbine. So with this solution we obtain the diagram represented on figure 6 for the test facility.

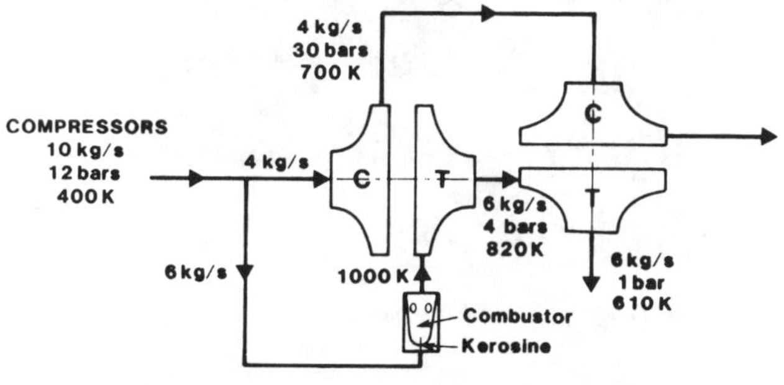

Fig. 6 – Diagram of the test facility.

The air of compressors is separated into 2 parts. One is compressed and goes to the test rig; the other one burns with kerosene in a combustor and then crosses 2 turbines.

The main advantage of this solution is that the duration is not limited because the air used is not taken from an air storage. But the drawback is the manufacture of the compressors and the turbines. Such a manufacture would be long and expensive. Anyway, this solution can be considered in the future, not at ONERA but at the CEPr to test the combustors of future aircraft motors with air mass flowrates larger than those of ONERA.

In summary, the solution chosen to obtain 4 kg/s of 40 bar -1000 K air is the high pressure storage with a heater (solution 2) because it is the least expensive way to produce non-vitiated air.

2.3. Description of the test rig

The main element of the test facility is the heater that allows the increase of the air temperature up to 1000 K.

2.3.1. The heater

The problem with heaters using air is that very large heat exchange surfaces are necessary in order to have a good efficiency because the conductivity of air is very low.

The technique developed for this heater is new.
There are nine parallel tubes with 2 collectors at each end of these tubes. Inside each tube there are 3 coaxial tubes because the principle of the heater is cold air flow between 2 annular flows of hot gases. Such a technology leads to a complicate design and manufacture but to an efficient heater and therefore to a reduced volume. The heater has a length equal to 2.50 m and a diameter of 0.80 m. Its weight is about 2 tons.

Figure 7 represents a half cross section of the heater and of one of the nine tubes.

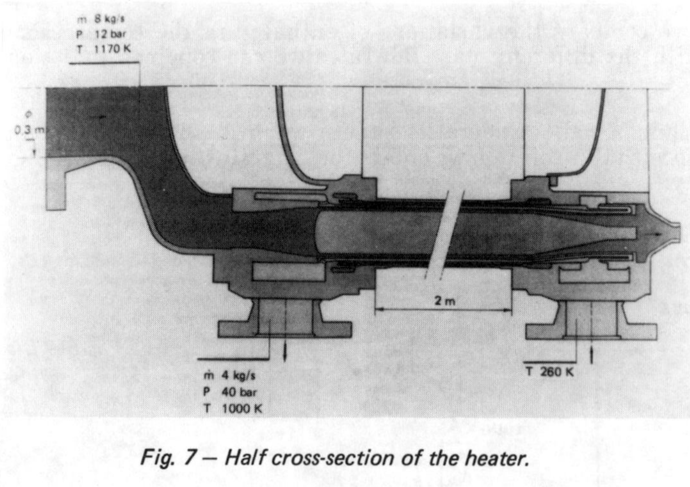

Fig. 7 — Half cross-section of the heater.

Inside the internal hot gas circuit, there is an obstruction which is only used to increase the velocity and thus, the heat exchange with the cold air without reducing the heat exchange surface.

In the collector for hot gases the effects of pressure and temperature have been separated by these 2 walls. One wall doesn't support pressure strength because the pressure is equal on its 2 sides but its temperature is high; the gas temperature is equal to 1100 K. The other wall is protected from convection by the first wall and has to support the pressure difference but with a reduced temperature level compared to the first one.

This was the main element of the test rig but another important element of this rig is the experimental combustor.

2.3.2. <u>The experimental combustor</u>

One of the combustor that will be tested on this test rig is representative of a part of the UDF TM combustor. A cross section of that single injector combustor is represented on figure 8.

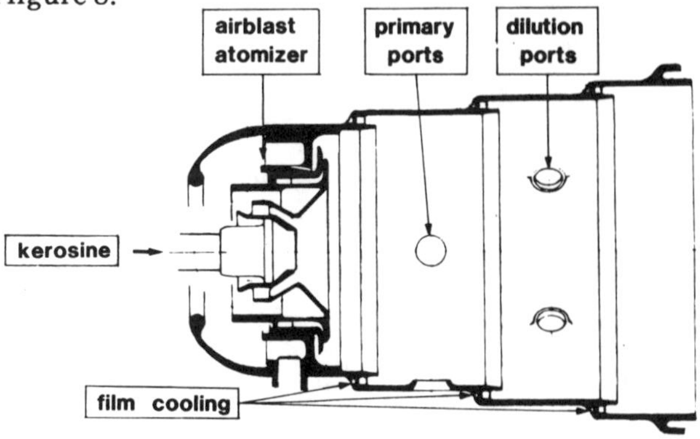

Fig. 8 – Cross-section of the experimental combustor.

The kerosene is injected in an airblast atomizer. The principle of such injectors is to have a swirl of kerosene between 2 swirls of air. This allows a good mixing of air and kerosene.

As in all combustors, there is primary air that participates in the combustion, dilution air to decrease the temperature of the combustion products and film cooling to protect the liner. The highest operating conditions for this combustor are: for the air 4 kg/s 1000 K 40 bar and for kerosene a fuel-air ratio equal to $20\ 10^{-3}$.

The temperature and the pressure are very high and under those conditions kerosene burns very quickly. Thus, if the combustion products are very close to the injector, it is possible for this injector to be destroyed.

The aim of these tests is to prove that this will not happen or if the injector is destroyed to find another convenient technology for the injector.

For such levels of pressure and temperature, the combustion efficiency is not a problem, but the resistance of the materials is.

2.3.3. The rest rig

Figure 9 represents a schematic view of the complete test rig with the heater, the experimental combustor and the combustor that delivers the hot gases for the heater.

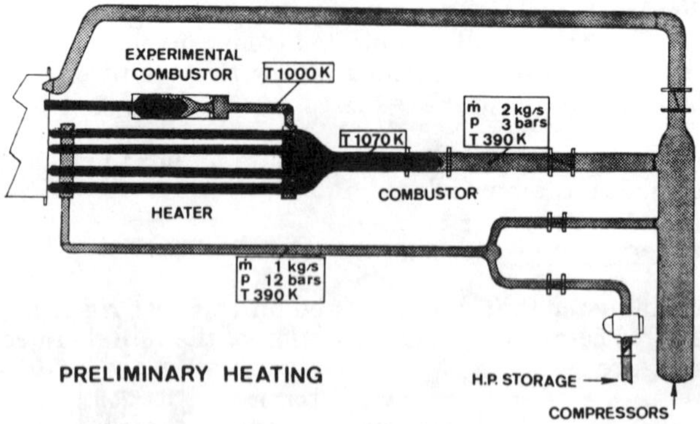

Fig. 9 – Diagram of the test rig.

There are 2 different sources of air, the compressors and the high pressure storage.

Air from the compressors is heated by a combustion with kerosene and goes through the heater.

Air from the storage is heated without vitiation in the heater and then feeds the experimental combustor.

But another circuit has been added because to reach stationary temperatures in the heater takes a long time and high pressurized air is limited. So, in order to save this air, we use air from the compressors to heat the walls of the heater. This is done for reduced air mass flowrates and pressures as indicated on figure 9 because the maximum air pressure is equal to 12 bar.

When the preliminary heating is finished the 2 different sources of air are commutated by limiting the pressure of the air from the storage to 12 bar.

Then, the air mass flowrates and the pressures of the cold air and of the hot gases are increased simultaneously so that the wall temperatures of the heater remain constant.

Figure 10 represents the operating conditions of the test rig. In fact there are several stages for the test because, for example, the experimental combustor has to be ignited for reduced pressure, temperature, air mass flowrate and equivalence ratio. This ignition occurs during the preliminary heating.

For all these different operations there are several air and kerosene valves that have to be regulated. All these valves are electrically controlled and a computer code allows an automatic positioning of these valves for a predetermined value of the air mass flowrate for the air valves and of the kerosene-air ratio for the kerosene valves.

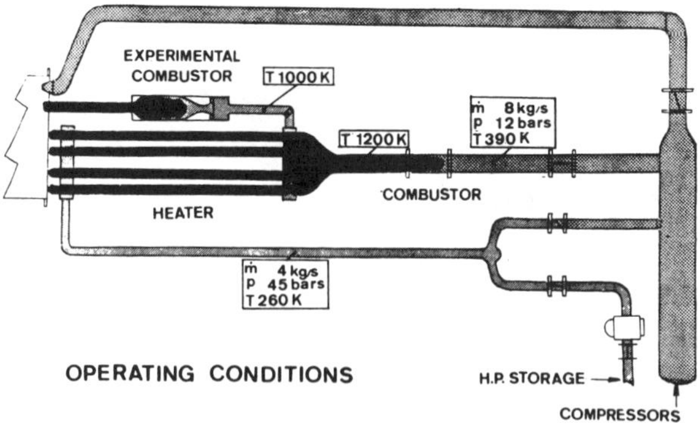

Fig. 10 – Diagram of the test rig.

The only human interferences during a usual test are the ignition of the 2 combustors and the control of the pressure of the air coming from the H.P. storage. But in the future even these operations will become automatic.

This automation of the experiments simplifies the tests but also allows identical experimental conditions for different tests. This is very useful for comparing and construing the experimental results with different injectors.

2.3.4. The instrumentation

The first interest of the experimentation presented here is to test the injector durability. Therefore, the main measurements are temperatures which are measured with thermocouples.

The use of a cooled endoscop inside the combustor allows the combustion and the injectors to be seen. A camera on this endoscop, a video-tape recorder and a T.V. give a picture of the combustion in the room where are the operators during the test.

When the durability of the injectors will be considered satisfactory, the second stage of the experiment will be the study of the combustion. For this a gas analysis device is generally used: a probe takes a sample of combustion products, this sample is continuously transferred to gas analysers. Infra-red absorption is used to measure CO and CO_2 concentrations, flame ionisation for C_xH_y and chemiluminescence for NO_x. These concentration measurements are used to calculate the equivalence ratio, the combustion efficiency, the temperature and the pollutant emission index.

These measurements are classical but are sufficient considering the aim of the study and overall its degree of emergency.

3. CONCLUSION

A particular study that begins at ONERA has been presented here but the test rig developed for this study is representative of future test rigs in the combustion field.

The new techniques used in this test rig are the heater but also the automation of the experiment. This automation has been possible because the valves used for kerosene and for air allow very accurate regulation even when problems are introduced into the system.

The heater is a very interesting tool for the study of future combustors with high pressure and temperature levels. It will be possibly also used for preheating the air with a hydrogen-oxygen combustion for studies on supersonic combustion.

LASER INSTRUMENTATION FOR PRODUCTION ENGINES

C.F. OLD

HARWELL LABORATORY

1. INTRODUCTION

This review begins by considering some of the problems currently limiting the development of production engines. We shall try to interpret these problems in terms of the fundamental physics and chemistry and assess how far available laser instrumentation can address the appropriate areas. The review is biased towards production engines since these not only pose particularly severe demands on instrumentation development, but may also show trends which are quite different from research engines. Much of the review will concentrate on spark ignition engines since the diesel engine is heavily influenced by spray behaviour which is treated in a later paper. In considering product performance, we shall restrict ourselves to investigations under steady state running. Although the study of transient operation is now just about within reach of current instrumentation, there is not as yet a substantial body of published experience.

2. CURRENT ENGINE PROBLEMS

Spark ignition engine development is driven by the twin demands of economy and emissions while retaining the necessary level of refinement to meet increasing market pressures for improved quality. The economy of the engine under steady state running can be described by the equation for efficiency of the ideal cycle which is $\eta = 1 - (\frac{1}{r})^{\gamma - 1}$ where r is the compression ratio and γ is the specific heat ratio of the air fuel mixture. Attempts to improve the efficiency by increasing the compression ratio quickly lead to limitation by auto-ignition (knock), while theoretically possible increases in γ are prevented by the flammability limits of lean mixtures and rapidly increasing cyclic variability. This phenomenon has a profound affect on refinement and smoothness as perceived by the driver.

Similarly, the two current approaches to emissions are based upon the use of either 3-way catalysts or alternatively lean burn running with an oxidation catalyst to remove residual traces of unburnt fuel. Again the second option encounters limits posed by the stability of lean running.

These considerations show that fundamental aspects of cyclic variability in ignition, flame propagation and auto-ignition, especially in lean mixtures, provide a common foundation for several aspects of technical performance.

In the case of diesel engines, the most pressing current problem is undoubtedly posed by impending legislation on emissions and particulates[1]. It is widely accepted that these aspects are controlled by the initial mixing of the fuel spray in the turbulent air flow of the combustion chamber. Closely allied to this however, is the auto-ignition behaviour of the diesel plume. There is thus some similarity with the spark ignition engine where the same process constitutes a limitation on achievable performance.

D. F. G. Durão et al. (eds.), Instrumentation for Combustion and Flow in Engines, 227–233.

3. LASER DIAGNOSTICS

Laser diagnostics are based upon the unique properties of laser radiation i.e. the intensity, bandwidth and coherence of the emitted light. These properties have enabled new techniques such as the non-linear spectroscopies, anemometry, holography and particle sizing. The unique advantages of laser diagnostics are found in the non-invasive high spatial and temporal precision. Against this however, optical access to the combustion chamber or engine manifold is always required thus perturbing to some extent the engine structure and processes such as heat transfer and gas flow. Although we shall concentrate most attention on the laser techniques, it must also be remembered that the conventional optical techniques such as Schlieren and ciné photography will continue to yield valuable insights as they have done in the past.

Probably the most well established technique is that of laser Doppler anemometry (LDA). This depends upon the passage of a small particle entrained in a fluid flow through a control volume formed at the intersection of 2 focused laser beams(2,3). The measurement of the modulation frequency of the light scattered by the particle gives a direct measurement of the fluid velocity component in the plane of the intersecting beams and normal to their bisector. An extension of the technique to measure the modulation depth of the scattered light signal provides an estimate of the size of the scattering particles. With well controlled optics, this technique (known as the visibility method) can be applied to measurement of droplets in a fuel spray or the two-phase in fuel flow in an inlet manifold. For measurement of velocity, the LDA technique is capable of operating in backscatter and will yield directional resolution with the aid of frequency shift. The particle sizing extension however must be operated in forward scatter and thus needs both an input and a signal collection window. An additional limitation is the restriction of the technique to point measurements. While this confers a desirable spatial precision, it also means that mapping of complex flows, such as those around an inlet valve, is cumbersome and time consuming. Furthermore, since it is difficult to achieve a sufficiently high signal rate to yield real quasi continuous information in real time, most LDA applications fall back on time integration methods. Detailed information about high frequency fluctuations in the flow is therefore lost and there has also been much discussion in the literature on how to separate low frequency turbulence from cycle to cycle variation.

The single point limitation of conventional LDA can now be circumvented by particle image velocimetry (PIV)(4). This method depends upon the formation of a well collimated sheet of intense laser light which is passed through the flow. Light scattered by particles moving in the plane of the sheet is collected either by high-speed camera or video in a direction orthogonal to the light sheet. PIV is not directionally sensitive, although the flow structure can usually be deduced by inspection. The two-dimensional velocity field may be determined from successive photographs of the particle distribution either by correlation methods or by an optical route based upon Young's fringes. A selected area of a double-exposed photograph is illuminated by an expanded laser-beam (usually a low-power HeNe beam). The interference pattern so formed is analysed to give the displacement of the particle images and hence the local 2-d flow vector in the plane of the light sheet. So far this method has mainly been applied to laboratory flows but application to running engines should be possible with the aid of appropriate fibre optics for beam input and signal collection(5).

In addition to the physical measurements of fluid velocity and particle size, chemical measurements of both species concentration and temperature are possible by laser spectroscopy. Perhaps the best known example of this is Coherent Anti-Stokes Raman Spectroscopy (usually known as CARS). Because of the use of pulsed lasers, this techniques offers extremely high temporal precision (typically 10-20 ms) which is capable of freezing the fluctuations in turbulent combustion chemistry. It also offers the advantage of a high intensity collimated signal beam. Direct measurements of species concentration and temperature are possible without the use of internal standards or calibration. Again however, the technique is only operable in forward scatter and data reduction can be extremely complex.

4. ENGINE BEHAVIOUR

Cyclic variability remains one of the major problems in engine development because of its effect on smoothness especially at part load and idle, and its rapid increase at leaner mixtures. It has also been suggested that it is closely linked with knock[(6)]. The sources of cyclic variability are unclear and many suggestions have been advanced. For example, Ziegler, de Soete, and Kalghatgi[(7-9)] have proposed that variability is introduced at ignition, while Müller suggests that it arises during the main flame propagation[(10)]. A third suggestion has been put forward by Lyon[(6)] who suggests that at least one input may be independent of the combustion process.

Ziegler and his colleagues base their conclusions on the extended lean flammability limit observed during comparisons of a conventional and a high current/short duration ignition system (Figure 1a). It is proposed that the initial improvement with the conventional ignition system is due to convection of the flame kernel by the flow away from the spark plug electrodes. Heat loss is thus reduced and the kernel grows at a greater rate. Fluctuations in mean flow would therefore be expected to lead to varying ignition efficiency in terms of the initial growth of the flame kernel. A counteracting influence is expected from the turbulence level. Higher turbulence should increase the heat loss and a higher spark energy is therefore required to achieve the same levels of ignition efficiency (Figure 1b). While the initial experiments were conducted in an optically accessed combustion cell, it was shown that the expected effects of modifications to the ignition system were realised in a conventional engine.

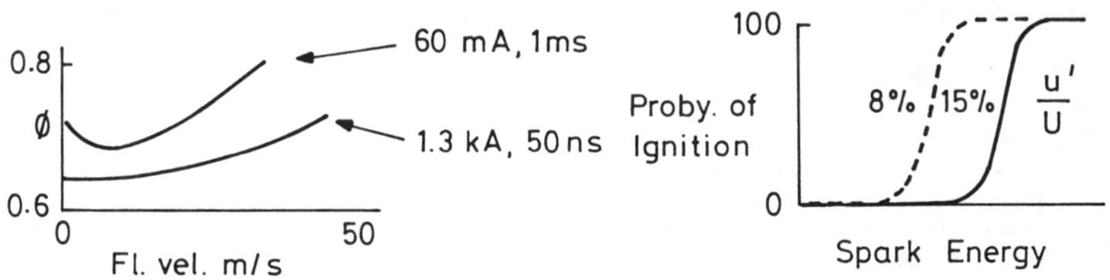

FIGURE 1. a) Dependence of lean flammability limit on mean gas velocity - after Ziegler et al[(7)]. b) Dependence of ignition probability on spark energy at different turbulence levels - ibid.

Further support for the generation of cyclic variability and ignition has been provided by the work of Kalghatgi[(9)]. Initial flame speed was measured by detecting the passage of the flame front across two narrow collimated laser beams which were placed at increasing distances from the spark plug. Measurements of the standard deviation of flame speed showed speed showed that variability was greatest in the immediate vicinity of the plug. Further away, the standard deviation fell to a lower and essentially constant level. It was suggested that this was because the flame kernel was easily perturbed by turbulent eddies when it was still very small near the electrodes. However, by the time it had grown to a larger size the overall flame envelope was much greater than the eddies and their perturbation was therefore averaged out.

Evidence for the generation of cyclic variability during the main flame propagation has been provided by Müller. Analysis of many pressure curves to derive mass fraction burnt (MFB) shows that the crank angle variability for a given burn fraction continues to increase throughout the main part of the combustion from 1-95% MFB. Furthermore, the variability itself increases at leaner air fuel ratios. In contrast, Müller claims that the variability is essentially constant over the interval 0-1% MFB, although measurements at these low levels must be near the limit of the sensitivity of the pressure measurements.

Yet a third theory is proposed by Lyon,[6] who showed that the timing of peak pressure (and thus, according to his combustion model, the level of peak pressure itself) could be correlated with small fluctuations in pressure 45 crank-angle degrees before ignition. No explanation for this observation was given, but presumably it relates to the temperature of the compressed mixture at ignition.

If we review these findings against the background of the current state-of-the-art of laser instrumentation, it is evident that existing instrumentation i.e. LDA, laser Schlieren and CARS, is adequate to probe many of the problems. Although there are still outstanding requirements such as, for example, 3-component LDA, the major need in this area would seem to be the co-ordinated application of techniques which are already available. For example, the application of coupled cyclically resolved Schlieren or LDA and pressure measurements in production engines should yield valuable information of direct use to the designer. However, although the experiments themselves are within the reach of the current technology, they will inevitably be very complex and make heavy demands on data acquisition, analysis and adaptation of a suitable engine. In consequence they will also be very expensive.

5. FLAME-PROPAGATION

The second area of engine behaviour which has recently come under close scrutiny is that of flame propagation because of its overriding importance in the development of lean burn technology. There is a need to evaluate possible improvements to flame propagation by control of fuel chemistry or in-cylinder air motion in order to realise the technological benefits of lean burn on emissions and economy. A major area of investigation concentrates on the reduced flame speed which can even lead to flame extinction in extreme cases. At low flame speeds, efficiency falls because the heat is released late in the thermodynamic cycle. When extinction occurs, the unburnt fuel is discharged along with the exhaust gases giving an unacceptable rise in hydrocarbon emissions. The experimental techniques deployed in this area are based on both large and small scale visualisation to determine the patterns of overall flame propagation and the detailed mechanisms of fuel entrainment and interactions at the cylinder wall.

Cuttler[11] has carried out an extensive photographic study of flame propagation in a side valve engine to obtain data on flame growth and associated pressure rise. Comparison of the volume of mixture entrained with rise of in-cylinder pressure indicated that mixture entrainment depends on engine speed and presumably therefore on the air motion and turbulence. In contrast, the progress of the combustion as indicated by the pressure rise depended only upon the air fuel ratio. This finding suggests that the combustion rate, as distinguished from entrainment rate, may possibly be limited by the chemistry. Improvements achievable by alterations to air flow and turbulence may therefore be limited.

Investigations later in the combustion process by Hayashi et al[12] have used laser shadowography together with high-speed ciné (100,000 fps) to elucidate the knock mechanism. These findings indicate that knock occurs by a spontaneous auto-ignition in the end-gas which generates an associated shock wave. This oscillates in the combustion chamber to generate the well-known associated knock frequency.

Fine scale structure of flame front propagation has been studied by several authors including Witze[13], zur Loye[14], Frank and Peters[15], and de Soete[8] by a planar imaging technique closely allied to PIV. The authors have exploited the disappearance of either oil or TiO_2 seeding in the burnt gas to delineate the boundaries between burnt and unburnt mixture.

While all these methods have been applied to research engines or rapid compression machines, there is as yet no information on their application to production engines. The advent of robust fibre optics would seem to offer a long awaited opportunity to set the limits to the extrapolation of laboratory results.

One example of a technique which has been successfully transferred from the laboratory to the test bed is laser Schlieren for flame speed measurement. The use of parallel detector beams to detect the expanding flame envelope was first deployed by Cole and Swords[16] in a single cylinder research engine. More recently, the same technique has been used to measure early flame velocities by Kalghatgi[9] and the Harwell group[17] in production engines.

One of the major drawbacks of simple detection of the flame envelope at a single point has been circumvented by the group at LAT Aachen by using an extensive array of optical fibres in the cylinder walls, cylinder head and piston top[18,19]. Although undoubtedly complex, these experiments have yielded valuable insight into the occurrence of knock and also into flame extinction in lean mixtures. Figure 2 shows the identification of the region of auto-ignition which spreads extremely rapidly with a pseudo flame speed in excess of 1000 m/sec. Use of the same technique at somewhat lower compression ratios and with lean mixtures shows how the flame front can, under certain conditions, fail to consume the full fuel charge, thus leading to discharge of unburnt fuel in the exhaust.

6. COMBUSTION CHEMISTRY

Control of auto-ignition in petrol engines, the optimisation of diesel ignition with fuels of decreasing quality, and the reduction of emissions in all reciprocating engines demand a knowledge of temperature and the distribution and concentration of individual species. Measurement of these parameters is particularly difficult in the hostile environment posed by an operating engine[20], but in recent years CARS has been successfully applied to temperature measurements in both diesel[21] and petrol[22] engines. A major problem for nitrogen thermometry in engines has been the pressure correction for motional narrowing. This problem has recently been solved[23]. As a result, it has been possible to obtain accurate end-gas temperature histories as shown in Figure 3. The narrow interval associated with flame transit arises because of the perturbation of the optical probe volume by the steep refractive index gradients at the flame front. It is thus now possible to derive valuable temperature information for the further development of auto-ignition models although it must be noted that the CARS technique remains subject to the limitations of forward scatter and single point measurement in a regime which is by its very nature highly turbulent and inhomogeneous.

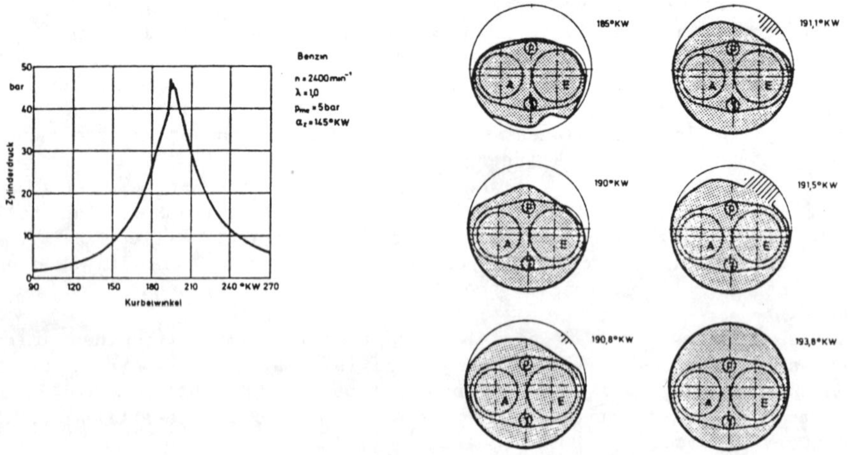

FIGURE 2: Definition of autoignition region using optical fibre array - after Knoche et al[18] and Klüting et al[19]

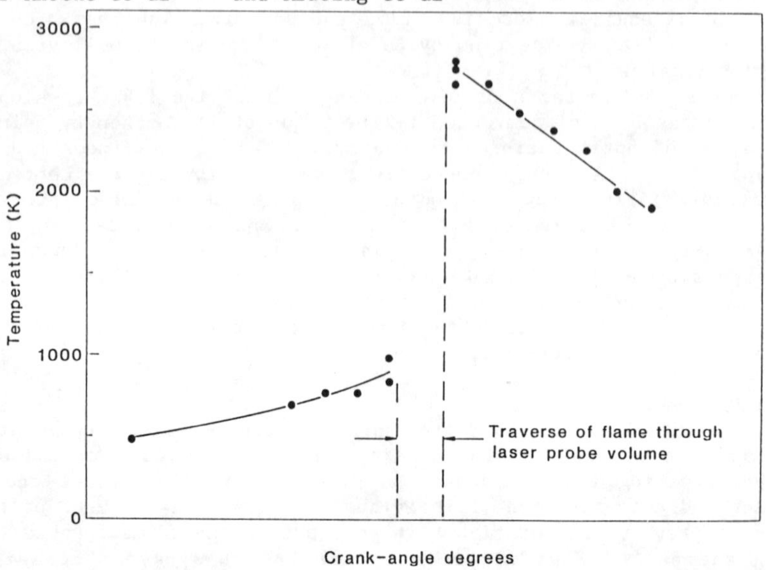

FIGURE 3: End gas temperature history in a firing production petrol engine - after Greenhalgh et al[22]

New developments currently in train seek to extend the application of CARS to species measurements. Previous laboratory experiments have shown that H_2O, CO_2, hydrogen and oxygen can be measured successfully. Such experiments are in principle also feasible in engines but interpretation is limited by the lack of detailed spectroscopic data, especially for complex fuel molecules. For the foreseeable future therefore, current developments such as those for measurements of air fuel ratios will be based on calibration techniques and comparisons made under different running conditions.

7. CONCLUDING REMARKS

At the time of writing, laser doppler and visualisation techniques exist in a satisfactory state to study many of the critical aspects of engine combustion. Progress hinges on their complex, detailed and expensive application to production engines.

Of the range of possible spectroscopic techniques, CARS can measure in-cylinder temperatures and can indicate the concentrations of the major combustion molecules. However, new spectroscopies are required to study the chemistry of pollutant formation and auto-ignition, both of which involve the study of much smaller concentrations of complex molecules.

ACKNOWLEDGEMENTS

Certain aspects of this paper are based upon work carried out at Harwell for a consortium comprising Elf UK, Ford, Jaguar and Lucas with the support of the Department of Trade and Industry.

REFERENCES

1. Cassaci C: LPG Symposium, Madrid, October 1985.
2. Durst F et al: Principles and Practice of Laser Anemometry, Academic Press, London.
3. Drain LE: The Laser Doppler Technique.
4. Third Inst. Symp. on Applications of Laser Anemometry to Fluid Mechanics, Lisbon, 1986, chapters 17 and 19.
5. Fujii H et al: ibid, paper 19.3.
6. Lyon D: Conf. on Petroleum-based fuels and automotive applications, I.Mech.E. London, November 1986.
7. Ziegler GFW et al: Conf. on Combustion in Engineering, vol I, I.Mech.E, Oxford 1983.
8. De Soete GC: ibid.
9. Kalghatgi GT: SAE.
10. Müller H: Conf. on Lean burn engines, VDI ber. 578, Wolfsburg, December 1985.
11. Cuttler DH: Cambridge Conference.
12. Hayashi T et al: SAE 841336.
13. Witze P: paper 30, this conference.
14. zur Loye AO et al: US DEn AC04-81, AL16338, 1985.
15. Franke C and Peters N: Engine Combustion, SFB224, Aachen, November 1985.
16. Cole JB and Swords MD: SAE 800043.
17. Old CF and Hutchinson P: Conf. on Practical limits of efficiency of engines, I.Mech.E. London, November 1986.
18. Knoche KF et al: as ref 15.
19. Klueting M et al: as ref 10.
20. Green RM and Lucht RP: SAND85-8621.
21. Kajiyama K et al: SAE 821036.
22. Greenhalgh DA et al: Autotech Conference, I.Mech.E, October 1985.
23. Greenhalgh DA : Agard Conference, Philadelphia, May 1986.

APPLICATION OF LASER ANEMOMETRY AND RAYLEIGH SCATTERING TO ENGINE FLOWS

C. ARCOUMANIS

Department of Mechanical Engineering
Imperial College of Science & Technology
London, England

1. INTRODUCTION

Following the extensive application of laser-based techniques to engine flows during the past decade and the parallel development of multidimensional computer codes for predicting the space-and time-resolved vector and scalar fields in the engine cylinder, the automotive industry is now better equipped to meet the ever increasing demands for higher fuel efficiency and lower exhaust emissions. The insight provided into the fluid mechanics of gasoline and diesel engines coupled with the development of fast and powerful computers, has also allowed reductions in the time required to evaluate new design concepts.

Laser-based techniques, despite their cost and required expertise, are offering advantages as research and development engineering tools which are no longer neglected by the traditionally conservative automotive community. To-day nearly all the research departments of the major engine manufacturers are equipped with laser instrumentation for testing new engine designs and/or obtaining data for their computer codes. Among the laser techniques, laser Doppler anemometry (LDA) has established itself as the main experimental technique for characterising complex internal flows, such as those in reciprocating engines, with and without combustion. LDA allows information to be obtained about all three velocity components, separately or simultaneously during an engine cycle, provided there is optical access to the combustion chamber, adequate concentration of seeding particles and instrumentation matching the flow characteristics. Laser Rayleigh scattering (LRS), on the other hand, provides information about scalar quantities, such as temperature/density or fuel concentration, which are directly related to the intensity of the scattered light from the fuel/air molecules. As an intensity-based rather than frequency-based technique, LRS requires particle-free flows, clean windows and more powerful lasers.

This paper discusses first the various ways of obtaining access to the engine cylinder through transparent windows and describes briefly the advantages and limitations of LDA and LRS for engine flows. In the second part, a number of applications of these techniques to research and production engines are presented and the most important conclusions are summarised.

2. EXPERIMENTAL SYSTEM

2.1 Flow Configuration

The most important requirement for the application of laser techniques to reciprocating engines is optical access to the combustion chamber; this should be provided with the minimum possible modification to the engine. Optical access can be obtained through the cylinder liner, the cylinder-head and the piston-bowl base. The choice of the window location and material depends on the engine type (gasoline or diesel), engine geometry and experimental technique to be employed.

D. F. G. Durão et al. (eds.), Instrumentation for Combustion and Flow in Engines, 235–253.

At Imperial College research started with an idealised engine, motored at 200rpm and equipped with a single axisymmetric inlet/exhaust valve, and a forward-scatter LDA system through a perspex cylinder liner and piston-bowl (1-3). This simple arrangement allowed parametric investigations to be performed on the effect of swirl, valve geometry, valve lift, compression ratio and piston-bowl geometry on the in-cylinder flow and provided detailed velocity data for the development and validation of two-dimensional computer codes (4,5). The experimental work was later extended to three-dimensional flows, with an off-centre, single valve (6,7) or two-valve cylinder heads (8,9) with and without induction swirl, over a wide range of engine speeds. Multi-dimensional computer codes were also developed, making use of LDA results as either boundary conditions at the valve curtain area or validation data near top-dead-centre (TDC) of compression (10-12).

The transition from model to production engine geometries was achieved through single-cylinder motored engines equipped with cylinder heads from production engines and incorporating realistic piston-bowl geometries. Optical access was provided, similarly to the model engines, through extension of the cylinder liner and piston by a perspex ring and piston-bowl. The use of perspex, however, imposes restrictions on the maximum permissible compression ratio and minimum crevice volume between the piston, cylinder wall and top piston ring due, respectively, to its mechanical properties and the need to prevent scratching of the perspex internal surface by the piston rings. For spark-ignition engine geometries with compression ratios up to 9, a perspex cylinder wall and/or piston-bowl can be used under motored conditions. However, at the higher compression ratios of diesel engines, deformation of the perspex piston crown has been observed (13); this can be prevented by throttling the induction air, which reduces the peak cylinder pressure, thus allowing engine operation at temperatures corresponding to compression ratios of ~19 (14). Justification for the use of motored engines has been provided by LDA measurements which showed that the flowfield before ignition is nearly identical to that under motoring conditions (15,16). This approach makes the experimental task easier since the additional complexities introduced by combustion, i.e. the use of solid seeding particles, window fouling by seeding or combustion generated particulates and shorter operating times due to increased thermal stresses, can be avoided.

In firing engines, the choice of the window material for optical access becomes even more important than in motored engines. Table 1 shows the properties of the most commonly used materials in comparison with those of plexiglass (perspex) and cast iron. For reasonably small windows, quartz (fused silica) has been extensively used in both gasoline and diesel engines. For larger windows or complete rings, however, the high tensile strength of sapphire offer advantages especially in diesel engines. The disadvantages are its high cost, hardness which makes machining difficult and the need for heating the sapphire external surface in order to minimize thermal stresses and prevent water condensation (17,18). This is especially important in skip-fired spark-ignition engines.

The positioning of the windows for optical access is also very important since different laser-based techniques have different directional requirements. LDA is probably the most versatile technique since the light which is scattered by the seeding particles can be collected at any angle between 0^0 and 180^0 without any effect on signal frequency (the signal intensity, however, varies with scattering angle). Contrary to LDA, LRS exhibits maximum intensity at 90^0 to the direction of propagation of the laser beam and its plane of polarisation and the positioning of the window for collecting the scattered light, relative to that of the incident light, becomes in this case very important.

In firing engines, optical access can be provided through windows located in the cylinder liner, cylinder head and/or piston bowl. These windows need to be small and preferably flush with the cylinder surface to prevent modification of the engine geometry (otherwise laser techniques are not truly non-intrusive (19)). According to Fig. 1a, full optical access through the cylinder head can be provided by inserting a large quartz or sapphire window and transferring the inlet and exhaust valves, as well as the fuel injector or spark plug, into the liner at mid-height of the clearance volume and flush with the cylinder wall. Additional small windows can be located around the periphery to allow the laser beam(s) to enter and exit the combustion chamber. Although the Sandia engine (20) with full access through the cylinder head is ideally suited for both LDA and LRS techniques and for instrumentation development in general, it is not representative of production engine designs and is limited to compression ratios of ~6 as a result of the need to accommodate the valves. Partial access can be obtained through a small window in the head with smaller than normal inlet and exhaust valves or, alternatively, a larger window with a single valve for both intake and exhaust (21).

Although the above arrangement provides full access into the chamber, it imposes collection of scattered light at 90^0 which is not useful for flow visualisation techniques usually employed to provide two-dimensional information about the flowfield and guide point techniques like LDA and LRS. A truly versatile research engine requires full or partial optical access through the head and access through the piston crown (Fig. 1b) or through a flat bowl base (Fig. 1c); the capabilities of single point or planar laser techniques can then be fully explored (21). However, even in this case there are certain limitations, namely that only the radial and swirl velocity components can be measured by LDA, the crevice volume is unacceptably long and realistic piston-bowls cannot be incorporated into the elongated pistons of these research engines without modification of the bowl base.

Full optical access into a disk-type combustion chamber can also be obtained through a quartz or sapphire cylinder liner but it is a very expensive method and there are a number of uncertainties and risks associated with this design (17). In principle, the measurement by LDA of all three velocity components is possible at any point in the engine cylinder but the curvature of the transparent cylindrical wall prevents detailed mapping of the flow by LRS and its visualisation by schlieren or shadowgraphy. Partial access can be provided by small windows in the cylinder liner which should be flush with the wall rather than flat and allow the beam(s) to approach as near as possible to the cylinder head. Access through the wall, however, does not imply necessarily access into the piston-bowl. For different than disk-type engine geometries, windows are also needed in the piston-bowl wall to correspond to the windows in the cylinder wall, so that their fields of view overlap near TDC. This method is, however, difficult and expensive and the alternative ways discussed previously for obtaining access into the piston-bowl should be pursued.

Summarising, there are no clear advantages with any of the designs discussed above. Cost, flexibility in stopping the engine and cleaning the windows, safety factors associated with a particular window design and material, running time for firing engines together with the experimental technique(s) to be employed, are the main factors to be considered in the design of a research engine.

2.2 Review of laser techniques.

The basic principles of LDA and LRS are well known (see for example (22-25)) and will not be repeated here. Only the basic components of typical systems for engine applications will be summarised below as a guide to prospective users.

2.2.1 Laser Doppler anemometry. Fig 2 shows the schematic of a typical single component forward-scatter LDA system for measuring instantaneous velocities in engine cylinders. LDA resolves the velocity vector in the direction perpendicular to the bisector of the two laser beams.

The most important components of an LDA system are listed below together with explanatory comments:

* laser-He-Ne (5-30 mW) can be used for low to medium speed in-cylinder flows under motored conditions, provided there is optical access in the forward direction, but a high power CW Ar-ion (>1W) is required for high speed flows (inlet valve flow) and flows with combustion.

* Frequency shift - rotating grating: Commercial diffraction gratings with standard motors and first order diffracted beams provide variable shifts up to 7MHz. If motored at high speeds, can provide shifts of 14MHz and higher. Frequency shifts of this order can be obtained by making use of the second order beams. Alternatively, bragg cells can be used: single bragg cells provide shifts of 40MHz which may be needed in certain flows but the resultant signal has to match the frequency bandwidth of the signal processing system. Double bragg cells with downmixing provide shifts of less than 10MHz.

* seeding - liquid and solid seeding particles are used in motored and firing engines, respectively, with a size range between 0.1 and 3μm. The range of solid particles for use in firing engines and their properties are given in (26).

* frequency counter - due to the wide variety of singal frequencies and signal-to-noise (S/N) ratios encountered in engine flows, counters should exhibit a flat reponse up to 100MHz and incorporate a burst detection circuitry, end of burst and validation circuitry in amplitude (hardware variable upper trigger level) and time (hardware or software 5/8 and 10/16 validation) domains.

In reciprocating engines, to obtain accurate information about the instantaneous flow velocity is not enough. It is the subsequent data analysis and interpretation which is important since the data reduction is complicated by the time (crank angle) dependent nature of the flow which makes difficult the separation of the mean flow from turbulence (27). The various ways of LDA data reduction are summarised below:

* ensemble-averaging: Instantaneous velocities within a small crank angle window (1-5°CA) are averaged over a number of engine cycles to provide a statistically adequate number of data points (~2000).

* smooth ensemble-averaging: Provided the flow is seeded heavily and the sampling rate is high, a smooth curve passes through the velocity data within an engine cycle which are averaged into 1°CA windows.

* conditional ensemble-averaging: Instantaneous velocity data are first divided into groups according to some conditioning parameter (flame speed/arrival time, peak cylinder pressure) and then the data from each group are ensemble-averaged.

* frequency filtering: Following a preliminary identification of the flow characteristics, a cut-off frequency is first selected and there the data are low-pass filtered to allow determination of the mean velocity within each engine cycle.

* time-averaging: Instantaneous velocity data are averaged over small successive crank angle windows in the engine cycle.

At present, there is no consensus among research workers about the method of data reduction in engine flows. There are two schools of thought in favour or against ensemble-averaging. Those against make use of smooth ensemble, frequency filtering and time averaging and aim at resolving cycle-to-cycle variations of the mean flow. This requires selection of a 'selective' parameter (e.g.

cut-off frequency) with adequate physical justification which is difficult to be provided in a flow where the time scales associated with the bulk flow variations are of the same order with the turbulent time scales. On the other hand, the supporters of ensemble averaging argue that before combustion starts any variations in the flow from one cycle to the next can only be due to turbulence since no other physical mechanism has yet been identified which can give rise to random variations in the mean flow**. In this respect, there is evidence that the flow entering the engine cyinder through the inlet valve is turbulent at all engine speeds, that any instabilities present during early industion at the shear layers do not persist through compression and that the turbulence generated in the induction system decays by TDC. After combustion any variations in the local flow around the spark plug at the time of ignition or in the ignition characteristics should be taken into consideration by conditioning the data upon flame arrival at the measurement location and then ensemble-averaging them within small crank angle windows (29). In addition of being unambiguous in its implementation and allowing direct comparison of data from different engines, ensemble-averaging allows *direct compatibility* between measurements and calculations. At present *only* ensemble-averaged results can provide boundary conditions and validation data for the development of multidimentional computer codes which solve the ensemble-averaged equations of motion.

The uncertainty about the proper method for separating mean flow from turbulence is far from resolved and will continue to occupy the minds and thoughts of the engine research community in the foreseable future. At present, the ensemble-averaging approach despite its inability to resolve the flow in individual cycles, seems to represent the most acceptable solution to modellers and experimentalists alike.

2.2.2 Laser Rayleigh scattering. Fig. 3 shows the schematic of a typical LRS system for engine applications. LRS provides a directionally sensitive signal whose wavelength is the same with that of the incident laser beam (elastic scattering) and its intensity proportional to the number density of the molecules in the cylindrical scattering volume and their scattering cross-section. The intensity of this signal in the absence of Mie scattering interference can be interpreted as density /temperature or fuel concentration in 'tailored' fuel/air gaseous mixtures. The most important components of an LRS system are listed below together with explanatory comments:

* laser- CW Ar-ion (1-4W) has been used to obtain information about fuel concentration (29,30). He-Ne lasers may also be used with photon correlators (31,32).

* optics - need for careful matching between the light transmission and collecting systems due to the directional sensitivity of scattered light (intensity is maximum at 90^0 to the direction of beam propagation and its plane of polarisation) and masking of reflected light from optics and windows (see error sources below).

* photodetector-due to the low Rayleigh scattering cross-section (relative to Mie scattering), very sensitive low-noise photodetectors are required with internal or external signal amplication.

* particle/noise filters - in addition to interference and electronic frequency filters to minimise stray light and white noise,respectively, software and/or hardware particle filtering is required (see below) for on-line and off-line discrimination between Mie and Rayleigh scattered light.

** Swirl center precession has been identified as a mechanism that under certain conditions gives rise to periodic variation in the position of zero swirl velocity with frequencies related to engine speed (28).

As an intensity-based technique, LRS is sensitive to interference from light of the same or different wavelength reaching the photodetector or internal variations in the response of the system. The various error sources associated with LRS are listed below:

* system's response
  - fluctuations in the laser power output
  - fluctuations in the input voltage to photodetector
  - drift in the electronics
  - electronic shot noise
  - limited response thime of the photomultiplier
  - dark noise in the photomultiplier
  - digitisation error in A/D converter

* bakcground radiation
  - ambient light
  - scattered light from optics
  - scattererd light from entrance/exit windows and cylinder walls.
  - flame radiation (chemiluminescent)
  - variation in the scattering cross-section of reactants, intermediates and products.

* particle interference
  - impurities in air stream
  - liquid fuel
  - combustion products (ie particulates)

Due to the difficulty to eliminate particles from the scattering volume or its vicinity before, during or after combustion, software particle filtering has been introduced (33,34) to discriminate between Mie and Rayleigh scattered light either on-line or off-line. The criteria for software filtering are the following:

- intensity of scattered light from particles is usually higher than the maximum intensity of the fuel/air mixture (amplitude filtering).
- initial rise in signal amplitude appears greater for particle scattering than molecular scattering (initial rise filtering).
- rate of fall in signal amplitude appers greater for particle scattering than molecular scattering (transient filtering).

Due to the need for adequate S/N ratio, the gaseous fuel to be used should be selected on the basis of its scattering cross section. In premixed flames, as in gasoline engines, careful matching is required between the scattering efficiency of reactants, intermediates and products. For example, propane gas, who is a good scatterer of light, exhibits at stoichiometric conditions a 23% difference in the scattering cross section between reactants and products (35) and, therefore, only a lean mixture of propane and air can be used. Alternatively, a correction factor should be calculated and applied to the Rayleigh signal as a function of the degree of reaction, fuel type and equivalence ratio (36). In diffusion flames (diesel engines) there is need to match the scattering cross-section of the fuel to that of air and the combustion products and this requires premixing the fuel with H_2 in order to reduce the total effective scattering cross-section of the mixture.

From the above, it becomes clear that the application of LRS to engines although promising is not straightforward. To circumvent the need for 'tailoring' the fuel, Freon gas has been frequently used to simulate gasoline (37) or diesel fuel (38) in motored engines, due to its high density and scattering cross-section, and to validate multidimentional computer codes for predicting scalar distribution in engine cylinders before combustion (39).

3. APPLICATIONS

In this section, examples will be presented of the application of LDA and LRS to engine flows in view of the role that these techniques have played in the understanding of the fluid mechanics of reciprocating engines. The presentation of the results follows the sequence of the unsteady flow processes in the engine cylinder starting with induction and proceeding towards the exhaust.

3.1 Laser Doppler anemometry

The flow in inlet manifolds is important as means of increasing volumetric efficiency (through reduction of pressure losses) and matching the flow generated by the inlet port. With helical ports , for example, the design of the manifold should be such as to generate secondary flows of the same rotation to that imparted to the flow by the helix. In addition, the flow distribution to individual cylinders of multi-cylinder engines may be different and this requires further investigation. Due to the complex geometry of production engine manifolds, idealised compact manifold junctions were used to simplify the experimental task (40) and provide validation data for a parallel computational programme (41). Although useful, these geometries are far from reality and experimental data are needed in production engine manifolds under operating conditions to allow development of multidimensional computer codes to cope with complex geometries and improve manifold design. The approach followed in (42) was to open ports in critical areas of a diesel engine manifold (i.e. in adjacent branches and in-between) and insert perspex windows which are nearly flush with the manifold wall. Using back scattered LDA, the flow in the three windows shown in Fig 4 was measured as a function of engine speed at various points across the window and found to vary in space and time within an engine cycle but to be fairly repeatable from one cycle to the next.

The flow at the exit of the inlet valve has been given recently a lot of attention (43-45) due to the need for detailed boundary conditions for multidimensional computer codes and, at the same time, identification of possible flow instabilities in the flow entering the engine cylinder. Depending on the velocity component to be measured, forward-or back-scatter LDA was employed to provide detailed information of the flow in the valve curtain area as a function of valve lift and distance from the cylinder head. Due to difficulties in measuring the valve flow in an operating engine, nearly all valve flow investigations were done under steady conditions by mounting the cylinder head in a test rig and sucking air through it at a constant mass flowrate. Fig 5 shows the three velocity components around the inlet valve of a prototype diesel engine incorporating a helical port (46). The validity of the steady flow assumption has been examined in (47) and found to be satisfactory for most of the valve opening period except during very early and late induction. An investigation into the effect of engine speed on the valve flow in the range 500-3250rpm (48) showed that as the speed increases, the flow inertia allows the fluid to enter the cylinder during the whole valve lift period whereas at low speed there is flow reversal through the inlet valve after BDC.

The in-cylinder flow during the induction stroke has been extensively investigated with LDA (27) and revealed a number of interesting features, some of which have been already included in computational models (49):

* Turbulence is generated in the shear layers of the unsteady annular jet entering the engine cylinder through the inlet valve.

* In the axial plane, a system of toroidal and ring vortices are formed during early induction which breaks down by bottom-dead-centre (BDC).
* The size of the main toroidal vortex scales with the instantaneous piston stroke during the first half of induction and then remains constant in size but decays in strength.
* Certain induction systems (e.g. shrouded valve (50), "radial" directed port (51)) give rise to a single toroidal vortex (tumbling, rolling vortex or barrel swirl) occupying the whole cylinder volume and rotating around an axis perpendicular to the cylinder axis (Fig. 6).
* In the tangential plane and depending on port geometry, a single, double or even more complicated vortex system is formed which is transformed by the time the inlet valve closes into a solid body type of rotation.
* The centre of swirl performs an off-axis helical motion with a processing frequency which is about five times the engine frequency (28).
* At BDC, the flow in the cylinder exhibits an axial stratification.

The flow during the compression stroke has been also extensively characterised by LDA in motored engines in order to establish the initial conditions for the subsequent combustion process. Some of the most important conclusions are summarised below:

* With the exception of the rolling vortex, no other motion in the axial plane persists through compression and the induction turbulence levels decay near TDC to levels of ~0.5 times the mean piston speed (52-54).
* The formation of a tumbling vortex, which survives through compression to the period of maximum compressive strain (~320^{0}CA), gives rise to increased turbulence levels and represents the most effective means of enhancing induction turbulence at the time of ignition.
* Swirl persists through compression exhibiting a gradual decay of its angular momentum and, in the presence of a piston-bowl, is transformed by squish and amplified in the smaller diameter bowl through conservation of its angular momentum. In the absence of a tumbling vortex and squish, swirl is not generally enhancing induction turbulence.
* Squish generates turbulence, which is confined in the region around the bowl edge, with levels determined by the bowl-entry diameter and bumping clearance.
* The flow pattern in the piston-bowl is mainly determined by the swirl-squish interaction (55), bowl eccentricity and bowl-base geometry (56).
* Turbulence intensities in disk chambers scale almost linearly with speed and are homogeneous and isotropic within 30%.

Due to the ability of LDA to measure flows with combustion, it has been exclusively used in spark-ignition engines to characterise the flow in the vicinity of the spark plug, ahead and behind the propagating flame front (57-61). The most important conclusions from these studies are summarised below:

* The propagating flame induces significant mean velocities ahead of the flame, in the direction normal to the flame surface.
* A relatively small increase of turbulence intensity has been observed in the vicinity of the flame front ahead of the flame. The effect of combustion on turbulence seems, however, to be speed dependent.
* There is a significant increase of turbulence intensity across the flame.
* Behind the flame, in the burned gas region, turbulence is rather homogeneous and isotropic away from the walls and decays repaidly.
* The wall boundary layer in motored engines is thin ($\leq$ 1mm), with its thickness decreasing with engine speed and increasing during combustion.

* The mean velocity in the vicinity of the spark plug just before ignition is correlated with the peak cylinder pressure.

It should be clear from the above listing of the most important conclusions from LDA measurements in motored and firing engines that not only a better insight into the fluid mechanics has been obtained but also quantitative information on the dependence of mean flow and turbulence on engine geometry and speed. This data has allowed multidimensional models of the cold flow to be validated and acquire their present predictive capability (62,63) and combustion sub-models to be developed.

3.2 Laser Rayleigh Scattering.

There are relatively few applications of LRS to engine flows and this is attributed to the sensitivity of the technique to light scattering from windows, optics and moving solid boundaries (i.e. valves, piston) and the need for a particle/droplet-free flow. The applications are divided into two groups dealing with motored and firing engines, respectively.

The difficulty to characterise diesel sprays in an operating engine has prompted attempts to simulate it by a transient gaseous jet (64-66). In an idealised motored engine, Freon-12 was injected through a nozzle inserted in the axisymmetric inlet/exhaust valve near TDC and its mixing with quiescent or swirling air was examined by LRS through a transparent (perspex) cylinder wall. Freon-12 was selected due to its high scattering cross-section and its high density relative to air. Initially, the engine was run without compression (64) and this allowed direct interpretation of the intensity of the Rayleigh signal as Freon concentration (mole fraction). Under compressing conditions, however, the time-varying pressure requires that the signal be analysed at the instantaneous pressure at which scattering takes place (66,67). Calibration made with different Freon/air mixtures (Fig 7) showed that signal intensity varies linearly with pressure. The instantaneous and ensemble-averaged Freon concentration (injection schedule) was quantified as a function of downstream distance from the nozzle (Fig 8) using the system of Fig 3 and found to exhibit a very sharp rise, followed by a relatively flat plateau whose level decreases with downstream distance and a relatively sharp fall. The degree of mixing with air was also quantified as a function of nozzle geometry, compression ratio, induction swirl and piston-bowl geometry and provided insight into the relative importance of the various parameters (68).

LRS was also used in a motored spark-ignition engine to provide time- and space-resolved measurements of vapour Freon-12 or Freon-113 concentration in the combustion chamber during the induction stroke, for both gaseous and liquid Freon injection into the intake port (69). The results showed significant differences in the space-resolved Freon concentration with a strong dependence on the location of the fuel injector and the intake valve lift.

To gain insight into the mechanism of turbulent flame propagation, a technique using imaged multipoint LRS has been developed and applied to an operating spark-ignition engine to measure turbulent flame thickness (70). The scattered light from the propagating flame front was imaged through a flat window in the cylinder head and a lens system onto a multichannel analyser. A typical density (intensity) profile through a flame is shown in Fig 9. The flame thickness was defined as the distance between the intercepts of the preflame and postflame densities with a straight line coinciding with the average maximum density gradient (Fig. 10). The mean flame thickness was found to increase with engine speed and the turbulent flame front to become more wrinkled. It was observed that islands of unburned mixture can be engulfed behind the flame front and this is more likely to happen as the turbulence intensity increases with engine speed.

Other areas where LRS is expected to find application in engine flows is the measurement of the space and time-resolved temperature in spark-igntion engines and the rate of vapourisation of fuel droplets. The technique may also be capable of resolving mixture non-homegeneities in the vicinity of the spark plug before ignition.

4. CONCLUSIONS

Over the last ten years the application of LDA to motored and firing engines has contributed to significant improvements in the understanding of the quantitative relationship between engine geometry and in-cylinder flow (27). LDA has also assisted in the development of multidimensional computer codes for predicting the flow and combustion in gasoline and diesel engines, by providing boundary conditions and validation data. LDA and Computational Fluid Mechanics (CFM) in their complementary role (71) are becoming standard engineering tools in automotive flow research and development. LRS, on the other hand, offers advantages as a scalar measuring technique in spark-ignition engines, where the flow is cleaner than diesels, and can be useful in providing validation data for computer methods attempting to predict the fuel concentration and temperature distributions in the engine cylinder. Although LRS is not expected to find the universal applicability of LDA, its application to engine flows may increase in the years ahead.

5. ACKNOWLEDGEMENTS

The author would like to thank his colleagues in the Fluids Section and especially those who co-authored the referenced papers for their technical assistance and useful discussions. The financial support provided for the work described in the Imperial College references by the Science and Engineering Research Council and the US Department of Energy, Office of Energy Utilization Research, Energy Conversion and Utilization Technologies Program is gratefully acknowledged.

6. REFERENCES

1. Morse A P, Whitelaw J H and Yianneskis Y : Turbulent Flow Measurements by Laser-Doppler Anemometry in Motored Piston-Cylinder Assemblies, J.Fluids Engng., V.101, 1979.
2. Arcoumanis C, Bicen A F, Vlachos N S and Whitelaw J H : Effects of Flow and Geometry Boundary Conditions on Fluid Motion in a Motored IC Model Engine, Proc.I.Mech.E., V.196, No.4, 1982.
3. Arcoumanis C, Bicen A F and Whitelaw J H : Effect of Inlet Parameters on the Flow Characteristics in a Four-Stroke Model Engine, SAE Paper 820750, 1982.
4. Ahmadi-Befrui B, Arcoumanis C, Bicen A F, Gosman A D, Jahanbaskch A and Whitelaw J H :Calculations and Measurements of the Flow in a Motored Model Engine and Implications for Open-Chamber Direct-Injection Engines, Proc. Symposium on Three-Dimensional Turbulent Shear Flows, AIAA/ASME Meeting, St.Louis, 1982.
5. Ahmadi-Befrui B, Gosman A D and Watkins A P : Prediction of In-Cylinder Flow and Turbulence with Three Versions of k-e Turbulence Model and Comparison with Data, Proc. Symposium on Flows in Internal Combustion Engines II, ASME Winter Meeting, New Orleans, 1984.
6. Vafidis C : Influence of Induction Swirl and Piston Configuration on Air Flow in a Four-Stroke Model Engine, Proc.I.Mech.E., V.198C, No.8, 1984.
7. Arcoumanis C, Bicen A F, Vafidis C and Whitelaw J H : Three-Dimensional Flow Field in Motored Four-Stroke Model Engines, SAE Trans., V.5.978, 1985.
8. Morse A P and Whitelaw J H : Measurements of the In-Cylinder Flow of a Motored Four-Stroke Reciprocating Engine, Proc.R.Soc.Lond., V.A377, 1981.
9. Bopp S, Vafidis C and Whitelaw J H : The Effect of Engine Speed on the TDC Flowfield in a Motored Reciprocating Engine, SAE Paper 860023, 1986.

10. Gosman A D, Tsui Y Y and Watkins A P : Calculation of Unsteady Three-Dimensional Flow in Model Motored Reciprocating Engine and Comparison with Experimental Data, Proc. 5th Int. Symposium on Turbulent Shear Flows, Ithaca, 1985.
11. Gosman A D : Multidimensional Modelling of Cold Flows and Turbulence in Reciprocating Engines, SAE Paper 850344, 1985.
12. Begleris P and Gosman A D : Analysis of In-Cylinder Flow Generation by Helical Ports, Proc. ASME Winter Meeting, Boston, 1987.
13. Arcoumanis C, Begleris P, Gosman A D and Whitelaw J H : Measurements and Calculations of the Flow in a Research Diesel Engine, SAE Paper 861563, 1986.
14. Arcoumanis C, Hadjiapostolou A and Whitelaw J H : Effect of Induction Air Throttling on the TDC Flowfield in Motored IC Engines, Proc. 2nd Int. Conference on Laser Anemometry, Glasgow, 1987.
15. Rask R B : Laser Doppler Anemometer Measurements in an Internal Combustion Engine, SAE Paper 790094, 1979.
16. Wigley G, Patterson A C and Renshaw J : Swirl Velocity Measurements in a Firing Production Diesel Engine by Laser Anemometry, Proc. Symposium on Fluid Mechanics of Combustion Systems, ASME Spring Meeting, Colorado, 1981.
17. Richman R M and Reynolds W C : The Development of a Transparent Cylinder Engine for Piston Engine Fluid Mechanics Research, SAE Paper 840379, 1984.
18. Witze P O and Baritaud T A : Influence of Combustion on Laser Velocimeter Signal Quality in a Spark Ignition Engine, Proc.Int. Symposium on Laser Anemometry, ASME-FED.Vol.33, 1985.
19. Amann C A : Classical Combustion Diagnostics for Engine Research, SAE Paper 850395, 1985.
20. Johnston S C, Robinson C W, Rorke W S, Smith J R and Witze P O : Application of Laser Diagnostics to an Injected Engine, SAE Trans., V.90, 1979.
21. Sinnamon J F, Lancaster D R and Steiner J C : An Experimental and Analytical Study Of Engine Fuel-Spray Trajectories, SAE Trans., V.89, 1980.
22. Durst F, Melling A and Whitelaw J H : Principles and Practice of Laser-Doppler Anemometry, Academic Press (London) Ltd., 2nd edition, 1981.
23. Drain L E : The Laser Doppler Technique, John Wiley & Sons Ltd., 1980.
24. van de Hulst H C : Light Scattering by Small Particles, Dover Publications Inc., New York, 1981.
25. Dyer T M : New Experimental Techniques for In-Cylinder Engine Studies, SAE Paper 850396, 1985.
26. Witze P O and Baritaud T A : Particle Seeding for Mie Scattering Measurements in Combusting Flows, Proc. 3rd Int. Symposium on Applications of Laser Anemometry to Fluid Mechanics, Lisbon, 1986.
27. Arcoumanis C and Whitelaw J H : Fluid Mechanics of Internal Combustion Engines-a Review, Proc.I.Mech.E., V.201, No.C1, 1987
28. Arcoumanis C, Hadjiapostolou A and Whitelaw J H : Swirl Center Precession in Engine Flows, SAE Paper 870370, 1987.
29. Witze P O and Martin J K : Cyclic-Variation Bias in Spark Ignition Engine Turbulence Measurements, Laser Anemometry to Fluid Mechanics II, eds. Adrian R J et al., LADOAN, 1986.
30. Arcoumanis C, Green H G and Whitelaw J H : Velocity and Concentration Measurements in a Model Diesel Engine, Experiments in Fluids, V.3, 1985.
31. Benhachmi D, Younes N, Yakout H, Emmerman P E and Goulard R : Rayleigh Thermometry with Low-Power Laser Sources, AIAA Paper 83-1554, Montreal, 1983.
32. Haumann J and Leipertz A : Time Resolved Concentration Measurements using Rayleigh Scattering Photon Correlation Technique, in Measurement Techniques in Heat and Mass Transfer, eds. Soloukhin R I and Afgan N H, Hemisphere, Washington, 1985.
33. Arcoumanis C : A Laser Rayleigh Scattering System for Scalar Transport Studies, Experiments in Fluids, V. 3, 1985.
34. Green H G : Characteristics of Steady and Unsteady Jets, Ph.D. Thesis, University of London, 1986.

35. Dandekar K V : Velocity and Density Measurements in Premixed Turbulent Flames, Ph.D. Thesis, Cornell University, 1982.
36. Namer I and Schefer R W : Error Estimates for Rayleigh Scattering Density and Temperature Measurements in Premixed Flames, Experiments in Fluids, V.3., 1985.
37. Kadota T : The Application of Laser Rayleigh Scattering to the Local Mixture Strength Measurements in SI Engine During Intake Stroke, SAE Paper 872151, 1987.
38. Arcoumanis C, Green H G and Whitelaw J H : The Application of Laser Rayleigh Scattering to a Reciprocating Model Engine, SAE Trans., V.2.949, 1985.
39. Arcoumanis C, Gosman A D, Whitelaw J H, Tsui Y Y and Wong K Y : Gaseous Simulation of Single-Hole Diesel Sprays, SAE Paper 872090, 1987.
40. Sierens R and Snauwaert P : Study of the Flow Pattern in Compact Manifold Type Junctions, ASME, J. Engng. for Gas Turbines and Power, 1987.
41. Dimitriadis C, Leschziner M A, Winterbone D E, Alexander G I and Sierens R : Computation of Three-Dimensional Flow in Manifold-Type Junctions, ASME-FED.Vol.28, 1985.
42. Arcoumanis C, Flamang P and Whitelaw J H : Flow in the Inlet Manifold of a Production Diesel Engine, To be presented at the 4th Int. Symposium on Applications of Laser Anemometry to Fluid Mechanics, Lisbon, 1988.
43. Bicen A F, Vafidis C and Whitelaw J H : Steady and Unsteady Air Flow through the Intake Valve of a Reciprocating Engine, J.Fluids Engng., V.107, 1985.
44. Arcoumanis C, Vafidis C and Whitelaw J H : Valve and In-Cylinder Flow Generated by a Helical Port in a Production Diesel Engine, J.Fluids Engng., V.109. 1987.
45. El Tahry S H, Khalighi B and Kuziak Jr. W R :Unsteady Flow Velocity Measurements around an Intake Valve of a Reciprocating Engine, SAE Paper 870593, 1987.
46. Arcoumanis C, Cossali E and Whitelaw J H : Valve Flow Generated by Helical Ports, To be presented at the 4th Int. Symposium on Applications of Laser Anemometry to Fluid Mechanics, Lisbon, 1988.
47. Arcoumanis C and Whitelaw J H : Are Steady Flow Inlet Boundary Conditions Valid for Engine Cylinder Calculations?, Proc. Congress on Modeling of Internal Combustion Engines, Valencia, 1987.
48. Arcoumanis C, Vafidis C, Vorropoulos G and Whitelaw J H : Correlation between Valve and In-Cylinder Flow in a Motored Reciprocating Engine, ASME Paper 87-FE-8, 1987.
49. Kent J C, Haghgooie M, Mikulec A, Davis G C and Tabaczynski : Effects of Intake Port Design and Valve Lift on In-Cylinder Flow and Burnrate, SAE Paper 872153, 1987.
50. Gosman A D, Tsui Y Y and Vafidis C : Flow in a Model Engine with a Shrouded Valve- a Combined Experimental and Computational Study, SAE Paper 850498, 1985.
51. Vafidis C, Vorropoulos G and Whitelaw J H : Effects of Intake Port and Combustion Chamber Geometry on In-Cylinder Turbulence in a Motored Reciprocating Engine, Proc. Symposium on Fluid Flow and Heat Transfer in Reciprocating Machinery, ASME Winter Meeting, Boston, 1987
52. Arcoumanis C, Bicen A F and Whitelaw J H : Application of Laser Doppler Anemometry to Motored Model Engines, Laser Anemometry in Fluid Mechanics, eds. Adrian R J et al, LADOAN, 1984.
53. Liou T M, Hall D A, Santavicca D A and Bracco F V : Laser Doppler Velocimetry Measurements in Valved and Ported Engines, SAE Paper 840375, 1984.
54. Hayder M E, Varma A K, Bracco F V : A Limit to TDC Turbulence Intensity In Internal Combustion Engines, AIAA J. of Propulsion and Power, V.1, No.4, 1985.
55. Arcoumanis C, Bicen A F and Whitelaw J H : Squish and Swirl-Squish Interaction in Motored Model Engines, J.Fluids Engng., V. 105, 1983.
56. Rask R B and Saxena V : Influence of Geometry on Flow in the Combustion Chamber of a Direct Injection Diesel Engine, ASME-FED.Vol.28, 1985.
57. Martin J K, Witze P O and Borgnakke C : Combustion Effects on the Preflame Flow Field in a Research Engine, SAE Paper 850122, 1985.

58. Hall M J , Bracco F V and Santavicca D A : Cycle-Resolved Velocity and Turbulence Measurements in an IC Engine with Combustion, SAE Paper 860320, 1986.
59. Hall M J and Bracco F V : A Study of Velocities and Turbulence Intensities Measured in Firing and Motored Engines, SAE Paper 870453, 1987.
60. Foster D E and Witze P O : Two-Component Laser Velocimeter Measurements in a Spark Ignition Engine, SAND 87-8817, 1987.
61. Foster D E and Witze P O : Velocity Measurements in the Wall Boundary Layer of a Spark-Ignited Research Engine, SAE Paper 872105, 1987.
62. Gosman A D, Tsui Y Y and Watkins A P : Calculation of Three-Dimensional Air Motion in Model Engines, SAE Paper 840229, 1984.
63. Gosman A D : Computer Modelling of Flow and Heat Transfer in Engines:Progress and Prospects, Proc. JSME Symposium on Diagnostics and Modelling of Combustion in Reciprocating Engines, Tokyo, 1985.
64. Arcoumanis C, Green H G and Wong K Y : Flow Structure Visualisation in Reciprocating Engines, Proc. 4th Int. Symposium on Flow Visualization, Paris, 1986.
65. Melton R B : Diesel Engine Fuel Injection Viewed as a Jet Phenomenon, SAE 710132, 1971.
66. Green H G and Whitelaw J H : Scalar Field Measurements in Reciprocating Engines, AIAA J. Propulsion and Power, 1987.
67. Green H G, Whitelaw J H and Wong K Y : Air-Fuel Mixture Characteristics of Reciprocating Engines, Comb. Sci. Tech., 1987.
68. Arcoumanis C, Whitelaw J H and Wong K Y : Gaseous Simulation of Diesel Sprays in Motored Model Engines, in preparation.
69. Kadota T : The Application of Laser Rayleigh Scattering to the Local Mixture Strength Measurements in SI Engine During Intake Stroke, SAE Paper 872151, 1987.
70. Smith J R : Turbulent Flame Structure in Homogeneous-Charge Engine, SAE Paper 820043, 1982.
71. Gosman A D : Application of LDA Measurements to Computational Fluid Mechanics, Proc.2nd Int. Conference on Laser Anemometry, Glasgow, 1987.

TABLE 1
Properties of materials

PROPERTY	CAST IRON	SAPPHIRE	QUARTZ	PLEXIGLASS
Tensile Strength (lbf/sq.in x 10^3)	60	40-58	7	10 (20^0C)
Coefficient of thermal expansion ($^0C^{-1} \times 10^{-6}$)	12	8	0.6	0.7
Melting Point ^{0}C	1200	2040	1700	80
Thermal Conductivity (Btu/h.ft.^{0}F)	26	14.5	0.8	1.3

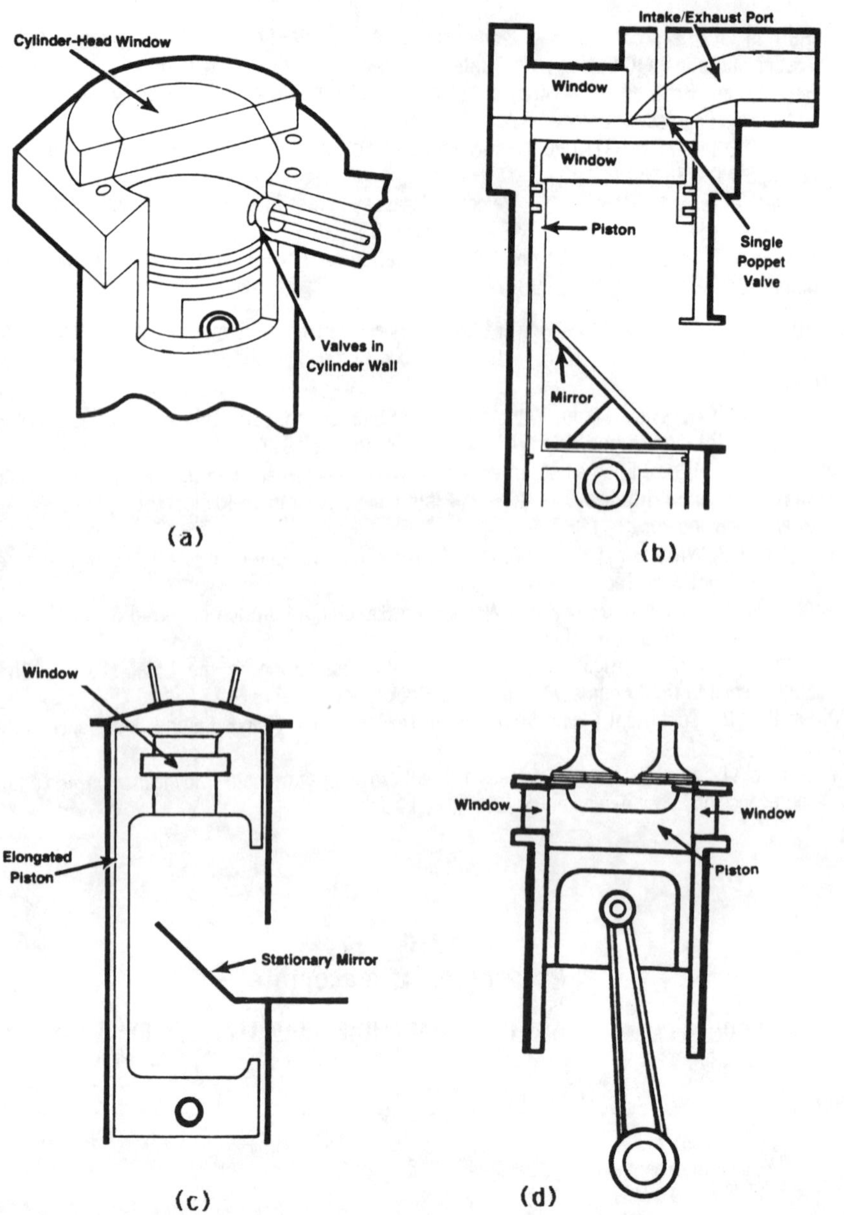

Fig. 1 Schematic diagrams of optical engines with (a) access through the cylinder head (b) access through the cylinder head and piston crown (c) access through the piston-bowl base (d) access through windows in the cylinder liner or through a transparent ring. (based on (19))

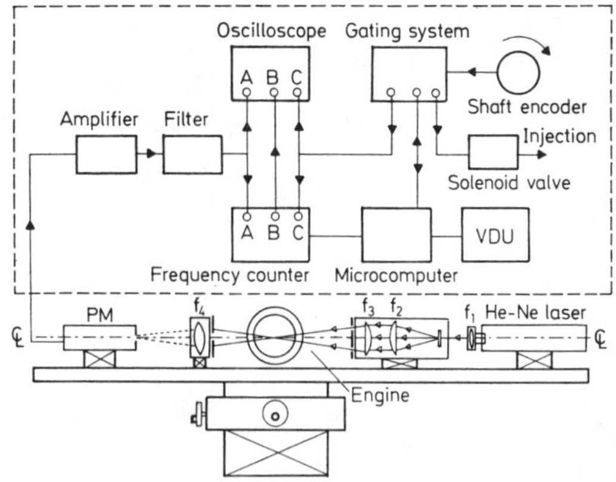

Fig. 2 Schematic diagram of LDA optics and signal processing system.

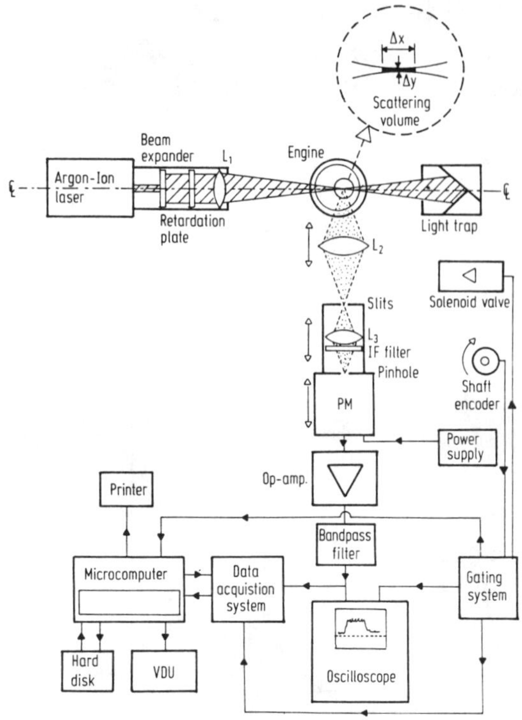

Fig. 3 Schematic diagram of LRS optics and signal processing system.

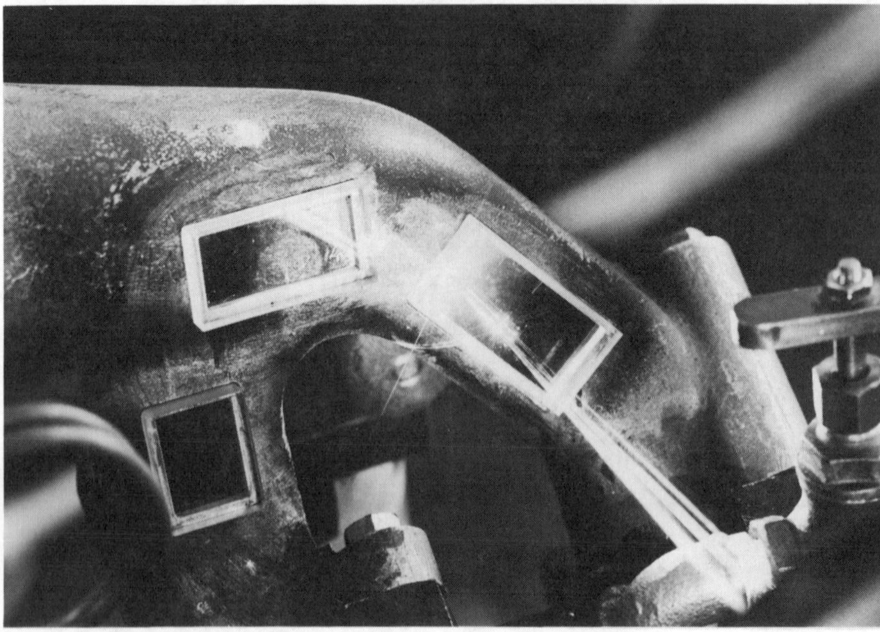

Fig. 4 Photograph of inlet manifold with perspex windows (42).

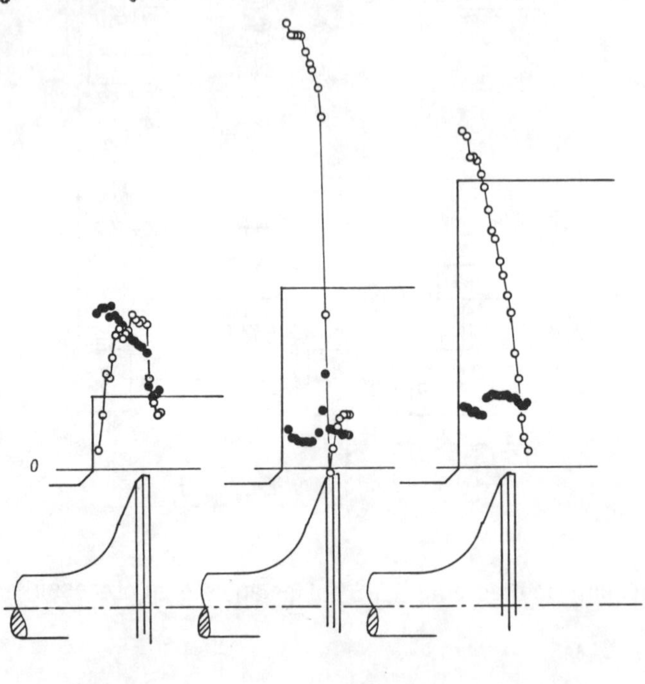

(a)

Fig.5 con't...

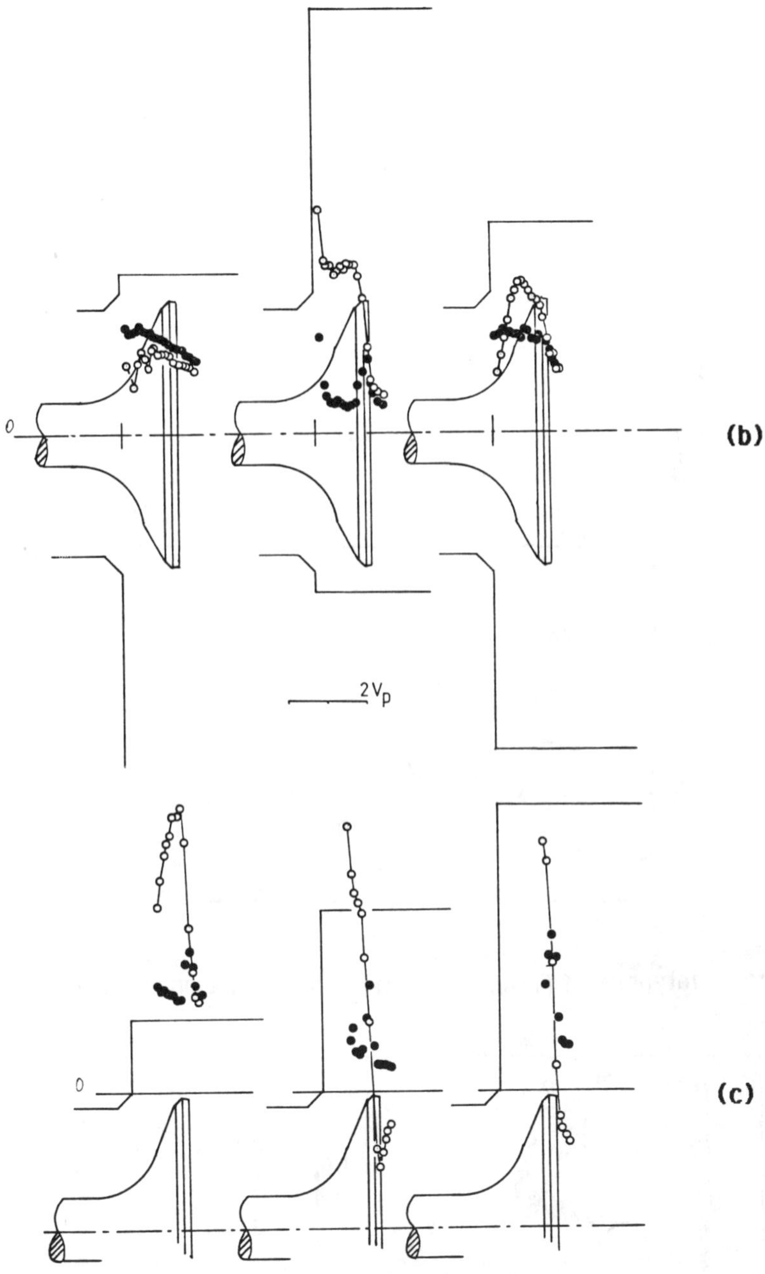

Fig. 5 Velocity distributions through the inlet valve of a prototype diesel engine under steady flow conditions (a) radial (b) swirl (c) axial velocity.

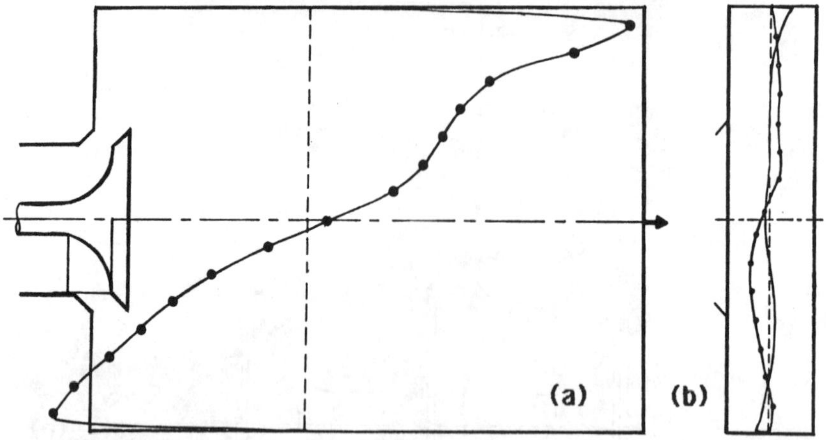

Fig. 6 Axial velocity profile indicating the formation of a rolling vortex at (a) inlet valve closing (b) top-dead-centre (50).

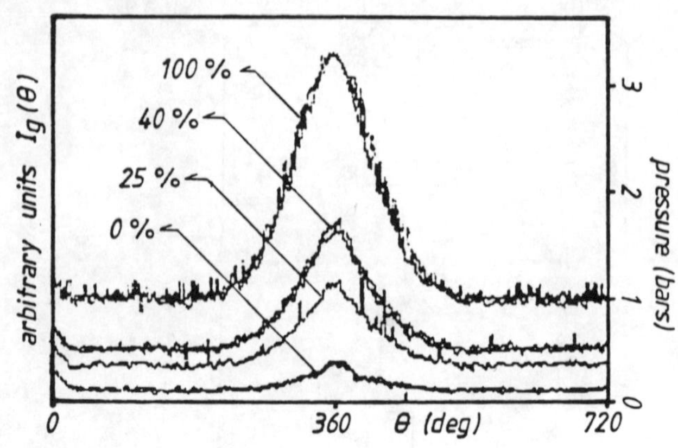

Fig. 7 Intensity of Freon/air mixtures as a function of crank angle (66).

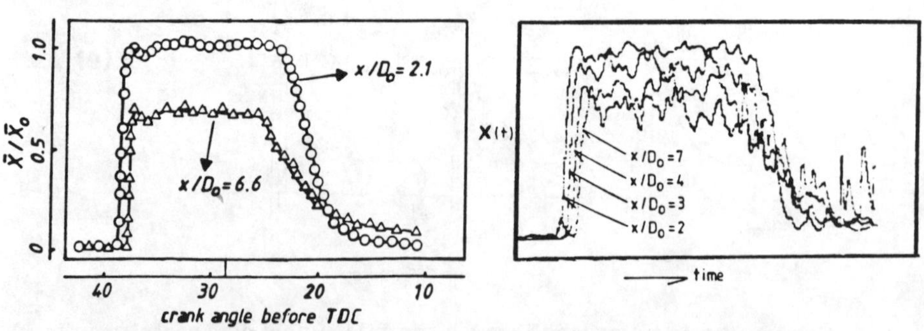

Fig. 8 Ensemble-averaged and instantaneous injection schedules as a function of downstream distance from the nozzle (66).

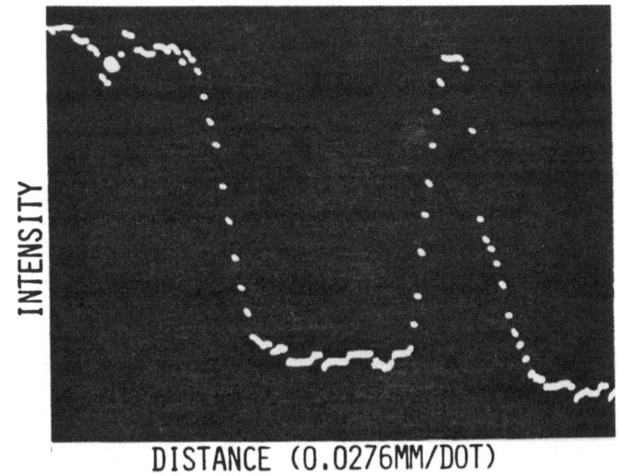

Fig. 9 Density (intensity) profile across the propagating flame (70).

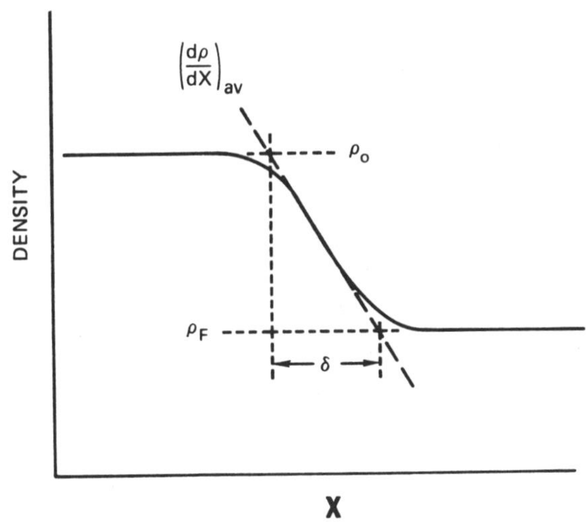

Fig. 10 Method of estimating flame thickness from the density profile across the flame (70).

INFLUENCE OF COMBUSTION ON LASER DOPPLER VELOCIMETER SIGNAL QUALITY IN A SPARK IGNITION ENGINE

P. O. Witze[1] and T. A. Baritaud[2]

[1]Sandia National Laboratories, Livermore CA, USA
[2]Institut Français du Pétrole, Rueil-Malmaison, France

ABSTRACT

Results are presented for an experimental investigation of the effects of premixed combustion on the accuracy and data rate of laser Doppler velocimeter measurements within a reciprocating engine. The question of interest is whether refractive index gradients at the boundaries of the burned gas region significantly defocus, deflect, or phase shift the incident laser beams, thereby disturbing the probe volume. It is shown that combustion has a negligible effect on both mean velocity and turbulence measurements, but can seriously degrade the signal quality, and thus the data rate. It is also shown that water condensation is a serious problem when a sapphire window is used, as compared to fused silica, and that the use of a laser line interference filter will significantly improve the data rate obtained from the burned gas region.

1. INTRODUCTION

The application of laser Doppler velocimetry (LDV) to reacting turbulent flows is generally faced with uncertainties about the effects of strong refractive index gradients on signal quality and the accuracy of velocity measurements. Hong, Jones and Weinberg (1) have shown that a moving interface can modify both the direction and phase of a traversing laser beam. In a dual-beam LDV system, either effect will cause the interference fringe pattern to move, creating an apparent velocity. The problem could be particularly severe during premixed homogeneous combustion in a spark ignition engine, since at elevated pressures the temperature interface across the flame has been shown by Smith (2) to be very thin, of the order of 0.1-0.4 mm. Dyer (3) has shown how the schlieren deflection of a single laser beam by a flame surface can be used to locate a transient flame front with excellent temporal resolution. Typically, this measurement is made far from the combustion zone, to take advantage of a long lever arm for the deflected laser beam. This flame/laser-beam interaction can only be detrimental to LDV, but it is not evident how serious the problem can be because the lever arm involved is much smaller. The theoretical analysis and experimental measurements of Hashimoto, et al. (4) conclude that the problem is serious for near-tangential incidence of the flame front with the laser beams, confirming earlier predictions (1). The experiment was performed by passing the incident LDV beams completely through a combustion bomb to a scattering surface outside the device, such that the lever arm was much greater than it would be for an in situ measurement. Their study emphasized the pseudo-velocities that refraction can create, but did not consider the effects of combustion on the LDV

D. F. G. Durão et al. (eds.), Instrumentation for Combustion and Flow in Engines, 255–266.

signal quality and data rate. This latter issue is of particular interest to us, because a previous study (5) reported serious difficulties in obtaining LDV measurements from the burned gas region. The loss in signal appeared to depend on the type of seeding particle used (aluminum oxide and titanium dioxide), and also suggested a sensitivity to the orientation of the two crossing beams with the flame surface.

The goal of the present study was to investigate the potential sources of LDV bias and signal quality deterioration caused by combustion. Because quantitative evaluation is difficult if the scattering source is random, the LDV signal was created by a controlled source. The scattering surface was located within the combustion chamber to identically reproduce the environment encountered during actual LDV measurements from particles. While this choice precluded the study of effects of combustion on particle scattering properties, it did permit isolation of the incident beam problems described above.

Early in the study it became evident that water vapor was condensing on the sapphire access window, totally blocking the incident laser beams during the expansion stroke. Setchell (6) noted this problem during his demonstration of Raman spectroscopy in an engine, and solved it by preheating the engine head and sapphire window to 180°C. Asanuma and Obokata (17) used a quartz window for LDV studies in a combusting engine, and also found it necessary to preheat the window to prevent fogging. In light of these problems, the first section of this paper is concerned with simply the transmission of a laser beam through a window during combustion.

The one other experimental detail studied was the need for a laser line interference filter to block the visible combustion emissions from the LDV signal. It is generally standard practice to use such a filter, but Rask (8) has succeeded in obtaining velocity measurements throughout the combustion period without one. We thus thought it would be useful to evaluate the benefits, if any, created by the use of a filter.

2. RESEARCH ENGINE

The measurements were made in the single cylinder Sandia research engine, which has been described in detail elsewhere (9). The unique feature of the engine is the location of the valves in the sidewall of the disc-shaped (right-circular-cylinder) combustion chamber, which permits optical access of the entire combustion chamber through a large window over the piston. There are two additional opposing windows in the sidewall of the clearance volume.

The engine was run using gaseous propane as the fuel, at stoichiometric conditions. To assure complete scavenging of residual combustion products, the engine was only fired every fifth cycle. LDV measurements were taken in the fourth and fifth cycles, to contrast motored and fired results. In this skip-fired mode, it is more necessary to heat the engine than to cool it. Water was used to preheat the engine cylinder and head, and electrical heat tape was used to preheat the intake air as it passed through a bed of copper beads. An electric motor was used to externally drive the engine, acting as a brake only during the fired cycles.

The 514.5 nm line of a 15 W argon-ion laser was used for all the experiments performed. The focal length of the LDV focusing lens was 450 mm, and the beam separation distance was 50 mm. The LDV signals were processed with a burst counter, pressure was measured using a water cooled piezoelectric transducer, and the crank-angle position was obtained from a digital shaft encoder with 0.2 crank-angle-degree (cad) resolution. The data were recorded serially by a PDP-11/34 computer.

3. WATER VAPOR CONDENSATION

It may not be readily apparent that water vapor can condense on surfaces in an engine cylinder upon the completion of combustion, when the gas temperature is far in excess of 2000 K. However, this can be a very real problem when windows are used to gain optical access to the combustion chamber. Condensation occurs in the thermal boundary layers because of the combination of the large water vapor content of the combustion products (12 percent mole fraction for stoichiometric propane/air) and the high combustion chamber pressure. The problem is enhanced if the engine is not fired every cycle, because of the increased time for heat transfer to cool the window surfaces between combustion events. In light of this dependence on heat transfer, the thermal diffusivity and, in particular, the thermal conductivity of the window material can be very important.

We have routinely used sapphire for our windows because of its much greater strength than fused silica, the next best candidate. However, sapphire has a room temperature thermal conductivity that is 18 times larger than that of fused silica, and we have recently come to realize that this property characteristic of sapphire is highly undesirable. The experiment to investigate this problem used a small mirror attached to a rod and positioned inside the combustion chamber at the LDV focal volume. One of the LDV beams was blocked, and a photodiode was positioned to catch the other, reflected beam. Because it was not desired to sense deflection of the beam from refraction, the full 8 mm aperture of the photodiode sensor was used.

Tests were performed using both sapphire and fused silica windows, for which the temperature of the window was varied by changing the temperature of the preheated intake air and the water circulating in the engine water jacket and head. Figure 1 summarizes the results, where the crank-angle phasing of the data have been adjusted slightly to account for the change in combustion rate brought about by the change in unburned gas temperature. It is seen that even the 135°C maximum preheat temperature we could achieve was not enough to fully prevent condensation on a sapphire window. In contrast, the fused silica window experienced no similar problems; for the 20°C results shown, there was considerable attenuation of the laser beam after the first fired engine cycle, but by the fifth combustion cycle the window was sufficiently heated such that no condensation occurred.

With the engine running at low speed (i.e., 100-200 rev/min), the water condensation on the sapphire could be observed visually. More dramatic, however, and clearly evident at any engine speed, was the effect of the water on the laser beam at the photodiode. With each fired engine cycle the beam "bloomed" to fill the entire aperture of the LDV collection lens, having a speckle-like pattern, and then slowly recovered to a beam-like quality as the water vaporized. Usually, a greatly reduced but similar effect could be seen in the first motored compression cycle following a fired cycle.

As a result of this transmission investigation, a fused silica window was used for the remainder of the reported study.

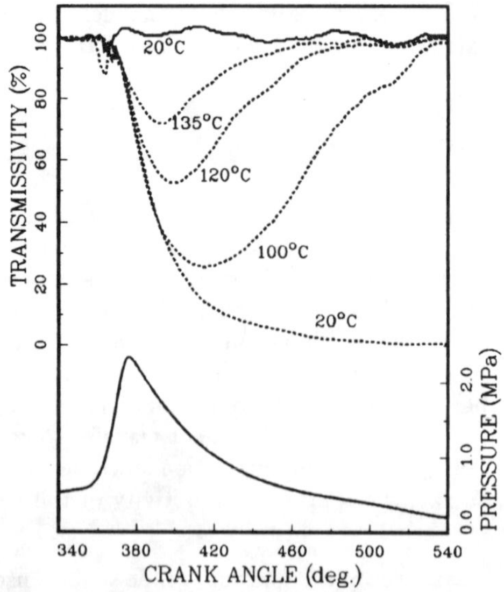

FIGURE 1. Loss in sapphire (dashed line) and fused silica (solid line) window transmission because of water condensation.

4. LASER BEAM REFRACTION BIAS

To investigate the effect of refraction-induced motion of the incident LDV laser beams by the flame, it was necessary to create a controlled Doppler signal source within the engine combustion chamber. After many unsuccessful attempts, a suitable LDV signal was finally obtained by scattering from a spinning rod of 6 mm diameter. The rod was painted black, and had twenty-four uniformly spaced lines scribed along the surface parallel to the spin axis. The rod was inserted into the engine through an oil-impregnated bronze sleeve bushing mounted in an access port in the side wall of the combustion chamber, as shown in Fig. 2. A small, directly-coupled electric motor was used to spin the rod. Because of the high pressures in the engine, it was necessary to place a collar on the rod to prevent it from moving axially during compression and combustion. This led to concern that the rod speed could slow down during combustion, due to increased friction between the collar and its seat on the bushing, so this area was machined for minimum contact area.

The overall performance of this source of Doppler signals proved to be quite satisfactory. The rod showed no evidence of slowing down during either compression or combustion. However, it was also found that the rod speed could not be exactly repeated from test to test, apparently because of binding within the bushing caused by imprecise alignment of the rod with the electric motor shaft. During each test, which typically lasted slightly more than one minute, the rotational speed of the rod would vary a small amount, not on a cycle-by-cycle basis, but rather as a drift throughout the test duration caused by friction heating within the bushing. This drift was the source of the ensemble averaged standard deviations in rod speed to be reported in this paper. Because the drift was slow compared to the cycle time, the comparison measurements between the motored fourth cycle and the fired fifth cycle are not biased by it, and thus are a true indication of the effects of combustion on LDV measurements.

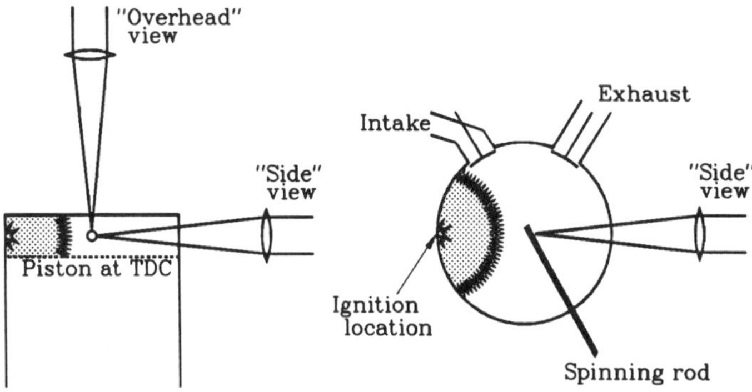

FIGURE 2. Optical configurations used to study velocity bias in LDV measurements caused by refraction-induced beam steering.

Two LDV configurations were used, both in direct backscatter, as shown in Fig. 2. The configuration observing from above the piston is our standard arrangement when measuring gas velocities from seed particles, except that we prefer to collect the scattered light at right angles to the incident optical axis, to avoid background noise from the laser flare off the window and piston surfaces.

Overhead-view results for the spinning rod are shown in Fig. 3, averaged from an ensemble of 35 motored and fired engine cycles. The data have been ensemble-averaged into windows having two cad resolution, with nominally 100 samples in each window average (although it will be shown that combustion can greatly reduce this number). Both the mean velocity and standard deviation are presented as the difference between the ensemble average and the time-averaged motored measurements, such that the results shown with the engine fired are the absolute bias errors in the mean velocity and turbulence intensity introduced by the combustion event. The ensemble-averaged motored engine data are shown to illustrate the accuracy of the measurements.

The mean velocity has been plotted using a very high resolution scale, and the results show that the data are not affected by combustion within the accuracy of the measurement. In contrast, the standard deviation in velocity shows a pseudo-turbulence occurring during combustion, with a maximum absolute value of approximately 1.0 cm/s occurring at the time of maximum rate of pressure rise. The 6-7 cad width of the pertubation is representative of the amount of cyclic variation observed in the combustion period. We are satisfied that the increase in standard deviation shown is not due to a change in the rotational speed of the rod, which conceivably could occur because of increased friction between the collar on the rod and the guide bushing. If a significant increase in friction was present, the mean velocity of the rod would be reduced, with the maximum change occurring at the time of maximum pressure. This condition is clearly not seen in the mean velocity measurements.

The particular result shown in Fig. 3 was obtained with the incident laser beams oriented to measure the velocity component normal to the flame surface. The result is atypical of the experiment, however, and in fact represents the most extreme case that we achieved out of a large number of tests. In general, when the LDV beams were introduced from above the piston, a combustion effect on the measured velocity was seldom evident; this observation is valid for all orientations of the two laser beams

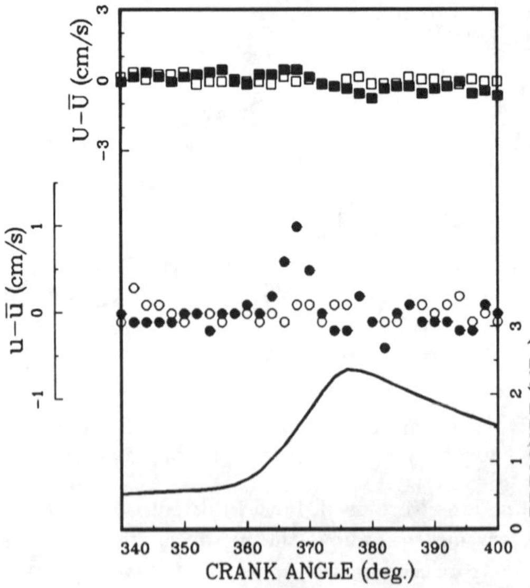

FIGURE 3. LDV measurements of the variations in mean velocity and standard deviation for motored (open symbols) and fired (closed symbols) cycles for a spinning rod observed from above the engine cylinder. $\overline{U}$ and $\overline{u}$ are the time-averaged values of mean velocity and standard deviation for the motored engine cycles.

relative to the flame surface. Even for the case shown, where it might appear there is an effect, in fact the bias error is negligible. The maximum perturbation in the standard deviation of approximately 1.0 cm/s would be insignificant to any gas velocity measurements made in an engine, where typical velocities are of the order of meters per second.

The nature of this combustion-induced pertubation is more clearly seen in Fig. 4, which shows the probability distribution of the individual velocity measurements for the crank-angle interval from 365 to 371 cad. The motored and fired distributions are essentially identical, except for a few outlying points in the fired case that were obviously affected by combustion. Only 5 points out of the total of 341 in the distribution lie in this region, but because of the squared behavior of the computation the standard deviation is perceptibly affected. However, in this instance the magnitude of the error is inconsequential, and in addition the identification of the statistical outlyers is readily apparent, such that they could be judiciously edited without fear of biasing the result.

The effects of combustion on the LDV signal are more pronounced when the incident laser beams are introduced from the side, which is the expected result because of the much longer path length from the window to the scattering surface. The data presented in Fig. 5 show a small but distinct pertubation in the mean velocity, and a larger increase in the standard deviation. The spark location was opposite the LDV access window, as diagrammed in Fig. 2, so that the flame surface was between the probe volume and the window from soon after the beginning of the pressure rise until peak pressure was reached. The effect of combustion on the measured velocities is seen to increase as the path length from the flame surface to the probe volume increased.

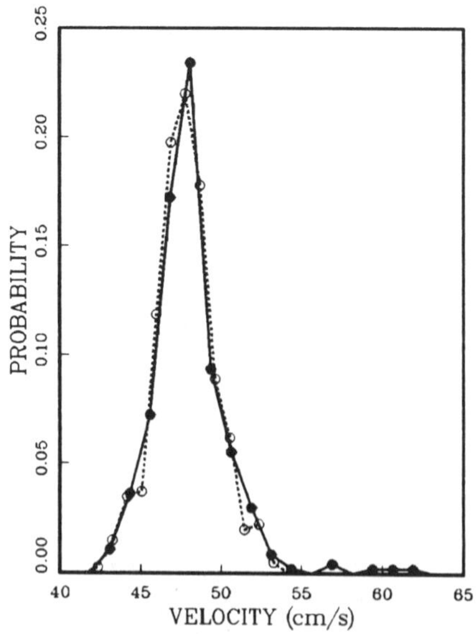

FIGURE 4. Probability distributions of motored (dashed line) and fired (solid line) velocities for a spinning rod observed from above the engine cylinder.

Unlike the results in Fig. 3, the measurements shown here are typical of all test results for this beam orientation. However, the conclusion derived from the data shown in Fig. 4 is the same as the previous case, in that the amplitude of the pertubations is still insignificant. For the maximum pertubations shown, the mean velocity has been biased by only 2.0 cm/s, and the standard deviation by only 1.3 cm/s.

The corresponding probability distributions for the region from 365 to 371 cad are given in Fig. 6, and show a significant distortion in the shape of the distribution with combustion. The distribution is not characterized by obvious statistical outlyers, so editing of the results to correct for combustion-induced bias errors would not appear to be feasible. The key point, however, is that the magnitudes of the bias errors do not warrant editing procedures.

While it is our conclusion that refraction-induced bias errors have a negligible impact on the accuracy of LDV measurements made in engines, the same cannot be said about the LDV data rate, which is shown in Fig. 7. The measurements made from above the piston show no change in data rate during flame passage, which occurs at approximately 365 cad.* The data rate does fall by 20 percent near the time of maximum pressure, which corresponds to the completion of combustion, after which it remains nearly constant throughout the remainder of the expansion stroke. We suspect that

*The large, low frequency oscillations in the data rate for the top window viewpoint were caused by engine vibration induced variations in the incident laser beam intensity. The vibration caused in-phase movement of the laser beam at the LDV beam splitter, resulting in amplitude modulation of the two beams exiting the splitter. This problem was discovered midway through the studies being reported, and was remedied for all measurements for which it was critical.

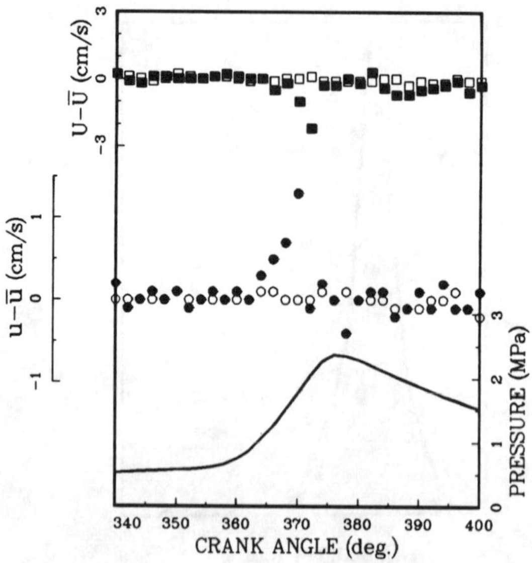

FIGURE 5. LDV measurements of the variations in mean velocity and standard deviation for motored (open symbols) and fired (closed symbols) cycles observed from the side of the engine cylinder.

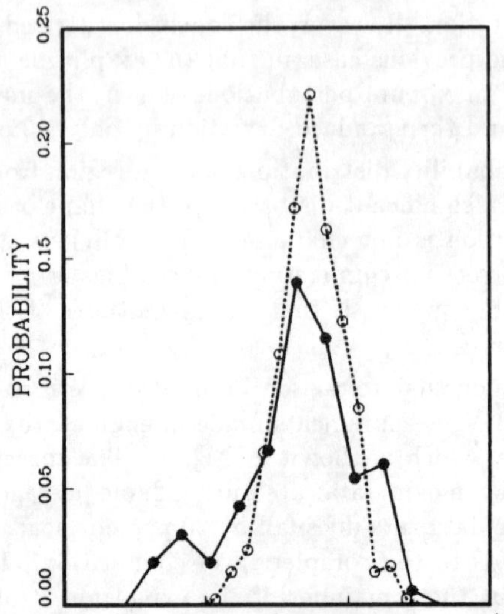

FIGURE 6. Probability distributions of motored (dashed line) and fired (solid line) velocities obtained looking from the side of the engine cylinder.

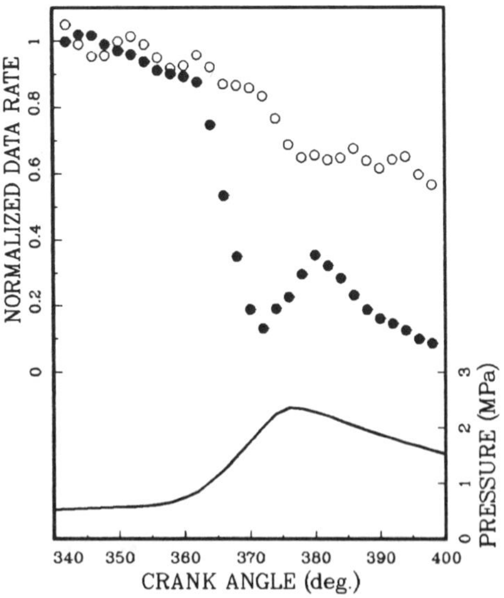

FIGURE 7. Comparison of LDV data rates between the "overhead" (open symbols) and "side" (closed symbols) views of the LDV optics.

this loss is caused by a reduction in signal quality related to thermal boundary layers at the window surfaces. In contrast to these results, the data rate obtained when viewing from the side shows a dramatic reduction from refraction during flame passage, followed by a recovery to a greatly reduced and continually decaying level during expansion.

We do not have a definitive explanation for the slow decay in data rate during compression, prior to combustion, nor for the greater losses in signal during the expansion stroke, but believe these may both be caused by refraction in the thermal boundary layer on the window surface. A previous study (10) has shown that the refractive gradients next to a window are large enough for shadowgraph visualization of swirling air motions under motoring conditions.

5. VISIBLE EMISSIONS

The combustion of a stoichiometric propane/air mixture produces a strong white-colored emission. When the mixture is richened, the color becomes increasingly orange, while when it is leaned it becomes partially clear, with patches of both blue and orange colors. Figure 8 shows the shape of the visible emission profile, measured with a photodiode. It is seen to be very similar to the pressure profile.

Qualitative measurements were made to determine the blocking effectiveness of a typical LDV collection aperture (130 microns), and a 10 nm full width at half-maximum laser line interference filter. Adding first just an aperture before the photodetector, the emission signal was reduced below the detectability limit, indicating at least a 100-fold decrease in the emission level. The sensitivity of the detector was then increased to achieve full-scale response to the light level passing through the aperture, and then the interference filter was added to the system. Once again the signal was reduced below the detectability limit.

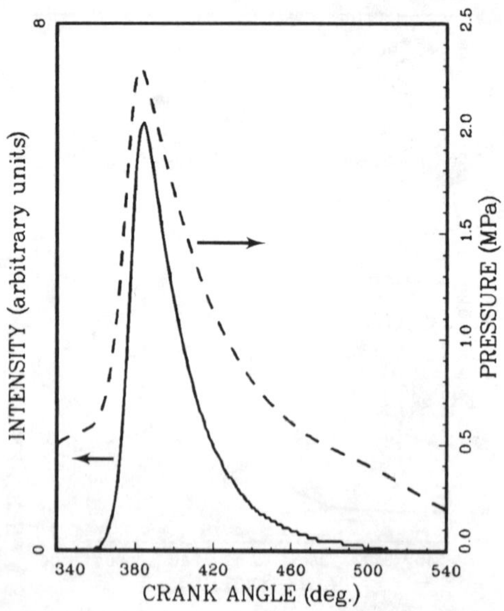

FIGURE 8. Profile of visible emissions from stoichiometric propane/air combustion, showing its similarity to the pressure history.

Because the laser line interference filter used transmits only 50 percent of the LDV signal, an experiment was performed to determine if the color filter was needed to make LDV measurements. The previously described spinning rod setup was used, in the top view configuration. This orientation does minimize the potential influence of visible emissions, but was chosen because it is the normal configuration used in our laboratory. Measurements made with and without the filter are compared in Fig. 9. The mean velocity is seen to be unaffected, whereas the standard deviation suggests that there are some problems with the signal quality. Even more important, however, is the effect the color filter has on the data rate, which is shown in Fig. 10. Although the color filter does reduce the signal intensity level by 50 percent, the data rate after the completion of combustion is twice that of the unfiltered test.

6. CONCLUSIONS

As a result of the described investigations on combustion-induced water condensation, beam refraction, and visible emissions, the following conclusions are offered regarding the application of LDV to reciprocating engines:

1) Velocity bias errors from flame-induced refraction of the incident laser beams are insignificant. However, higher data rates can be achieved if the optical path length through the postflame gases is minimized.
2) Sapphire is not a suitable window material for post-combustion studies, because excessive preheat temperatures must be maintained to avoid water condensation on its surface.
3) Although the use of a laser line interference filter before the LDV photodetector reduces the intensity of the scattered signal by 50 percent, the data rate during combustion is significantly improved because of rejection of emission noise.

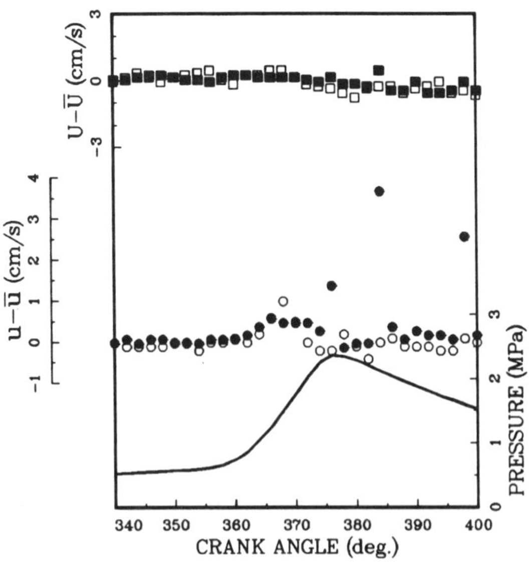

FIGURE 9. Comparisons of LDV measurements of mean velocity and standard deviation made with (open symbols) and without (closed symbols) a laser line interference filter.

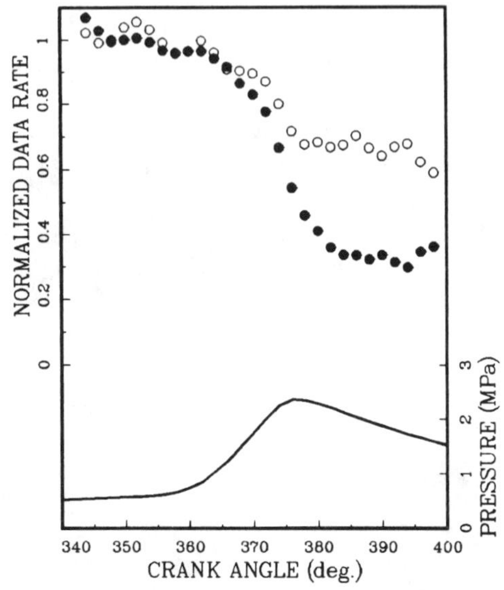

FIGURE 10. Comparison between LDV data rates obtained with (open symbols) and without (closed symbols) a laser line interference filter.

ACKNOWLEDGMENT

This work was performed at the Combustion Research Facility of the Sandia National Laboratories, and was funded by the U.S. Department of Energy, Energy Conversion and Utilization Technologies Program, the Motor Vehicle Manufacturers Association, and the Groupement Scientifique Moteurs, France.

REFERENCES

1. Hong N-S, Jones A R, Weinberg F J: Doppler Velocimetry within Turbulent Phase Boundaries. Proceedings of the Royal Society A, 353, 77-85, 1977.
2. Smith J R: Turbulent Flame Structure in a Homogeneous-Charge Engine. Transactions of the SAE, 91, 150-164, 1982.
3. Dyer T M: Characterization of One- and Two-Dimensional Homogeneous Combustion Phenomena in a Constant Volume Bomb. Transactions of the SAE, 88, 1196-1216, 1979.
4. Hashimoto M, Hamamoto Y, Ohkawa H, Sata 0: Laser Doppler Velocimetry - Effect of Refractive Index Change at Flame Front. Bulletin of the JSME, 26, 1908-1913, 1983.
5. Witze P 0, Martin J K, Borgnakke C: Conditionally-Sampled Velocity and Turbulence Measurements in a Spark Ignition Engine. Combustion Science and Technology 36, 301-317, 1984.
6. Setchell R E: Initial Measurements within an Internal Combustion Engine using Raman Spectroscopy. Sandia National Laboratories Report SAND78-1220, 1978.
7. Asanuma T, Obokata T: Gas Velocity Measurements of a Motored and Firing Engine by Laser Anemometry. Transactions of the SAE, 88, 401-409, 1979.
8. Rask R B: Laser Doppler Anemometer Measurements in an Internal Combustion Engine. Transactions of the SAE, 88, 371-382, 1979.
9. Johnston S C, Robinson C W, Rorke W S, Smith J R, Witze P 0: Application of Laser Diagnostics to an Injected Engine. Transactions of the SAE, 88, 353-370, 1979.
10. Witze P 0, Vilchis F R: Stroboscopic Laser Shadowgraph Study of the Effect of Swirl on Homogeneous Combustion in a Spark-Ignited Engine. Transactions of the SAE, 90, 979-992, 1981.

REFRACTIVE INDEX MATCHING FOR LDV MEASUREMENTS NEAR WALLS AND IN COMPLEX GEOMETRIES

J.C.F. PEREIRA

Instituto Superior Técnico
Dept. Mech. Eng.
Av. Rovisco Pais
Lisbon - Portugal

1 - INTRODUCTION

Very often the velocity characteristics in the near wall region of many flows are of great relevance to the understanding of transport processes that are relevant to engineering applications. Flows with heat transfer in curved smooth or rough walls comprising multiple recirculation regions or flows of polymer solutions which exhibit strong drag reducing effects are flow examples where detailed characterization of the velocity problems is required close to the walls. Flows with complex geometries with or without internal solid walls or flows in packed or fluidized liquid-solid beds are examples where velocity measurements are difficult to obtain. Several measuring techniques can be used, the non intrusive LDA is in many situations the most adequate technique. However, its applicability can be limited by the changes of position of measuring volume which occurs when light is passed through solids or fluids of different refractive index.

To avoid this difficulty a suitable fluid should be selected with the same refractive index of the transparent material used to build the solid walls or the objects placed inside the flow stream. The matched index of refraction should be maintained through the experiments, so temperature control may constitute an additional problem but the method increases the accuracy and applicability of LDV measurements in continuous or discontinuous isothermal flow phases.

This paper reviews some basic refractive index matching techniques used for eliminating the problems of distortion of the optical paths through curved surfaces in liquid flows. The choice of a fluid, whose index of refraction matches that of the curved walls or objects placed inside the flow, is advised based in theoretical and practical considerations. In addition several examples of flow measurements made using index matched facilities will be presented giving special emphasys to the velocity measurements in a time dependent flow in a piston cylinder assembly.

In section two of this paper some theoretical concepts for selection of an appropriate fluid or fluid mixtures with a desired refractive index are outlined together with some practical sugestions. Section three reviews some of recent applications of refractive index matched. Section four presents LDV measurements of mean axial and radial velocities of a time dependent laminar flow in a piston/cylinder assembly, using a matched index of refractation flow system.

D. F. G. Durão et al. (eds.), Instrumentation for Combustion and Flow in Engines, 267–284.

2 - SELECTION OF WORKING FLUIDS AND TRANSPARENT MODELS

2.1 - Refractive-Index Correction Equations

When a light beam enters an optically denser medium, the beam is refracted according to the law of refraction (better known as Snell's law, Jenkins and White, 1976), in which the angle of incidence (α_1) and the angle of refraction (α_2) are related through the refractive indexes of the media, n_1 and n_2

$$n_1 \sin\alpha_1 = n_2 \sin\alpha_2 \tag{1}$$

This rule is applied to the light passage through several layers system as illustrated in Figure 1. This figure shows the position changes occuring

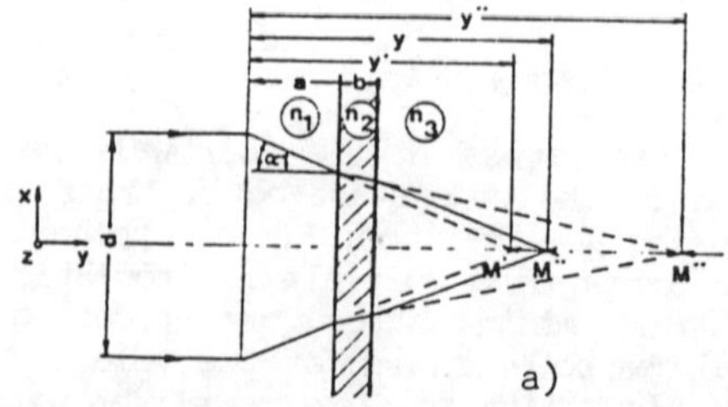

Figure 1. Change of angle between two light beams by light passage through a three-layer system.

due to refractive index differences. The true intersection position of the laser beams inside the medium n_3 is given by:

$$y = [y' + b\,[1 - \frac{\cos\alpha_1}{\sqrt{(\frac{n_2}{n_1})^2 - \sin^2\alpha_1}}]]\;\frac{1}{\cos\alpha_1}\sqrt{(\frac{n_3}{n_1})^2 - \sin^2\alpha_1} \tag{2}$$

For curved interfaces the situation is much more complex as illustrated in Figures 2. The interface acts like a poor lens, bending the rays towards or away from the axis dependending on whether n_w is greater than n_1 or not. The amount of bending also depends on the height of the intersection above the horizontal axis. Figure 2-a shows the real intersection of two beams. The non-linear behaviour of the position of beam intersections may originate large errors if one wants to measure close to a curved wall (as represented in Figure 2-b)). Furthermore if the intersection angles with

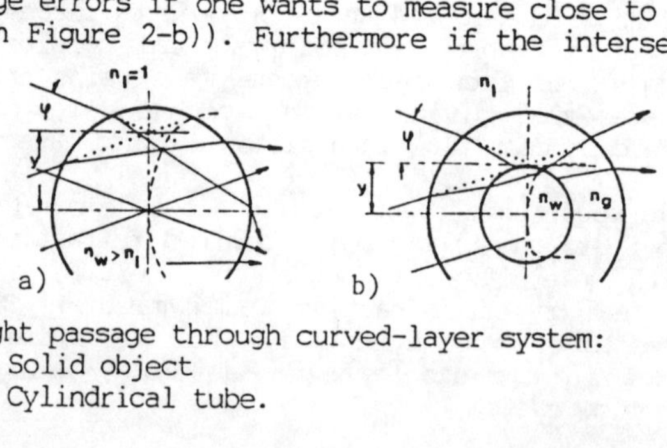

Figure 2. Light passage through curved-layer system:
a) Solid object
b) Cylindrical tube.

the surface normal are greater than the critical angle no beams enter into medium n_w. The height r_a above which no rays enter is given by

$$r_a = R_i \frac{n_w}{n_1} \qquad \text{for} \qquad (n_1 > n_w) \tag{3}$$

Computation of the position of intersection of two rays may be complicated, but it can be given by refractive-index correction equations such as those of Bicen (1982). Figure 3-a illustrates the determination of control-volume position for measurements of axial U-velocity component. The relative distances from the centreplane to the true intersection

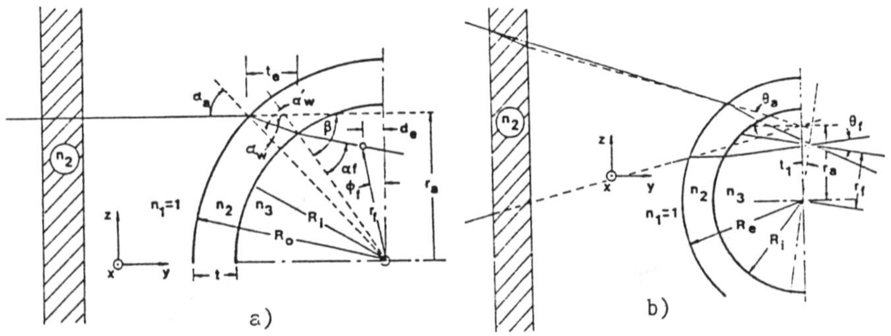

Figure 3. Determination of the control volume position
a) Measurement of the axial velocity component;
b) Measurement of the radial velocity component.

control volume position is given by:

$$r_f = R_i \sin \beta - (R_i \cos \beta - d_e) \tan (\beta - \alpha_f) \tag{4}$$

and

$$\phi_f = \sin^{-1} \left(\frac{d_e}{r_f}\right) \tag{5}$$

being

$$\alpha_a = \sin^{-1} \left(\frac{r_a}{R_o}\right) \quad ;$$

$$\alpha_w = \sin^{-1} \left(\frac{n_1 r_a}{n_2 R_o}\right) \quad ; \quad \alpha'_w = \sin^{-1} \left(\frac{n_1}{n_2} \frac{r_a}{R_i}\right)$$

$$\alpha_f = \sin^{-1} \left(\frac{n_1 r_a}{n_3 R_i}\right) \quad ; \quad \beta = \alpha_a + \alpha'_w - \alpha_w$$

$$de = \sqrt{R_o^2 - r_a^2} - y$$

$$t_e = \sqrt{R_o^2 - r_a^2} - R_i \cos \beta$$

$$y = y_1 \sqrt{\left(\frac{n_3}{n_1}\right)^2 - \sin^2 \alpha'} \Big/ \cos \alpha'$$

$$y_1 = y_2 + te \left(1 - \cos \alpha' \Big/ \left(\sqrt{\left(\frac{n_2}{n_1}\right)^2 - \sin^2 \alpha'}\right)\right)$$

$$y_2 = y_o - R_o + \sqrt{R_o^2 - r_a^2}$$

$$Y_o = R_o \cdot \cos \alpha' \Big/ \sqrt{\left(\frac{n_3}{n_1}\right)^2 - \sin^2 \alpha'} - t \left(1 - \frac{\cos \alpha'}{\sqrt{\left(\frac{n_2}{n_1}\right)^2 - \sin^2 \alpha'}}\right)$$

$$\alpha' = \sin^{-1} \left(\frac{\sin \alpha}{n_1}\right)$$

$$\alpha = \tan^{-1} \left(\frac{D/2}{L_f}\right)$$

where D and L_f are respectively the distance of the incident laser beams and the lens focal lenght.

Figure 4a) shows the determination of the control volume position for the measurement of the radial V-velocity component. The real intersection point is given by r_f and θ_f:

$$r_f = r_a \frac{n_1}{n_3}$$

$$\theta_f = \sin^{-1} \left[\frac{n_1}{n_2} \frac{r_a}{R_i} \cos \alpha'\right] - \sin^{-1} \left(\frac{n_1}{n_2} \frac{r_a}{R_o} \cos \alpha'\right) -$$

$$- \sin^{-1} \left(\frac{n_1}{n_3} \frac{r_a}{R_i} \cos \alpha'\right) + \sin^{-1} \left(\frac{r_a}{R_o} \cos \alpha'\right) \qquad (6)$$

As an example to show the changes in control volume position due to refractive index mismatched let us suppose that one would be interested to measure with one component LDV system the axial and radial velocity components along the simmetry plane inside a cylindrical tube with t = 5 mm thickness and R_i = 45 mm inner radius. The distance between the focusing beams was 50 mm and the lens focal distance was 150 mm. The material of the container box and cylindrical tube was Duran glass with refractive index n_2 = 1.4718. The refractive-index correction equations calculate the relative distances d_e and d_r from the real position of the control volume to the location where the beams would intersect if the refractive index of the fluid and walls was matched. The dimensions of the scattering volume were:

diameter at $1|e^2$ intensity = 0.108 mm
length = 0.99 mm.

Figures 4a) and 4b) show the displacements "d_e" and "d_r" corresponding to changes in refractive index from $n_2|n_3$ = 0.86 to $n_2|n_3$ = 1.05; ($n_1 = n_3$ and n_2 = 1.4718). Up to $r|R_i$ = 0.5 very small deviations occur. However for $r/R_i \geq 0.75$ and in the range $0.95 \leq n_2|n_3 \leq 1.05$ control volume displacements d_e and d_r equal to 0.4 mm were detected.

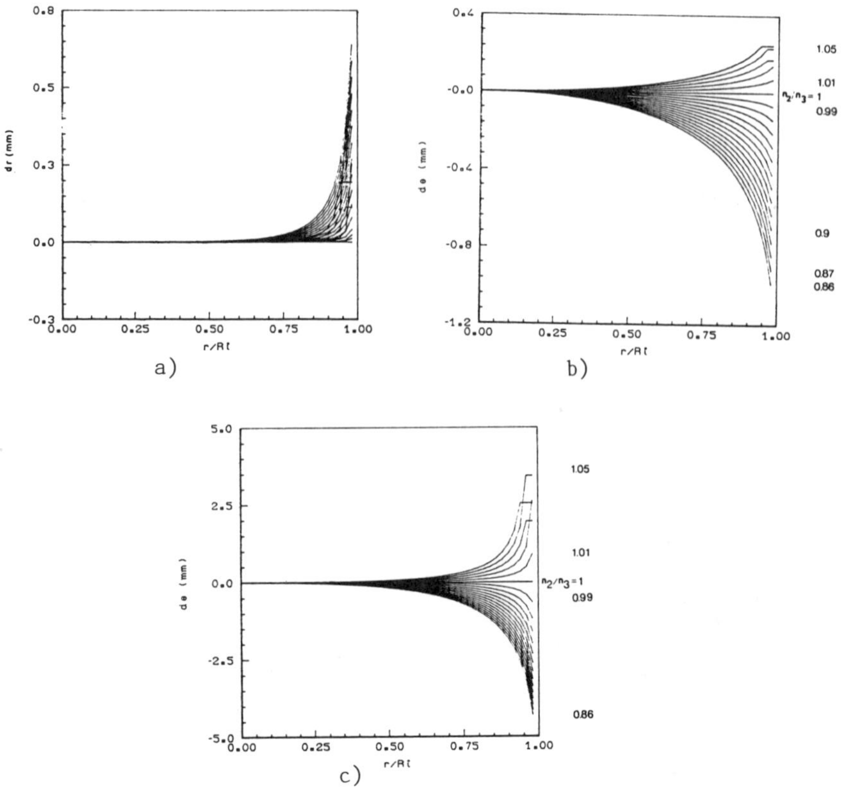

Figure 4. Computation of the control volume position for $n_1 = n_3$ as a function of n_2/n_3.
a) Measurement of axial velocity component;
b) Measurement of axial velocity component;
c) Measurement of radial velocity component.

So measurements of the radial velocity component are very sensitive to displacements of the control volume position as it is shown in figure 4c). This figure shows that a change of 1% in refractive index deviates the intersection beams position of 1.0 mm at r/R_i = 0.98.

2.2 - Basic Theoretical and Practical Considerations on the Calculation of the Refractive Index

Several empirical equations exist to predict refractive index of mixtures, see e.g. Elphick et al (1982), Kleine (1983), Perry (1973). The following are a summary of three existing methods.

Method I: Constancy of the specific refractivity:

The specific refractivity is defined by $K_d = (n - 1)/\rho$, being ρ the density that can be replaced as a function of volumetric concentration, C, for each fluid in the mixture. As an example let us consider only two fluids denoted by 1 and 2. The resulting equation for a two-component mixture is:

$$(\bar{n} - 1) = C_1 (n_1 - 1) + C_2 (n_2 - 1) \qquad (7)$$

and it allows the determination of the refractive index of the mixture n.

Method II: Correlation between density and refractive index:

The Eykman equation correlates density with refractive index for a single component fluid by:

$$\rho = \frac{(n^2 - 1)}{A (n + 0.4)} \qquad \text{A is a constant} \qquad (8)$$

yielding, for a mixture of two fluids:

$$\frac{\bar{n}^2 - 1}{A (\bar{n} + 0.4)} = C_1 \frac{n_1^2 - 1}{A (n_1 + 0.4)} + C_2 \frac{n_2^2 - 1}{n_2 + 0.4} \qquad (9)$$

and solving equation (9) for $\bar{n}$ allows the determination of the refractive index of the mixture.

Method III: Lorentz-Lorenz equation:

The refractive index (n) of a transparent medium depends on the dielectricity ε by

$$n = \sqrt{\varepsilon} \qquad (10)$$

where $$\varepsilon = \frac{1 + 2 \,.\, 4\pi N \,.\, \alpha/3}{1 - 4\pi N.\, \alpha/3}$$

being N the number of molecules per unit volume and α the molecular polarizability. Defining a molar refractivity A,

$$A = \frac{4}{3} \pi.\, N_L\, \alpha \qquad (11)$$

where N_L is the Loschmidt number and introducing Boyl's law $\frac{N_L}{N} = \frac{RT}{P} = \frac{M}{\rho}$, n can be given by:

$$n = \frac{\sqrt{M/\rho + 2A}}{\sqrt{M/\rho - A}} \tag{12}$$

For a mixture of two fluids equation (12) may be applied in the form:

$$P_M = \frac{\overline{M}}{\overline{\rho}} = \frac{\overline{n}^2 + 1}{\overline{n}^2 + 2} \tag{13}$$

to compute $\overline{P}$, $\overline{\rho}$ and $\overline{M}$ the following equations are used:

$$\overline{P}_M = \Sigma\ P_{M_j}\ f_j \tag{14-a}$$

$$\overline{M} = \Sigma\ M_j\ f_j \tag{14-b}$$

$$\overline{\rho} = \Sigma\ \rho_j\ \frac{V_j}{V} \tag{14-c}$$

$$f_j = \frac{N_j}{N} = \frac{V_j\ M_j}{\Sigma\, V_j\ M_j} \tag{14-d}$$

$$V = \Sigma\ V_j \tag{14-e}$$

where V_j is the volume concentration of each fluid in the mixture.

The temperature dependence of the refractive index, n, can be determined by differentiation at an initial temperature to:

$$n = n_0 + \left(\frac{\partial n}{\partial T}\right)_{T_0} \Delta T = n_0\ (1 - K_T\, \Delta T) \tag{15}$$

being the temperature coefficient given by

$$K_T = \frac{\dfrac{3\ AP}{2T_0\ M}}{\left(1 - \dfrac{A\rho}{M}\right)\ \left(1 + \dfrac{2A\rho}{M}\right)} \tag{16}$$

with good accuracy the molecular refractivity is determined by summation of the atomic refractivities of the atoms forming a compound, see Perry (1973). From equation (15) the refractive index decreases with the temperature. The constant K_T may be obtained either theoretically or experimentally using a refractometer. Most organic liquids have temperature coefficients close to $K_T = -4.5 \times 10^{-4}\,{}^oC^{-1}$. Although equations (12) and (15) give the values of the refractive index of the fluid mixtures very often the fluids are not 100% pure and so the chemistry formula is unknown. So experiments should be performed to match exactly the refractive index of the transparent solid materials with the working fluid. A simple experience can be conducted like it is illustrated in Fig. 5. Half laser beams passes through the curved solid wall and the other half through the fluid. If refractive index is matched only a single light spot appears on a wall placed away from the text section.

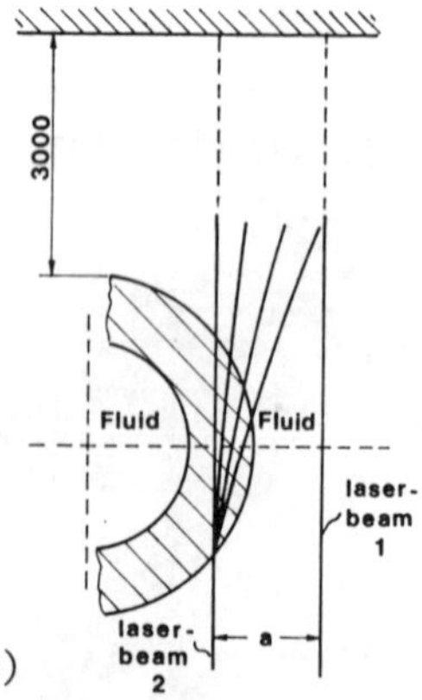

a)

Figure 5. Experimental verification of refractive index matched.

Table I lists some of the most used transparent solid materials materials and Table II gives indication of the refractive index-values for some liquids.

Solid	Index of Refraction
Plexiglax (Perspex)	1.48 - 1.49
Pyrex class	1.473/4
Duran 50	1.471/3
Polystyrene	1.59

TABLE I - Refractive indeces of some transparent glass materials

SOME LIQUIDS THAT MAY BE USED WITH PYREX AND DURANGLASS TUBES
TABLE II

Liquid	n_D	ρ	μ (cP)
Decolin trans	1.4696	0.87	2.4
Glycerol	1.470	1.2563	3.800
Geraniol	1.473 pure 1.494 techn.	0.87	7.95
Trichloroethylene	1.4735	1.4556	0.55
Isophorone	1.474	0.923	2.4
Dimethylsulphaxid	1.4783	1.1	2.3
Furfurylalcohol	1.485	0.93	
Terpineol	1.485	0.93	
Terpene Solvent (Deponol®)	1.477	0.87/89	
Turpentine Oil	1.465-1.478	0.87/89	1.487
Castor Oil	1.477	0.961	986
Olive Oil	1.468	0.918	84
Soybean Oil	1.4729	0.927	70
Light Fuel Oil	1.463	0.83	5.5
Light Fuel Oil +35 vol% Geraniol	1.472		5.35
Light Fuel Oil +8 vol% Perchloroethylene	1.473		3.75
Turpentine +			
60 vol% Trichloroethylene	1.474		0.68
19 vol% Xylene	1.474		1.25
22.5 vol% Toluene	1.474		1.21
14 vol% Benzene	1.474		1.33

The choice of the working fluid is very often limited by the value of its viscosity but many other liquid properties may affect its choice e.g. mixability with other liquids, stability, fugacity, flammability, explosivity and corrosive action on metals, plastics and scalings are examples of important properties of interest in the choice of liquid mixtures.

Table III lists some of the experimental work performed using refractive index matching in single phase flows. Refractive index matching proved to be very useful when local velocity information is required close to curved walls and it extend the range of applicability of LDA technique to complex flow geometries. For two phase flows, Nouri et al. (1987) reported a detailled study of liquid-solid flow in a horizontal pipe. The fluid used were dibutylphthalate that due to its high viscosity was limited to studies up to Re = 1500. For higher Re, a mixture of tetraline and turpentine was used that matched the refractive index of diakon (plexiglass) particles.

Authors Year	Flow Configuration	Material of Solid	Fluid	Main Finding
Durst et al 1979	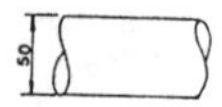 Horizontal pipe	Glass	Light fuel oil + 30% by volume Palatinol ~ 25^oC	Velocity measurements close to wall
Elphick et al 1982	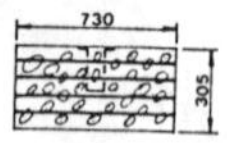 Model heat exchange	Pirex rods 12.7 mm diameter	Xylene + 44% volume Varsol 20^oC $\nu = 9 \times 10^{-6}\ m^2/s$	$n = C_{xy}\ \lvert -6 \times 10^{-5}\ T + 5.9 \times 10^{-2} \rvert - 4.5 \times 10^{-4}\ T + 1.45$ C_{xy} = % volume of Xylene
Durst and Loy 1985	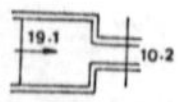 Sudden contraction in a pipe	Glass	Light fuel + 30% by volume Palatinol. 25°	Velocity measurements close to the wall $\mu = (5.2 - 0.16\ T)^{x10^{-3}}$ $20^o \le T \le 25^oC$ $\rho = 869 - 26.74\ T$ $20^o \le T \le 29^oC$
Srivener Lugrin 1986	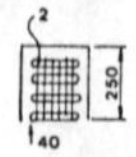 Model of porous cell	Pirex glass $n = 1.4728$	$n_1 = 1.4364$ $n = 1.4969$ $\sim^2 58\%$ by weight oil solvent and lubrificant $T = 20^oC$	$\Delta T = 1^oC \nRightarrow \Delta n = 10^{-3}$ $\mu = 7\ C_p$ Velocity measurements in complex flow geometry
Horvay and Leuckel 1984	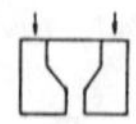 Flow in a swirl chamber	Perspex $n = 1.5$	Tetraline + 40% by weight of turpentine 25°	Measurements in three different swirl chamber configurations $T = 25^oC$ n-terpentine = 1.441 n-tetraline = 1.554 $\nu_{mix} \sim 1.6$ cSt.
Nowshiravani et al 1982	Flow in a rod bounder 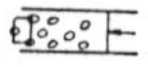	Pirex glass $n = 1.474$	Sohio MDI-57 oil ($n = 1.482$) 28.6% mineral seal oil $n = 1.455$ $T = 23^oC$	$\nu_{mix} = 11.57$ cs $\rho = 824\ Kg/m^3$ Velocity measurements in complex flow geometry
Dybbs and Edwards 1984	Complex rod bundle porous medium 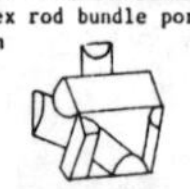	48.5% by weight Dow Corning 556 ($n = 1.46$) 51.5% Dow Corning 550 $n = 1.495$	Pirex $n = 1.474$ $28^o.1$	$n_{mix} = 1.474$ $\nu = 43.3$ cs $\rho = 1030\ Kg/m^3$
Varty and Currie 1984	Flow in tube bundles 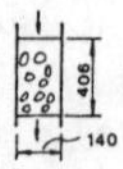	78.7 volume Tetrachloroethylene + 21.3% Freon-113.	Borosilicate Glass $n = 1.4708$ $T = 23$	$n = 1.5 - 0.0417\ C$ C = % Volume Freon - 13 $0 \le C \le 35\%$ $T = 23^oC$

Table III - Previous experimental work using refractive index matching in single phase flow in simple or complex flow geometries

3 - LDA-MEASUREMENTS OF TIME DEPENDENT LAMINAR FLOW AT A SUDDEN TIPE EXPANTION.

3.1 - Apparatus and experimental procedure

The piston cylinder assembly used for this study is schematically shown in Figure 6. The fluid used was a mixture of dibutylphthalate and disel oil, that had the same refractive index (± 0.001) of the Duran glass (1.4718) used to build the cylinder, tube and tank walls of the test section.

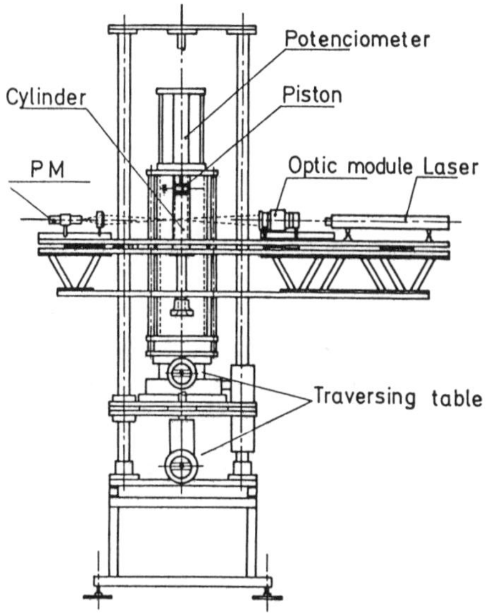

Figure 6. Sketch of test section.

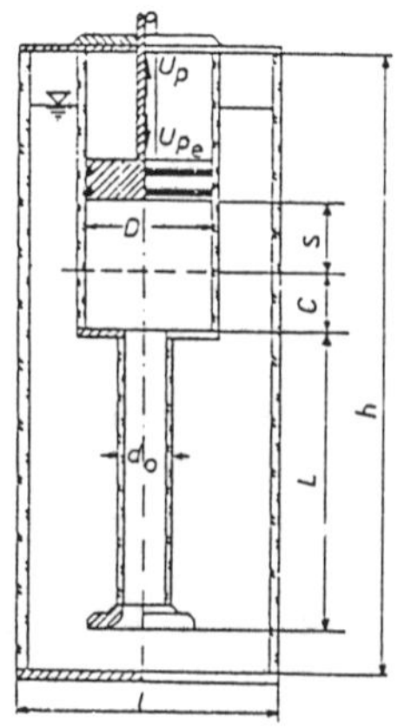

Figure 7. Piston/cylinder geometry.

The fluid entered and left the cylinder through a centrally located, square-edged orifice connected to a flared inlet tube. The flat head piston was driven at a constant speed by a very precise volumetric pump which was electronically controlled.

The geometry of the system shown in Figure 7 is specified by:

Main cylinder diameter	$D = 45$ mm
Orifice diameter	$d_O = 19$ mm
Tube lenght	$L = 200$ mm
Piston stroke	$S = 0$ to 150 mm
Clearance	$C = 0$ to 150 mm
Containment box height	$h = 600$ mm
Containment box width	$l = 200$ mm
Intake piston vel.	$U_p = 1$ to 19.5 mm/s
Exhaust piston vel.	$U_{pe} = 27$ mm/s

The test section, was constructed from Duran glass and aluminum, and all parts were machined to very close tolerance regarding parallelness of walls, surface roughness, manufacturing of step corners, etc.

The clearance gap between the piston and the cylinder walls (0.1 mm) was sealed withfour rubber "0" -rings, that allowed the piston to be traversed without leakage of oil. The test section was placed on the top of a three-dimensional traversing table which allowed the measuring position to be located within ± 0.05 mm in the three coordinate directions.

The two beams were focused by a biconvex lens with a focal lenght of 150 mm. Forward scatterd light was collected through a lens of focla lenght 300 mm and focussed on the pinhole (150 m) of a photomultiplier. Characteristics of the optical system, e.g. Durst et al. (1981), are tabulated below.

- o Intersection half-angle between beams 0 = 9.44
- o Fringe spacing l = 1.927 m
- o Diameter of scattering volume at $1/e^2$ intensity d = 108 m
- o Lenght of scattering volume at a/e^2 intensity l = 990 m
- o Number of fringes N_f = 56

3.2 - Signal processing

The Doppler signals were amplified and band-pass filtered to remove the low frequency pedestal and high frequency electronic noise. The signals were passed to an oscilloscope to be displayed and to the input of a frequency-tracking demodulator BBC-Goertz L5301, for processing. From the tracker output, a DC-voltage corresponding to the shift frequency was subtracted. This analog output of the frequency tracker, which was proportional to the instantaneous Doppler frequency, was then digitized using a 12-bit ADC and read directly into a minicomputer. The block diagram in Figure 8 shows the entire signal processing system.

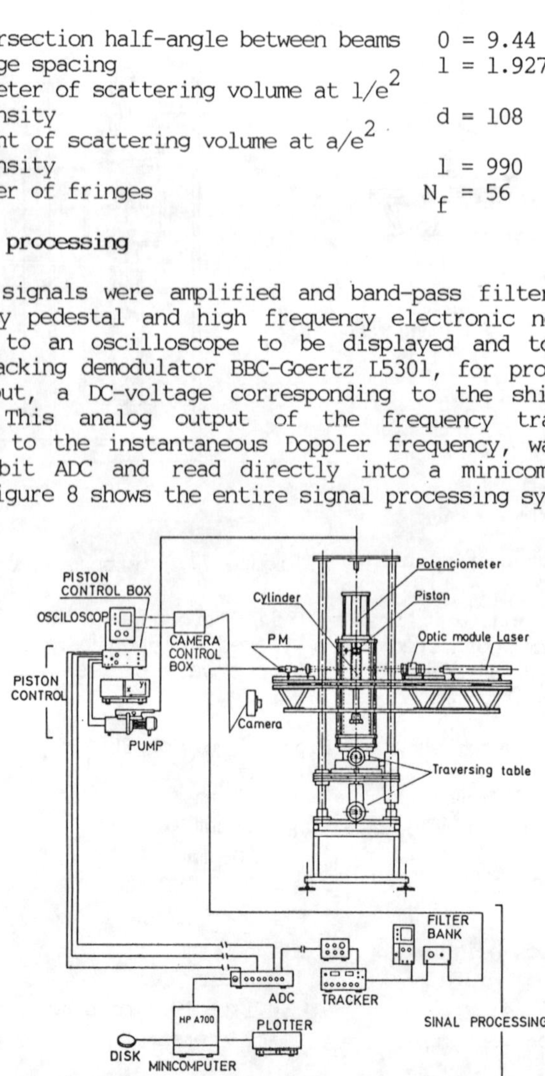

Figure 8. Block diagram of Data Acquisition and Processing System for LDA-Measurements.

Simultaneous to the axial velocity records, information on the instantaneous piston position was stored on a second channel of the ADC in order to allow velocity information to be obtained as a function of space and piston location. The ADC was externally triggered at a frequency of 175 Hz using a pulse generator. The trigger pulses were, however, first passed through a gating circuit of the piston control box which thus allowed acquisition of data in prescribed intervals of the piston movement. Using this system the velocity information over a full cycle could be obtained either by recording the velocity continuously over one full cycle, or by sampling at a succession of piston positions, by adjusting the gating circuit.

3.3 - Accuracy of the Results

Many sources of experimental error are small due to the laminar character of the flow, such as turbulence-induced velocity bias or gradient broadening of turbulence quantities due to the finite size of the control volume. Other errors were reduced by appropriate signal conditioning and/or calibration. For instance, subtracting the DC equivalent of the shift frequency from the tracker output, and amplifying the signal by a factor of 8, allowed a substantial increase in the resolution of the analog-digital conversion. Calibration of the entire signal processing system with sine wave signals, reduced small systematic errors and the resulting uncertainty of the recorded velocity was estimated to be $\pm$ 0.1 mm/s. THis emphasis on measurement accuracy was necessary due to the very low velocities (< 1 mm/s) occuring in some regions of the cylinder.

Errors due to averaging over the control volume area were minimized by using a shorter focal length lens, resulting in a smaller control volume (d = 0.108 mm, l = 0.99 mm). Crank angle broadening, i.e. the error due to the finite length of the averaging window, was found to be negligible (< 1%) for window sizes less than 1 mm. The window was further reduced to 0.2 mm in recording the impulsive start of the piston.

The refractive index matching was performed to within n = $\pm$ 0.001, and was maintained by controlling the temperature to within $\pm$ 2.5^{o}C. Estimates of the position uncertainty were possible by performing wall-to-wall profiles and then comparing the external traversing length to the cylinder diameter. These agreed to within approximately 0.05 mm.

Finally, the piston was driven by a very precise volumetric pump ($\pm$ 0.1 mm/s) at a maximum speed of 27 mm/s.

3.4 - Results

Examples of LDV measurements for the impulsively started flow corresponding to Clearance C = 40 mm (C/D = 1.78) and piston velocity Up = 11.9 mm/s (Re_D = 98) are shown in Figures 9 and 10. Figure 9 shows measured axial and radial velocity vectors for different piston positions during the transient motion of the piston. The corresponding streamline contours, obtained by integration of the axial profiles are shown in Figure 10.

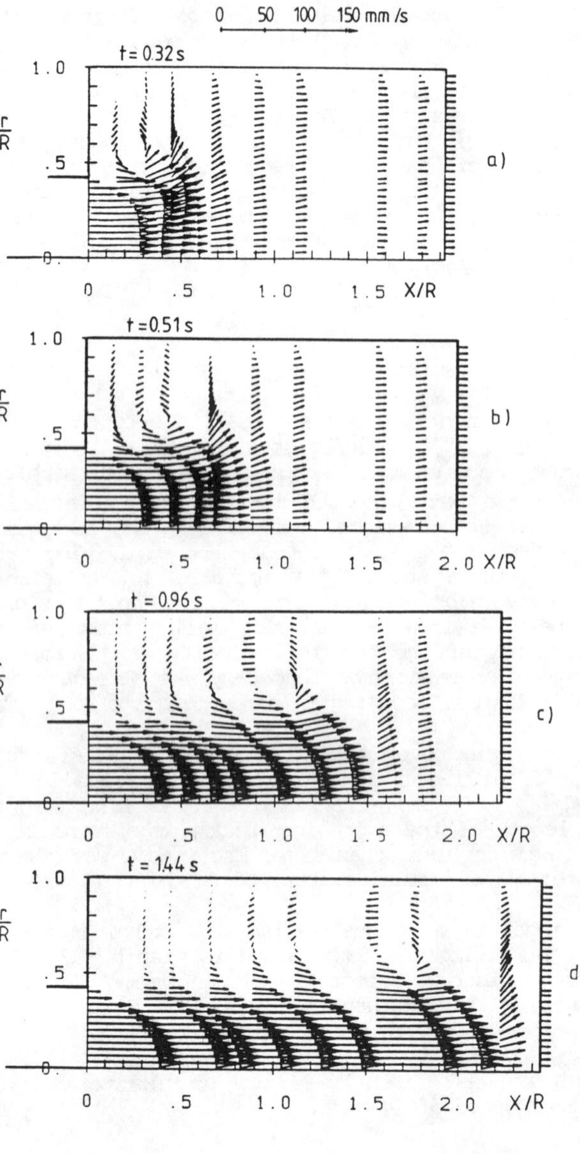

Figure 9. Measured velocity vectors at differenct piston positions, C = 40 mm; U_p = 11.9 mm/s.

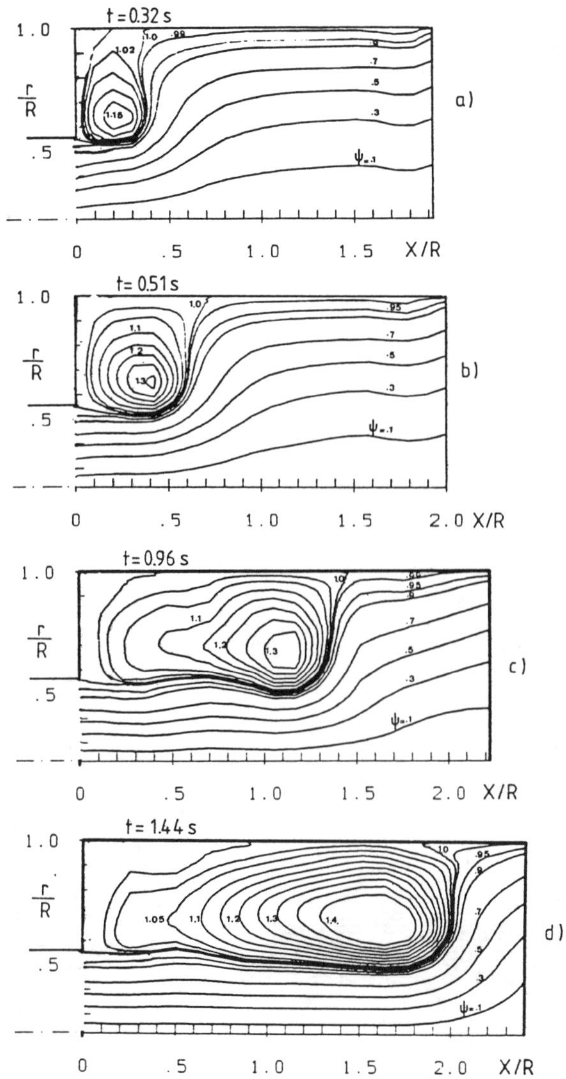

Figure 10. Streamline contours at different piston positions, C = =0 mm; U_p = 11.9 mm/s.

When the piston starts to move, the initially thin boundary layer separates at the sudden expansion and vortical fluid rolls up. At this time instant the axial velocity profile exhibits a maximum close to the tube wall, the position of which recedes from the wall as the thickness of the boundary layer increases. A similar evidence of the displacement of the maximum velocity (peak) from the center line was reported by Didden (1979). After t = 0.32 s as shown in Figure 10 a) this occurence is still present. The inlet jet flow spreads to fill the cylinder space. With further piston displacement the vortex is convected and the unsteady recirculation region is characterized by a moving reattachment point as shown in Figure 10 b) and c). The maximum velocities inside the recirculation region are of the order of magnitude of the piston velocity as Figure 9 b) and c) show. For t = 1.44 s (Figure 10 d)) the vortex suffers an eliptical deformation with the occurence of weak recirculating region and a presence of a counter rotation vortex with very small velocity (0.1 mm/s) inside this recirculation region.

Figures 11 and 12 show the velocity vectors and streamlines for the second case investigated (C = 16.5 mm; U_p = 11.9 mm/s; Re_D = 98). For comparison purposes the piston velocity was the same as the previous flow case value of the clearance was reduced to C = 16.5 mm. A comparison between Figure 9 and 11 shows the effect of the clearance volume on the flow structure. As the flow visualization has revealed the vortex collides with piston and strong negative velocities (twice the piston velocity) arise inside the recirculation regions. A secondary vortex appears in the corner of the sudden axpansion (se Figure 12). The appearance of this localized vortex is due to separation of the shear layer caused by the main recirculating flow on the wall.

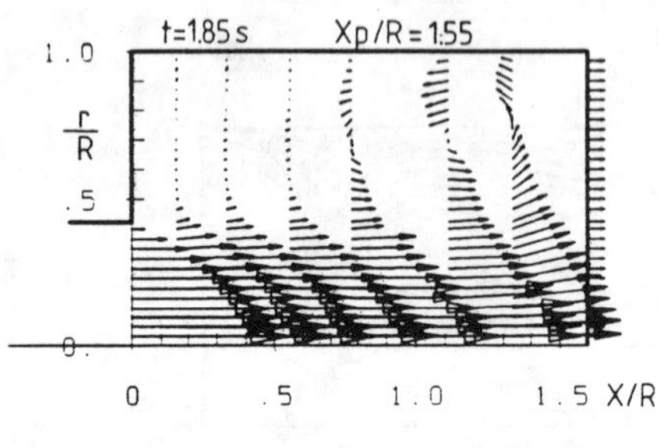

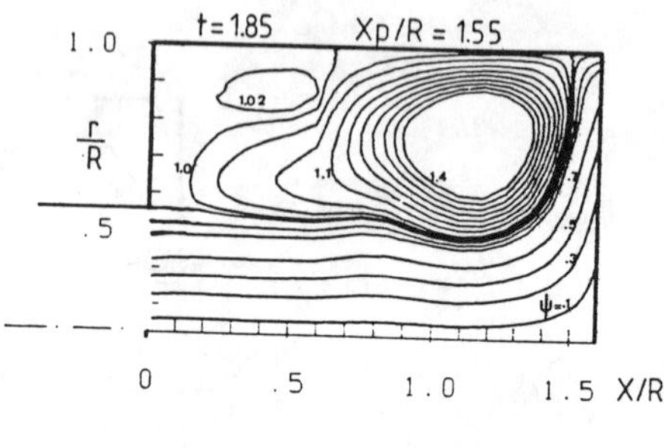

Figure 11. Measured velocity vectors at different piston positions, C = 16.5 mm; U_p = 11.9 mm/s.

Figure 12. Streamline contours at different piston positions C = 16.5 mm; U_p = 17.8 mm/s.

The results show that the piston velocity has smaller influence on the flow structure than the clearance volume, as was reported by Gorman et al (1978), and Morse et al (1979) for motored piston cylinders assemblies.

4 - CONCLUDING REMARKS

Refractve-index matching may solve many problems encountered in LDV measurements near curved walls or complex flow geometries. The temperature control of the working fluid is a very important factor for the maintanance of the refractive-index value.

LDV measurements of the unsteady flow in a piston/cylinder assembly were obtained using refractive index matching. The results show that the impulsively started flow displays a very complex structure comprising of many recirculation regions. The clearance volume was proved to be, rather than the piston velocity, the parameter that has more influence on the flow structure.

References

Bicen, A.F. (1982) Refraction Correction for LDA Measurements in Flows with Curved Optical Boundaries. TSI Quarterly, 8, No. 2.

Didden, N. (1979) On the Formation of Vortex Rings: Rolling-Up and Production of Circulation. J. of Appl. Math. and Phys. (ZAMP), Vol. 30, pp. 110-115.

Durst, F. and Zare, M. (1975) Laser-Doppler Measurements in Two-Phase Flows. Proc. Symp. on the Accuracy of Flow Measurements by Laser-Doppler Methods. Copenhagen.

Durst, F., Keck, T. and Kleine, R. (1979) 6th Symp. on Turbulence in Liquids, Chem. Engng. Dept., University of Missouri-Rolla.

Durst, F., Melling, A. and Whitelaw, J.H. (1981) Principles and Practice of Laser-Doppler-Anemometry. Academic Press, New York, 2nd Edition.

Durst, F. and Loy, T. (1985) Investigations of Laminar Flow in a Pipe with Sudden Contraction of Cross Sectional Area. Computers and Fluids, Vol. 13, Nr. 1, pp. 15-36.

Edwards, R.V. and Dybbs, A. (1984) An Index Matched Flow System for Measurement of Flow in Complex Geometries. Proc. 2nd Int. Symp. on Applications of Laser Anemometry to Fluid Mechanics, Lisbon.

Elphick, I.G., Martin, W.W. and Curie, I.G. (1982) Application of LDA to high Reynolds Number Cross Flow. Proc. 2nd Int. Symp. on Applications of Laser Anemometry to Fluid Mechanics, Lisbon.

Gosman, A.D., Melling, A., Watkins, A.P. and Whitelaw, J.H. (1978) Axisymmetric Flow in a Motored Reciprocating Engine, Proc. Inst. Mech. Eng., 198, 11.

Horvay, M. and Leukel, W. (1984) LDA Measurement of Liquid Swirl Flow In Converging Swirl Chamber with Tangential Inlets. Proc. 2nd Int. Symp. on Applications of Laser Anemometry to Fluid Mechanics, Lisbon.

Jenkins, F.A. and White, H.E., Fundamentals of Optics, 4th Edition, McGraw-Hill, 1976.

Kleine, R. (1983) Messungen in Verbrennungssystemen und in Fluissigkeits-Stromungen - Brechungsindexanfassung, Nurzlehrgang-Laser-Doppler-Anemometrie - LSTM, University Erlangen - Nurnberg, 28 Februar - 3 Marz.

Morse, A.P., Whitelaw, J.H. and Yanneskis, M. (1979) Turbulent Flow Measurements by Laser-Doppler Anemometry in Motored Piston-Cylinder Assemblies. Trans. of the ASME, Vol. 101, pp. 208-216.

Nouri, J.M., Whitelaw, J.H. and Yianneskis, M. A Refractive-Index Matching Technique for Solid/Liquid Flows. Laser Anemometry to Fluid Mechanics - III, Editors Adrian et al, LADOAN - IST - Lisbon - Portugal, pp. 335-346.

Perry, R.H. (1973) Chemical Engineers Handbook. McGraw-Hill

Scrivener, O. and S. Lizon and S.L. Lugrin (1986) Velocity Measurements Inside 3-D Porous Media Using Laser Doppler Anemometry.

Varty, R.L. and Currie, I.G. Index-Matched Laser-Anemometer Measurements of Cross-Flow in Tube Bundles. 2nd Int. Symp. on Applications of Laser Anemometry to Fluid Mechanics, Lisbon.

PROBLEMS AND SOLUTIONS FOR SPRAYS IN PISTON ENGINES

C.F. OLD

HARWELL LABORATORY

1. INTRODUCTION

Two of the most important criteria for assessing the performance of the modern diesel engine, namely economy and emissions, are strongly influenced by the atomisation of the fuel spray and its subsequent interactions with the ambient air during the early stages of injection. It is widely considered that any effective control of diesel engine combustion will be based on an improved predictive understanding of the momentum, concentration and temperature fields for both the liquid and gas phases. The experimental study of these parameters is difficult because of the complex interaction of several processes i.e. the sequence comprising atomisation, evaporation, mixing with the surrounding air, and auto-ignition. The problem is further complicated by high-speed transience, particularly in high-speed automotive engines, the droplet density and momentum, and frequently by the small physical scale of the engine.

2. GENERAL DESCRIPTION OF SPRAY STRUCTURE

Many attempts have been made to derive a satisfactory description of the general features of injected sprays. These studies lead to a picture which suggests the existence of a dense central core surrounded by a much larger and more dispersed mixing region. (Figure 1) Sato[(1)] carried out an extensive examination on sprays under a range of conditions using high-speed photography. He concluded that the main jet region was not significantly affected by increased ambient temperature but the mixing zone was substantially reduced by faster evaporation. Typical values of Sauter mean diameter (SMD) for the droplets in the mixing region were about 32-40 μm. Double flash photography gave information about the velocity distribution which showed a positive and approximately linear correlation between velocity and droplet size in the mixing region. Smoke tracers were used to establish air flow patterns around the spray during injection and indicated that the ambient air was displaced outwards and away from the nozzle at the end of the spray cloud. However, the flow was in the reverse direction i.e. into the spray in a lateral direction over much of the plume length. An extensive parametric study yielded an empirical description of spray growth.

A more detailed study to establish the effect of nozzle geometry on the near field region was carried out by Arai and co-workers[(2)]. These experiments used a model system in which water was continuously injected at pressures of 1-500 bar into ambient pressures of 1-100 bar. Small additions of electrolytes to the water permitted the use of an electrically conducting grid to define the breakup length. At low ambient pressures the breakup length was observed to show hysteresis at Reynolds numbers (Re) below 10^4. As the ambient pressure increased, the breakup length became single valued and further increases in Re number led to a progressively shorter breakup length and larger spray angle. (Figure 2).

D. F. G. Durão et al. (eds.), Instrumentation for Combustion and Flow in Engines, 285–293.

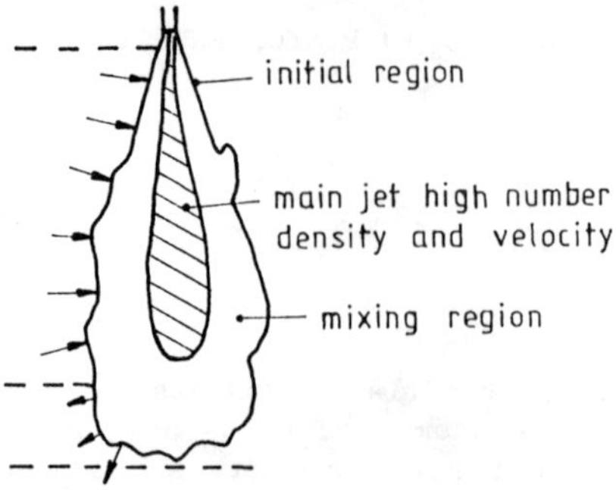

FIGURE 1: General description of spray structure - after Sato[1]

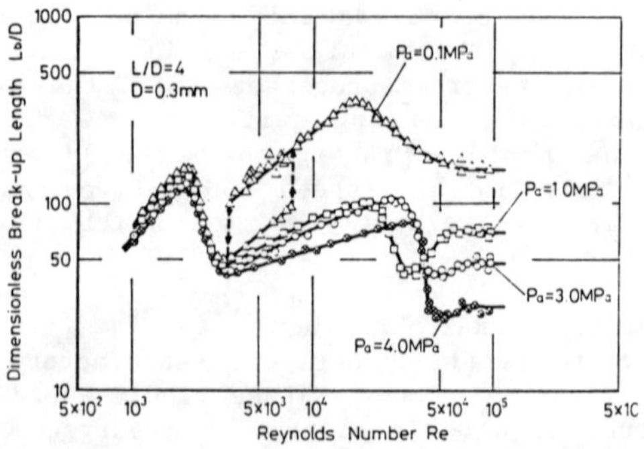

FIGURE 2: Dimensionless break up length as function of Reynolds number - after Arai[2]

The effect of the nozzle geometry was investigated over the range of diameters from 0.3 - 3 mm and length/diameter aspect ratio (L/D) from 1-50. Over much of the range of Re the nozzle aspect ratio had comparatively little effect. However, beyond values of 5×10^4, the higher aspect ratio nozzles (L/D = 40-50) increased the dimensionless breakup length by a factor of 2-3. (Figure 3). Effects of nozzle geometry and injection pressure on spray angle are shown in Figure 4. Spray angle increases with injection pressure relatively independent of nozzle geometry up to about Re $\sim 5 \times 10^4$ when dependence on nozzle geometry suddenly increases. For intermediate values of aspect ratio (L/D = 4-10), sharp peaks are observed in the dependence of spray angle on injection pressure.

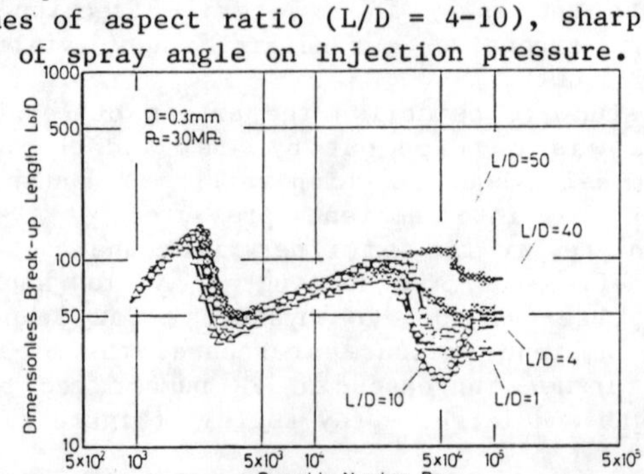

FIGURE 3: Effect of nozzle geometry on break up length - after Arai[2]

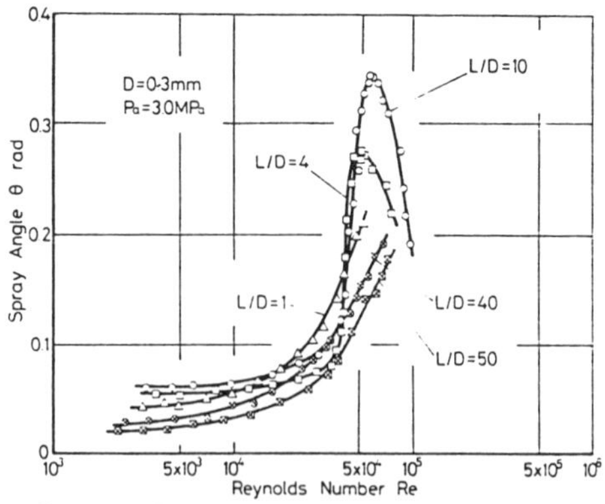

FIGURE 4: Effect of nozzle geometry on spray angle - after Arai[2]

Studies of the behaviour of combinations of commercial injectors and pumps have been carried out by Yule et al[3]. Bursts of fuel are injected into quiescent or well controlled flowing CO_2 at temperatures up to 400°C and pressures up to 50 bar. Measurements of flow rate, penetration and spray angle are made as a function of time while the spray structure is recorded by means of spark shadow photography with exposures down to 0.1 μs or high-speed ciné at up to 11000 fps. Results for penetration at room temperature are shown in Figure 5 for a range of injection pressures from 5-50 bar. The droplet sizes in the mixing zone were estimated to decrease from 100 μ at 5 bars to 20 μ at 50 bars. Figure 6 shows the effect of an increase of ~100K in ambient temperature. The observed decrease in spray angle was attributed largely to the reduction in gas density at the higher temperatures. It is interesting to note that significant cyclic variability was observed.

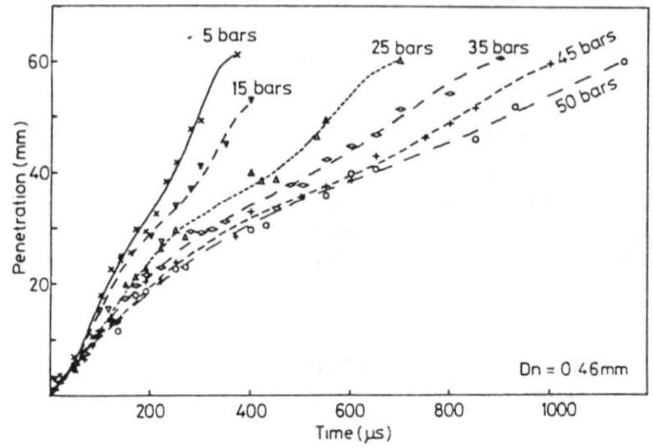

FIGURE 5: Effect of injection pressure on spray penetration - after Yule[3]

Although many extensive descriptions have been given for the behaviour of unconfined sprays, it must always be remembered that the most important region for automotive engine applications is the near field zone immediately after the nozzle. This region contains the highest droplet densities and velocities and is therefore the most difficult to study experimentally. An attempt to investigate the fundamental details of spray breakup was mounted by Ruiz and Chigier[4] using single shot injection of a non-combusting fluid at pressures of 80-800 bar into a high-pressure cell at up to 60 bar and 400°C. It was postulated that spray breakup could be due to either aerodynamic interaction at the surface of the liquid or, alternatively, due to instabilities at the free surface arising from the radial velocity components induced by turbulence or cavitation. A facility to include a coaxial air pulse was provided to regulate the surface shear velocity. The breakup mechanism was studied using 0.5 μs flash photography with final magnifications up to 100.

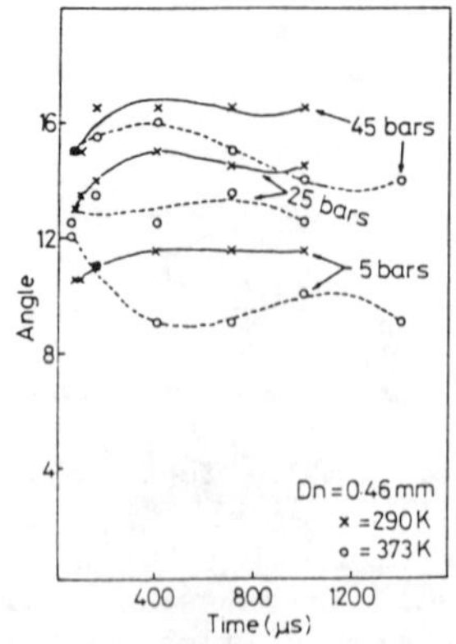

FIGURE 6: Effect of ambient temperature on spray angle - after Yule[13]

Studies of the discharge coefficient showed that cavitation occurred over most of the operating range. Variations in ambient pressure affected the spray angle but did not alter the initial part of the spray. In contrast, higher injection pressures increased the breakup in the near field region thus reinforcing the association of breakup with cavitation. Changes in the relative velocity at the air liquid interface altered the spray angle but only weakly. It was suggested that the shear velocity accounted primarily for the growth of instabilities. Ruiz and Chigier concluded that the initial surface perturbations were generated by cavitation in the nozzle and amplified by interfacial shear stresses. The effect of the surrounding gas density was thus mainly to influence the atomisation of perturbations on the liquid surface.

More recently, the increasingly stringent legislation on diesel emissions has led to intensified efforts to probe the breakup region in detail. Cavaliere et al[5] have used pulsed emissions from a Nd-YAG laser to illuminate the dense core of a diesel spray injected at up to 50 bar into a blow-down tunnel providing air flows up to 500°C and 10 m/s. Initial results taken at room temperature and pressure suggest that the spray starts in a very dispersed form 0.15 ms after injection but becomes more concentrated in the period 0.9 ms. This result is perhaps contrary to intuitive expectation and highlights the need to probe the near field region in detail.

3. OPTICAL DIAGNOSTICS FOR SPRAYS

The small scale and high velocities and densities favour non-invasive optical methods for quantitative measurements in the near field region[6].

The most well established method for particle sizing by laser optics is probably the laser diffraction technique[7-9]. This depends upon the small angle scattering of a beam of parallel laser light passing through a cloud of droplets. An ensemble average of all the droplets encountered is obtained covering a size range of approximately 2 to 1000 μ. The method is independent of the exact position of the droplets and therefore provides an integrated result for the volume of the spray illuminated by the beam. It is thus limited by line of sight averaging, although this disadvantage has been mitigated to some extent by tomographic techniques as demonstrated by Yule[10]. Because the laser diffraction technique is independent of the position of the scattering droplet, the results are also independent of droplet velocity. However, an attempt to incorporate time resolution has been made by Watson[11] but it should be noted that this time dependence relates to the changes in transient drop size of the spray and not to the velocities of the individual particles. The time resolution is provided by increasing the capability of the data collection equipment.

The limitations of the diffraction technique include its restriction to forward scatter and its dependence on very accurate measurement of the angular intensity distribution in the forward direction. The latter requirement renders it particularly sensitive to changes in refractive index of the transmitting medium i.e. it is restricted to isothermal application. The same criterion demands also that the window access to the measurement volume be of very high quality and unperturbed by for example, liquid films.

Despite these restrictions for engine applications, careful application of the laser diffraction technique has yielded valuable insights in a number of laboratory simulations and specially adapted engine experiments. Tabata[12] used laser diffraction to measure the effects of fuel viscosity and surface tension on droplet size in a spray injected into an optically accessed high pressure chamber (Figure 7). The results showed that the drop size fell as injection pressure increased and it was noted that there was a lower threshold limit of injection pressure to develop a fine spray. This lower limit increased within increasing fuel viscosity and surface tension. It was also noted that at low injection pressures the drop size fell with increasing ambient pressure but this trend reversed at higher injection pressures (Figure 8).

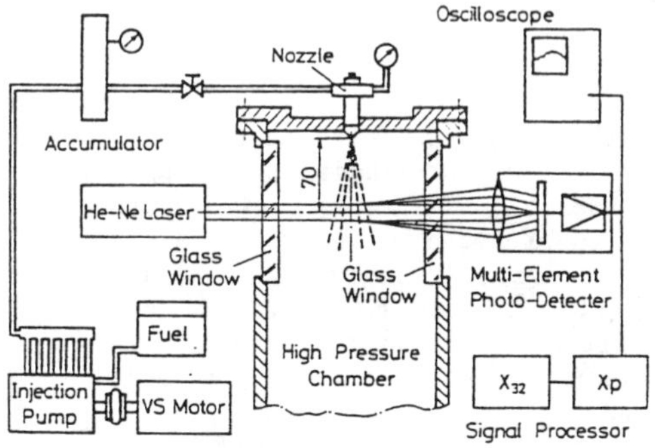

FIGURE 7: Laser diffraction measurements of droplet size dependence on fuel properties - after Tabata[12]

Numerous workers have attempted to overcome the limitations posed by the volumetric averaging by adaptating the spatially resolved LDA technique[13]. Two approaches are now in common use, namely the visibility technique and the phase Doppler method.

Both techniques depend upon the detailed examination of the Doppler burst generated when a droplet passes through the probe volume at the intersection of the input laser beams. This provides a point measurement with additional information on velocity of a single particle over the size range ~1-2000 μ. At present, neither method will operate in full backscatter although there is evidence that the phase doppler method has greater potential than the visibility technique in this respect. Signal processing is also an aspect of major concern and efforts are currently being made to push up the signal processing rate beyond the kHz range in an attempt to provide real time information on transient sprays.

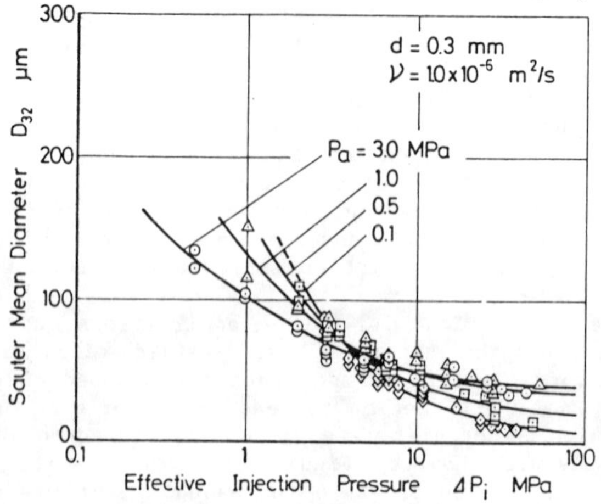

FIGURE 8: The effect of injection pressure on droplet size - after ibid

The visibility technique depends upon the measurement of the modulation depth of the doppler burst using a single signal collector which is usually placed off axis in an approximately forward scatter direction. The phase Doppler technique on the other hand relies upon the measurement of the phase difference in the signals collected by two detectors and is at its most sensitive at larger scattering angles. Table 1 shows a comparison of the main features of the two techniques.

Opinion is growing amongst workers in the particle sizing field that the phase doppler technique has advantages over the visibility method (14-16).

So far, the visibility method has been applied by Maeda (17) to measurements in the peripheral region of a diesel spray at up to 120 bar. These measurements however, are restricted to regions approximately 150 mm below the nozzle, i.e. in the more diffuse mixing region of the spray. Similar work is currently in progress at RWTH Aachen, where Dibelius is carrying out studies on diesel sprays injected into a nitrogen filled cell at up to 50 bar pressure and 700°C (18).

The difficulties in applying the visibility and phase Doppler techniques to dense regions of the spray in the near field region have stimulated Beretta and his colleagues (19,20) to explore a new technique known as polarisation light scattering. This exploits the size dependent difference in the intensities of vertically and horizontally scattered polarised light. The intensity of the scattered light, Q_{ij} is given by

$$Q_{ij} \ 1 \ N\overline{C}_{ij}(\alpha,\theta)$$

where Q = intensity; i,j is the vertical or horizontal orientation, N = no. concentration, α = size parameter $\pi d/\lambda$, θ = scattering angle.

TABLE 1. COMPARISON OF VISIBILITY PHASE-DOPPLER TECHNIQUES

Visibility	Phase-Doppler
Non-linear with diameter	Linear with diameter
Size range is complex function of optical geometry	Linear with slit separation insensitive to slip shape
2 colour probe volume for signal validation	Single colour (cheaper)
Dynamic range of electronics must be large; e.g. 1000:1	Electronics phase-matched
Complex even for LDA experts	Simple for LDA experts
Droplet diameter must be < probe volume to avoid multiple occupancy	Droplet diameter ~ probe volume
Lower limit 0.5/1.0 μm	Lower limit < 10 μm only for spherical particles?
Not backscatter	Perhaps backscatter for metal spheres

If a model is assumed for the size distribution of the droplets, then the ratio γ of the horizontally and vertically polarised components, Q_{HH} and Q_{VV}, respectively can be used to derive the size parameter while the absolute intensity of the scattered light gives the average number concentration and hence the volume fraction of liquid in the spray. As can be seen from Figure 9 the relationship between γ and the scattering angle is complex but the authors propose that a region can be identified where a reliable monotonic relationship between γ and droplet size can be established. This has been used successfully in a dense spray to measure the population of small droplets below 30 microns diameter.

More recently, Beretta et al have extended this technique to develop the method of spectral ensemble light scattering (SELS)(21) which employs selected wavelengths to derive the size distribution explicitly. This eliminates the need to assume size distribution before analysis of the scattering data. The method has been compared with the visibility technique in an oil flame with very promising results as shown in Figure 10.

4. SPRAY CHEMISTRY

Study of this aspect of fuel sprays is even less advanced than the examination of the physical behaviour of the dense near field region. Parameters of particular interest include measurements of air fuel ratio and temperature profiles during spray auto-ignition. Knoche(22) has recently published measurements of air fuel ratios in the mixing region at ignition in a large optically accessed test cell using spontaneous Raman spectroscopy at up to 60 bar and 600°C.

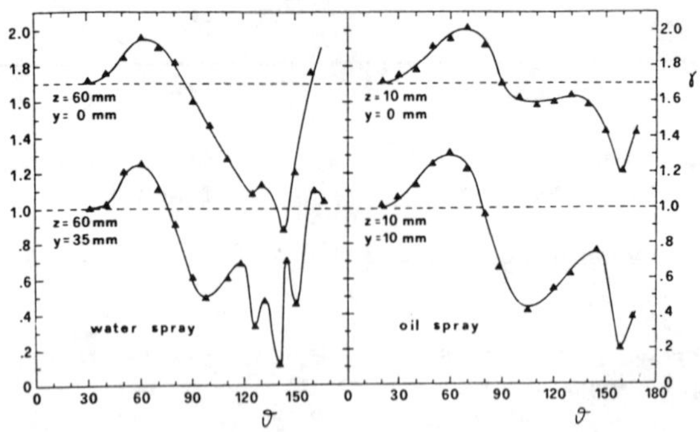

FIGURE 9: Dependence of γ on scattering angle - after Beretta et al[19]

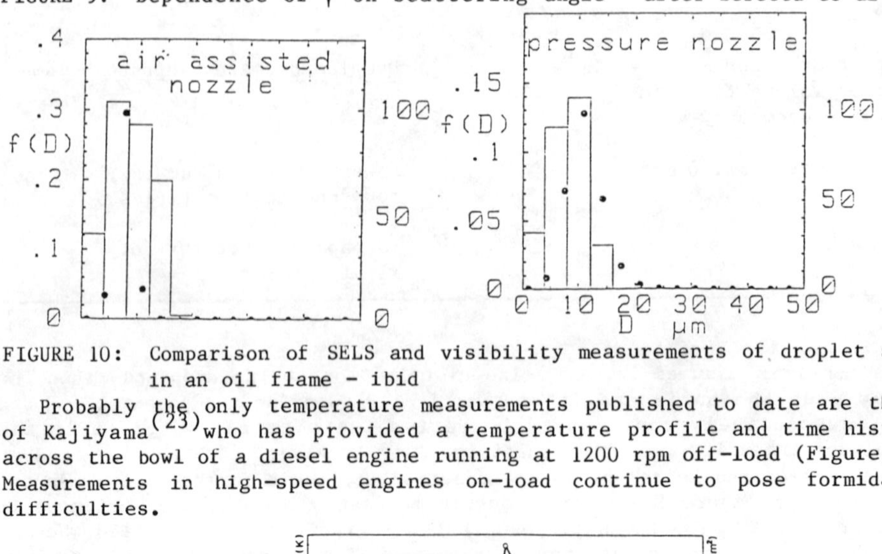

FIGURE 10: Comparison of SELS and visibility measurements of droplet size in an oil flame - ibid

Probably the only temperature measurements published to date are those of Kajiyama[23] who has provided a temperature profile and time history across the bowl of a diesel engine running at 1200 rpm off-load (Figure 11). Measurements in high-speed engines on-load continue to pose formidable difficulties.

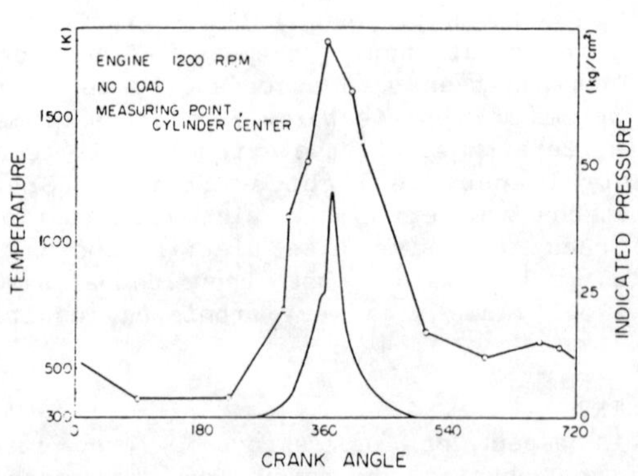

FIGURE 11: CARS temperature profile across a diesel engine combustion chamber - after Kajiyama[23]

5. CONCLUDING REMARKS

This literature review indicates that most of the essential physical parameters in the mixing zone are accessible by a variety of techniques and especially by those based on laser optical methods. There are promising prospects for studying the breakup region close to the nozzle, but there is still a great deal of work to be done in this area.

From the chemical point of view, it is possible that the measurements of the temperature field in the mixing zone and perhaps also in the dense core would be accessible by CARS. Applications of CARS to other combustors such as coal furnaces posing environments of comparable difficulty have been reasonably successful and auger well for applications to diesel sprays. Measurement of species concentrations however remains exceptionally difficult and it will be some time before such data can be successfully obtained from practicable combustors.

REFERENCES

1. Sato GT: ICLASS-85, Inst. Energy, London, 1985.
2. Arai M et al: ibid.
3. Yule AJ et al: ibid.
4. Ruiz R and Chigier N: ibid.
5. Cavaliere et al: Int. Symp. Heat and Mass Transfer in Engines, Dubrovnik, 1987.
6. Azzopardi BJ: Int. Jnl. Heat & Mass Transfer, 22, 1245, 1979.
7. Dobbins RA et al: ATAA Jnl., 1, 1982, 1963.
8. Chigier NA: Prog. Energy and Combustion Sci., 3, 175, 1977.
9. Swithenbank J et al: Prog. Astro. Aero, 53, 421, 1977.
10. Yule AJ et al: 18th Int. Symp. on Combn., 1981.
11. Watson DJ: ICLASS-85, Inst. Energy, London, 1985.
12. Tabata M et al: ibid.
13. Drain LE: Int. Conf on Advances and Applications of Laser Anemometry, Manchester, 1985.
14. Bacchalo WD and Houser MJ: 3rd Int. Conf on LDA in Fluid Mechanics, Lisbon, 1986.
15. Lightfoot NS and Negus CR: ibid.
16. Hemsley D: - private communication.
17. Maeda M et al: ibid.
18. Dibelius G et al: Motorische Verbrennung, SFB224, RWTH Aachen, 1985.
19. Beretta F et al: Combustion Sci Tech., 36, (1 and 2), 19, 1984.
20. Beretta F et al: 20th Int. Symposium on Combustion, p.1249, 1984.
21. Beretta F et al: 21st Int. Symposium on Combustion, Munich, 1986.
22. Knoche KF et al: Motorische Verbrennung, SFB224, RWTH Aachen, 1985.
23. Kajiyama K et al: SAE 821036.

DISPERSED TWO-PHASE FLOWS, ITS EXPERIMENTAL INVESTIGATION AND NUMERICAL PREDICTION

F. DURST AND T. BÖRNER

Lehrstuhl für Strömungsmechanik, Universität Erlangen-Nürnberg, Egerlandstr. 13, 8520 Erlangen, FRG

1. INTRODUCTION

The measurement of particle velocities, size distribution and concentration has become of considerable interest to environmental control and protection agencies as well as to engineers and scientists working on pollution problems. Furthermore, there are many flows in natural and industrial environment in which suspended particles influence or even dominate heat and mass transfer processes. Detailed studies of these transfer processes require suitable measuring techniques to provide information on particle velocity, size distribution and concentration and on variations of these quantities in time and space. Hence, the occurence of two-phase or multiphase flows in different fields of engineering is large and their properties differ due to the wide variety of phase combinations and due to the large number of possible flow regimes. In recent years considerable progress has been made understanding the physics of some of these regimes of dispersed two-phase flows. This understanding was achieved by means of detailed experimental investigations that will shortly be summarized in section 2 of this paper. The results of these studies entered two-phase flow modelling refining numerical prediction procedures for dispersed two-phase flows.These are looked at in section 3.

The need for local measurements to examine spatial distribution of size and concentration, and the necessity to have instantaneous records to study transient variations, impose special instrumentation requirements. Instruments for this purpose, based on light scattering from particles are presently under development in many laboratories. Some of them are highlighted in this paper and their employment to measure flow and particle properties is described. The development and optimization of instrumentation to measure two-phase flow properties is greatly facilitated by physically correct descriptions of the generation of light scattering phenomena from particles. These descriptions are provided in this paper with special emphasis being given to the laser-Doppler instrument (section 2). The generation of laser-Doppler signals in a dual-beam LDA-system can be readily explained by light scattering from particles penetrating a region in space common to both incident light beams. This region is usually referred to as the measuring volume of the LDA-system although signals may result from outside of this volume. These additional signals are usually not considered in laser-Doppler anemometry and are eliminated in actual measurements by designing the receiving optics to only collect light from the crossing region of the two incident light beams, e.g., see Durst and Whitelaw (1).

For the prediction of particulate two-phase flows, two approaches are possible, discussed in section 3. One treats the fluid phase as a continuum and the particulate second phase as single particles. This approach, which predicts the particle trajectories in the fluid phase as a

D. F. G. Durão et al. (eds.), Instrumentation for Combustion and Flow in Engines, 295–339.

result of forces acting on particles, is called the Lagrangian approach. Treating the solid as a continuum using the Eulerian form of the continuum, momentum and energy equations for the fluid and particle phases, is referred to as the Eulerian approach to predict two-phase flows.

Both approaches are discussed in the present paper and their basic equations for the particle and fluid phases as well as their numerical treatment are presented in section 3. Particular attention is given to the interactions between both phases and their mathematical formulations. The resulting computer codes are discussed.

As long as the particles of a particulate two-phase flow are small enough, they will follow the fluid motion and the whole flow can be treated as a homogeneous flow with appropriately changed physical properties. This kind of particulate two-phase flow, which is basically in its treatment a single phase flow, will not be considered here. In the present paper it is presumed that the particles are large enough for a slip velocity to exist between the two phases and it is the task of two-phase flow prediction procedures to yield information of the effects of this slip velocity on the relative motion of both phases.

Most of the experimental work on particulate two-phase flows has dealt with the flow of round single particles in gas or liquid on one hand and on the other hand, with the flow of high concentration suspensions, where only integral parameters are taken to characterize the flow properties. An example of the latter is the measurement of the pressure loss of a particulate flow in a pipe and its correlation with the total mass-flow rate of the solid and fluid phase. This kind of investigation of integral parameters has been forced onto studies of particulate two-phase flows by lack of suitable measuring techniques to yield more refined information. Only during the last few years new experimental techniques have emerged that permit local information to be reliably obtained on both phases. Laser-Doppler anemometry, for example, allows the local measurement of the velocity, size and concentration of particles as well as the measurement of the fluid velocity. In this way, local flow information can be obtained in the entire flow domain.

Such detailed information of many relevant parameters of particulate two-phase flow is absolutely necessary to improve the assumptions introduced into two-phase flow prediction procedures to close the set of equations that describe this flow. Through detailed experimental studies, it is also possible to see whether information on the flow around single particles can be transferred to particulate flow systems.

Section 4 emphasizes the close link required between experimentally and numerically working fluid mechanics that is necessary for research work of the fluid described here.

2. EXPERIMENTAL INVESTIGATION OF DISPERSED TWO-PHASE FLOWS USING LASER-DOPPLER ANEMOMETRY

2.1 Introduction to Laser-Doppler Anemometry

It is important to point out here, that laser-Doppler anemometry always requires scattering particles to yield signals from which Doppler information can be obtained. These particles may occur naturally in a flow or may deliberately be seeded into the flow. Depending on the particle size, one can presume for small particles to follow the flow ($R_p = 1$) /5/, or for larger particles to have a destinct slip velocity as shown in Fig. 2.1, where R_p is the ratio of rms-particle fluctuating velocity to the corresponding rms-velocity of the fluctuating fluid.

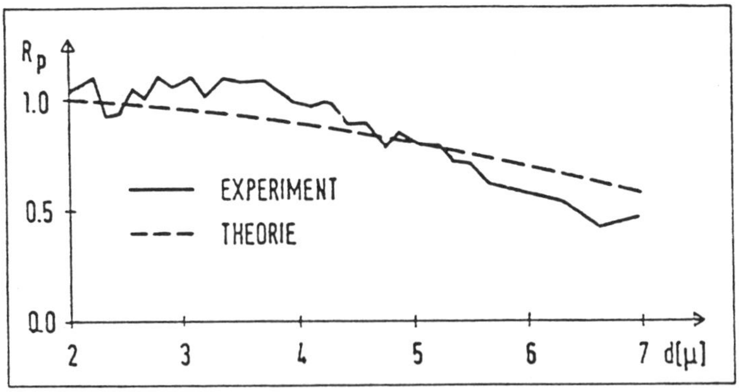

FIGURE 2.1. Ratio of rms-values of fluctuating velocity of particles to fluid /5/

The physical existence of an interference fringe pattern in the common region of two crossing laser-beams has been the basis of explanations of LDA-signal properties for a long time, /2/ and /3/. A physically correct description of LDA-signals, e.g. of their generation and their properties are given below.

2.1.1 The generation and detection of LDA-signals. A number of different models have been proposed in the literature to explain and physically understand laser-Doppler signals. Among these the fringe model has been the most popular one. Using wave theory, the Doppler model is the physically most correct description of the signal generation. The frequency shift due to a moving source and to a moving observer are the basic principles of the Doppler model. This model is readily understood by considering an experiment with an aeroplane. Figure 2.2 shows a transmitter that emits waves of constant frequency in a certain direction which is given by the unity vector $\{1\}_i$. An aeroplane, that contains a receiver and a transmitter for the particular waves, penetrates the emitted "wave beam". It is known from the basic considerations of the Doppler effect that the receiver in the aeroplane will record a wave frequency different from the one transmitted if the aeroplane has a velocity component in the direction of the unity vector $\{1\}_i$. This frequency shift is given by the equation:

$$\nu_p = \nu_o \left\{ 1 - \frac{\{\hat{U}\}_i \{n\}_i}{c} \right\} . \tag{2.1}$$

The transmitter in the aeroplane is in a second step employed to emit waves in all directions. Since the receiver's wave has the frequency ν_p the transmitter can only emit waves of this particular frequency.

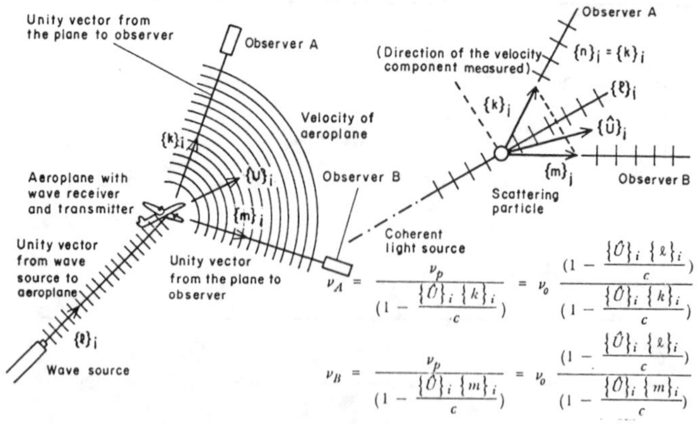

FIGURE 2.2. 'Doppler model' to explain LDA-signal frequency

Two observers placed at two different positions (A) and (B) will record additional frequency differences due to the fact that the transmitter in the aeroplane represents a moving wave source. The two frequencies can be calculated and taking their difference frequency yields after linearization $\{\hat{U}\}_i \ll c$:

$$\nu_D = \frac{\nu_o}{c}\{\hat{U}\}_i(\{k\}_i - \{m\}_i) = \frac{1}{\lambda_o}\{\hat{U}\}_i\{n\}_i \,. \qquad (2.2)$$

The last equation indicates that velocity measurements can be made by recording the frequency difference of the two waves observed at positions (A) and (B).

If the aeroplane is replaced by a small tracer particle the same derivations apply, since that particle fulfils the action of the receiver and the transmitter in the aeroplane. This is readily understood by the following simple considerations. If a small particle penetrates a laser beam a certain amount of light is scattered by it. Thus, the particle acts as "transmitter". However, before that action can take place the particle must have acted as a "receiver" of the light wave, i.e. as far as the frequency of the scattered light wave is concerned, the scattering action of the particle can be understood as the receiving and sending of light waves.

2.1.2 Fringe Model for Laser-Doppler Signals. In Figure 2.3 the principles of the interference fringe model, which was proposed by Rudd /2/ and vom Stein and Pfeifer /3/, are indicated. This model provides an easily visualized picture of a varying light intensity pattern in the measuring volume and this is a basis for both a qualitative and quantitative analysis of a number of features of laser-Doppler signals. Two coherent light beams having plane wave fronts - as would be the case in the waist region of two Gaussian light beams - intersect at an angle 2φ. This yields a pattern of plane interference fringes. The fringe spacing Δx is proportional to the wavelength of light and inversely proportional to the sine of the half-angle between the two beams. As a particle moves through the region of interference fringes it will scatter light whose intensity will vary according to the light intensity variations inside the common region of the two beams. These light intensity variations are detected by a photomultiplier and it can easily be shown that the frequency of the resulting signals is exactly that determined from the analysis based on Doppler considerations.

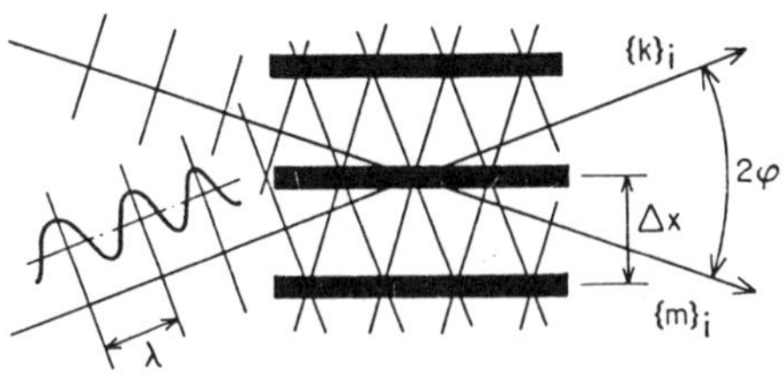

Fringe distance:

$$\Delta x = \frac{\lambda}{2 \sin \varphi}$$

Signal frequency:

$$\nu = \frac{\hat{U}_\perp}{\Delta x} = \frac{2\hat{U}_\perp \sin \varphi}{\lambda}$$

$$|\{k\}_i - \{m\}_i| = 2 \sin \varphi \qquad \hat{U}_\perp = |\{\hat{U}\}_i| \cdot \cos a$$

$$\nu = \frac{2\, \hat{U}_\perp \sin \varphi}{\lambda} = \frac{\{\hat{U}\}_i}{\lambda} (\{k\}_i - \{m\}_i)$$

FIGURE 2.3. 'Fringe model' to explain features of LDA-signals

Considering laser-Doppler signals as created by an interference fringe pattern permits signal quality conclusions on the basis of coherence properties of the two light beams and on their relative light intensities. If such information is available, it is very helpful to set up optimum optical arrangements. The fringe model was in this respect a significant step forward to introduce the measuring technique and to provide a basis for the optimized usage of laser-Doppler anemometry.

When fringe model considerations are combined with the Gaussian light intensity distribution of the interfering beams, a more complete description of LDA signals is obtained. Using $1/e^2$ of the peak intensity to define the signal starting and ending points, its boundaries, the "effective probe volume" is ellipsoidal in shape. Thus the probe volume, at least for a differential Doppler or interference fringe system, can be visualized as a set of parallel fringe planes occupying an ellipsoidal volume at the beam intersection point. The short axis of the ellipsoid is primarily dependent on the beam waist size and varies only weakly with beam intersection angle. However, the long axis increases rapidly as beam angle decreases.

The effects of improper beam intersection can also be readily visualized with the fringe model. In this case spherical rather than plane wavefronts interfere. The resulting fringe surfaces are distorted and are no longer parallel planes, e.g. see ref. /3/. Particles with the same velocity, but crossing different parts of the probe volume, will therefore produce different Doppler frequencies. Erroneous measurements will result in principle, but the errors are small in practice.

2.2 Two-Phase Flow Application for LDA

2.2.1 Light scattering principles. It is well known that the amplitudes of LDA-signals depend on the employed laser power and on the properties of the photomultiplier used in the detection optics. Furthermore, the amplitude and modulation depth of LDA-singals also depends on the size of the scattering particles, the optical properties of the particle material, the wavelength of the employed laser radiation, the angle between the two incident light beams, and the size and location of the receiving aperture. Various authors have employed computer programs to predict the strength and modulation depth of LDA-signals on the basis of Mie's (1980) scattering theory, e.g. see Dändliker and Eliason /6/, Adrian and Early /7/, Cherdron, Durst and Richter /8/. Comparisons between predictions and experiments clearly showed that the formula described by Mie /9/ yields correct predictions of amplitude properties of LDA-signals when correctly applied to both laser beams of an LDA-optical system and taking into account the complete arrangement of the transmission and receiving optics, e.g. see Durst and Eliason /10/, Durst and Umhauer /11/, Adrian and Orloff /12/.

The aforementioned investigations clearly demonstrated that Mie's scattering theory can be employed to correctly predict the amplitude of LDA-signals and their dependence on various influencing parameters. Irrespective of this finding, simplified models were proposed on the basis of fringes inside the measuring volume of LDA-systems. The LDA-signal amplitude was proposed to be proportional to the total light flux striking a scattering particle, e.g. see Farmer /13/, inside the fringe system. Hence, this model suggests the modulation depth of LDA-signals to be only dependent on the particle diameter relative to the fringe spacing. The dependence of the modulation depth on the optical properties of the particle material, on the size and location of the light collecting aperture etc., is not accounted for by this model. This clearly indicates that its validity is limited to special scattering properties of particles and to very special optical arrangements. Nevertheless, its simple explanation of the dependence of the signal modulation depth on particle size yielded a wide acceptance of this model as being generally applicable.

The limitations of the fringe model to predict amplitude characteristics of LDA-signals have been pointed out by a number of authors. Durst /12/ suggests the fringe model to only heuristically explain the modulation depth dependence of LDA signals on particle size. More detailed considerations are provided by Adrian and Orloff /10/, Robinson and Chu /13/ and Roberds /14/ who pointed out the limitations of the fringe model to predict amplitude properties of LDA-signals. These authors show that the signal amplitude and signal visibility can be correctly treated by the fringe model in the scalar diffraction theory limit, but only if the scattered light is collected in the forward direction with a large light collecting aperture. Under other conditions, fringe model considerations will yield results that differ from those that are obtained experimentally.

The origin of light scattered from larger particles compared to the fringe-spacing and particles being greater than the LDA measuring volume and its interpretation has been the objective for further research activities carried out in many laboratories.

Some basic ideas for LDA-velocity measurements in two-phase flows were brought forward by Durst and Zaré /17/ who showed that the light waves produced by two laser beams, reflected by the smooth surface of a large body, interfere and produce fringes in space. The location and shape of the interference pattern is dependent on the arrangement of the incident laser beams, the shape of the body, and its location. Furthermore, the rate at which the fringes cross a detector in space is linearly related to the transverse velocity component of the reflecting body perpendicular to the symmetry line between the two incident beams.

(a) Reflecting Spheres. Two laser beams, reflecting from a spherical particle, interfere and produce a non linear fringe system. The shape and spacing of the fringes are functions of the angel between the incident laser beams and their wave length, as well as the sphere size and the direction of observation. The fringes are, of simple shape, however - namely, parallel planes - in the backward direction. The fringe separation distance in this region, for large rations L/R and small angles φ, can be formulated as:

$$\Delta x = [1 + (\frac{2n_1}{n_2} - 2)\frac{L}{R}] \frac{\lambda}{2n_1 \sin\varphi} \tag{2.3}$$

This relationship indicates that the radius of the sphere R can be obtained by measuring the fringe distance Δx in the backward direction. The authors have used a double element photodiode, with elements spaced 2 mm apart, to obtain information of the sphere diameter through phase measurements between the two detected signals. Examples of such measurements are given by Durst and Zaré /17/.

As shown in Figure 2.4a, when the sphere moves across the two laser beams, the fringes appear to originate from a point along the direction of one of the beams, move around the entire space and then disappear into a point along the direction of the second beam.

The rate at which fringes cross any point, when produced by a moving nondeformable sphere, is the same at all points in the surrounding space, and is equal to the rate of fringe appearance at the source line or the rate of their disappearance at the sink line. The Doppler frequency detected at the source, resulting from the movement of the sphere, can be formulated as:

$$f_D = \frac{2(U_\perp \cos\beta \pm U_\parallel \sin\beta)\sin\varphi}{\lambda} \tag{2.4}$$

The angle β is a function of L/R and φ. For large values of ratio L/R and small angles β, equation (2.4) tends towards the universal equation of laser-Doppler anemometry for single phase measurements. In the case of a deforming particle, the rate of change of fringes would not remain the same all over space and this fact allows measurement of the particle deformation by using proper multidetection systems.

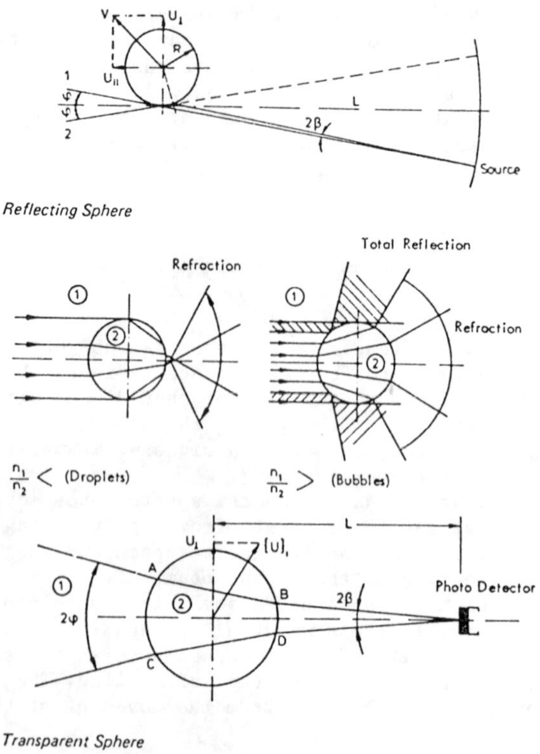

FIGURE 2.4. Configuration for spherical particles

(b) Transparent Spheres. As the light beam passes through a transparent particle, the relative refractive index of the particle plays a role in the characteristics of the propagated light. Two cases are distinguishable, Figure 2.4b $n_1/n_2 < 1$; the light propagates from an optically lighter medium to a denser medium. All of the light incident on the sphere passes through it and the sphere acts as a positive lens. An example of this case occurs in the measurements of water drops falling in air. $n_1/n_2 > 1$; the light propagates from an optically denser medium to a lighter medium. The sphere acts as a negative lens, and at a certain region total reflection occurs on the sphere surface. Such a case happens in the measurements of gas bubble-liquid flows and insoluble liquid-liquid flows. The partly or totally reflected light from the interface of the spherical particle can be treated in a manner similar to part 2.4a.

When two beams are transmitted through a sphere, the shape of the fringes produced by the beams is linear in the vicinity of the optical axis in the forward direction, and for large ratios L/R and small angles φ one may show the fringe distance to be:

$$\Delta x = \left[1 + \frac{2(L-R)}{R \cdot \cos\varphi} \right] \cdot \frac{\lambda}{2\tan\varphi} \tag{2.5}$$

For constant geometrical and physical conditions, the fringe distances are functions of sphere radius R. Consequently, spherical particle size measurements can be performed using a double photo-detector as discussed below.

When transmitted through a moving particle, each of the beams is refracted at both interfaces (entering and leaving) and therefore Doppler-shifted twice. The general motion of the sphere and its surface movements all contribute to the rate of passage of the interference fringes. For a simple case of a solid sphere with no rotational movements (so that the four points A, B, and D in Figure 2.4b have equal velocities), the Doppler frequency detected in the forward direction can be formulated as:

$$f_D = \frac{2U_\perp(\sin\varphi - \sin\beta)}{\lambda} \tag{2.6}$$

For large ratios L/R and small beam crossing angle β, equation (2.6) tends towards the universal equation of laser-Doppler anemometry for single phase movements.

2.2.2 Optical systems for two-phase flow measurements. Some optical systems have been proposed in the literature for two-phase flow investigations to allow for simultaneous measurements of fluid- and particle-velocity, for particle size distribution and local particle concentration. A selection of configurations all involving laser-Doppler anemometry is given below.

In sufficiently dilute suspensions the following light scattering characterisitics of particles are extracted and evaluated for:

frequency	velocity detection
amplitude	particle sizing
relative signal modulation	particle sizing
signal counts per time	local mass flow rate
signal time	residence time (size)
signal symmetry	particle trajectory
coincidence on multiple detectors	particle sizing
phase shift of frequencies on multiple detectors	particle sizing

The above signal properties require extensive processing equipment for full analysis. This is why some researchers reduce the complexity of the two-phase flows by selecting mono-disperse particle-dispersions for basic two-phase flow investigations.

Börner et al. /18/ suggested the application of laser-Doppler anemometry in a conventional single component dual beam mode LDA-system combined with a digital signal evaluation scheme for basic gas-solid flow investigations. Amplitude and relative signal modulation of a Doppler-event were evaluated to yield information on the signal belonging to phase 1 or phase 2. Figure 2.5 gives an overview on the data-processing equipment and the typical signal properties for small and large particles.

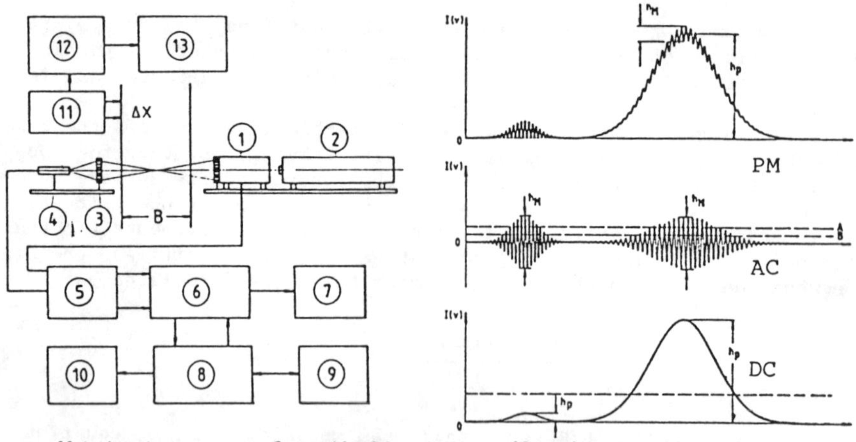

1 bragg cell backward scatter module (f=300mm)
2 15mW He-Ne-laser
3 collecting lens (f=250mm)
4 photomultiplier (pin hole Ø=300µm)
5 variable filter bank
6 transient recorder (2 x 20 kHz)
7 oscilloscope
8 mini-computer (DMA)
9 plotter
10 graphics terminal
11 differential pressure transducer
12 y/t-plotter
13 integrating volt meter

FIGURE 2.5. Discrimination based on signal amplitude and its relative modulation.

Using the above mentioned LDA-system and signal evaluation procedure glass particles of a uniform size ($\bar{d}_p$=108 µm) were investigated in a gas flow also seeded with magnesia particles of less than 3 µm diameter to indicate the gas-flow velocity.

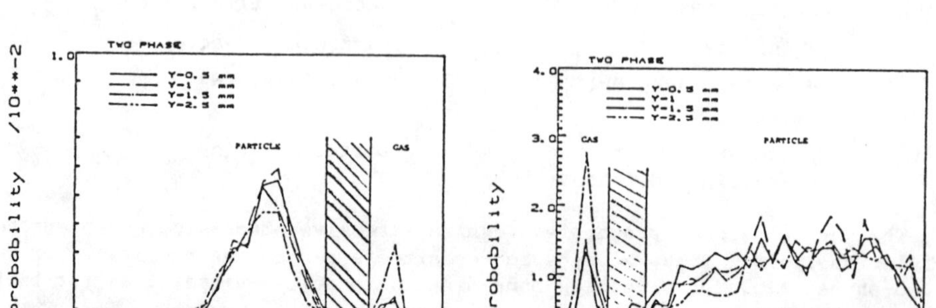

FIGURE 2.6. Distributions of relative signal modulation (visibility) and signal amplitude in a two-phase model flow.

The distributions of signal amplitude and relative signal modulation (visibility) based on 1000 Doppler events in Figure 2.6 show clearly that neither visibility nor amplitude allows satisfactory separation of the information obtained from both phases. It could be shown that a combined discrimination scheme based on amplitude and visibility yielded satisfactory separation of the individual phases.

Lee et al. /19/ presented experimental results of a droplet/ gas flow employing two schemes evaluating the lower part of the particle size distribution and the large particles separately with a reference beam type laser-Doppler anemometer. In Figure 2.7 the method to size small particles is demonstrated. By means of digital data processing the centrical particle trajectories within the measuring volume were selected by requesting a predetermined number of fringes. The light intensity in the region of the probe volume where this fringe number occurred was assumed to be uniform. This allowed direct scattering light intensity evaluation for particle sizing and also to use this information to separate the flow properties of fluid and particle phases.

Intensity distribution
in measuring volume

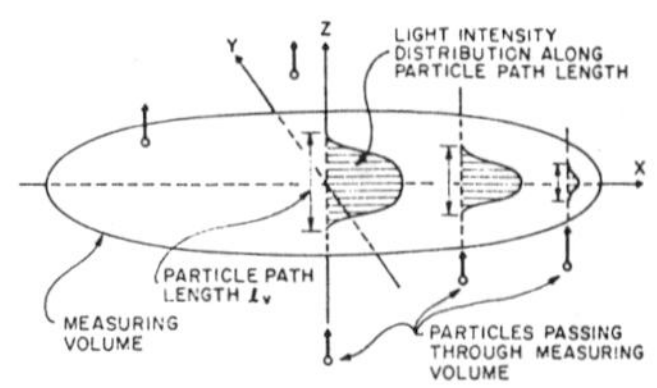

Doppler signal

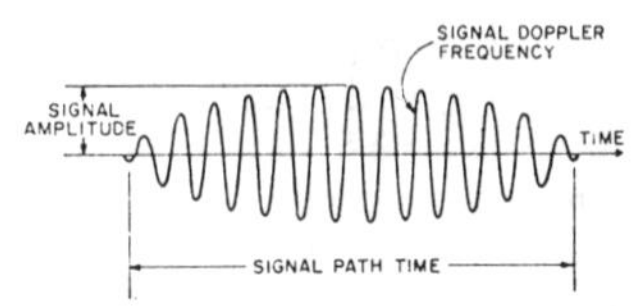

Electronically selected
aerea of measuring volume
for amplitude discrimination

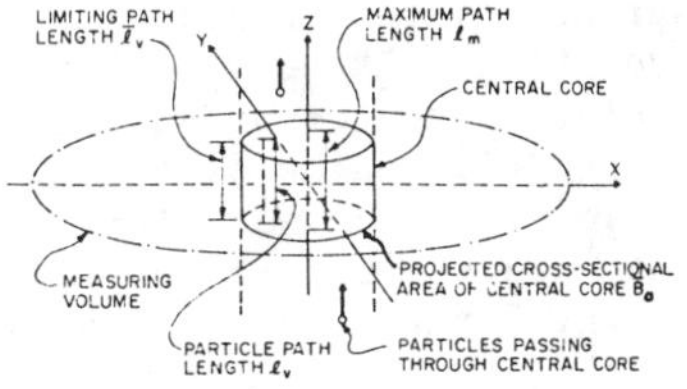

FIGURE 2.7. Scheme to evaluate central particle trajectories by Lee et al. /19/.

For larger particles Lee et al. /19/ suggest to use the residence time of a particle that can be detected by a signal drop-out due to the blockage of the reference beam as shown in Figure 2.8a. This blockage time is only characteristic for the size of the particle if the trajectory is centrical. The central passage was assumed to have happened only if a Doppler-event at the end of the blockage time was obtained.

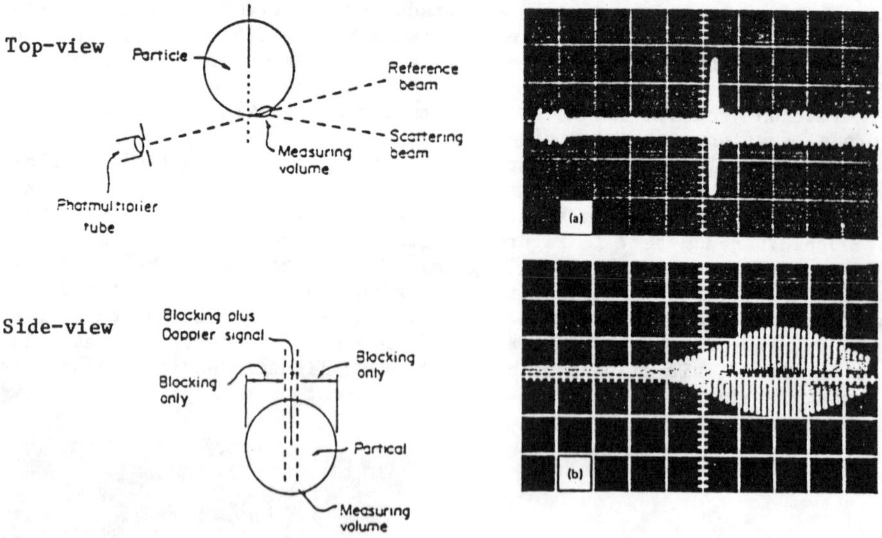

FIGURE 2.8. Set-up and signal characteristics of large particles, Lee et al. /19/.

The data processing speed was increased by the authors using a specially designed LDA-counter processor.

Transient time evaluation of a particle passing a LDA-measuring volume together with signal symmetry requirements has been used by Brankovic et al. /20/ investigating the rise of gas bubbles in still liquids. Figure 2.9 gives an overview on the employed test-facility. A dual beam LDA-system produced bubble signals as depicted by the top trace in the figure after optimizing the receiving optics with respect to signal amplitude and signal modulation.

Appropriate filtering allowed digital evaluation of signal frequency and residence time. It was shown that the high frequency contribution in the signal center corresponded to the bubble velocity, whereas the time between the side peaks of the low frequency signal contribution indicated the bubble entering or leaving the measuring volume respectively. The technique proved to be effective for the bubble size range of .05 to 1.2 mm, e.g. for gas bubbles in water being sufficiently spherical.

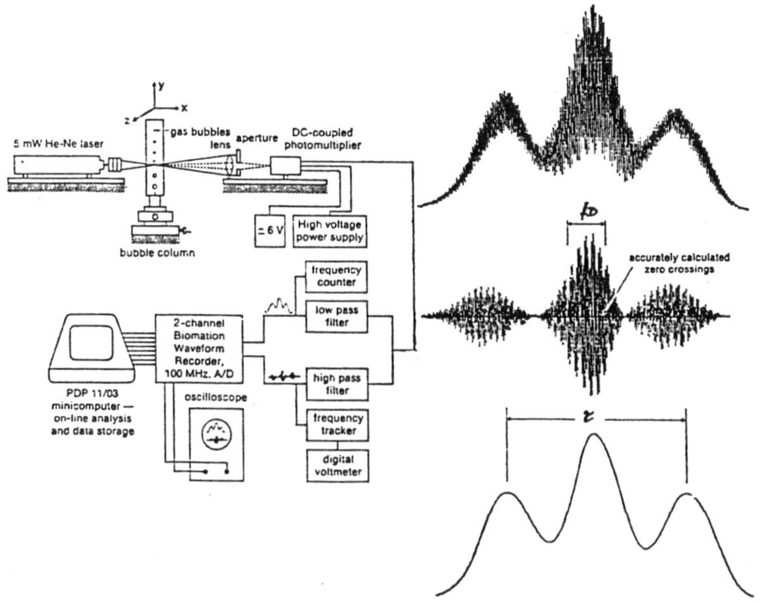

FIGURE 2.9. Bubble velocity and bubble size measurement, Brankovic et al. /20/.

The two techniques described below employ scattered light intensity for particle sizing together with laser-Doppler velocity detection. To this end it is required to make sure that particles get illuminated with exactly the same light power.

Durst and Ruck /4/ and Ruck /5/ employed a white light source to illuminate particles in the set-up shown in Figure 2.10 while superimposing a laser-Doppler probe volume to select particle trajectories that go through a defined intensity field. These particle signals provide a velocity information (LDA-frequency) as well as size information (amplitude of white light). Typical size distributions for small particles can be obtained as shown in Figure 2.1 characterizing their particular properties to follow the fluctuations of a turbulent flow.

Modarres et al. /21/ superimposed a green and a blue LDA-measuring of different size volume from an argon ion laser concentrically to select an area within the measuring volume of Gaussian intensity distribution to be used for particle sizing ensuring a defined illuminating light power in the case of the large probe. The two pairs of beams were orientated perpendicular to one another to allow for simultaneous two-component measurement. As shown in Figure 2.11 coincidence of LDA-events in both measuring volumes was requested to allow further processing of the information.

For the data reported however, narrow particle size distributions were chosen to describe the basic behaviour of a two-phase flow free jet and overcome the inherent sizing problems.

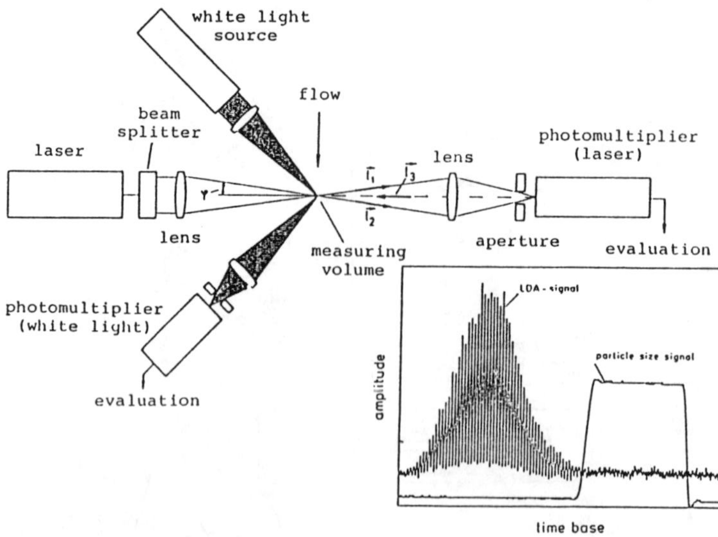

FIGURE 2.10. Velocity and size detection by means of LDA and superimposed white light source, Ruck /3/.

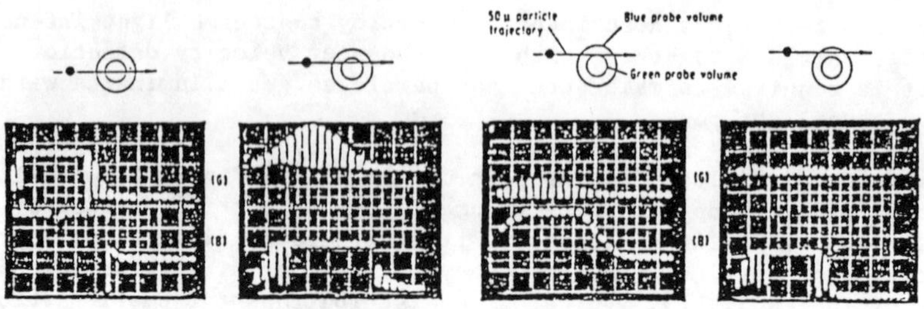

FIGURE 2.11. Amplitude evaluation requesting signal coincidence from two concentric measuring volumes /21/.

Most recent developments in particle velocity and size detection of dispersed two-phase flows employ multiple scattered light detectors. The digital data processing for evaluation is presently improved and will be carried out on LDA-counter processors. Saffmann et al. /22/ defined an experimental set-up as shown in Figure 2.12 to analyse and quantify the phase difference of a Doppler-event observed from two detectors oriented in space. It could be shown that this phase difference is linearly proportional to the particle diameter. The potential of this technique was first discussed by Durst et al. /17/ and Durst /23/.

Experimental arrangement

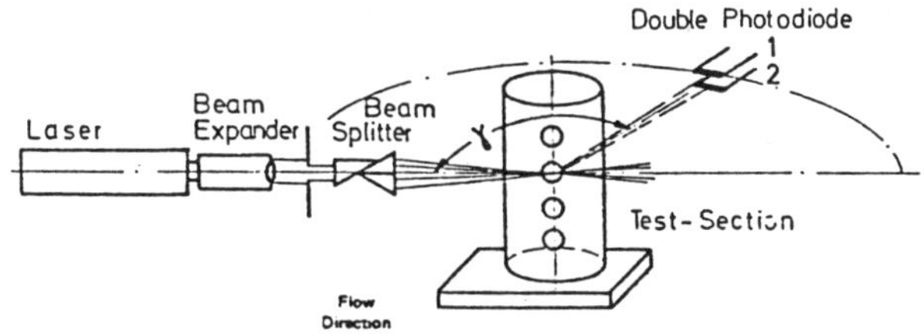

Geometrical definition of set-up

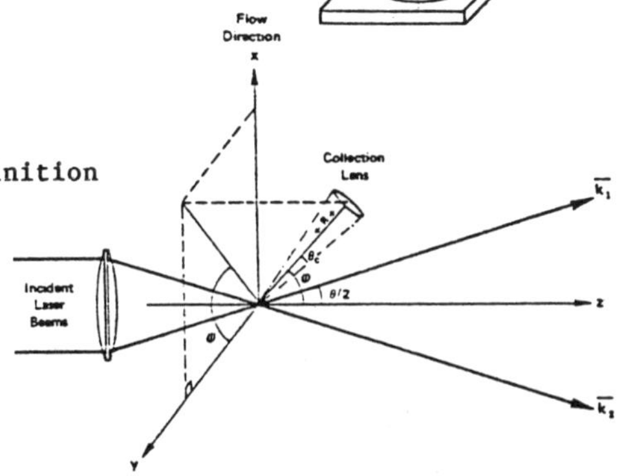

FIGURE 2.12. Phase-Doppler set-up for velocity detection and particle sizing.

The selected operating conditions for two detectors in space are controlled by the light scattering intensity distribution and the phase shift effect controlled by the space angle PHI. Separation of the detectors in space predetermine the particle size range investigated. The noisiness of the phase-information strongly relates to the collecting angle THETA determined by the aperture of the receiving optics as shown in Figure 2.13.

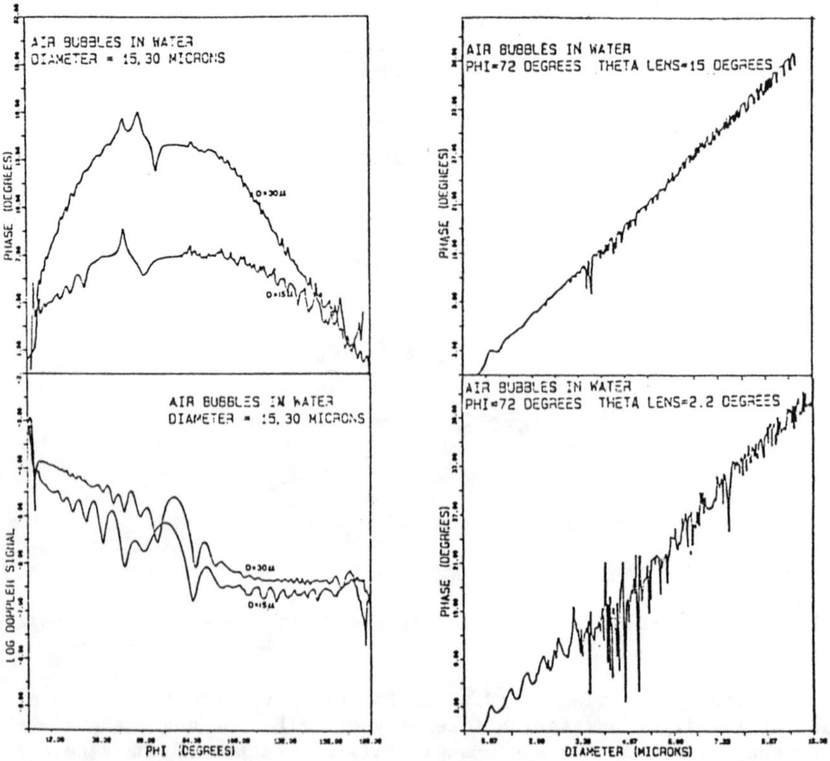

FIGURE 2.13. Phase-Doppler optimization by Saffmann et al. /20/.

2.3 Detailed Experimental Results of Gas/Particle Flow Measurements with LDA

The motion of particles in a turbulent flow is dominated by individual local effects that should not be lumped together and correlated in an integral form. Detailed local measurements can separate out local features of the particle motion, and detailed quantitative investigations of particulate two-phase flows will allow for appropriate modelling of the fluid turbulence, the drag forces (interaction of fluid and particles), the particle/wall interactions and the particle/particle interactions independently. To be able to study these interactions a two-phase channel flow with three separate inlets in parallel was carefully set up as specified in the flow geometry in Figure 2.14. With this flow geometry, it was felt that a large variety of parameters and their effects on particulate two-phase flows could be studied. For example, it was expected that the rate of dispersion of the particulate center flow will be strongly dependent on the flow rate ratios of the inner and the outer channels and this will also be influenced by the particle fluid interactions. Figure 2.14 gives the geometrical details and dimensions of the test section that was designed to allow all features of the spreading particle jet to be studied. The coordinate system later referred to, is defined in the Figure and the measuring planes as well as the resolution of the local measurements is indicated.

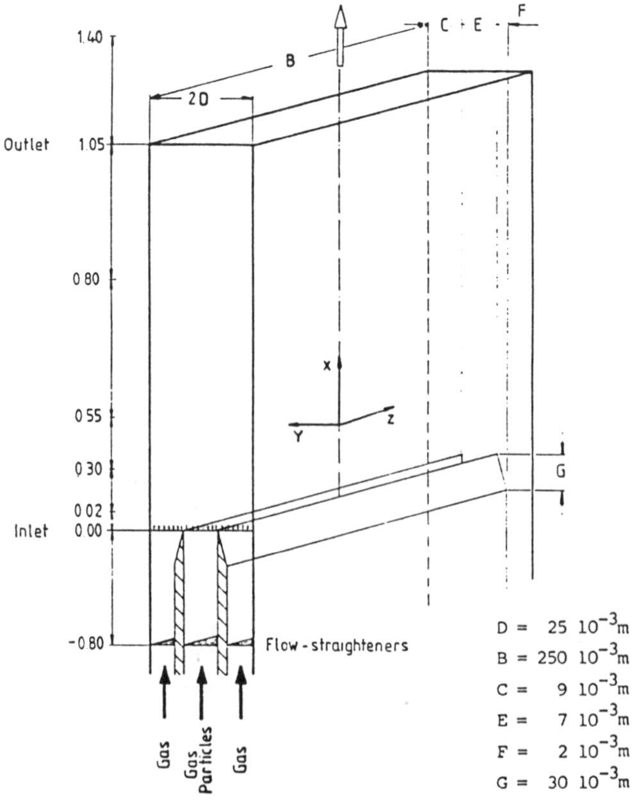

FIGURE 2.14. Geometry of confined two-phase jet flow.

2.3.1 Experimental set-up. The aforementioned test facility consisted of a closed loop that was set up to supply an upward gas-flow in a flat channel of 250 x 25 mm cross-sectional area and 3.20 m in total length. The gas flow was driven by a blower providing a maximum flow rate of 0.5 m^3/s at a differential pressure of 25 10^4 Pa. The temperature of the flow was kept constant to 32°C throughout the measurements. The total flow rate was measured by the differential pressure across an orifice also controlling a pneumatic regulating valve. For the flow-configuration considered in this study the total flow was split into three individual flows led separately into the test-section over flow-straighteners.

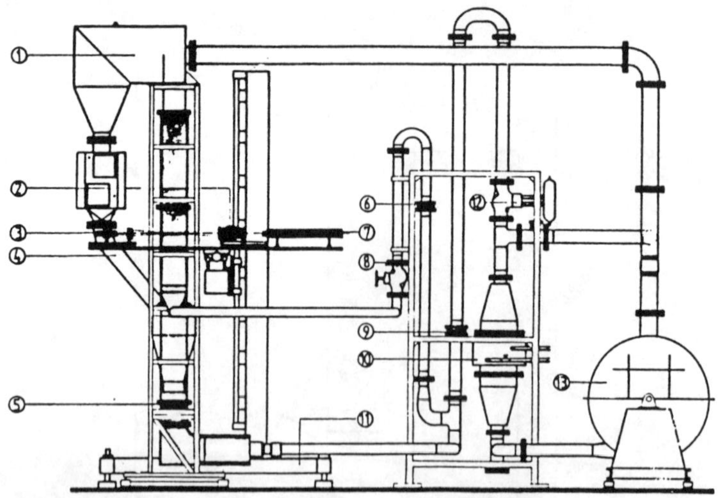

1 Cyclone
2 Emmitting LDA-Optics
3 Receiving LDA-Optics
4 Particle Recirculating Loop
5 Test Section
6 Orifice for Gas Flow Rate Measurements
7 He-Ne Laser (25 mW)
8 Regulating Valve
9 Orifice for Gas Flow Rate Measurements
10 Heat Exchanger
11 Traversing Mechanism for Laser
12 Regulating Valve
13 Blower

FIGURE 2.15. Overall view of the experimental facility /24/.

The flow passing through the central channel was homogeneously supplied with spherical glass particles across the width B of the channel using a calibrated vibration feeding device. After a developing length of 0.8 m for each of the channels the separating plates ended with the outlet walls between the three flows being shaped like knife edges; this is shown in Figure 2.14. This permitted a confluence of the three flows without flow separation and a minimum of wake effects.

All measurements in the above described test section were collected in the center plane of the wider part of the channel ($z = 0$ m) and at different downstream stations corresponding to $x = 0.0, 0.02, 0.3, 0.550, 1.05$ m. Traversing across the smaller width of the channel were performed within $+12.5 \leq y \leq -12.5$ mm, as indicated in Figure 2.14. The small tics shown at the inlet level were the measuring locations for each profile and these indicate that 25 local positions were chosen for one profile.

In the experiments described here, the flow field for two experimental test-cases were carefully studied. Single phase flows and corresponding particulate two-phase flows were studied. For the single- as well as for the two-phase flow situation the ratio of the gas velocities in the two outer channels and the gas velocity in the central channel was varied according to the values in Table 2.1.

TABLE 2.1. Flow conditions for studied test-cases.

	Test case 1		Test case 2	
average gas velocity in outer channels:	$\bar{U}^{o}_{g}$ = 8.20	m/s	5.00	m/s
average gas velocity in central channel:	$\bar{U}^{c}_{g}$ = 9.65	m/s	9.65	m/s
average gas velocity:	$\bar{U}$ = 8.60	m/s	6.30	m/s
flow rate in central channel:	$\dot{V}_g$ = 1.689 10^{-2}	m³/s	1.689 10^{-2}	m³/s
flow rate in both outer channels:	$\dot{V}_g$ = 3.69 10^{-2}	m³/s	2.25 10^{-2}	m³/s
mass flow rate for particles:	$\dot{m}_p$ = 3.210 10^{-3}	kg/s	3.210 10^{-3}	kg/s
mass flow rate for gas:	$\dot{m}_g$ = 6.081 10^{-2}	kg/s	4.450 10^{-2}	kg/s
void fraction central channel:	α = 7.25 10^{-5}		7.25 10^{-5}	

The density of the glass particles was 2780 kg/m³ whereas the density of the gas was 1.1307 kg/m³ with a dynamic viscosity of 18.6 10^{-6} kg/ms and a kinematic viscosity of 1.6 10^{-5} m²/s.

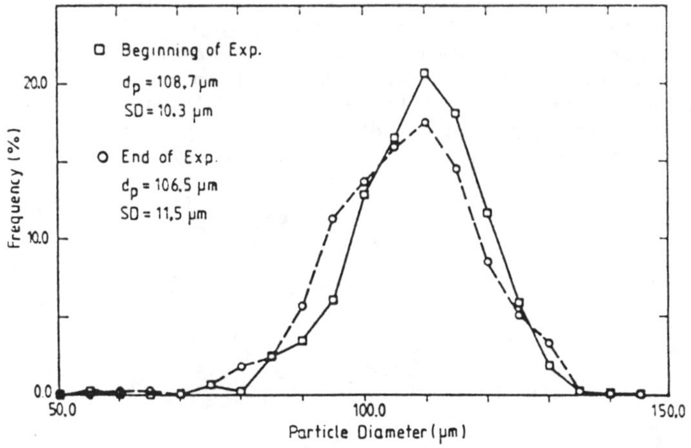

FIGURE 2.16. Particle size distribution.

To ensure constant particle size distribution throughout each of the experiments, samples of particles were taken and analysed before and after the experiment. In Figure 2.16 it is demonstrated that the change in mean particle size the increase in the broadening of the distribution due to wear of the particles is very small.

The influence of the knife edge geometry (see Figure 2.14) can be regarded as negligible if only distributions of the rms-value of the downstream velocity fluctuation show up in the measure. Far downstream of the inlet section due to turbulent mixing, a typical channel velocity profile developed in both flow cases described here.

2.3.2 Results. The inlet conditions given in Table 2.1 were maintained constant throughout the two experiments. The loading in the central channel was μ = 0.168 kg/kg which corresponds to a volume fraction of $\alpha = 7.25\ 10^{-5}\ m^3/m^3$.

Figure 2.17 gives an overview over the results for test-case 1 providing a central air-particle flow mixing with the outer flows with approximately the same average velocity from the outer channels. Gas (+) and particle mean velocities (•) were measured at different downstream stations, and rms-values of the gas and particle velocity fluctuations and the local volume fraction of particles are displayed from left to right in the three columns of Figure 2.17; the inlet flow being described by the bottom profiles, the outlet flow by the top profiles of each column. The scales on the right hand side give from top to bottom the volume fraction, the rms-value of the velocity fluctuations, and the mean velocity.

The following features of the flow can be deduced from the velocity data. At the inlet the particles are in the average slower when compared to the gas phase with slip velocity of $\bar{U}_S$ = 1.45 m/s on the center line. The average gas velocity in the inlet-section is lower than the average particle inlet velocity. This is maintained throughout the channel and this is the reason why the particles are being decelerated along the entire channel. Due to their interaction with the flow and due to their inertia they reach a center line slip velocity at the outlet of 1.18 m/s which is more than twice their corresponding terminal velocity. In the boundary layer close to the channel wall particles are on the average faster than the air flow due to their turbulence transport from the channel centre and due to their high inertia.

For the particulate two-phase flows the features of the fluid rms-velocity fluctuations agree with those of single phase flows except the local level of turbulence turns out to be higher. This may be due either to some unknown turbulence producing effects of the particles, or to still some cross-talks between the velocity information of the two phases caused in the data processing stage.

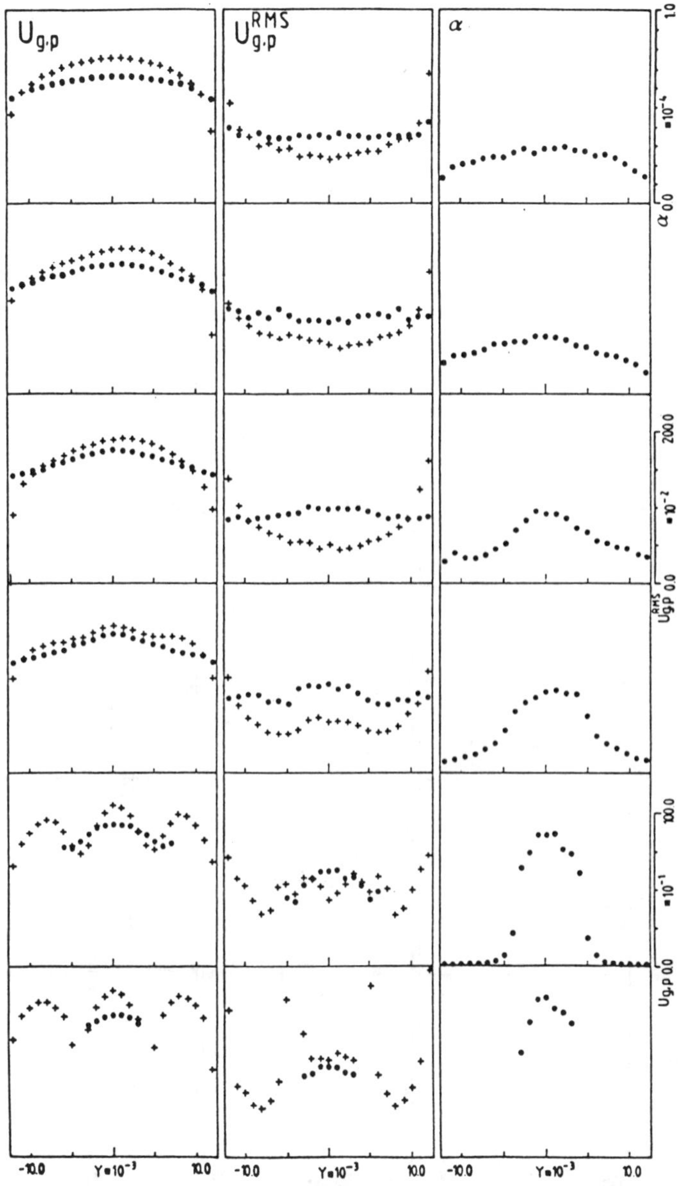

FIGURE 2.17. Two-phase gas(+)-, particle(•)-velocities and α (test-case 1)

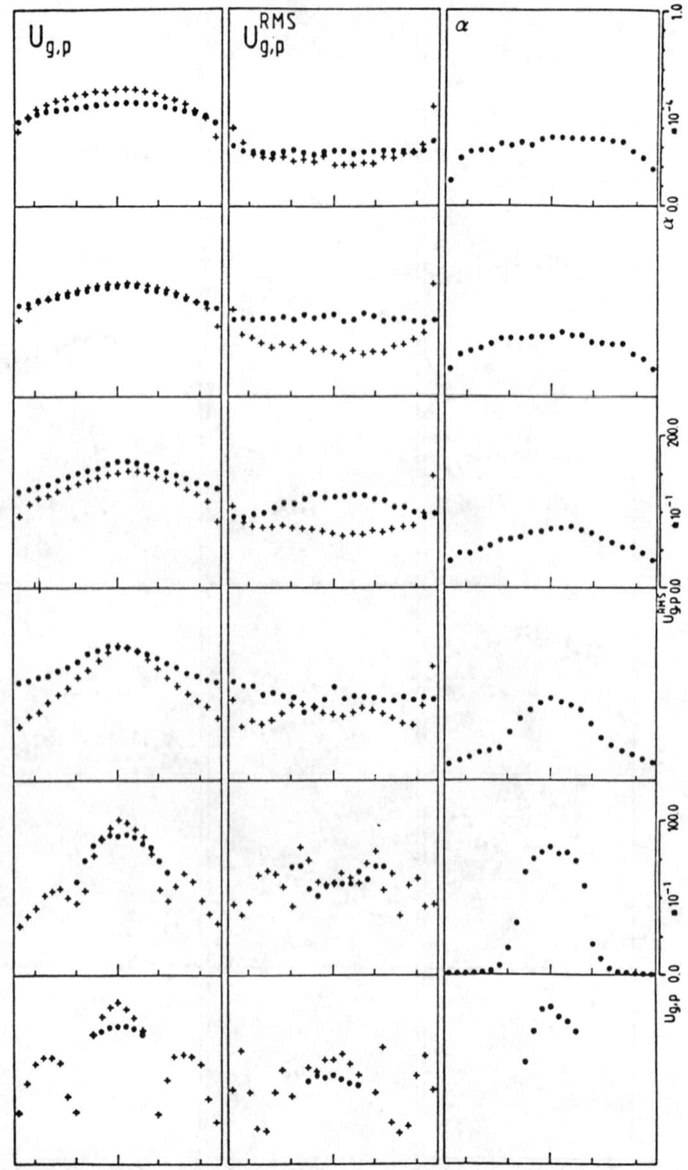

FIGURE 2.18. Two-phase gas(+)-, particle(●)-velocities and α (test-case 2)

The following most remarkable features of the investigated particulate two-phase flows should be listed. The rms-value of the turbulent velocity fluctuation of the gas phase is higher in the inlet plane than that of the particles. This agrees with the assumption that the particles only "respond" to fluctuations of the gas phase and they do this with a reduced amplitude. Further downstream, at a distance of x = 0.1 m the turbulence intensity of the gas phase reaches a minimum, but rises to a maximum at x = 0.8 m and then drops again. This effect can be explained by the fast decay of turbulence in the wider part of the channel following the inlet plane and the retarded influence of the near wall boundary layer being developed at both of the outer walls. The particle rms-value of the particle velocity fluctuations stays very high to 0.5 m downstream of the inlet plane. This is due to the particle inertia, but the present measurements may also be slightly influenced by the data aquisition and the signal processing procedures. Studies to check this are presently being carried out.

The measured volume fraction of the particles indicate their spreading from the central jet into the whole downstream channel. The inlet profile in Figure 2.17/18 is not fully symmetric and indicates that the particle feeding device may not yield a uniform distribution and that the developing length of 0.8 m is insufficient to even out this defect of the particle concentration distribution. The spreading of the particles downstream of the inlet plane seems to take place very rapidly showing that the redistribution of particles in the larger channel is completed after 0.5 m already. This conclusion from a first impression of the data is however not correct, since the mean velocity and the turbulence intensity profiles have not reached their fully developed state. The counting procedure to determine local particle concentration is very sensitive to the signal intensity. This may be the reason why the points close to the wall are generally very low. Unexpectedly fast mixing and spreading of the particles can be explained by relatively high inlet particle fluctuating velocity and particle inertia which are the most important parameters controlling the dispersion rate.

Figure 2.18 gives an overview of the results for test-case 2 providing gas-flows to the outer channels that are distinctly slower than the central gas velocity. With these inlet conditions it can be observed that the average gas velocity in the inlet plane is lower than the average particle velocity. This is the reason why particles flow faster across the entire channel at x = 0.1 m downstream of the inlet plane. Gravity and particle-fluid interactions decelerate the particles and in the outlet plane x = 1.05 m the gas-phase average velocity is higher than the corresponding velocity of the particles. The slip velocity is $\bar{U}_S$ = 0.93 m/s at the center line and turns out to be again negative near the wall very similar to the experimental data of Lee and Durst /25/ and Tsuji et al. /26/. The resultant maximum slip velocity is distinctly lower than in test-case 1. This shows that the flow is not fully developed.

2.4 Conclusion Based on Experiments

It could be shown that experimental techniques are available that are capable to explore basic two-phase flow properties in model flows with a single particle size being present and also more complex flows with distributions of particle sizes. Beside velocity information of the phases involved in the flow the local volume concentration of the disperse phase is an important quantity. In the above study this quantity has been extracted evaluating particle number density and converting it with the mean particle velocity and the total mass flux into the required local concentration. The employed evaluation is not justifiable for all two-phase flows namely not for those with high turbulence intensity.

It has been indicated that complex two-phase flow phenomena that are difficult to measure have to be treated in separate test stages to identify their dominating features, e.g. turbulent particle dispersion, particle/fluid interaction, particle/wall interaction. To enlarge the present knowledge of these flow features individual experiments have to be identified to give better insight into simple particulate two-phase flows. This will allow to subsequently better understand the complex flow-situations.

3. NUMERICAL PREDICTION OF DISPERSED TWO-PHASE FLOWS

3.1 Introduction to Numerical Prediction Procedures

In the last 10 years, the numerical treatment of single-phase flow problems has reached a level which permits calculations for laminar as well as turbulent flows. With the increasing capacity of available computers, systems of nonlinear, coupled, partial differential equations can be solved for exceedingly complex flow problems. For two-phase flows similar computations can be carried out, but it is necessary to solve additional transport equations for the second phase and to handle those interaction terms which treat the mass, momentum and energy exchanges between the two phases /23/.

Particulate two-phase flows can be dealt with theoretically or numericallly in two ways, and there are two corresponding approaches to describe the transport equations for the particulate phase. The Langrangian approach treats the fluid phase as a continuum and predicts the trajectory of a single particle in the fluid flow as a result of various forces acting on the particle. By assuming different starting positions of the particles and following their trajectories, a solid-fluid flow can be simulated. As the locations of the particles are known, the mass, momentum and energy transfers of the solid to the fluid phase and vice versa can be calculated.

Treating the particle phase as a continuum too, and solving the appropriate equations for the fluid and the particle phases, makes up the basic feature of the Eulerian approach. The volume concentration of each phase has to be introducecd and the continuity equation for both phases solved. The back-influence of the solid particles onto the fluid has then to be proportional to the void fraction of the solid phase.

The equations take into account the particle/fluid interactions through the drag force, based on a single particle motion, and the gravity effect. The influence of fluid turbulence on the particles, but the influence of the particles on fluid flow turbulence are mostly neglected. It is also assumed that there are no particle/particle interactions and that the interactions of the particles on the wall can be negelected. This indicates that predictions are nowadays mostly carried out for low particle concentrations.

In the following sections the basic equations and the two solution procedures for both the computational approaches to predict particulate two-phase flows are presented. As the continuous phase is treated in the same way for both approaches, the fluid equations and their numerical solutions will be explained first. The different treatments of the motion of the dispersed phase and its interaction with the fluid phase are then given, first for the Eulerian method and then for the Lagrangian method.

3.1.1 Basic equations and solution procedures. The basic equations for a single-phase Newtonian fluid are well known (e.g. ref. /28/) and are therefore only summarized here for the case of a stationary, two-dimensional, axially symmetric laminar or turbulent flow. With these assumptions for the flow, the continuity equation can be written as follows:

$$\frac{\partial}{\partial x}(\rho u) + \frac{1}{r}\frac{\partial}{\partial y}(r\rho v) = 0 \tag{3.1}$$

and the momentum equations for the u and v velocity components of the fluid phase as:

$$\frac{\partial}{\partial x}(\rho u^2) + \frac{1}{r}\frac{\partial}{\partial y}(r\rho u v) = -\frac{\partial p}{\partial x} + 2\frac{\partial}{\partial x}\left(\mu\frac{\partial u}{\partial x}\right) + \frac{1}{r}\frac{\partial}{\partial y}\left[r\mu\left(\frac{\partial u}{\partial y}+\frac{\partial v}{\partial x}\right)\right] \tag{3.2}$$

$$\frac{\partial}{\partial x}(\rho u v) + \frac{1}{r}\frac{\partial}{\partial y}(r\rho v^2) = -\frac{\partial p}{\partial y} + 2\frac{\partial}{\partial y}\left(\mu\frac{\partial v}{\partial y}\right) + \frac{\partial}{\partial x}\left[\mu\left(\frac{\partial v}{\partial x}+\frac{\partial u}{\partial y}\right)\right] + \frac{2\mu}{r}\left(\frac{\partial v}{\partial y}-\frac{v}{r}\right) \tag{3.3}$$

For a turbulent flow, the laminar viscosity μ is replaced by an effective viscosity:

$$\mu_{\text{eff}} = \mu + \mu_t \tag{3.4}$$

and hence, the present approach to compute the fluid motion uses an eddy viscosity concept for the Reynolds shear stress.

Furthermore, to yield a local description of the turbulent momentum transport in the fluid, the turbulent viscosity μ_t is described by the turbulence energy k and the turbulence dissipation rate ε, for which transport equations have to be solved. Details of this so-called k-ε model have been given by e.g. Rodi /29/. The turbulence model formulated in reference /29/ for single-phase flow has not been changed for the two-phase flow calculations presented in this report. This is justified due to the low concentration of particles treated in the present work.

For the calculations of a single-phase channel flow the boundary conditions are normally specified in the following manner:

Inlet: u, v are prescribed

Exit: $$\frac{\partial u}{\partial x} = \frac{\partial v}{\partial x} = 0 \tag{3.5}$$

Wall: $u = v = 0$

Symmetry axis: (pipe axis) $\frac{\partial u}{\partial y} = 0, \quad v = 0$

These equations for single-phase flows are also taken as a starting point for the transport equations of the continuous phase of the particulate two-phase flows treated in this report. In any continuation of the present study, they have to be changed in such a way that the presence of the particles and their effects on the fluid phase are taken into account in a physically more correct way.

In particulate two-phase flows the particle phase occupies a definite volume, and this has to be taken into account in formulating the transport equations for the fluid phase. Only the part of the incoming volume which is not taken by the particle phase enters the continuity and momentum equations for the fluid phase. In the Lagrangian approach the volume of the particles is neglected, so that the continuity equation of the fluid does not have to be changed and the different terms in the fluid momentum equation do not have to be weighted by the corresponding volume concentrations. With this simplification, only the momentum equations have to be solved for each particle class in the Lagrangian approach. For the Eulerian method, a volume concentration of the fluid and the particle phase will be defined and a continuity equation solved for every phase. In addition to this, the different terms in the momentum equations are weighted by the volume concentration.

In the equations for the continuity and momentum transport of both phases, interaction terms occur. Since it is assumed that there is no phase change, only the momentum transfer between the phases has to be considered. This transfer of momentum yields additional terms that have to be included in the momentum equation of the fluid phase. It is assumed here that they will be dependent on the velocity difference between the two phases. Of course, the momentum transfer from the particle phase to the fluid phase has to be opposite to the corresponding momentum transfer from the fluid to the particles.

In the Lagrangian approach, principally the gain or loss of particle momentum is calculated and included as a sink or source term in the fluid momentum equation. The drag force between the two phases is calculated in the Eulerian approach and added with opposite signs to the momentum equations of both the phases. Similarly, other terms that describe interactions between the phases can be added to the momentum equations.

Generally, in particulate two-phase flows, the particles have a density different from that of the fluid. Therefore the gravity term has to be included in the momentum equations of the continuous phase. The resulting pressure in the fluid equations is the static pressure, which also takes into account buoyancy effects. As the gravity term will be dependent on the particle concentration, its magnitude changes with location in space. The above mentioned additional momentum terms in the fluid momentum equations model the back-influence of the particles on the fluid.

In the Lagrangian approach, Newton's law of motion is used to find the velocity of every particle in the flow domain. Integration of the velocity with time gives the particle path, so that the particle velocities and locations can be used to get the momentum transfer terms. In the Eulerian approach, the void fraction of the particles in every point of the flow domain is solved from a particle continuity equation, rather than the individual particle locations being determined. Together with the particle velocities, which are known as a function of space the corresponding transfer terms can be derived in the Eulerian method.

The numerical solution procedures of the extended transport equations of the continuous phase are the same for both the approaches and the basic ideas are presented below. The calculation procedure is based on the TEACH code, which has been described in detail by Patankar and Spalding /30, 31/.

The partial differential equations for the mass and the momentum are solved by using finite differences, derived by integrating the differential equations over control volumes. For this an orthogonal grid is applied, where the two velocities and the pressure are stored in staggered positions (Figure 3.1). The necessary linearizaton of the nonlinear equations is performed by hybrid differences to get an implicit finite difference scheme.

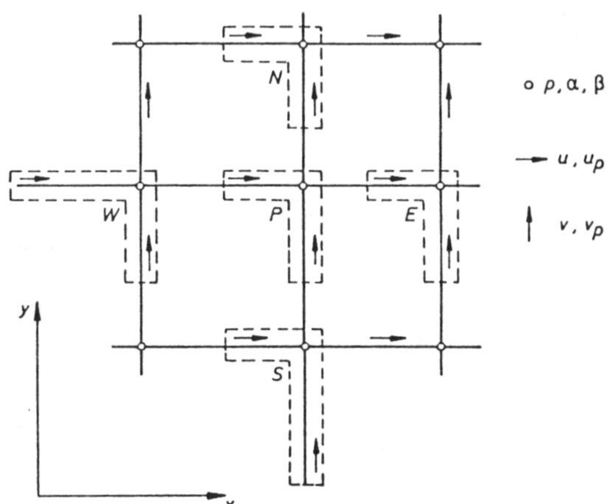

FIGURE 3.1. Staggered grid and control volume.

Because of the elliptic nature of the partial differential equations, an iterativ solution procedure is employed. One starts with distributions for the velocity and pressure fields, and calculates the fluid velocities from the momentum equation. Thereafter, the solution of a pressure correction equation (poisson equation), which has been derived from the continuity equation, yields corrected values for the velocity and the pressure fields, so that the continuity equation is satisfied. With the corrected values, the momentum equations are solved again and the whole procedure is repeated until convergence is achieved. This was assumed to be the case when the normalized residuals at each point, summed over the whole calculation domain, were each smaller than 0.005.

The resultant finite difference equations are solved using a tri-diagonal matrix algorithm. The velocities and the location or the "void fraction of the particles" are obtained by solving the corresponding equations after a specific number of interactions of the fluid solution procedure has been carried out. With the computed information on the particles, the momentum transfer terms can be calculated and introduced in the fluid equations before the solution procedure is repeated. Details on how this has been performed for both approaches will now be shown.

3.1.2 The Eulerian approach. The basis of the Eulerian approach is the treatment of the dispersed phase as a continuum. As there are two "fluids" present, the definition of a volume concentration of each phase, the so-called void fraction, is necessary and will be explained first. After the derivation of the additional terms in the fluid continuity and momentum equations, the corresponding transport equations for the particle phase are established. Finally the numerical procedure for the solution of the dispersed two-phase flow problem is explained.

Continuity equations: The Eulerian approach is based on the assumption that there are two continua in two-phase flow. At one point in the flow domain and at a specific time, however, only one fluid can be present. As with a numerical procedure, it is not possible to resolve every point in time or space; it is necessary to average over a specific time and space. Definition of a volume concentration gives the fraction of a control volume which is occupied by the first or the second phase.

The void-fraction of the dispersed phase is:

$$\alpha = \frac{V_p}{V} \tag{3.6}$$

As stationary flow problems only are being dealt with, it is assumed that this variable is not changing with time, and an integration with time is thus not necessary. In order to get a physically reasonable meaning for the void-fraction, the ratio of the control volume and the particle volume should be much larger than unity; otherwise the concept that the dispersed phase can be treated as a continuum is not justified. The control volume over which the void-fraction is averaged is the same as that for the numerical control volume approach. The transport of mass or momentum has now to be weighted by this void-fraction, to take into account the fact that both the phases are flowing through the control volume.

As there is no mass exchange between the two phases, two separate continuity equations exist:

$$\frac{\partial}{\partial x}(\beta \rho u) + \frac{1}{r}\frac{\partial}{\partial y}(\beta r \rho v) = 0 \tag{3.7}$$

and for the dispersed phase:

$$\frac{\partial}{\partial x}(\alpha \rho_p u_p) + \frac{1}{r}\frac{\partial}{\partial y}(\alpha r \rho_p v_p) = 0 \tag{3.8}$$

The particles will have constant density, so that the continuity equation of the dispersed phase can be written as:

$$\frac{\partial}{\partial x}(\alpha \rho_p u_p) + \frac{1}{r}\frac{\partial}{\partial y}(\alpha r \rho_p v_p) = 0 \tag{3.9}$$

Momentum equations: The momentum equation for the continuous phase has to be weighted by the volume concentration of the fluid. In addition to this, the momentum transfer term, based on the drag force between the two phases, has to be taken into account. As this term appears in the momentum equation of the dispersed phase with opposite sign, it will be mentioned here and discussed in detail when the particle momentum equations are derived.

The fluid momentum equation in the x direction (with gravity) is:

$$\begin{aligned}\frac{\partial}{\partial x}(\beta\rho u^2) + \frac{1}{r}\frac{\partial}{\partial y}(\beta r\rho u v) \\ = -\beta\frac{\partial p}{\partial x} + 2\frac{\partial}{\partial x}\left(\beta\mu\frac{\partial u}{\partial x}\right) \\ + \frac{1}{r}\frac{\partial}{\partial y}\left[\beta r\mu\left(\frac{\partial u}{\partial y}+\frac{\partial v}{\partial x}\right)\right] \\ - \beta\rho g - 0.75\,\alpha\beta c_D\frac{\mu}{d^2}Re_p(u-u_p)\end{aligned} \tag{3.10}$$

The fluid momentum equation in the y direction is:

$$\begin{aligned}\frac{\partial}{\partial x}(\beta\rho u v) + \frac{1}{r}\frac{\partial}{\partial y}(\beta r\rho v^2) \\ = -\beta\frac{\partial p}{\partial y} + 2\frac{\partial}{\partial y}\left(\beta\mu\frac{\partial v}{\partial y}\right) \\ + \frac{\partial}{\partial x}\left[\beta\mu\left(\frac{\partial v}{\partial x}+\frac{\partial u}{\partial y}\right)\right] \\ + \frac{2\beta\mu}{r}\left(\frac{\partial v}{\partial y}-\frac{v}{r}\right) - 0.75\,\alpha\beta c_D\frac{\mu}{d^2}Re_p(v-v_p)\end{aligned} \tag{3.11}$$

The starting point for the momentum equations of the dispersed phase is Newton's law of motion for a particle in a fluid, which is, for the x direction:

$$m^p\frac{du}{dt} = c_D\left(\frac{\pi d^2}{4}\right)\cdot\frac{1}{2}\rho|\Delta u|\Delta u \tag{3.12}$$

Changing the total by the substantial derivative and dividing by the particle volume results in:

$$\rho_p\left(u_p\frac{du_p}{dx} + v_p\frac{du_p}{dy}\right) = \frac{3}{4}c_D\frac{\rho}{d}|\Delta u|\Delta u \tag{3.13}$$

Introduction of a particle Reynolds number and multiplication by the void-fraction gives:

$$\left(\alpha\rho_p \frac{\partial u_p}{\partial x} + \alpha\rho_p v_p \frac{\partial u_p}{\partial y}\right) = \frac{3}{4}\alpha c_D \frac{\mu}{d^2} Re_p \Delta u \tag{3.14}$$

The use of the continuity equation for the dispersed phase results in the momentum equation for a suspension of particles flowing with a slip velocity u in a fluid:

$$\frac{\partial}{\partial x}(\alpha\rho_p u_p^2) + \frac{1}{r}\frac{\partial}{\partial y}(\alpha r u_p v_p) = \frac{3}{4}\alpha c_D \frac{\mu}{d^2} Re_p \Delta u \tag{3.15}$$

To obtain consistent momentum equations for a flowing suspension, the following has to be taken into account:

(1) The pressure variation, which can occur in the flowing fluid, acts also on the dispersed phase.

(2) The gravity force on the particles: as gravity is also introduced into the fluid momentum equation, buoyancy effects result from the coupling of the two phases by the pressure.

(3) The momentum transfer term has to be symmetric with respect to the volume concentration of both the phases, as two continua are being dealt with. For this reason, instead of the particle void-fraction, the product of both the volume concentrations has be be used.

(4) The second derivatives in the momentum equations of a Newtonian fluid are simulating the molecular momentum transport in a laminar flow. The additional momentum transport, resulting from turbulent fluctuations, is modelled very often with the eddy viscosity concept. This turbulent momentum transport, which is derived from the Reynolds equations, also has a physical meaning in the momentum equations of the dispersed phase. For this reason, the second derivatives also will be taken into account in the dispersed phase for simulation of the particle turbulent fluctuations. As the modelling of diffusion phenomena with the Lagrangian approach is not possible in such a way, the influence of the turbulence on the particles has not been considered here ($\sigma = 10^{30}$).

Finally, the momentum equations for the dispersed phase can be written in the following way. The dispersed phase momentum equation in the x direction is:

$$\begin{aligned}
&\frac{\partial}{\partial x}(\alpha\rho_p u_p^2) + \frac{1}{r}\frac{\partial}{\partial y}(\alpha r\rho_p u_p v_p) \\
&= -\alpha\frac{\partial p}{\partial x} + 2\frac{\partial}{\partial x}\left(\alpha\frac{\mu}{\sigma}\frac{\partial u_p}{\partial x}\right) \\
&\quad + \frac{1}{r}\frac{\partial}{\partial y}\left[\alpha r\frac{\mu}{\sigma}\left(\frac{\partial u_p}{\partial y} + \frac{\partial v_p}{\partial x}\right)\right] \\
&\quad - \alpha\rho_p g + 0.75\,\alpha\beta c_D \frac{\mu}{d^2} Re_p(u - u_p)
\end{aligned} \tag{3.16}$$

and the dispersed phase momentum equation in the y direction is:

$$\frac{\partial}{\partial x}(\alpha\rho_p u_p v_p) + \frac{1}{r}\frac{\partial}{\partial y}(\alpha r \rho_p v_p^2)$$

$$= -\alpha\frac{\partial p}{\partial y} + 2\frac{\partial}{\partial y}\left(\alpha\frac{\mu}{\sigma}\frac{\partial v_p}{\partial y}\right)$$

$$+ \frac{\partial}{\partial x}\left[\alpha\frac{\mu}{\sigma}\left(\frac{\partial v_p}{\partial x} + \frac{\partial u_p}{\partial y}\right)\right] + \frac{2\alpha\mu}{r\sigma}\left(\frac{\partial v_p}{\partial y} - \frac{v_p}{r}\right)$$

$$+ 0.75\,\alpha\beta c_D \frac{\mu}{d^2} Re_p(v - v_p) \tag{3.17}$$

From the momentum equations for the two phases, it can be seen that, for $\alpha = 0$, the fluid momentum equation transforms to a single-phase flow equation and the particle momentum equation approaches the equation of motion for a single particle.

The drag coefficient in the momentum transfer term, which is based on the motion of a single particle, is expressed by the relationship:

$$c_D = \frac{24}{Re_p}(1 + 0.15\,Re_p^{0.687}) \tag{3.18}$$

which approximates quite well to the standard drag curve for a sphere. The experimental data for this correlation have been obtained by measuring the drag coefficient of a single sphere in a uniform flow. The question as to how far it is possible to extend such results also for flow conditions where turbulence, non-uniform fluid velocity or multiparticle motion have roles will not be discussed here.

Besides the drag term introduced, additional forces, acting on the particles, are possible. A non-uniform fluid velocity causes migration of particles /30/ in the same way as the result of particle spinning /31/. With increased concentration, there will be some diffusion-like interaction of the particles. These effects are not taken into account, as the influence of turbulence on particles and vice versa will be neglected.

The system of six partial differential equations can now be solved together with the equation:

$$\beta = 1 - \alpha \tag{3.19}$$

to obtain the seven independent variables u, v, u_p, v_p, α, β and p). As already described, the solution of the mass and momentum equation of the continuous phase gives the fluid velocities and the pressure. The particle velocities are obtained from the particle momentum equation and the particle void-fraction by solving the particle continuity equation.

The fluid boundary conditions are the same as those described already. For inlet and exit, the particle boundary conditions are the same as for the fluid phase.

As no second derivatives occur in the particle transport equations ($\sigma = 10^{30}$) and $v_p = 0$ is assumed for the symmetry axis and wall, no conditions are necessary for u_p and α for those boundaries; so we have:

inlet: u_p, v_p, α prescribed

exit: $\frac{\partial u_p}{\partial x} = \frac{\partial v_p}{\partial x} = \frac{\partial \alpha}{\partial x} = 0$ (3.20)

3.1.3 The Lagrangian approach. The Lagrangian treatment of the particulate phase combined with the Eulerian approach for the fluid was first proposed by Migdal and Agosta /28/. They showed that the particles could be treated as sources of mass, momentum and energy in the fluid flow field. The elliptic form of the equation for the conservation of the general variable ϕ in axisymetrical coordinates is as follows:

$$\frac{\partial}{\partial x}(\rho u \phi) + \frac{1}{r}\frac{\partial}{\partial y}(r \rho v \phi) = \frac{\partial}{\partial x}\left(\Gamma \frac{\partial \phi}{\partial x}\right) + \frac{1}{r}\frac{\partial}{\partial y}\left(r \Gamma \frac{\partial \phi}{\partial y}\right) + S_\phi + S_\phi^p \quad (3.21)$$

Equation (3.21) contains the additional source term S_ϕ^p which represents the net efflux of ϕ into the fluid phase owing to the particle-fluid interaction. This source term is calculated by solving the Lagrangian equation for the corresponding particle variable ϕ^p. This is the essence of the so-called PSI cell (particle-source in cell) approach, which was used by Sharma /29/ for the computation of gas-particle flows in orifices and venturis, by Crowe et al. /30/ for gas-droplet flows, by Sharma et al. /31/ for the computation of swirling gas-particle flows in pulverized coal classifiers, and by Lockwood et al. /32/ for the prediction of pulverized coal-fired furnaces.

Inspection of equation (3.21) shows that the solution procedure for the fluid can be the same as for the single-phase flow or for the above Eulerian treatment of the particulate phase. We may neglect mass transfer and use time-averaged variables for the turbulent flow; then the equations for the conservation of mass and axial/radial momenta are discretized and solved according to the TEACH numerical solution procedure. The standard two-equation k-ε turbulence model, together with the isotropic eddy viscosity for the computation of gas-particle flows in orifices and venturis, by Crowe et al. /36/ for gas-droplet flows, by Sharma et al. /37/ for the computation of swirling gas-particle flows in pulverized coal classifiers, and by Lockwood et al. /38/ for the prediction of pulverized coal-fired furnaces.

Inspection of equation (3.21) shows that the solution procedure for the fluid can be the same as for the single-phase flow or for the above Eulerian treatment of the particulate phase. We may neglect mass transfer and use time-averaged variables for the turbulent flow; then the equations for the conservation of mass and axial/radial momenta are discretized and solved according to the TEACH numerical solution procedure. The standard two-equation k-ε turbulence model, together with the isotropic eddy viscosity hypothesis, is employed for the prediction of the turbulence transport.

Calculation of the particle trajectories by solution of the equations of motion of the particles, as well as the calculation of the particle source terms, is performed in the same iteration loop after the solution of the fluid flow equations, in order to handle the interaction between the fluid and the particles.

Velocities and trajectories of particles: According to the TEACH numerical solution code for single-phase flow, the flow domain is subdivided into a finite number of control volumes (cells) forming an orthogonal grid for the numerical calculation. The particles are introduced at a finite number of starting locations with a finite number of particle sizes at each starting location. If the total mass flow rate of particles is $\dot{m}_p$, if the mass fraction of particles starting at a location j is f_j and if the mass fraction of particles of class i with diameter d_i is f_i, the number flow rate of spherical particles of class i at starting location j is:

$$\dot{N}_{ij}^{p} = \frac{6\dot{m}^{p}}{\pi \rho_{p,i} d_i^3} f_i f_j \tag{3.22}$$

The equation of motion for the particle of class i starting at location j is:

$$\begin{aligned} m_i^p \frac{dV_{ij}^p}{dt} &= \frac{1}{2} c_D \rho A_i^p |V - V_{ij}^p|(V - V_{ij}^p) \\ &+ m_i^p g \left(1 - \frac{\rho}{\rho_{p,i}}\right) \end{aligned} \tag{3.23}$$

where m_i^p is the mass of a single particle. The influence of the Reynolds number ($Re_p = \rho d |V-V_{ij}^p|/\mu$) on the drag coefficient of spherical particles is satisfactorily described by /39/

$$c_D = \frac{24}{Re_p}(1 + 0.15\, Re_p^{0.687}) \tag{3.24}$$

over a wide range of Reynolds number (approximately $1 < Re_p < 10^3$).

For the calculation of the particle trajectories, the method of Sharma /35/ is used. If the fluid velocity is constant during short time intervals Δt, equation (3.23) is solved analytically to give:

$$\begin{aligned} V_{ij}^p &= V - (V - V_{ij0}^p) \exp\left(-\frac{\Delta t}{\tau}\right) + g\tau \left(1 - \frac{\rho}{\rho_{p,i}}\right) \\ &\times \left[1 - \exp\left(-\frac{\Delta t}{\tau}\right)\right] \end{aligned} \tag{3.25}$$

where V_{ij0}^p is the particle velocity at the beginning of the time increment Δt, and τ is the particle characteristic time defined by:

$$\tau = \frac{4\rho_{p,i} d_i^2}{3\mu c_D Re_p} \tag{3.26}$$

The particle position after the time interval Δt is determined from:

$$x = x_0 + (V_{ij}^p + V_{ij0}^p)\frac{\Delta t}{2} \tag{3.27}$$

Note that equation (3.23) can be also solved using a Runge-Kutta integration, as was suggested by Lockwood et al. /38/.

Particle momentum source term: The transfer of momentum between the phases is due to the drag force. If we consider only one class of particle and only one starting location, and assume a constant particle diameter along the trajectory, the particle momentum source term for any arbitrary control volume can be written as:

$$S_{ij}^{p} = \dot{N}_{ij}^{p}\, \rho \frac{A_{i}^{p}}{2} \int_{t_{in}}^{t_{out}} c_{D}(V - V_{ij}^{p})|V - V_{ij}^{p}|\, dt \tag{3.28}$$

where t_{in} and t_{out} are the times when the particle enters and leaves the control volume. However, the calculation of the source terms is more efficient if we multiply equation (3.23) by N_{ij}^{p} dt and integrate it over the residence time of the particle in the control volume $\Delta t_{R} = t_{out} - t_{in}$:

$$S_{ij}^{p} = \dot{m}_{ij}^{p}(V_{ij,out}^{p} - V_{ij,in}^{p}) - \dot{m}_{ij}^{p} g \left(1 - \frac{\rho}{\rho_{p,i}}\right) \times (t_{out} - t_{in}) \tag{3.29}$$

The mass flow rate of the particles along the trajectory is:

$$\dot{m}_{ij}^{p} = m_{i}^{p} \cdot \dot{N}_{ij}^{p} \tag{3.30}$$

The momentum source term resulting from all particle trajectories crossing the control volume is obtained from:

$$S_{cell}^{p} = \sum_{i,cell} \sum_{j,cell} S_{ij}^{p} \tag{3.31}$$

Inspection of equation (3.29) shows that the particle momentum source term is composed of two parts: the change of momentum of the particles, and the influence of all external forces acting on the particles. In our case, all other forces acting on particles, except gravity, are neglected. Some other authors using the Lagrangian approach, such as Sharma /35/, Crowe et al. /36/ and Sharma et al. /37/ define the particle source term as the net efflux of particle momentum represented by the first term on the right-hand side of equation (3.29). This can lead to an erroneous solution of the equations for the fluid flow, because the interaction of particles and the fluid is realized through the drag force and not only through the change of the particle momentum. When particles reach their settling velocities their momentum remains constant, but the exchange of momentum with the surrounding fluid owing to the gravity force still exists. It has to be noted here that the second term on the right-hand side of equation (3.29) is very often much larger than the first one.

Void-fraction: The void-fraction, which is used in the Eulerian approach, has to be calculated also by the Lagrangian method, to compare the results of both the approaches.

For the arbitrary control volume V and a single particle trajectory, the void-fraction is given by:

$$\alpha_{ij} = \frac{\dot{m}^{p}_{ij}(t_{out} - t_{in})}{\rho_{p,i} V} \tag{3.32}$$

For all the particle trajectories crossing the control volume, the following equation is obtained:

$$\alpha = \frac{1}{V} \sum_{i,\text{cell}} \sum_{j,\text{cell}} \frac{\dot{m}^{p}_{ij}}{\rho_{p,i}} (t_{out} - t_{in}) \tag{3.33}$$

The Lagrangian approach requires a relatively large number of starting locations for the particles in order to obtain a reasonably smooth void-fraction profile. This is due to the Eulerian definition of the void-fraction, which assumes that particles can be treated as a continuum.

3.2 Comparison of Experiments and Predictions

The modelling of the motion of heavy particles in turbulence is one of the most difficult problems for the prediction of turbulent two-phase flows. If an Eulerian approach for the treatment of the particulate phase is used, the mathematical formulation is relatively simple, except it is necessary to evaluate a dispersion coefficient for the particles. This is usually done by specifying a Schmidt number which relates fluid and particle diffusion coefficients.

Here the dispersion of particles was based on an analysis of Al-taweel and Carley /40/ first employed by Schönung /41/ using an eddy viscosity concept. A fluctuation ratio between the two phases is obtained by the particle response in an oscillating fluid. The frequency is assumed to be the frequency of the energy containing eddies.

This method is not correct for heavy particles because of their inertia. In this case particle disperion is influenced by upstream particle turbulence rather than local fluid turbulence. This problem can be solved only if some additional equations are solved for the particle properties, like turbulence energy and turbulence scales and if the disperion coefficient is calculated directly from these local particle quantities. In addition the heavy particles experience the so-called effect of crossing trajectories due to the existence of body forces like gravity. This effect causes the particles to cross turbulence eddies, lose their velocity auto-correlation and disperse less than fluid elements /42/. At present the only possibility to take into account all these effects is to apply a Lagrangian simulation of heavy particle motion in a randomly generated fluid turbulence for which the fluctuating kinetic energy and scales are known.

For the prediction of the turbulent fluid flow field the model of Durst et al. /24/ is used. The model is based on a single phase flow k-ε model of turbulence with additional sources and sinks due to the presence of particles and their interactions with the fluid. The sources of mean momentum in both longitudinal and transverse directions, sources of kinetic energy of turbulence k as well as its dissipation ε are calculated from the ensemble averages of a large number of Lagrangian particle trajectories.

The motion of particles in fluid turbulence is predicted by means of a Lagrangian stochastic-deterministic model (LSD) /43/. The numerical solution of fluid equations provides the fields of mean velocity components as well as of k and ε. All three components of the fluctuation velocity are calculated from known k, ε and turbulence production terms by using an algebraic shear stress turbulence model described by Launder /44/. The calculation of the particle trajectories from the inlet to the outlet of the flow domain is performed by a numerical solution of instantaneous particle momentum equations

$$\frac{d\tilde{u}_i^p}{dt} = \frac{1}{\tau^p} (\tilde{u}_i - \tilde{u}_i^p) + g_i \qquad (i = 1,2,3) \tag{3.34}$$

where index i denotes any of the three coordinate directions and g the acceleration due to the gravity acting only in longitudinal direction (i = 1). Fluid velocity fluctuations are generated as Gaussian random numbers for all three directions corresponding to local turbulence intensity of the fluid and added to local mean velocity components to obtain instantaneous values $\tilde{u}_i$. The instantaneous values are kept constant during the fluid Lagrangian integral time scale T_L corresponding to the eddy lifetime, which is calculated from an approximate expression:

$$T_L = C_T \frac{k}{\varepsilon} \tag{3.35}$$

Effective eddy size is given by

$$L_L = U_g^{RMS} T_L \tag{3.36}$$

The numerical solution of the equation (3.34) allows to directly take into account the relative motion between the eddy and the particle. That gives the possbility to stop the eddy-particle interaction and generate new fluid fluctuation velocity components either when the eddy life time is over or when the particle crosses the eddy. Mean particle flow properties are calculated by taking ensemble averages for the same control volumes as for the numerical solution of the fluid equations from a large number of trajectories. The LSD model contains a coefficient C_T in the expression for T_L (3.36). The value of C_T was optimized by means of comparisons of the predictions with the experimental results of Snyder and Lumley /42/ for the case of particle dispersion from a point source in grid generated turbulence and the value $C_T = 0.3$ was accepted.

A good agreement of predictions for the dispersion for all particles sizes investigated by Snyder and Lumley is illustrated in Fig. 3.2. The LSD model was also applied for the simulation of the experiments of Wells and Stock /45/ under similar conditions but with additional electrostatic field acting on the glass particles. The agreement of the predictions with their experiment using the same $C_T = 0.3$ was again very good /46/. On the basis of the predictions of particle dispersion in fully developed turbulent pipe flow in the same publication /45/ it was possible to explain a much faster dispersion of heavy particles than the small one reported in the experiments of Arnason /47/. It was pointed out that initial conditions for the particles are very important for heavy particle dispersion especially within short dispersion times.

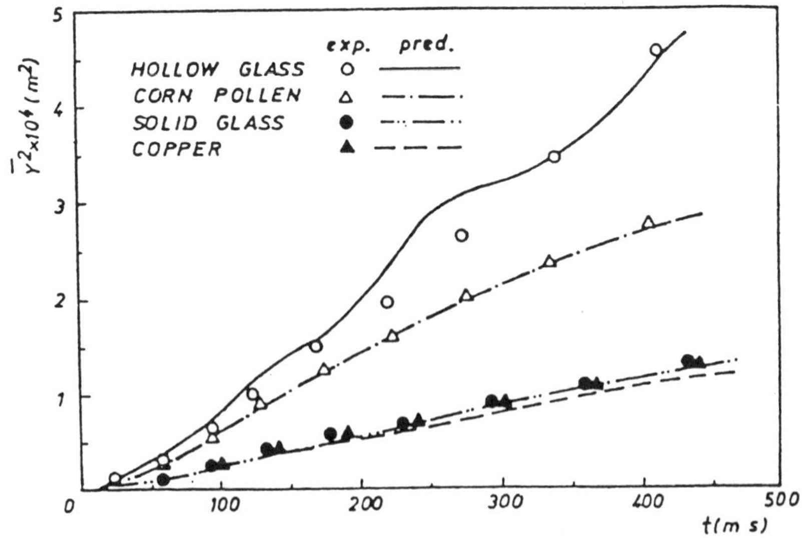

FIGURE 3.2. Comparison of LSD predictions with experiments of Snyder and Lumley /42/.

For the prediction of the gas-particle two-phase flow in the experimental test cases described above a solution scheme modified from the TEACH program was used. The calculations of the particle motion, followed by averaging to obtain mean and fluctuation particle velocities, volume concentration and the source terms, were incorporated into the iteration loop after solving the fluid flow equations. Thus the interaction of the phases was taken into account in a converged solution. For the predictions of the two-phase flows in the test cases, 2000 particle trajectories were calculated to obtain mean values and the source terms for the fluid equations. In order to save computer time the particle calculations were performed only after the solution of fluid equations was almost converged. The boundary conditions for the fluid equations were taken the same as in single-phase flow prediction, e.g. with universal wall functions and zero gradients at the symmetry plain and at the outlet. Measured mean velocity, turbulence intensity and particle volume fraction distributions are specified at the inlet cross section to the flow field. Preliminary calculations showed that the inlet transverse particle fluctuation velocity had a large influence on the dispersion rate. Hence, the measured value of 0.38 m/s was specified as an inlet condition to produce the results to be shown in this paper. At solid walls ideal reflections of the particles were assumed as the boundary conditions.

The prediction procedure gives all necessary data for the comparison with experiments. Figure 3.2 gives an overview for the estimation of the validity of the predictions. Two vertical columns of plotts related to two cross sections for the channel, one at the near field of jet development (x = 100 mm) and one at the outlet (x = 1050 mm), both for the test case 2, which was estimated to be more complex than the case 1. The agreement of predicted mean fluid velocity profiles with the experiments at both cross sections is very good, although there is some faster decay of the central jet at x = 100 mm, probably due to the sharp edges of the separation plates between the three channels, producing some transverse fluid velocity in the direction to the symmetry plain, which was not taken into account as an inlet condition for the predictions. Another reason for this discrepancy may be some numerical diffusion, which was not seriously checked at present stage of the predictions. Predicted mean particle velocities ($\bar{U}_p$) agree very well with the measured ones at the outlet, but at x = 100 mm the predictions give higher values near the walls. The explanation is in fact for the calculation of particle trajectories the fluid velocity in near wall control volumes was kept constant instead of interpolating it down to zero at the wall. In addition a unique particle size of 108 μm was used for the predictions, thus neglecting some smaller size particles accommodating faster to local fluid velocity and producing lower local ensemble slip velocity being negative in the near field of the jet development. The predicted profiles of particle volume concentration x show slower dispersion at x = 100 mm, but faster near the outlet. The largest differencies are observed close to the walls and this may be due to the larger error in counting particles crossing the measuring volume where a certain fraction of the scattered light is reflected out of the forward direction. Fluid fluctuation velocity distributions U_{RMS} agree with the experiments except near the symmetry plain at the outlet where the predicted values are too low. The plotts at the top of Figure 3.3 show the comparisons of measured and predicted particle fluctuation velocity U_p^{RMS} distributions. Note that this is the first model that predicts any turbulence of the particles and in that case the agreement with experiments at x = 100 mm is very good. However, at the outlet the predicted values are as low as the half of the measured ones. This large difference cannot be explained by any simple physical argument and it needs more attention for future work both in the experiments and in the predictions.

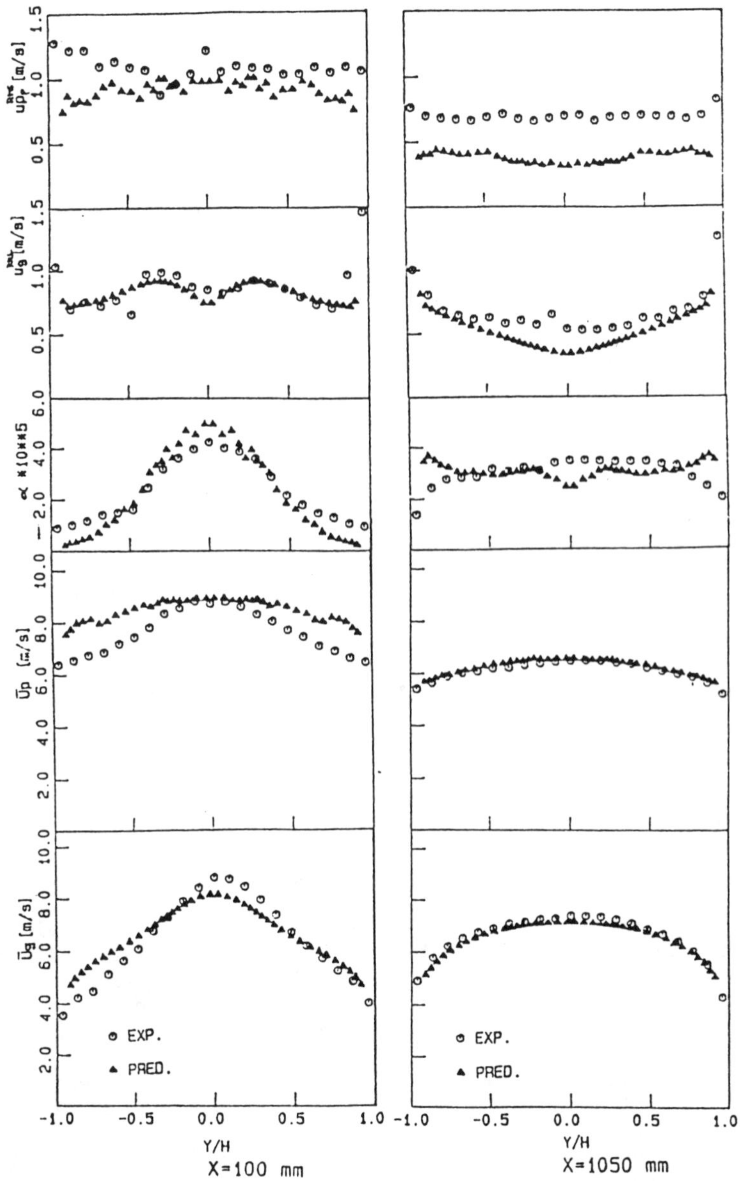

FIGURE 3.3. Comparison of predictions and experiment

3.3 Conclusions

The motion of heavy particles in turbulent two-phase flows is very complex. Both the experiments and the predictions show that there is no equilibrium of the momentum exchange between the phases and no universal integral similarity criteria can be specified for the kind of flows investigated in this work. Heavy particle dispersion is dominated by inlet conditions and due to their inertia and mixing the mean difference velocity of the phases may be negative near the wall in an upward flow. The high particle spreading rate results in very frequent wall contacts. They are not yet completely understood because of the complex fluid flow in that region and of the highly non equilibrium particle motion with reflections at the wall being dependent also on particle and wall materials.

It is encouraging that in some aspects the predictions give good agreement with the experiments. The physical models employed here describe the chosen dispersed tubulent two-phase jet flow. Still, at this stage of the investigation further development is important to understand experimental results of more complex flow situation.

4. COOPERATION OF EXPERIMENTAL AND NUMERICAL WORK IN FLUID MECHANICS

The two phase flow prediction models have shown their ability to predict some very interesting phenomena experimentally observed in upward vertical wall bounded two-phase flow: The center line mean difference velocity between the phases is larger than the particle free fall velocity and near wall slip velocity may be negative, that means particles can be faster than the fluid.

One of the most important and not yet completely resolved problems is the modelling of the influence of fluid turbulence on particle motion. It can be shown that particle dispersion remains of interest only in the majority of applications, a few integral scales of turbulence, or a few particle characteristic response times or both of them. In the above test cases, the particle motion is strongly influenced by the inlet conditions and much less by the local fluid turbulence. That is why the "dispersion coefficients" in the Eulerian approach should be evaluated from transport equations for the particle turbulence energy and particle turbulence scale rather than the mean flow and mean particle velocity fields. If a Lagrangian approach is used, the stochastic-deterministic modelling seems to be promising as a starting point for the future work in this field.

Continuous work was input into experimental investigations of two-phase flows as well as further modelling of the observed phenomena. In many of these instances the direct collaboration of experimentalists and computational working fluid mechanists was necessary.

Work-shops /48, 49/ were organized to regularly meet and evaluate and improve the abilities of the existing codes against predictions of other research groups.

The work-shops gave an opportunity to see and compare different prediction procedures applied to the same experimental test cases. The agreement between the predictions and experiments was satisfactory and encouraging for the future work to improve the models.

It can be estimated as a very useful event for the researchers experienced in two-phase flow predictions and experimentation. The discussions on papers presented during the work-shop, as well as on prediction procedures and results in comparison with the experiments, were very fruitful and enabled the exchange of ideas between the participants.

5. NOMENCLATURE

Section 2:

c	velocity of light
f_D	Doppler frequenz
$\{k\}_i$	unit vector of light wave
$\{l\}_i$	unit vector of incident light beam
L	distance from particle
n	refractive index
$\{m\}_i$	unit vector of light wave
$\{n\}_i$	sensitivity vector of anemometer, $\{n\}_i = \{k\}_i - \{m\}_i$
R	particle radius
$\{\hat{U}\}_i$	instantaneous velocity
β	laser beam crossing angle
λ_o	wave length of light source
ν_o	frequency of light source
ν	resultant frequency
ν_D	Doppler frequency
Δx	fringe spacing on screen in distance L
φ	space observer angle

Section 3

x	axial coordinate
y	radial coordinate
r	radial coordinate
ρ	density of fluid
u	axial velocity of fluid
v	radial velocity of fluid
p	pressure
μ	molecular dynamic viscosity of fluid
μ_{eff}	effective viscosity ($u_{eff} = \mu+\mu_t$)
μ_t	turbulent viscosity
k	turbulent energy
ε	dissipation
α^*	void fraction of dispersed phase ($\alpha^* = \alpha/\alpha_o$)
V	control volume
V_p	volume occupied by dispersed phase

α	void fraction of fluid phase
ρ_p	density of particles ($\rho_p^* = \rho_p/\rho$)
u_p	axial particle velocity
v_p	radial particle velocity
g	acceleration due to gravity
c_D	drag coefficient
d	particle diameter
Re_p	particle Reynolds number
m^p	mass of a particle
t	time coordinate
Δu	velocity difference ($\Delta u = u-u_p$)
	constant for turbulent mixing of particles
$\dot{m}^p$	total mass flow rate of particles
ϕ	general variable
Γ	exchange coefficient
S_ϕ	source term
S_ϕ^p	source term representing interaction of particulate phase
f_j	mass fraction of particles, starting at location j
f_i	mass fraction of particles with diameter d_i
$\dot{N}_{ij}^p$	number flow rate of spherical particles with diameter d_i at starting location j
V	fluid velocity
V^p	particle velocity
A^p	projection area of a particle ($A = \pi d^2/4$)
Δt	time interval
τ	particle characteristic time
x	particle position
D	pipe diameter
γ	loading of particles

Indices

i	diameter class d_i
j	starting location
0	time $t = 0$

6. ACKNOWLEDGEMENTS

The authors gratefully acknowledge the support of the Deutsche Forschungsgemeinschaft, DFG and the International Bureau of the Kernforschungsanlage, Jülich.

The engagement of all the students and researchers involved in this work within their collaboration at LSTM-Erlangen is very much appreciated here.

Thank you also to Frau I. Paulus for her expert typing of the manuscript.

7. REFERENCES

1. F. Durst and J.H. Whitelaw: Optimization of Optical Anemometers. Proceedings of the Royal Society, London, series A324 (1971) pp. 157-181.
2. M.J. Rudd: A New Theoretical Model for the Laser-Doppler Meter. Journal of Physics E: Scientific Instruments, Vol. 2 (1969) pp. 723-726.
3. H.D. von Stein, H.J. Pfeiffer and B. Lehmann: Vorrichtung zur Messung der Geschwindigkeit eines elektrische Wellen streuenden Objektes. Deutsche Patentanmeldung P1673403.5-52 (8.11.1967).
4. F. Durst and B. Ruck: Effective Particle Size Range in Laser-Doppler Anemometry. Experiments in Fluids, 5, pp. 305-314 (1987).
5. B. Ruck: Untersuchungen zur optischen Messung von Teilchengrößen und -geschwindigkeit mit Streulichtmethoden. Dissertation Univ. Karlsruhe (1981).
6. R. Dändliker and B. Eliason: A Theoretical Analysis of Laser-Doppler Flowmeters. Optica Acta Vol. 21, No. 2 (1974) pp. 119-149.
7. R.J. Adrian and W.L. Earley: Evaluation of LDV Performance using Mie Scattering Theory. (In reference /19/) (1976) pp. 426-454.
8. W. Cherdron, F. Durst and G. Richter: Computer Programs to Predict the Properties of Scattered Laser Radiation. SFB 80/TM/121 (1978) Sonderforschungsbereich 80, University of Karlsruhe.
9. G. Mie: Beiträge zur Optik trüber Medien, speziell kolloidaler Metalllösungen. Ann. Physik 25, IV Folge, Nr. 3 (1908) pp. 377.
10. F. Durst and B. Eliason: Properties of Laser-Doppler Signals and their Exploitation for Particle Size Measurements. Proceedings of the LDA-Symposium, University of Denmark (1975).
11. F. Durst and H. Umhauer: Local Measurement of Particle Velocity, Size Distribution and Concentration with a Combined Laser-Doppler Particle Sizing System. Proceedings of the LDA-Symposium, University of Denmark (1975).
12. R.J. Adrian and K.L. Orloff: Laser Anemometer Signals: Visibility Characteristics and Application to Particle Sizing. Journal of Applied Optics, Vol. 16 (1977) pp. 677-684.
13. W.M. Farmer: Measurement of Particle Size, Number Density and Velocity using a Laser Interferometer. Journal of Applied Optics, Vol. 11 (1972) pp. 2603-2609.
14. F. Durst: Scattering Phenomena and their Application in Laser-Doppler Anemometry. Journal of Applied Mathematics and Physics (ZAMP), Vol. 24, No. 4 (1973) pp. 619-643.
15. D.M. Robinson and W.P. Chu: Diffraction Analysis of Doppler Signal Characteristics for a Cross Beam Laser Doppler Velocimeter. Journal of Applied Optics, Vol. 14 (1975) pp. 2177-2181.

16. D.W. Roberts: Particle Sizing using Laser Interferometry. Journal of Applied Optics, Vol. 16 (1977) pp. 1861-1865.
17. F. Durst and M. Zare: Laser-Doppler Measurements in Two-Phase Flows. Proceedings of the LDA-Symposium (1975), University of Denmark.
18. T. Börner, F. Durst and E. Manero: LDV Measurements of Gas-Particle Flows and Digital Data Processing. Paper No. 4.5, Proc. Third Int. Symp. on Applications of LDA to Fluid Mechanics, Lisbon, Portugal (1986).
19. S.L. Lee and J. Srinivasan: An LDA-Technique for Simultaneous Velocity and Size Measurement of Large Spherical Particles in a Two-Phase Suspension Flow. Int. J. of Multiphase Flow, Vol. 8, No. 1 (1982) p. 47.
20. A. Brankovic, T. Börner and W.W. Martin: The Measurement of Mass-Transfer Coefficients of Bubbles Rising in Liquids Using Laser-Doppler Anemometry. Laser Anemometry in Fluid Mechanics, Published by Ladoan-Instituto Superior Tecnico, 1096 Lisbon, Portugal (1984).
21. D. Modarres, H. Tan and S. Elgobashi: Two-Component LDA Measurement in a Two-Phase Turbulent Jet. AIAA Journal, Vol. 22, No. 5 (1984).
22. H. Saffmann, P. Buchhave and H. Tanger: Simultaneous Measurement of Size, Concentration and Velocity of Spherical Particles by a Laser-Doppler Method. Proc. Second Int. Symp. on Applications of LDA to Fluid Mechanics, 8.1 (1984).
23. F. Durst: Review-Combined Measurements of Particle-Velocity, Size Distribution and Concentration. J. of Fluid Eng., Vol. 104 (1982) p. 284.
24. D. Milojevic, T. Börner and F. Durst: Prediction of Turbulent Gas-Particle Flow Measured in Plane Confined Jets. PARTEC 86-Contribution, Nürnberg, April 16-18, 1986
25. S.L. Lee and F. Durst: On the Motion of Particles in Turbulent Duct Flows. Int. J. Multiphase Flow, 8, No. 2, pp. 125-146.
26. Y. Tsuji, A. Morikawa and H. Shiomi: LDV Measurements of an Air-Solid Two-Phase Flow in a Vertical Pipe. J. of Fluid Mech., 139, (1984) pp. 417-434.
27. F. Durst, D. Milojevic and B. Schönung: Eulerian and Lagrangian Predictions of Particulate Two-Phase Flows: A Numerical Study. Appl. Math. Modelling, vol. 8 (1984) pp. 101-115
28. W.F. Hughes and E.W. Gaylord: Basic Equations of Engineering Science. Schaum's Outline Series, McGraw-Hill, New York (1964).
29. W. Rodi: Turbulence Models and their Application in Hydraulics. IAHR-Section, Experimental and Mathematical Fluid Dynamics (1980).
30. S.V. Patankar and D.B. Spalding: A Calculation Procedure for Heat, Mass and Momentum Transfer in Three-Dimensional Parabolic Flows. Int. J. Heat Mass Transfer (1972) 15, pp. 1787.
31. S.V. Patankar: Numerical Heat Transfer and Fluid Flow. McGraw-Hill, New York (1980).
32. P.G. Saffman: The Lift on a Small Sphere in a Slow Shear Flow. J. Fluid Mech. (1965) 22, pp. 385.
33. S.I. Rubinow and J.B. Keller: The Transverse Force on a Spinning Sphere Moving in a Viscous Fluid. J. Fluid Mech. (1961) 11, 447.
34. D. Migdal and D.V. Agosta: A Source Flow Model for Continuum Gas-Particle Flow. Trans. ASME, J. Appl. Mech. (1967) 34E, pp. 860.
35. M.P. Sharma: Numerical and Experimental Study of Gas-Particle Flows in Orifices and Venturis: Application to Flowmeter Design. Ph.D. Thesis, Washington State University (1977).

36. C.T. Crowe, M.P. Sharma and D.E. Stock: The Particle Source-In-Cell (PSI-Cell) Model for Gas Droplet Flows. Trans. ASME, J. Fluids Engng (1977) 325.
37. M.P. Sharma, D.K. Cornelius, J.G. Rice and D.R. Dougan: Numerical Computation of Swirling Gas-Particle Flows: Application to Pulverized Coal Classifiers. ASME Winter Annual meeting, Nov. 1980, Paper No. 80-WA/HT-31.
38. F.C. Lockwood, A.P. Salooja and S.A. Syed: A Prediction Method for Coal-Fired Furnaces. Combust. Flame (1980) 38, 1.
39. R. Cliff and W.H. Gauvin: Motion of Entrained Particles in Gas Streams. J. Chem. Eng. (1971) 49, pp. 439.
40. A.M. Al-Taweel, J.F. Carley: Dynamics of Single Spheres in Pulsated Flowing Liquids. A.I.Ch.E.-Symposium Series, 67, 1971.
41. B. Schönung: Numerische Simulation teilchenbeladener vertikaler Rohrströmungen. Ph.D.-Thesis, University of Karlsruhe, 1983.
42. W.H. Snyder and J.L. Lumley: Some Measurements of Particle Velocity Autocorrelation Functions in a Turbulent Flow. J. Fluid Mech. (1971) Vol. 48, part 1, pp. 41-71.
43. D. Milojevic: Lagrangian Stochastic Deterministic Model (LSD) for the Prediction of Turbulent Fluid-Particle Flows. Presented at 2nd Work-shop of Two-Phase Flow Predictions, Erlangen, FRG, (1985).
44. B.E. Launder: Second-Moment Closure: Methodology and Practice in Turbulent Models and Their Applications. Direction des Etudes et Recherches d'Electricite de France (1984).
45. M.R. Wells and D.E. Stock: The Effects of Crossing Trajectories on the Dispersion of Particles in a Turbulent Flow. J. Fluid Mech. (1983) vol. 136, pp. 31-62.
46. D. Milojevic: Lagrangian Stochastic-Deterministic (LSD) Predictions of Particle Dispersion in Turbulence. To be published.
47. G. Arnason: Measurement of Particle Dispersion in Turbulent Pipe Flow. Ph.D. Thesis, Washington State University, Pullman (1982).
48. F. Durst, T. Börner, M. Arnal and H. Zeisel: Work-Shop on Two-Phase Flow Predictions. LSTM-Report No. 57/N/84, Lehrstuhl für Strömungsmechanik, Univ. Erlangen-Nürnberg (1984).
49. D. Milojevic, T. Börner and F. Durst: Second Work-Shop on Two-Phase Flow Predictions. KFA-Jülich, Bilateral Seminars of the International Bureau, Jülich (1985).

TEMPERATURE MEASUREMENTS BY COHERENT ANTI-STOKES RAMAN SCATTERING IN INTERNAL COMBUSTION ENGINES *

ROBERT P. LUCHT

COMBUSTION RESEARCH FACILITY, SANDIA NATIONAL LABORATORIES, LIVERMORE, CA, USA 94550

1. INTRODUCTION

Broadband, coherent anti-Stokes Raman scattering (CARS) measurements of unburned gas temperatures have been performed in an internal combustion engine (Lucht et al., 1987). The engine was fueled with n-butane and air and operated under both knocking and non-knocking conditions. Temperatures were determined by fitting calculated CARS spectra to experimentally obtained, cycle-averaged nitrogen CARS spectra. The estimated accuracy of the CARS temperatures is 5%, and was presently limited by long-term drift in the broadband dye laser frequency spectrum. The CARS temperatures are compared with the predictions of a heat release model for the engine (Ferguson et al., 1987).

CARS was also used to measure temperatures in the boundary layer near the cylinder head wall of a propane-fueled engine (Lucht and Maris, 1987). A lens-prism arrangement was used to traverse the probe volume perpendicular to the cylinder head wall while maintaining the alignment of the CARS optical system. The spatial resolution of the measurements was approximately 25-50 μm, and CARS signals were obtained at distances within 25 μm of the wall. Temperature profiles were measured 0.6, 2.7 and 4.2 msec after peak pressure was reached. The thickening of the boundary layer during the piston expansion stroke was clearly evident.

2. TEMPERATURE MEASUREMENTS IN THE KNOCKING ENGINE

We have performed broadband, coherent anti-Stokes Raman scattering (CARS) temperature measurements in the unburned gas region of an internal combustion engine operated under both knocking and non-knocking conditions. The motivation for performing the measurements arose from the need to understand the chemical kinetic processes which lead to autoignition of the unburned gases. Rapid heat release resulting from autoignition is responsible for engine knock, which currently limits the compression ratio and thermal efficiency of production automobile engines. In the last few years, great progress has been made in developing fundamental chemical kinetic models of the autoignition of hydrocarbon fuels (Smith et al., 1984; Leppard, 1985; Cox and Cole, 1985; Pitz and Westbrook, 1986; Dimpelfeld and Foster, 1986). The temperature history of the unburned gas is a critical input to such chemical kinetic models.

The CARS measurements are compared with the results of a heat release model for the engine. Heat release models are used to predict the thermodynamic state of the burned and unburned gases in the engine, and have been employed in chemical kinetics studies of autoignition to predict the temperature history of the unburned gas (Meintjes, 1985; Dimpelfeld and Foster, 1986). However, we know of no previous comparison of heat release model predictions with experimental measurements of temperature in engines. In this paper and in a companion paper which describes the heat release model in more detail (Ferguson et al., 1987), the temperature predictions of the heat release model are compared with CARS temperature measurements.

---

* Work supported by the United States Department of Energy, Division of Energy Conversion and Utilization Technologies, and by the Office of Basic Energy Sciences, Division of Chemical Sciences.

D. F. G. Durão et al. (eds.), Instrumentation for Combustion and Flow in Engines, 341–353.

The first CARS measurements in an internal combustion engine were reported by Stenhouse et al (1979), who performed narrowband, scanning CARS measurements in an engine fueled either with propane or gasoline. Temperature was determined from nitrogen CARS spectra, and propane spectra were also obtained. Klick et al. (1981) performed the first broadband CARS measurements in an engine, using a unique noncollinear phase-matching technique (Marko and Rimai, 1979). Rahn et al. (1982) performed scanning CARS measurements in the Sandia optical engine. They determined temperature from nitrogen CARS spectra, and acquired hydrogen and carbon monoxide spectra. They also demonstrated nonresonant background subtraction via polarization analysis (Rahn et al., 1979) and in-situ referencing (Farrow et al., 1982) in an engine. Alessandretti and Violini (1983) performed broadband CARS measurements in a gasoline-fueled engine using collinear phase-matching. Kajiyama et al. (1983) performed collinear, broadband CARS measurements in a diesel engine.

The measurements reported here differ from those of previous investigations primarily in the manner in which the spectra obtained from the engine are analyzed to obtain temperature. Specifically, collisional narrowing of the nitrogen CARS spectra is shown to be a significant effect throughout the temperature and pressure regime of interest in this study. With the exception of Kajiyama et al. (1983), collisional narrowing was ignored in previous CARS investigations in engines. We use the methodology of Koszykowski et al. (1985) to calculate collisional narrowing effects. In addition, the nonresonant susceptibility of n-butane was recently measured (Farrow et al., 1987) and, consequently, the nonresonant susceptibility of the unburned gases as a whole is known very accurately. We also found that careful consideration of the spectral response of the intensified diode array detector was necessary for accurate determination of temperatures.

The Sandia optical engine has been described in a number of publications. The configuration used for this experiment is discussed in detail by Smith et al. (1984). The engine was fueled with a stoichiometric n-butane/air mixture. The premixed charge was introduced into the chamber through a shrouded intake valve on the side of the piston chamber. Because of the valve shrouding, the gases were introduced with high swirl. The flow was highly turbulent because of the small size of the valve. Four spark plugs, spaced at 90 degree intervals around the chamber, ignited the mixture at top-dead-center (TDC), which corresponded to 360 crank angle degrees (CAD). Flame front propagation was dominated by swirl-induced convection, and the unburned gases were confined in the center of the chamber.

The engine was operated under both knocking and non-knocking conditions. The intake manifold was heated to 433 K and pressurized to 1.55 atm for knocking operation. Operating conditions for the non-knocking case were identical except that the intake manifold was pressurized to only 1.27 atm. The engine was skip-fired, with one fired cycle followed by two non-fired cycles. Skip firing enhanced cycle-to-cycle repeatability because residual gases from a fired cycle were almost completely eliminated from the chamber before the next fired cycle. The standard deviation in peak pressure for the fired cycles was approximately 1%, and the time at which autoignition occurred was also very repeatable.

Optical access for the CARS beams was provided by two diametrically-opposed fused silica windows. The windows were 10 mm in diameter and 10 mm thick. The cylinder head was a 75 mm diameter, 25 mm thick sapphire window which provides optical access for flow visualization.

The CARS apparatus is shown in Figure 1. The pump beams for the CARS measurements were provided by the 532-nm second harmonic of a Nd:YAG laser (Quanta-Ray DCR2). Part of the 532-nm laser radiation was also used to pump a broadband dye laser centered at 607 nm. The dye laser oscillator was transversely pumped, and the amplifier was longitudinally pumped. A three-dimensional phase-matching scheme was used to combine the beams at the CARS probe volume. The beams were focused at the center of the chamber using a single 200-mm-focal-length lens. The diameter of the beams on the lens was approximately 7 mm. The crossing angle of the beams was approximately 0.05 radians (full angle), and the longitudinal spatial resolution was less than 2 mm. The polarization of the pump and Stokes beams was horizontal.

The laser beams and the CARS signal beam were recollimated upon exiting the engine chamber using a 200-mm-focal-length lens. The pump and Stokes laser beams were spatially separated from the CARS signal beam using a three prism arrangement. The CARS beam was focused onto the entrance slit of a 0.85-m double spectrometer (Spex 1402). The polarization of the CARS beam was rotated for highest grating efficiency. A relay lens was used to magnify the spectrum at the exit plane of the spectrometer by a factor of four, and focus it onto an intensified diode array detector (Princeton Applied Research Corp. Model 1420). The resultant spectral resolution was approximately 0.7 cm^{-1}. Typically, CARS spectra were averaged on the diode array for eight Q-switched laser shots, and four or five spectra were acquired at a given crank angle. Temperatures reported in this paper are thus determined from spectra averaged for 30-40 laser shots.

Nitrogen CARS spectra acquired from the engine were normalized by dividing by a nonresonant background spectrum generated from oxygen. The normalization corrected the nitrogen spectra for the spectral profile of the dye laser and the nonuniformity of pixel response on the detector. Two 50 mm prisms were used to direct the pump and Stokes beams through an oxygen-filled cell. Oxygen was used rather than argon for this measurement because of its higher breakdown threshold. Calculations showed that the contribution to the CARS susceptibility from the oxygen Q-branch at 1550 cm^{-1} was negligible relative to the nonresonant susceptibility at the nitrogen Q-branch frequency, 2330 cm^{-1}.

The CARS spectra of nitrogen were analyzed using the Sandia CARS fitting code (Farrow et al., 1982; Farrow et al., 1985) to determine gas temperature. Theoretically calculated spectra were fit to the experimental spectra using a nonlinear least squares optimization routine, with temperature as a fitting parameter. At the temperatures of interest in this study, 400 to 1100 K, accurate determination of temperature is difficult because the first vibrational hot band of nitrogen is very weak. Thus, temperature information is contained primarily in the shape and width of the fundamental band. It is therefore very important that the physics of processes which affect the shape and width of the fundamental are well-modeled.

Figure 2 shows an experimental CARS spectrum obtained in the unburned gas just prior to autoignition at a measured pressure of 21 atm. The spectrum shown is the average of thirty laser shots. A calculated CARS spectrum, fit to the experimental spectrum, is also shown. Agreement between theory and experiment, in terms of the shape and rotational structure of the fundamental band, and the relative intensity of the hot band, is excellent.

Determination of the resonant CARS susceptibility for nitrogen at the pressures and temperatures encountered in the unburned gases in the engine requires accurate treatment of collisional narrowing and knowledge of the nitrogen linewidths. Collisional narrowing of the nitrogen spectra is significant at the pressures encountered in the engine. Collisional narrowing was first shown to be of importance in high pressure combustion diagnostics by Hall et al. (1980). In our investigation, collisionally narrowed spectra were calculated using a methodology developed by Koszykowski et al. (1985). The method is computationally efficient, in that only one G-matrix inversion is required per spectrum, and rigorous, in that the rotational transfer model for the G-matrix does not have to be compromised in order to obtain computational efficiency. Koszykowski et al. (1985) use an exponential gap model for rotational transfer. The rotational transfer model is optimized so that calculated isolated linewidths match the measurements of Rahn and Palmer (1986). An exponential gap model approach was also investigated by Hall et al. (1980) and Kataoka et al. (1983), but calculations were so time-consuming that it was impractical for routine analysis of spectra.

The temperature and rotational state dependence of self-broadened nitrogen linewidths have been measured by Rahn and Palmer (1986), and the results of the measurements have been incorporated into the Sandia CARS code. For the engine measurements, it is assumed that the broadening contribution for all gases is the same as that for nitrogen. Farrow et al. (1987) found that the broadening contribution due to n-butane was 1.5 times that due to nitrogen. Because the fuel mass fraction was only 0.031 for the stoichiometric gas mixture however, nitrogen linewidths were only 2% greater than in pure nitrogen.

The effect of uncertainty in the nonresonant susceptibility on the accuracy of temperatures determined from CARS spectra has been discussed by Hall and Boedecker (1984) and Bedue et al. (1984). In the engine, the effect of the nonresonant susceptibility on temperature accuracy is significant because of the high pressures. For an isolated Raman line, the peak third-order susceptibility is nearly independent of pressure, whereas the nonresonant susceptibility is proportional to pressure. Thus, the ratio of resonant to nonresonant intensities is inversely proprtional to the square of pressure for an isolated line. For the Q-branch of nitrogen, the lines are heavily overlapped at high pressure and the peak susceptibility increases with pressure. As the temperature increases, however, the degree to which lines are overlapped decreases, and the nonresonant background becomes more important in determining temperature from CARS spectra.

Farrow et al. (1987) recently measured the nonresonant susceptibility of n-butane using a high resolution CARS system. The spectrum of 5% nitrogen in n-butane in a room temperature cell was acquired, and the nonresonant susceptibility and linewidth broadening contribution of n-butane were used as fitting parameters for the spectrum. The nonresonant susceptibility of n-butane was found to be 12 times that of nitrogen. In addition, the off-resonant susceptibility due to strong Raman resonances near 2900 cm^{-1} also contributes to the real part of the susceptibility. For polarization-aligned CARS, the total contribution to the real part of the susceptibility at the nitrogen Q-branch wavelength was 15 ± 2 times the nonresonant susceptibility of nitrogen.

Thus, although the mole fraction of n-butane was only 0.031, the nonresonant susceptibility of the unburned gases was significantly affected by the n-butane contribution. When the spectrum shown in Figure 2 is analyzed assuming that the nonresonant susceptibility of the unburned gases is equal to that of nitrogen, the fitted temperature is in error by +110 K. Changing the nonresonant susceptibility by 10% was found change the fitted temperature by 30 K for nitrogen spectra at 20 atm and 1000 K. Errors due to uncertainty in the nonresonant susceptibility of the gas in the cylinder increase with increasing pressure. Polarization techniques to suppress the nonresonant background (Rahn et al., 1979) could be used to eliminate potential errors posed by uncertainties in the nonresonant background. However, the resonant CARS signal intensity is reduced by a factor of sixteen for a such a polarization arrangement.

The predominant sources of scatter in the temperature measurements in this experiment were shot-to-shot noise and long-term drift in the dye laser spectrum. For a single laser shot, the standard deviation of nonresonant background intensity for each pixel on the diode array dectector was nearly 20% at a resolution of 0.7 cm^{-1}, and about half that amount at a resolution of 3 cm^{-1}. This spectral noise in the shot-to-shot spectra prevented the determination of accurate temperatures from single shot CARS spectra. All temperatures reported here were determined from spectra which had been averaged for 30-40 laser shots. The cycle-to-cycle repeatability of the Sandia engine is excellent. The cycle-to-cycle fluctuations in the bulk temperature in the engine should be correlated with cycle-to-cycle fluctuations in engine pressure, which are on the order of 1%. Thus, at 360 CAD (TDC), cycle-to-cycle fluctuations in the bulk temperature should be less than 10 K.

The temperature in the CARS probe volume may also vary due to turbulent fluctuations in the unburned gases. Smith (1980) measured the standard deviation of nitrogen density in the unburned gases, and deduced that the standard deviation of temperature was approximately 30-40 K for a non-skip-fired engine. In the non-skip-fired engine, cycle-to-cycle fluctuations in pressure and bulk temperature were twice as high as in the skip-fired engine. Therefore, we can estimate that in the skip-fired engine, fluctuations in temperature in the CARS probe volume due to turbulence are approximately 20-30 K. Such small deviations from the mean temperature will not significantly bias the temperature determined by averaging CARS spectra on the detector.

Random errors in determination of the average temperature may arise from dye laser spectral noise and long-term drift, from cycle-to-cycle and turbulent fluctuations in unburned gas temperatures, and from shot noise on the detector. Long-term drift in the dye laser spectrum contributed an estimated random error of 30-40 K. It is difficult to assess the random error associated with shot noise on the detector, spectral noise in the dye laser spectrum, and cycle-to-cycle and turbulent fluctuations. However, the fact that averaged spectra taken successively at the same point in the engine cycle typically yield temperatures which are equal to within 10-20 K suggests that random errors due to factors other than long-term drift in the dye laser spectrum are small. The estimated total random error in the measurements is therefore dominated by long-term drift in the dye laser spectrum. The estimated total error for the measurements, corresponding to a single standard deviation for a single data point, is 50-60 K (5-6%).

The results of the CARS measurements of temperature in the unburned gas region of the engine operated under knocking and non-knocking conditions is shown in Figure 3. The predictions of the heat release model for the Sandia engine are also shown. The heat release model, and the comparison with the CARS measurements, are discussed in detail by Ferguson et al. (1987). The bulk gas temperature is the average temperature of the unburned gas in the cylinder, and the core gas temperature is the average temperature of the unburned gases exclusive of the boundary layer at the walls of the cylinder. Before TDC, the agreement between the CARS measurements and the core gas temperature predicted by the heat release model is excellent. It must be emphasized here that the CARS temperature measurements and the heat release predictions were not normalized with respect to each other, and that the good agreement is not the result of varying undetermined parameters in the heat release model (Ferguson et al., 1987).

After TDC, the CARS temperature is systematically higher than the calculated core gas temperature. The measured CARS temperatures could be higher than the temperatures predicted by the heat release model due to exothermic reactions in the unburned gas. The n-butane/air mixture is assumed to be non-reacting in the heat release model. If a significant fraction of the butane breaks down prior to autoignition, significant heat release may occur in the unburned gas.

3. TEMPERATURE PROFILES IN THE BOUNDARY LAYER

Historically, the state of gases near walls in internal combustion chambers has been of great interest to engine researchers. The temperature profile of gases near walls is important, first of all, because of heat transfer considerations (Woschni, 1967; Greif et al., 1979). The ability to accurately predict heat transfer using multidimensional modeling would be an important tool for the design of efficient engine chambers (Diwakar, 1984; Ikegami et al., 1986). Engine heat transfer is an especially important issue for adiabatic or ceramic engines.

Secondly, the possible influence of the thermal boundary layer on pollutant emission from engines has long been recognized. For many years, for instance, it was believed that the primary source of unburned hydrocarbon emissions from engines was the so-called "quench layer," a region of cool gas near engine walls. Although it is now believed that the main sources of unburned hydrocarbon emissions from engines are engine crevices, the temperature profile near the crevices will certainly affect the level of emissions (Lyford-Pike and Heywood, 1984).

Furthermore, the thermal boundary layer is a calculated quantity in most engine heat release calculations. The shape and thickness of the boundary layer has a strong influence on the predicted difference between bulk (all gases in the chamber) and core (gases outside the boundary layer) temperatures in the engine (Ferguson et al., 1987). Nitric oxide emissions from an engine chamber are most sensitive to peak temperatures in the chamber, and these temperatures will be calculated incorrectly in heat release codes unless the thermal boundary layer is accurately modeled.

Despite the great interest in the thermal boundary layer in engines, there have to our knowledge been no previous quantitative measurements of the temperature profile near a wall in an operating internal combustion engine. Lyford-Pike and Heywood (5) measured thermal boundary layer thicknesses in a homogeneous-charge, spark-ignition engine using Schlieren photography. Their study was noteworthy in that they obtained measurements for a number of different locations in the engine and for virtually the entire engine cycle. Farrell and Verhoeven (1987) recently used speckle interferometry to measure temperature profiles in the boundary layer of a motored engine.

We measured temperature profiles in the boundary layer of the Sandia engine using CARS. For these studies, the engine was fueled with a stoichiometric propane/air mixture. As in the measurements in the knocking engine, the premixed charge was introduced into the chamber through a shrouded intake valve on the side of the piston chamber, the mixture was ignited using four spark plugs, and the engine was skip-cycled. The intake manifold was heated to 373 K and pressurized to 1.27 atm.

The cylinder head was a stainless steel disk with a convex lower surface which projected into the clearance volume. The geometry of the disk and engine chamber is shown in Figure 4. The center of the disk was 7 mm lower than the edge of the disk, and the radius of curvature of the convex surface was 70 mm.

A three-dimensional phase-matching scheme was used to combine the beams at the CARS probe volume. The beams were focused at the center of the chamber (radially) using a 200-mm-focal-length lens. The diameter of the beams on the lens was approximately 7 mm. The crossing angle of the beams was approximately 0.05 radians (full angle), and the measured longitudinal spatial resolution was better than 2 mm. The polarization of the pump and Stokes beams was horizontal.

The laser beams were moved axially within the chamber using the prism and lens arrangement shown in Figure 4. By moving the lens and the prism closest to the lens on each side of the chamber the same distance vertically, the CARS probe volume could be moved vertically without misaligning the optical system. Dial indicators with a precision of approximately 10 μm were used to measure the positions of the prisms and lenses. The position of the CARS probe volume with respect to the cylinder head surface was determined by translating the 532 nm beams (with the laser in the long pulse mode) until they just started to be blocked by the surface, as indicated by scattered light off the surface. The beams were then moved further until they were completely blocked. The distance between these two positions was approximately 50 μm, which represents an upper limit for the spatial resolution. The spatial resolution was probably closer to 25 μm because the CARS signal depends on the square of the pump beam intensity. The absolute positions for the start of beam blockage and complete blockage were reproducible to approximately 10 μm from day-to-day. The position of the CARS probe volume was referenced to the midway point.

CARS spectra were obtained at three different crank angle locations. For the measurements reported here. spark ignition occurred at 360 CAD and peak pressure occurred at 376 CAD. The temperature profile was measured at 378. 385, and 390 CAD, or 0.6, 2.7 and 4.2 msec past the time of peak pressure. The occurrence of peak pressure indicates that the combustion process in the bulk gases is completed. Our measurement point was at the radial center of the chamber. Because the flames burn inward from the four spark plugs. the engine gases probed at 378 CAD were freshly burned.

Figures 5 and 6 show typical spectra and theoretical fits for the burned gases at 378 CAD, 1.93 mm from the wall. and at 385 CAD and 25 μm from the wall, respectively. The spectra have been averaged for thirty laser shots. There is a great deal of scattered light in the spectrum shown in Figure 6. At this position both the pump and Stokes beams are partially obscured by the cylinder head wall. For this spectrum a vertical offset was used as one of the fitting parameters in the CARS code to account for the scattered light. For spectra acquired further from the wall scattered light was negligible.

The measured temperature profiles near the wall at 378, 385, and 390 CAD are shown in Figures

7 The data points shown on the plots are temperatures determined from CARS spectra which were averaged for approximately 30 laser shots. The data is compared with profiles calculated from power law expressions for the boundary layer temperature profile. The temperature profile $T(z)$ is calculated from the expression

$$T(z) = T_w + (T_c - T_w)(z/\delta)^n \qquad \text{for} \qquad z < \delta \tag{1}$$

and

$$T(z) = T_c \qquad \text{for} \qquad z \geq \delta \tag{2}$$

where T_w is the wall temperature, estimated to be 400 K, T_c is the freestream core temperature, δ is the boundary layer thickness, and z is the distance from the wall.

The solid lines in Figure 7 are convolutions of computed power law temperature profiles with a function describing the spatial response of the CARS probe volume. As a first estimate of the effect of the finite extent of the CARS probe volume on the measured temperature profile, the spatial response of the CARS probe volume was assumed to be Gaussian. Furthermore, it was assumed that temperatures, rather than intensities, were averaged within the volume. By comparing the convolved profiles with the power law profiles, we concluded that the spatial resolution of the measurements is sufficient to accurately measure the temperature profile at distances further than 50 μm from the wall.

The parameters n and δ were adjusted to give reasonable fits to the experimental data shown in Figure 7. The value $n = 1/4$ seemed to give the best fit for all three cases. The temperature profiles clearly show that the boundary layer thickness increases as the expansion stroke proceeds. The measured boundary layer thicknesses at 378, 385 and 390 CAD were 0.25, 1.0, and 1.5 mm, respectively.

Curiously, the core temperature is equal to within experimental error at 378 CAD and at 385 CAD, but drops significantly at 390 CAD. A possible explanation for the near equality of the core temperatures at 378 and 385 CAD involves radial stratification of temperatures in the burned gases as the combustion process proceeds. The first-burned gases in the engine, near the cylinder wall in this case, will reach higher temperature than gases at the center of the chamber, where the measurements are performed, because of compression due to combustion. Consequently, the burned gas will be radially stratified with the lowest temperature in the center of the chamber just after combustion ceases. Turbulent transport in the burned gases will tend to equalize temperatures in the burned gases, increasing the temperature at the center of the chamber. The temperature at the center of the chamber will therefore be determined by the competition between expansion of the gases and heat transfer to the walls, which will tend to lower the overall core temperature, and turbulent transport, which will tend to raise the temperature at the measurement point.

4. SUMMARY

Broadband CARS has been used to measure temperatures in the unburned gas region of a knocking internal combustion engine. The measurements show good agreement with heat release model predictions for the engine, and demonstrate our ability to extract accurate temperatures from collisionally narrowed, high pressure nitrogen spectra. We have also measured temperature profiles in the thermal boundary layer of a propane-fueled engine. The spatial resolution of the measurements is approximately 25 μm normal to the wall, and CARS spectra have been obtained 25 μm from the wall. Power law expressions have been compared with the CARS data.

5. ACKNOWLEDGEMENTS

This research was performed at the Combustion Research Facility and supported by the United States Department of Energy, Division of Energy Conversion and Utilization Technologies, and by the Office of Basic Energy Sciences, Chemical Sciences Division. The experiments in the knocking engine were performed in collaboration with Bob Green and Rich Palmer of Sandia, Rich Teets of the General Motors Research Laboratories, and Colin Ferguson of Purdue University. The boundary layer measurements were performed in collaboration with Mark Maris of Computer Genetics. For both experiments, the technical assistance of Dennis Sparger is gratefully acknowledged.

6. REFERENCES

Alessandretti, G.C. and Violino, P. (1983). Thermometry by CARS in an automobile engine. *J. Phys. D: Appl. Phys.* **16**, 1583-1594.

Bedue, R., Gastebois, P., Bailly, R., Pealat, M., and Taran, J.P. (1984). CARS measurements in a simulated turbomachine combustor. *Combust. Flame* **57**, 141-153.

Cox, R.A. and Cole, J.A. (1985). Chemical aspects of the autoignition of hydrocarbon-air mixtures. *Combust. Flame* **60**, 109-123.

Dimpelfeld, P.M. and Foster, D.E. (1986). Predictions of autoignition in a spark-ignition engine using chemical kinetics. *SAE Trans.*, Paper No. 860322.

Diwakar, R. (1984). Assessment of the ability of a multidimensional computer code to model combustion in a homogeneous-charge engine. *SAE Paper No. 840230, SAE Trans.* **93**, 2.85-2.108.

Farrell, P.V. and Verhoeven, D.D. (1987). Heat transfer measurements in a motored engine using speckle interferometry. *SAE Paper No. 870456.*

Farrow, R.L., Mattern, P.L., and Rahn. L.A. (1982). Comparison between CARS and corrected thermocouple temperature measurements in a diffusion flame. *Appl. Opt.* **21**, 3119-3215.

Farrow, R.L. and Rahn, L.A. (1985). Interpreting coherent anti-Stokes Raman spectra measured with multimode Nd:YAG pump lasers. *J. Opt. Soc. Am. B* **2**, 903-907.

Farrow, R.L., Lucht, R.P., Clark, G.L. and Palmer, R.E. (1985). Species concentration measurements using CARS with nonresonant susceptibility normalization. *Appl. Opt.* **24**, 2241-2251.

Farrow. R.L., Lucht. R.P., and Rahn, L.A. (1987). Measurements of the nonresonant third-order susceptibilities of gases using coherent anti-Stokes Raman spectroscopy. *J. Opt. Soc. Am. B* **4**, 1241-1246.

Ferguson, C.R., Green, R.M., and Lucht, R.P. (1987). Unburned gas temperatures in an internal combustion engine: II. Heat release computations. *Combust. Sci. Tech.*, accepted for publication.

Greif, R., Namba. T., and Nikanham, M. (1979). Heat transfer during piston compression including side wall and convection effects. *Int. J. Heat Mass Transfer* **22**, 901-907.

Hall. R.J., Verdieck, J.F., and Eckbreth, A.C. (1980). Pressure-induced narrowing of the CARS spectrum of N_2. *Opt. Comm.* **35**, 69-75.

Hall, R.J. and Greenhalgh, D.A. (1982). Application of the rotational diffusion model to gaseous N_2 CARS spectra. *Opt. Comm.* **40**, 417-420.

Hall, R.J. and Boedeker, L.R. (1984). CARS thermometry in fuel-rich combustion zones. *Appl. Opt.* **23**, 1340-1346.

Ikegami, M., Kidoguchi, Y., and Nishiwaki, K. (1986). A multidimensional model prediction of heat transfer in non-fired engines. *SAE Paper No. 860467.*

Kajiyama. K., Sajiki, K., Kataoka, H., Maeda, S., and Hirose, C. (1982). N_2 CARS thermometry in a diesel engine. *SAE Trans.* **21**, Paper No. 821036, 3243-3251.

Kataoka, H., Maeda, S., Hirose, C., and Kajiyama, K (1983). A study for N_2 coherent anti-Stokes Raman spectroscopy thermometry at high pressure. *Appl. Spectrosc.* **37**, 508-512.

Klick, D., Marko, K.A., and Rimai, L. (1981). Broadband single-pulse CARS spectra in a fired internal combustion engine. *Appl. Opt.* **20**, 1178-1181.

Koszykowski, M.L., Farrow, R.L., and Palmer, R.E. (1985). Calculation of collisionally narrowed coherent anti-Stokes Raman spectroscopy spectra. *Opt. Lett.* **10**, 478-480.

Leppard, W.R. (1985). A detailed chemical kinetics simulation of engine knock. *Combust. Sci. Tech.* **43**, 1-20.

Lucht, R.P. and Maris, M.A. (1987). CARS measurements of temperature profiles near a wall in an internal combustion engine. *SAE Paper No. 870459.*

Lucht, R.P., Green, R.M., Palmer, R.E., Teets, R.E., and Ferguson, C.R. (1987). Unburned gas temperatures in an internal combustion engine: I. CARS temperature measurements. *Combust. Sci. Tech.*, accepted for publication.

Lyford-Pike, L.J. and Heywood, J.B. (1984). Thermal boundary layer thickness in the cylinder of a spark-ignition engine. *Int. J. Heat Mass Transfer* **27**, 1873-1878.

Marko, K.A. and Rimai, L. (1979). Space- and time-resolved coherent anti-Stokes Raman spectroscopy for combustion diagnostics. *Opt. Lett.* **4**, 211-213.

Meintjes, K. (1985). End-gas pressure and temperature in the CFR knock rating engine: calculations with an engine simulation. General Motors Research Memorandum 57-1040.

Pitz, W.J. and Westbrook, C.K. (1986). Chemical kinetics of the high pressure oxidation of n-butane and its relation to engine knock. *Combust. Flame* **63**, 113-133.

Rahn, L.A., Zych, L.J., and Mattern, P.L. (1979). Background-free CARS studies of carbon monoxide in a flame. *Opt. Comm.* **30**, 249-252.

Rahn, L.A., Johnston, S.C., Farrow, R.L., and Mattern, P.L. (1982). CARS thermometry in an internal combustion engine. In *Temperature: Its Measurement and Control in Science and Industry, Vol. 5*, American Institute of Physics, New York, pp. 609-613.

Rahn, L.A. and Palmer, R.E. (1986). N_2 Raman linewidths for CARS measurements of temperature and major species concentrations in flames. Submitted to *J. Opt. Soc. Am. B.*

Smith, J.R. (1980). Temperature and density measurements in an engine by pulsed Raman spectroscopy. *SAE Trans.* **89**, Paper 800137, 808-816.

Smith, J.R., Green, R.M., Westbrook, C.K., and Pitz, W.J. (1984). An experimental and modeling study of engine knock. *Twentieth Symposium (International) on Combustion*, The Combustion Institute, Pittsburgh, Pennsylvania, pp. 91-100.

Stenhouse, I.A., Williams, D.R., Cole, J.B., and Swords, M.D. (1979). CARS measurements in an internal combustion engine. Appl. Opt. **18**, 3819-3825.

Woschni, G. (1967). A universally applicable equation for the instantaneous heat transfer coefficient in the internal combustion engine. *SAE Paper No. 670931, SAE Trans.* **76**, 3065-3083.

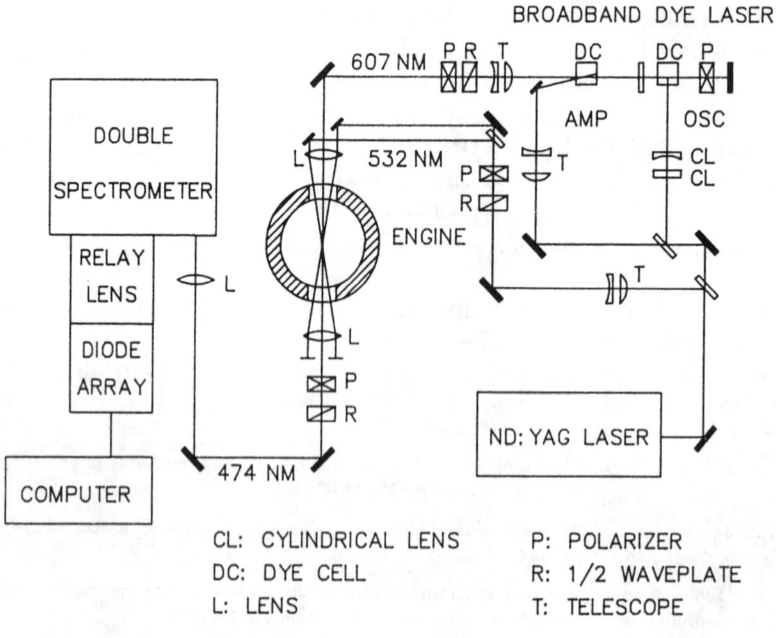

Figure 1. Experimental apparatus for the CARS measurements.

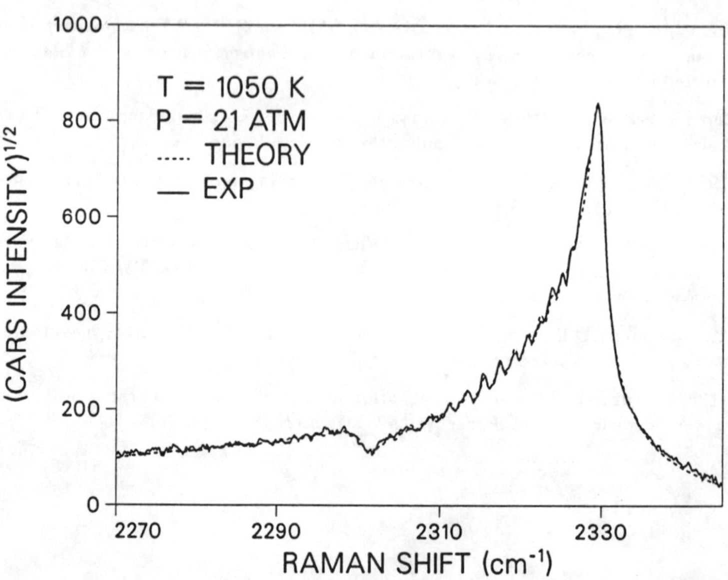

Figure 2. Experimental nitrogen CARS spectrum and the theoretical fit to the spectrum. The spectrum was obtained at 370 crank angle degrees (CAD) in the knocking engine just prior to autoignition. The methodology of Koszykowski et al. (1985) was used to calculate the theoretical spectrum.

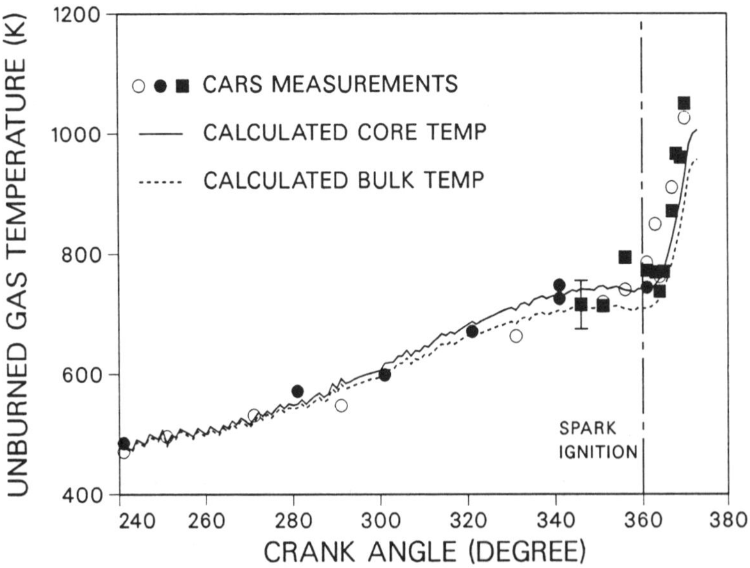

Figure 3. Comparison of CARS temperature measurements and the predictions of a heat release model (Ferguson et al., 1987) in the knocking engine. The different symbols on the CARS measurements represent data acquired on different days; the CARS data represented by the solid squares were normalized using nonresonant spectra obtained from the parallel cell. In the heat release model, the core gas temperature is the average temperature of the unburned gas in the engine outside the boundary layer, and the bulk gas temperature is the average temperature of the unburned gas including the boundary layer.

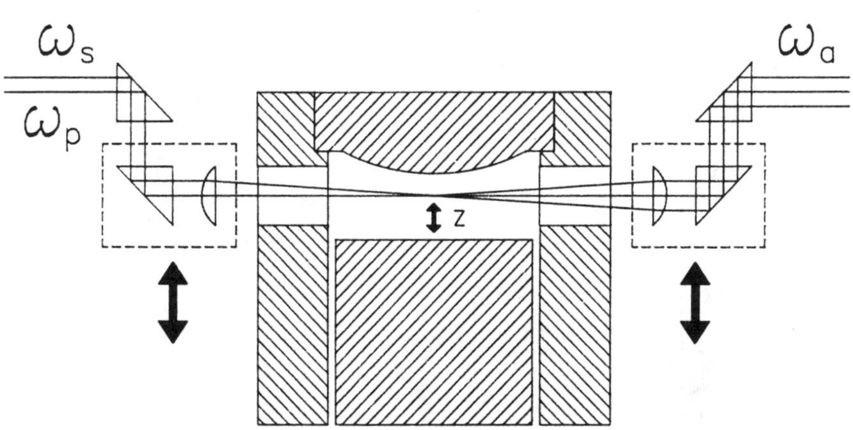

Figure 4. Convex cylinder head, and lens-prism apparatus for moving the CARS probe volume normal to the cylinder head surface. The frequencies for the pump and Stokes beams, and the CARS signal are ω_p, ω_s, and ω_{as}, respectively.

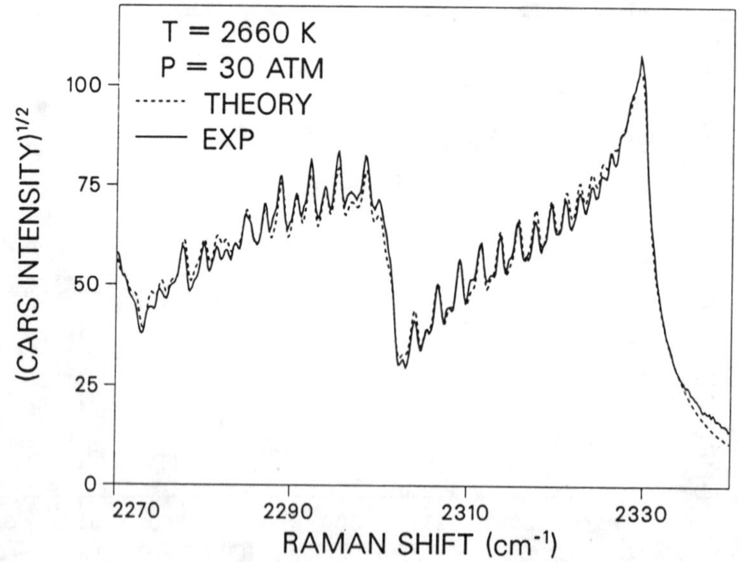

Figure 5. CARS spectrum and theoretical fit at 378 CAD, 1.93 mm from the wall. The spectrum is the average of approximately 30 laser shots. The temperature determined from the fit is 2660 K.

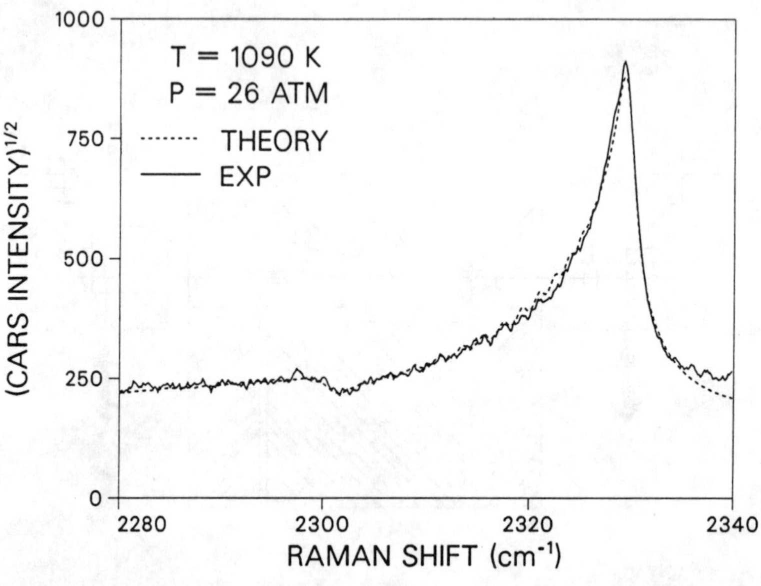

Figure 6. CARS spectrum and theoretical fit at 385 CAD, 0.025 mm from the wall. The spectrum is the average of approximately 30 laser shots. The temperature determined from the fit is 1090 K.

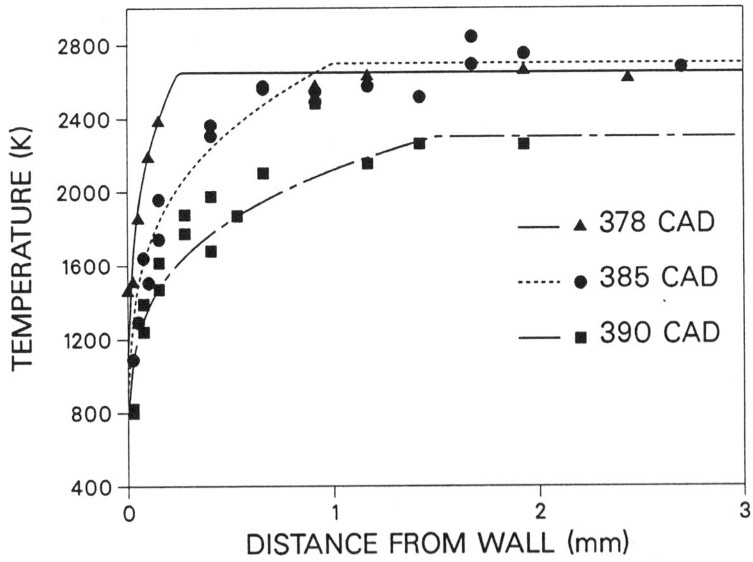

Figure 7. CARS measurements and power law calculations of the temperature profile at 378, 385, and 390 CAD. The lines are the convolved power law expressions.

S.I. ENGINE ION PROBE DIAGNOSTICS

C.G.W. SHEPPARD and E-S. A.A. IBRAHIM

DEPARTMENT OF MECHANICAL ENGINEERING, LEEDS UNIVERSITY

1. INTRODUCTION

At Leeds, as elsewhere, there is currrently a great deal of interest in the effects of turbulence on flame propagation within spark ignition engines (1,2). Such data are also required to verify an engine combustion mathematical model (3,4) which incorporates Bradley's data (5) for turbulent burning velocity. Leeds engine flame propagation data has previously relied on colour cine photography (1); this method is expensive and requires comprehensive optical access. The present study was initiated to investigate the possibility of using multiple ion probes (i) to augment or replace the photographic technique and (ii) in an investigation of knock.

Electrostatic probes were probably first introduced into petrol engines by Schnauffer (6); he used up to 24 probes to define the growth of the flame front. Subsequently others have used such probes in engines, e.g. refs (7-9), generally adopting applied voltages in the range 100 - 200 volts. All experienced cyclic variation, which made analysis difficult. Arrigoni et al (10) designed a special probe having three sensing electrodes to measure local flame speed and direction, they also claimed to measure flame thickness. Ali and Bradley (11), like Arrigoni, noted a variation in peak ionization with mixture strength; they suggested that this effect could be used in an engine management system. However, Mortara and Canta (12) claimed that ion current signal resolution was inadequate for engine control purposes. Witze et al (13) used electrostatic probe output to conditionally sample laser doppler velocimetry data; using laser shadowgraph photography, they also noted that the onset of ionization at an ion probe coincided with its engulfment by the visible flame front.

The present work reports some preliminary experiments to determine a suitable probe voltage, to confirm the coincidence of onset of ionization at a probe with the arrival of the visible flame front and to establish criteria for conditional sampling. Then, an assessment is made of a method using conditionally sample ion probe data to map flame progress in a single cylinder engine. Finally experiments using ion probes and high speed colour photography under engine knocking conditions are reported.

2. EXPERIMENTAL EQUIPMENT

Some preliminary work was conducted using a conventional Ricardo E-6 variable compression ratio, single cylinder research engine (4). Spherical "single" electrostatic probes were used, with the engine block acting as the second electrode. The spherical shape presented the same aspect to the flame front regardless of its direction of travel. The probe material was platinum 40% rhodium - chosen for its high melting

D. F. G. Durão et al. (eds.), Instrumentation for Combustion and Flow in Engines, 355–364.

point, high work function and corrosion resistance. Probe sphere sizes used were 1.1 mm formed at the end of a 0.5 mm wire and 0.8 mm at the end of a 0.1 mm wire. Probes, were insulated using 0.8 mm i.d. and 2 mm o.d. alumina tube. In the Ricardo engine three probes were used simultaneously, positioned in the cylinder on a diametric line passing through the spark plug location. The probes were at 19, 43 and 65 mm from the spark plug, itself 7 mm from the cylinder wall in the 76.2 mm bore. The probes were fixed in a spacer interposed between the block and cylinder head, a piston of increased crown height being used to compensate for the spacer thickness (14).

The bulk of the work was undertaken using a modified Ruston single cylinder, side valve engine of 76.2 mm bore and 82.5 mm stroke. The basic engine and instrumentatioon have been described in earlier publications (1,2,14). The cylinder head used is shown diagrammatically in Fig.1. The cylindrical insert in this head was either perspex, for filming combustion, or steel, for multiple ion probe work. This proved necessary because electrical interference prevented the use of more than one ion probe in perspex. The metal insert accommodated ion probes at radial distances of 5, 10, 15, 20 and 32 mm, there were 12 circumferencially equi-spaced probes at the outer radius. The insert could be rotated to reposition the probes. Since only six channels to the computer were available for ion probes at any one time, only two probes in the outer circle could be used simultaneously. One of these probes was always chosen to be in a "reference" position, to allow conditional sampling - as described later. In the flame propagation study, side ignition was adopted and a special "lipped" gasked was used to cut-off the valve chamber for about 18 degrees either side of top dead centre (1). In the knock study, the spark plug was located in the valve chamber and a normal flat gasket was used.

An ion probe signal processing unit was designed and built to interface between the probes and the eight channel multiplexed analogue to digital converter (ADC) feeding the VAX-8600 mini computer. This unit housed eight identical circuits which allowed adjustment of probe potential of 15 volts DC; separate amplification of each circuit allowed full use of the range of the ADC for maximum sensitivity for each probe. The data aquisition system was driven by an Hohner 360 pulse per revolution shaft encoder. This and the other (pressure, spark timing, crank angle indicator) instrumentation is fully described elsewhere (14).

For the high speed photography a Hitachi 16 HM cine camera was used, generally with a set speed of 8,000 frames per second. The film was 16 mm Eastman Ektachrome video news film (type 7250 tungsten balance).

3. PRELIMINARY EXPERIMENTS

Initial tests were conducted to determine the effects of probe size, polarity and voltage on ion current and indicated flame arrival time. The Ricardo engine was used, operating at 2000 rpm, with a compression ratio of 9 and an equivalence ratio of 0.82. For all three probes the ion current (i) increased linearly with applied voltage, (ii) increased with probe size, and (iii) proved higher with positive than negative polarity (14) - the latter effect was expected given the higher mobility of the negative (electron) charge carriers (15,16). Ion probe current also increased with distance from the spark plug, as a result of the higher temperature (and associated ionization level) as combustion proceeds (14). For all subsequent tests the larger probe size was selected, as the smaller probes proved to be insufficiently robust. Since ion current signal to noise ratio proved satisfactory, a negative probe bias was adopted; it was reasoned that a probe collecting positive ions would give a truer indication of flame arrival time than one collecting the more mobile electrons.

The effect of applied voltage on recorded flame arrival time at a fixed probe, that 43 mm from the spark plug, is shown in Fig.2. The data are those averaged over many cycles and those for cycles selected on the basis of nearly identical pressure records. Earlier apparent flame arrival times are indicated for higher applied voltage. Since the voltages used were too low for significant ionic wind effects (16,17), the earlier flame arrival was considered to be associated with increased probe "sphere of influence" at higher voltages. The noted differences in flame arrival time are only of the same order as the resolution of the shaft encoder used. Nevertheless, since the signal to noise ratio proved adequate at the 3 volt applied potential difference, it was considered sensible to adopt this voltage in all subsequent experiments.

A number of films were taken of combustion in the Ruston engine, using a perspex cylinder head insert with a single electrostatic probe. In all films the onset of ionization at the probe proved coincident with the arrival of the visible flame front, within the limits of the accuracy imposed by the shaft encoder and the camera framing rate; this is in accord with the earlier findings of Witze (13).

4. CONDITIONAL SAMPLING

The Ruston engine was fitted with the steel insert housing multiple ion probes and subjected to prolonged running at 1400 rpm with a stoichiometric mixture and a compression ratio of 8, ignition advance was fixed at 30 degrees before top dead centre (TDC). Cylinder pressure data and signals from five fixed ion probes were recorded over many cycles. The data were subject to considerable cyclic variation, as shown in Fig.3 - where a near guassian distribution in flame arrival time at the "reference" position is evident. It can also be seen that flame arrival time is inadequate to define a particular cycle in this engine; for a given flame arrival time (FAT) there is quite a spread in the recorded cylinder peak pressure (PP). A similar spread in the crank angle at which peak pressure (PPCA) was developed, and in indicated mean effective pressure (IMEP) was also noted (14). This is thought to be related to different flame growth patterns in the cycles recorded. Shown in Fig. 4 are cylinder pressure and reference position ion probe signals for typical slow, average and fast burning cycles. Each ion probe trace shows the characteristic spark interference (at a crank angle of 30 degrees before TDC) and, some continuing low level noise until a sharp increase in ion current is experienced when the flame first touches the probe. The current then falls as the flame passes over the probe, which is then immersed in a post flame region in which recombination takes place. Then a further rise in ion current occurs, reaching a peak coincident with that of cylinder pressure and associated (ionization inducing) gas temperature. The earlier flame arrival with faster pressure development is clear. Earlier combustion generates higher peak cylinder pressure and correspondingly greater ionization levels.

Shown in Table 7.1 are cylinder pressure data for a number of cycles selected to have very similar values of PP at a given FAT at the reference probe. It is clear, that for cycles selected to have similar FAT and PP, the values of IMEP and PPCA are essentially the same. It can be further seen from the tabulated data that, for cycles of similar reference probe

FAT and PP, the FAT at other ion probe positions are also nearly identical. In fact the complete ion records at these other positions are nearly coincident (14). Nevertheless some slight variation existed. This might be due to,

(i) Minor variation in flame front wrinkles relative to the probe on cycles of nevertheless similar overall flame growth.

(ii) Limitations imposed by the experimental technique. It proved impossible to reduce data sampling interval below 1 degree crank angle using the existing ADC, when multiplexed for eight channels at engine speeds above 750 rpm.

The variation of FAT at probe 4 proved greatest. This probe was that nearest the spark. The flame development in the early stages is likely to be rather more sensitive to cyclic variation in turbulence than in the later stages of combustion (3,4).

Shown in Fig.5 are ion probe FAT's for a large number of probe positions, collected with the head insert rotated twelve times and collated on the basis of similar peak cylinder pressure and FAT at the reference probe. The data so generated compare favourably with contours of flame position drawn from a high speed photographic record for a similar cycle (14). Also shown in the diagram are the relative positions of the ion probe placements.

5. PROBE RESPONSE TO KNOCK

The compression ratio of the Ruston engine was increased to 10 and the ignition advanced to 38 degrees before TDC, with all other conditions as reported previously, in order to operate with intermittent knock. For this work the spark plug was re-located to the valve chamber, and five ion probes were used in the positions shown in Fig.6. Resultant pressure and ion probe current records are shown in Fig.7, for typical normal and knocking cycles. It can be seen that,

(i) The initial rate of pressure rise (and hence the initial combustion rate) was greater in the normal combustion than in the knocking case. As a result the flame was able to traverse the entire chamber in a time shorter than the end gas spontaneous induction period, so that no knock occurred.

(ii) The flame arrival times at the individual probes are mostly earlier in the normal cycle, again suggesting a greater combustion rate. The exception to this was probe 4; this probe showed an earlier increase in ion current in the knocking cycle.

(iii) In the knocking cycle, all ion probes experienced a sharp increase in ion current at the crank angle corresponding to the sharp increase in pressure accompanying the onset of knock. This suggests that the earlier response of probe 4 in the knock cycle was probably due to a different flame growth pattern rather than to spontaneous ignition initiating at this point.

(iv) Under knocking conditions an oscillaton in ion current occurred. The oscillation frequency, at about 5kHz in this engine, matched that in the pressure record.

Shown in Fig.8 are tracings of successive positions of the flame front, taken from a colour cine film of the combustion event in the Ruston engine under similar operating conditions.* This shows the flame development relative to the ion probe positions, with spontaneous ignition occurring

*The films were shown at the meeting, colour printing was not possible in these Proceedings.

in the end gas region occupied by probes 4 and 5. The crank angle at which autoignition occurs is similar to that noted in the ion probe signals of Fig.7. In Fig.8, multiple end gas autoignition centres are evident. This occurred in most filmed knock cycles, with the position of the first ignition site(s) proving random within the end gas region (14). Further films were taken with the compression ratio of the engine raised to 11. These exhibited rather more severe knock, with multiple autoignition sites distributed over half of the piston crown area. In the frame following that in which autoignition was noted, a marked increase in yellow colouration was observed throughout the combustion space. This was presumed to be associated with the carbon formation reported in earlier studies (18). The increased ionization levels noted with knock might be related to the increase in carbon concentration.

6. SUMMARY

(i) For accurate recording of flame arrival time at an ion probe in an engine, the voltage applied to the probe should be biased negative - at the lowest level consistent with good signal to noise ratio.

(ii) Ion probes are useful for assessing cyclic dispersion in engines.

(iii) The response of a single ion probe may not always be sufficient to identify like cycles. However, used in conjunction with other information (e.g. a pressure record or another ion probe remote from the first), an ion probe provides a reliable and inexpensive means of effecting conditional sampling.

(iv) Using conditional sampling, it is possible to collate data from a restricted number of probes (rotated to many positions) in order to build up an accurate picture of flame development.

(v) In-cylinder ion probes have a characteristic response to knock; they may therefore be used as knock indicators.

7. TABLES

TABLE 7.1 Conditional Sampling Parameters and Flame Arrival Times (°BTDC) at the reference probe and other locations)

REF. PROBE	PP (atm)	IMEP (atm)	PPCA (°ATDC)	PROBE NO. 1	2	3	4
	45.4	6.9	6	10	8	11	20
	44.4	7.1	7	11	9	12	21
13	43.8	7.0	6	11	8	11	19
	44.4	7.0	7	12	7	10	20
	45.6	6.9	5	12	7	12	17
	42.8	7.4	9				
11	44.3	7.5	9	(not recorded)			
	43.0	7.0	8				
	42.8	7.5	10				
	42.6	7.7	11				
	41.4	7.9	11				
8	43.0	7.8	10	(not recorded)			
	42.6	7.8	11				
	41.7	7.7	10				

8. REFERENCES

1. Al-Mamar, F.N.A.A.; Hynes, J. and Sheppard, C.G.W. : Turbulence And Combustion Rate In A Dual Chamber S.I. Engine, Comb. Sci. and Tech. 45, 85, 1986.
2. Al-Mamar, F.N.A.A.; Desoky, A.R.A.A. and Sheppard, C.G.W. : Combustion In The Divided Chamber Spark Ignition Engine, paper C46/83, "Combustion in Engineering", I.Mech.E., 1983.
3. Bradley, D.; Lawes, M; Hynes, J. and Sheppard, C.G.W. : Limitations To Turbulence Enhanced Burning Rates In Lean Burn Engines, submitted to meeting, "International Conference on Combustion in Engines", I.Mech.E., 1988.
4. Hynes, J. : Turbulence Effects on Combustion in Spark Ignition Engines, Ph.D. Thesis, University of Leeds, 1986.
5. Abdel-Gayed, R.G.; Bradley, D. and Lawes, M. : Turbulent Burning Velocities : A General Correlation In Terms Of Straining Rates, to be published in Proc. Roy. Soc. 1987.
6. Schnauffer, K, SAE Journal 34, 17, 1934.
7. Warren, J.A. and Hinkamp, J.B. : New Instrumentation For Engine Combustion Studies, SAE Transactions 64, 665, 1956.
8. Higashino, I. : Flame Propagation In Spark Ignition Engines, Bulletin JSME 4, 15, 1961.
9. Harrow, G.A. : Some Applications Of Basic Combustion Research To Gasoline Engine Development Problems, SAE Paper 680765, 1965.
10. Arrigoni, V.; Calvi, F.; Cornetti, G.M. and Pozzi, U. : Turbulent Flame Structure As Determined By Pressure Development And Ionization Intensity, SAE Paper 730088, 1978.
11. Ali, M. and Bradley, D. : Electrostatic Probe Diagnostics In Gasoline Engines, Paper C68/83, "Combustion in Engineering", I.Mech.E., 1983.
12. Mortara, W. and Canta, M. : Engine Stability Sensors, SAE Paper 830428, 1983.
13. Witze, P.O.; Martin, J.K. and Borgnakke, C. : Conditionally - Sampled Velocity And Turbulence Measurements In A Spark Ignition Engine, Comb. Sci. and Tech. 36, 301, 1984.
14. Ibrahim, E-S. A.A. : Flame Propagation And Knock In S.I. Engines, Ph.D. Thesis, University of Leeds, 1987.
15. Lawton and Weinberg, F.J. : Electrical Aspects of Combustion, Clarendon Press, 1969.
16. Sheppard, C.G.W. : Electrical Phenomena In Flames, Ph.D. Thesis, University of Leeds, 1971.
17. Bowser, R.J. and Weinberg, F.J. : The Effect of Direct Electric Fields On Normal Burning Velocity, Combustion and Flame, 18, 296, 1972.
18. Miller, C.D. : Roles of Detonation Waves and Autoignition in Spark Ignition Engine, SAE Transactions 1, 98, 1947.

9. ACKNOWLEDGEMENTS

The support of SERC and the Egyptian government are gratefully acknowledged, as is the encouragement and advice of Derek Bradley.

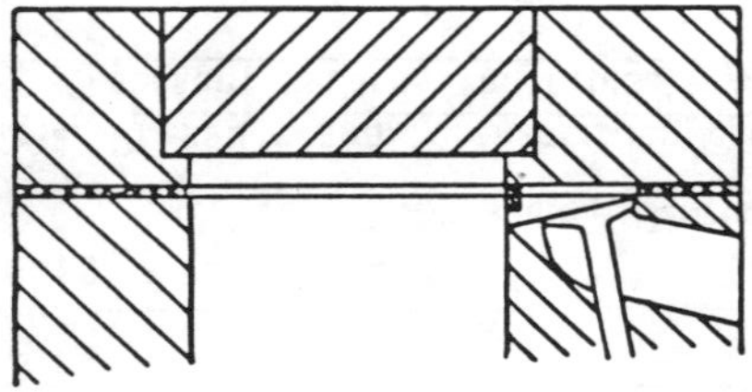

Fig 1 Diagrammatic representation of Ruston engine

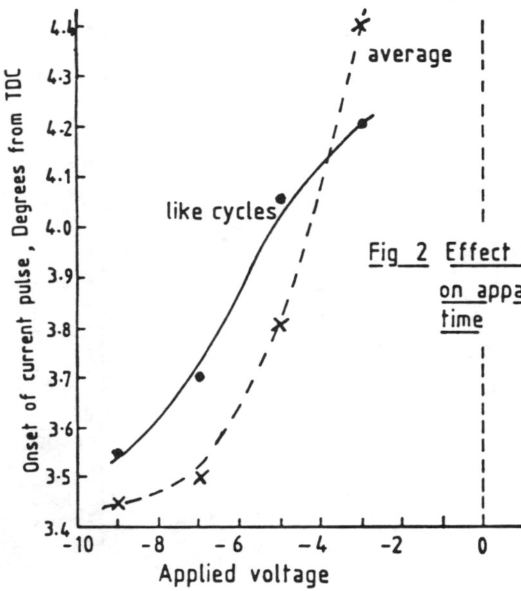

Fig 2 Effect of applied voltage on apparent flame arrival time

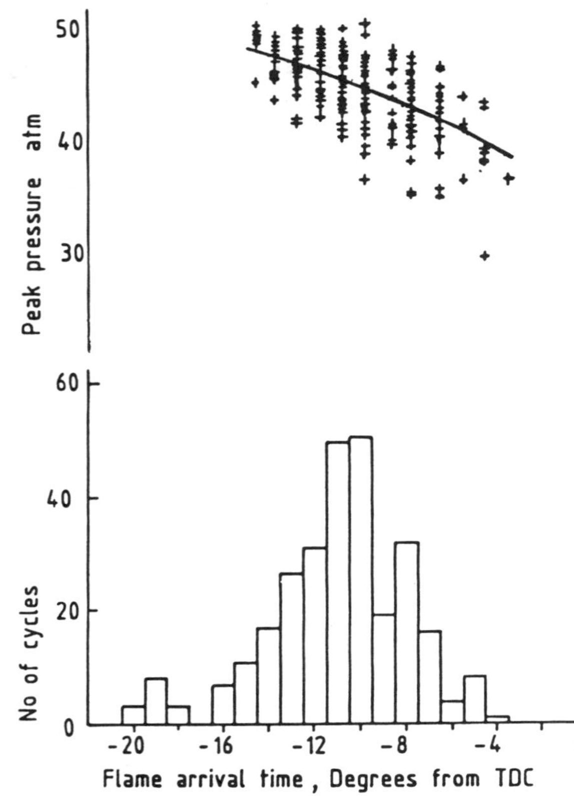

Fig 3 Cyclic variation in peak cylinder pressure and flame arrival time

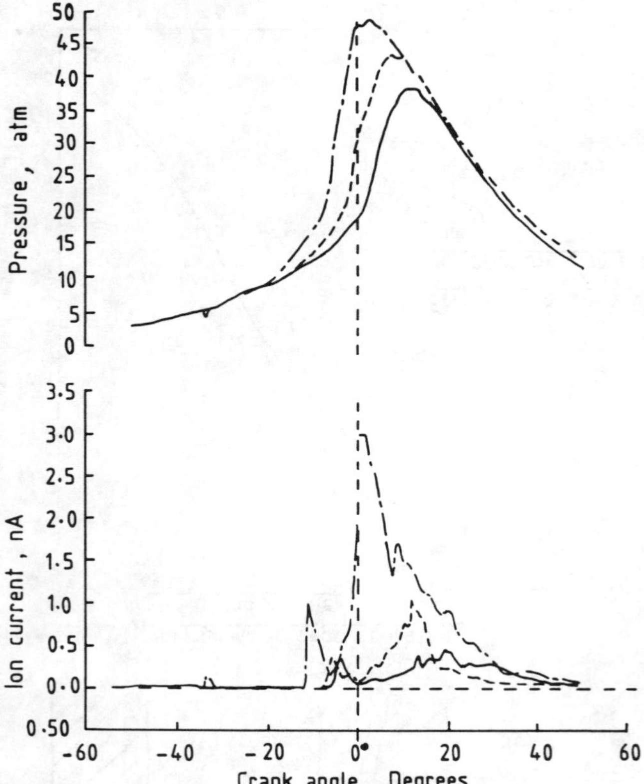

Fig 4 Pressure and reference ion probe current records for typical cycles

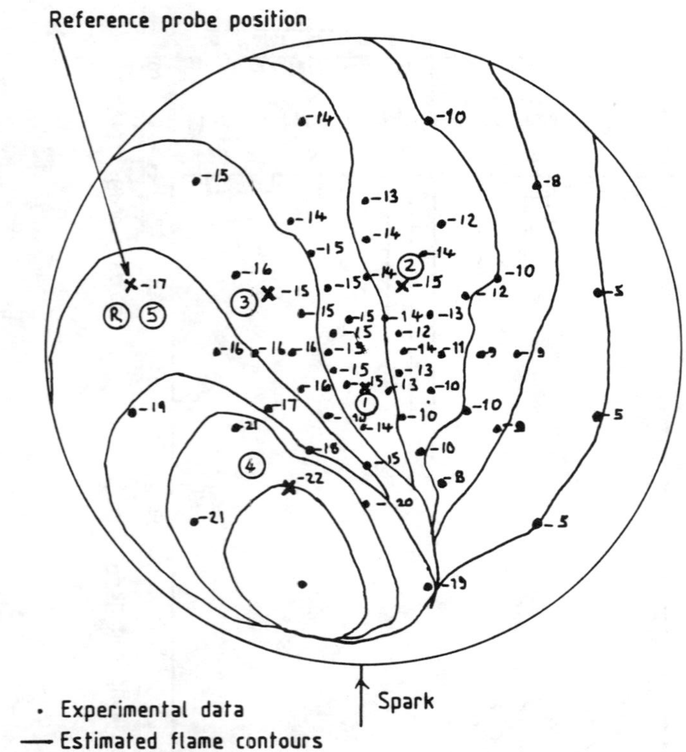

Fig 5 Flame progress diagram

(crosses mark ion probe relative positions in initial orientation of cylinder head inserts)

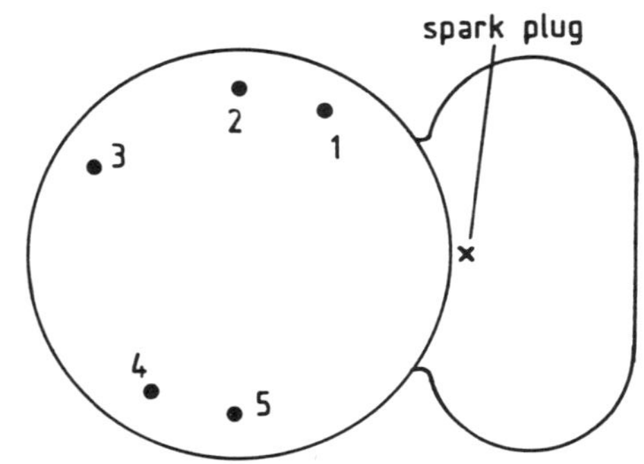

Fig 6 Position of ion probes during knock experiments

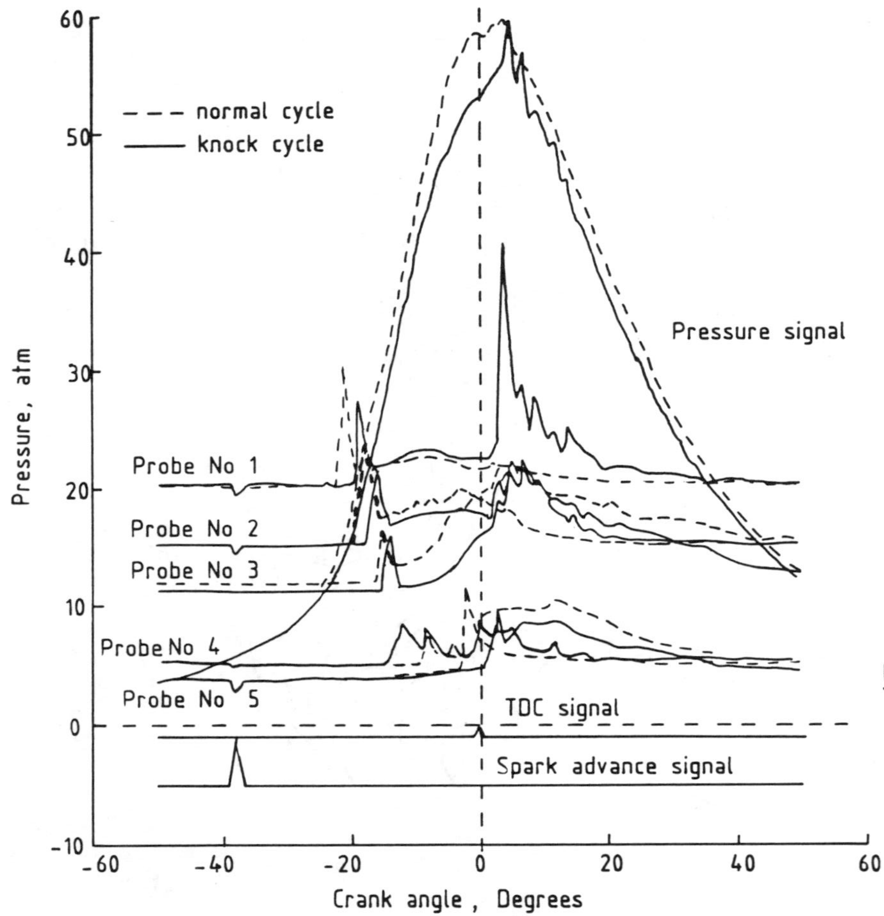

Fig 7 Cylinder pressure diagrams and multiple ion probe records for typical normal and knock cycles

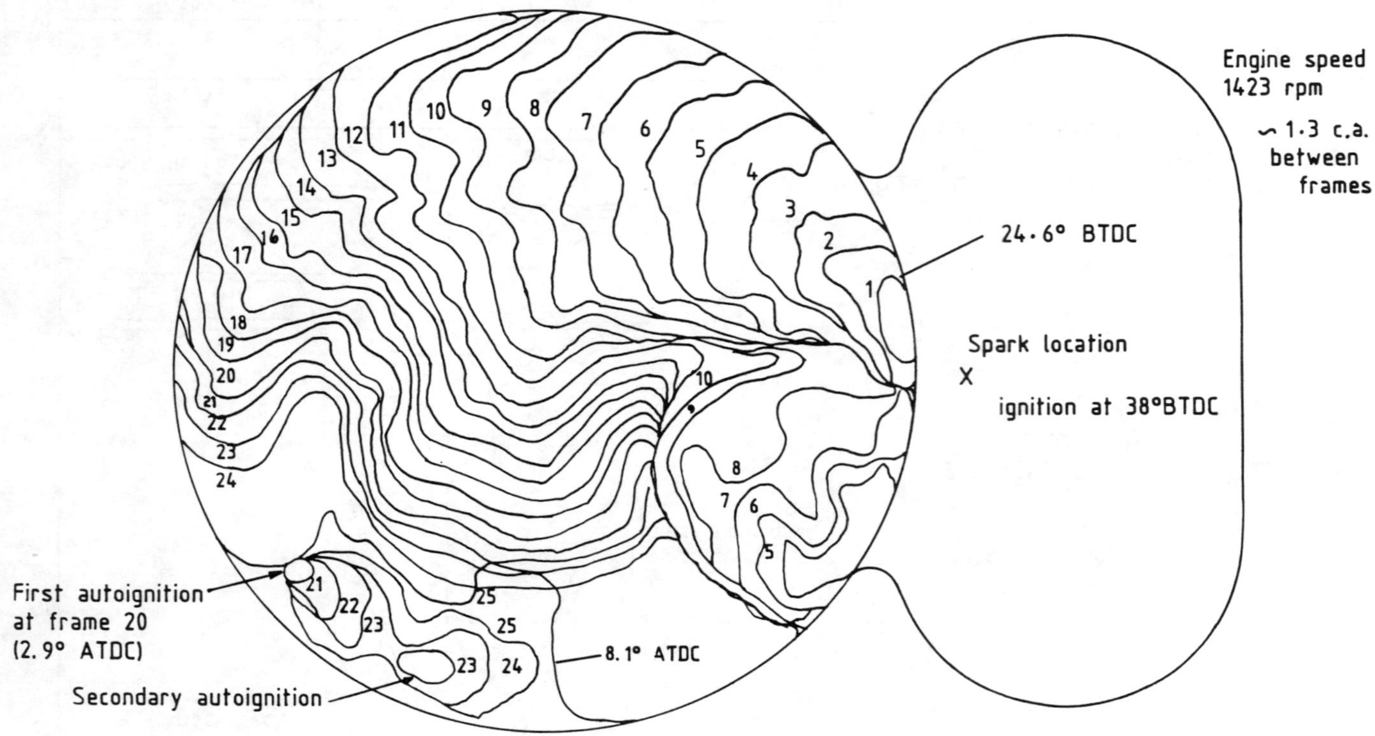

Fig 8 Successive flame positions traced from cine film of a knock cycle

THE EFFECT OF MAXIMUM VALVE LIFT ON TDC FLOW FIELD IN A HIGH SWIRL PISTON ENGINE

M. HAGHGOOIE AND J.C. KENT

Ford Motor Company
Research Staff
P. O. Box 2053
Dearborn, Michigan 48121-2053

ABSTRACT

The in-cylinder flow velocities in a motored engine were measured near top-dead-center (TDC) of compression using a single component laser doppler anemometer (LDA). The engine had a high-swirl intake port and a pancake chamber. The maximum intake valve lift was selectable via different rocker arm ratios. The results indicate that for this cylinder head, the angular velocity of the swirling flow increases with decreasing valve lift, but the rms fluctuations are generally insensitive to the valve lift except at small lifts. For all lifts studied, the mean swirl is rigid-body up to r/R = 0.75; the swirl after TDC of compression decays monatonically; and the rms fluctuations are higher in the central region of the cylinder than the wall region. The tangential velocities and their rms fluctuations were found to vary linearly with engine speed over the range investigated (200 to 1000 RPM).

INTRODUCTION

Background

The predominant role of in-cylinder fluid motion on the combustion process is well recognized (1,2). Continuing research efforts are aimed at understanding the nature of flow inside the combustion chamber and its influence on combustion. The motivation for the research is improved engine performance: increased maximum power, while maintaining good fast-burn characteristics and idle stability. One of the desired results is a correlation of flow and combustion as affected by engine design and operating variables. Such information is useful to the engine designer, and is also needed by the researcher for developing engine simulation models.

To achieve an adequate level of understanding of piston engine flow phenomena, it is imperative to conduct experiments with well-defined engine design variations. Some related flow studies in piston/cylinder systems have been reported, in which various simplifications have been used to facilitate application of diagnostic measurement methods (3-6). Because of the complex dependence of in-cylinder flow on the intake port and combustion chamber geometry, systematic changes in these variables are required for application of the results to production engines.

Literature Review

A number of flow measurements in motored engines have utilized hot-wire anemometry, see for example references (6-9). This measurement method has high frequency response and provides a continuous record of the observed velocity vs time. Both are advantageous for detailed analysis of turbulence

D. F. G. Durão et al. (eds.), Instrumentation for Combustion and Flow in Engines, 365–376.

characteristics such as correlation time and power spectrum. However, the application of hot-wire anemometry to in-cylinder flow measurements is difficult. A hot-wire is insensitive to the flow direction, and therefore, the signal is rectified when flow reversals occur. Furthermore, hot-wire calibration depends on the temperature and density of the gas and on the flow orientation, all of which undergo significant temporal variations in the cylinder of an engine (6,10).

Experiments in engine flow have also been done using laser doppler anemometry, LDA, (3-5, 11). This technique is well suited for in-cylinder use, because its measurements are unaffected by temperature and density. Moreover, it is non-intrusive, has good spacial resolution, and is directionally sensitive. Despite these advantages of LDA, accurate flow velocity measurements require proper seeding and appropriate collection optics (12). Since an LDA measures the velocity of discrete seed particles, it is not able to provide a continuous velocity record.

Two approaches have been used for measurements of in-cylinder flow: ensemble average (4,11) and cycle resolved (13,14). In ensemble averaged method the velocity data at a given location and crank angle window are collected over many cycles. While in cycle resolved technique the seed concentration is increased so that the velocity data can be collected and analyzed for individual cycles. At present, general agreement has not been reached on the significance of either method for measuring piston engine turbulence and its correlation with the combustion rate. We have chosen to employ the unambiguously defined ensemble average method for the present investigation.

The present study was motivated by observations of increased combustion rate when the intake valve lift was reduced; Stivender (15) and Oforah (16) hypothesized that increased intensity of small scale turbulent fluctuations due to high inlet velocities (restricted flow area) caused the observed increase in combustion rate. Davis et al. (17), reported a similar change in experimental burn rate with valve lift and developed an engine flow simulation model that predicted increased turbulence intensity for reduced valve lift.

Because of the direct influence of valve lift on the velocity of the flow entering the cylinder, it is of importance to understand its effect on in-cylinder fluid motion at the time of ignition, shortly before TDC of compression.

No measurements of in-cylinder turbulence are available for the engines used in references (15-17). However, measurements have been reported on the effect of valve lift on turbulence in two model engines. Arcoumanis et al. (4), using LDA, showed relatively little effect of valve lift on the rms velocity fluctuations at TDC for a centrally located valve. Daneshyar and Fuller (6) employed hot-wire anemometry and a cycle resolved measurement method. They observed that for a shrouded valve, the turbulence intensity decreased about 20% as the valve lift was reduced by 50%. These different flow observations associated with valve lift changes could be due to the specific flow geometries (or possibly the measurement techniques).

In this paper, we describe an LDA setup which was used for measuring the flow velocities near TDC of compression in a motored engine equipped with a cylinder head similar to a production head. Experimental results are reported on the nature of the swirl inside the cylinder, the effect of maximum intake valve lift on the flow field, and the scaling of mean and rms velocities with engine speed.

EXPERIMENTAL SETUP

The experimental setup consists of an engine with optical access, a seeding system, LDA optics, a counter processor, and a computer for storing and analyzing the data. These items, shown schematically in Figure 1, are described in the following.

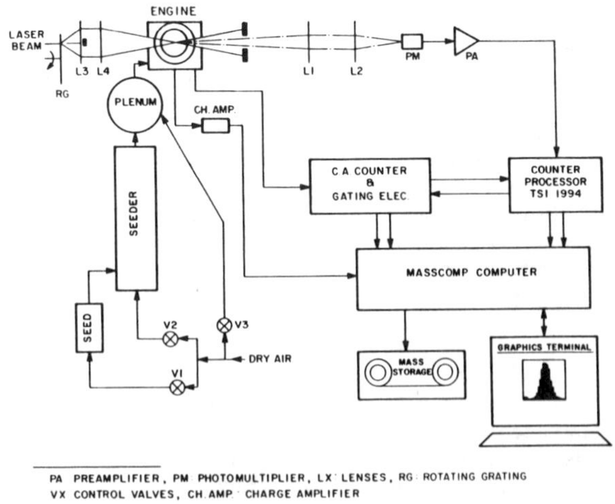

FIGURE 1. Schematic of experimental setup.

Engine

A single cylinder Waukesha engine was modified to provide optical access into the combustion chamber and establish a measurement environment which is free from lubricating oil. The absence of oil is required to reduce fouling of the optical window and engine wear which would result from accumulation of abrasive seed particles in the oil. The optical access is via a glass ring inserted between the cylinder and cylinder head. Dynamic sealing of the piston-cylinder interface is accomplished by a single continuous teflon-filled bronze ring. This arrangement proved to be satisfactory for oil-free operation. To moderate friction and compression-generated heat, the engine head and cylinder wall are cooled by water flow.

The valve guides, rocker arms, and pushrods were lubricated sparingly to minimize the introduction of oil into the cylinder with the intake flow. Specially designed grooved intake and exhaust valve guides with o-ring seals were used for isolating the valve guides from the intake and exhaust ports, thus preventing the seed particles from entering the valve guides and seepage of the lubricating oil into the ports. The grooves collect any seed and wear particles present in the valve guide. Without these precautions, the seed particles passing around valve steam mixed with the oil and formed a thick paste which caused the valves to stick.

Diffferent intake valve lifts were achieved, with a fixed cam profile, by changing the pivot point of the rocker arm. Using a specially designed rocker arm assembly, maximum valve lifts of 100% (10 mm) and 75%, 50%, 25% of full lift could be selected. The engine had a pancake-shaped combustion chamber, bore and stroke of 101.6 mm, and a compression ratio of 8:1. The cylinder head incorporated a high swirl intake port. The engine was also equipped with a shaft encoder (1 degree crank angle resolution) for recording the occurance of valid velocity measurements. The cylinder pressure was continuously monitored by a Kistler transducer for detecting ring leakage or valve sticking.

Optics

A single component LDA system was employed, using forward scatter to increase the detectivity of the small (1-2 micron) seed particles. The laser beam (514.5 nm) from an Argon Ion laser (Lexel model 95), typically operated at 500 milliwatts power, was focused onto a rotating grating (TPD type H) which provides beam splitting and frequency shifting of 5 MHz. The equal intensity first order diffracted beams from the rotating grating were first made parallel and then focused onto the measurement location by a 371.3 mm focal length lens. The beam crossing half-angle was 3.79 degrees, producing a measurement volume of 1.5x0.1 mm in length and diameter respectively. The collection optics consisted of collecting lenses and a pin hole of 0.2 mm diameter. The scattered light intensity was measured with a photomultiplier (EMI 990 7B) located behind the pin hole and analyzed with a counter processor (TSI 1994B) to obtain the period of the Doppler signal.

The laser as well as input and collection optics were assembled on an optical bench mounted on a milling machine table. Thus the probe volume could be located within the measurement chamber with an accuracy of 0.025 millimeter in all three dimensions. The input beams entered the combustion chamber through the cylindrical window. Due to the window curvature, the beam crossing angle changes as the probe volume is traversed along the bore diameter. Therefore, a beam tracing algorithm was used to calculate the correct beam crossing angle and position of the measurement probe.

Seeder

An essential component of an LDA setup is a seeding mechanism which generates appropriate seed particles for measuring the fluid velocity. First, the seed particles must follow the fluid motion with reasonable accuracy. This means that they must have appropriate size and density for the given fluid under investigation. And second, they must be visible to the collection setup, i.e. they must have large scattering cross section in the direction of detection. A discussion of seed particle generation, velocity fidelity, and detection with various optical setups is given elsewhere (12).

The seed particles used in these experiments were alumina powder with a nominal diameter of 1 micron. The alumina particles are introduced into the intake air stream through a two-stage fluidized bed as shown in Fig. 1. In the first stage the alumina powder is fluidized and coarsely deagglomerated by passing dry air through the powder. Remaining agglomerates are broken up in a second stage bed, which contains 100| brass beads in a column 10 cm diameter by 30 cm high. This bed is fluidized by air flowing through a porous plate at the bottom. To break up large air bubbles traveling upwards through the column, six wire mesh screens (0.5 mm x 0.5 mm) are dispersed within the column. The performance of the seeder was checked by measuring the particle velocity lag as it passes through a converging nozzle (12). And it was found that the size of particles are less than 2 microns in diameter, which is adequate for resolving the flow fluctuations present in IC engines (18).

The output of the seeder is mixed in a plenum with the main intake air which has been dried to a dew point of -100F using a drier (Anderson Industrial model MPS-50). The use of dry air is essential if the measurements are to extend into the expansion stroke. With undried air, the condensation of water vapor during the expansion becomes a strong light scattering source which overwhelms the scattered light from the seed particles.

Data Acquisition

To interpret the in-cylinder velocity data, it is necessary to record the measurements as a function of engine crank angle and spatial coordinates. The measurement location is set by positioning the milling machine table such that the probe volume coincides with the desired location, while the corresponding crank angle is recorded by a pulse counter. The crank angle pulses produced by the shaft encoder are counted by a 12 bit counter which is reset at TDC of intake.

The Doppler signals from the photomultiplier are fed to a TSI counter processor via a UHF amplifier (TSI model 10099). The counter is set to remove the low frequency component of the signal (pedestal) and count the time for eight complete cycles of the Doppler burst. The data-ready pulse from the counter processor is then used to store the burst period and the contents of the crank angle counter through a parallel digital I/O (MASSCOMP PI16) in a MASSCOMP computer. To confine the measurements into a predefined crank angle interval, a gating signal is produced by comparing the contents of crank angle counter with a pair of preset 12 bit binary numbers - set by switches - which correspond to the lower/upper bounds of crank angle window. This gating pulse is used to enable the counter processor.

RESULTS AND DISCUSSION

The velocity measurements were carried out in the single cylinder LDA engine motored at 500 RPM with wide open throttle. The measurements were made along a bore diameter at the center of the clearance height (7 mm from the cylinder head). The tangential component of the velocity was measured in radial steps of 2.5 mm over the crankangle window of +-30 degrees around TDC of compression. At a given position, at least 10,000 Doppler bursts are stored (on average, 166 measurements per degree crank angle) over many cycles. Occasionally the number of measurements was increased by a factor of ten, but no significant difference in the measured quantities was observed. Fig. 2 shows a typical histogram of velocity distribution at $r/R = 0.25$ and 30 degrees before TDC of compression.

Although the signal processor imposes predetermined conditions on an input signal in order to be considered as a valid Doppler burst, some noise signals do pass through the system and are measured as valid bursts. For the experiments reported here, the frequency of noise to signal ratio is on average less than 0.2% for $r/R < 0.95$ and increases to about 10% as the probe volume approaches the cylinder wall. This increase is mainly due to scattered light from the cylinder wall entering the detection system. And since the input laser beams are frequency shifted, this results in a Doppler signal from the stationary wall, which is easily identified and discarded.

Velocity distributions at a given crank angle θ and radius r are used to calculate the mean (ensemble average) velocity $U(\theta,r)$ and the rms velocity fluctuation $u'(\theta,r)$. Figure 3 represents a typical plot of the mean velocity vs r/R for $\theta = 30$ degrees before TDC of compression. The inset shows an schematic of the cylinder head. The vertical bars in Fig. 3 are not measurement errors; they represent the rms $u'(\theta,r)$. Plotted in this figure are results of several measurements obtained on different days. Barring systematic errors, the accuracy of data is evident by repeatability of the measurements. The estimated uncertainties of the mean and rms velocities are about 4% and 12% respectively. From Fig. 3 it is clear that the swirl has the characteristics of a rigid body rotation up to $r/R = 0.75$. For $r/R > 0.75$ the effect of wall shear is evident.

The turbulent boundary layer thickness δ was estimated using

$$\delta/x = 0.37*(U*x/\nu)^{-0.2}$$

which is derived for a turbulent boundary layer on a semi-infinite plate (19). In this expression x is the distance from the leading edge of the plate, ν is the kinematic viscosity, and U is the free stream velocity. We have employed x = U*dt, with U taken as the tangential velocity outside of the boundary layer (10 m/sec in Fig. 3), and dt is the time from intake valve closing to the time of measurement (30 msec in Fig. 3). The calculated boundary layer thickness is 9.7 millimeters and are indicated in Fig. 3, which is in good agreement with the measured data. Bopp et al. (20) also observed rigid body behavior for in-cylinder swirling flow. However, the boundary layer is not evident in their data. This could be due to the lower swirl ratio in their engine and hence smaller U*dt which would lead to a thinner boundary layer. The application of above equation to their data results in an estimated boundary layer thickness of 3 to 4 millimeters, depending on the engine RPM.

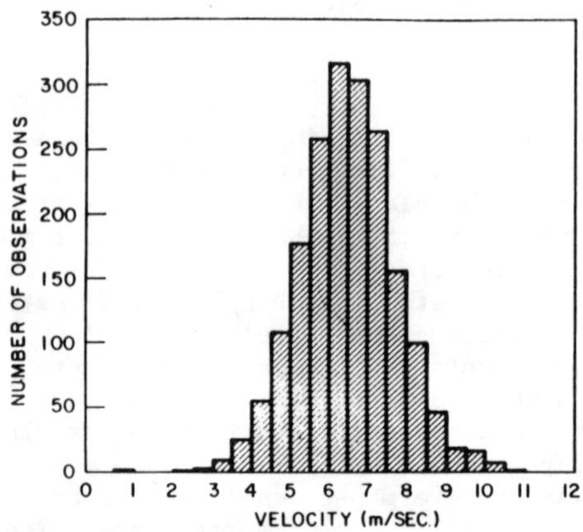

FIGURE 2. Velocity distribution probability at r/R = 0.5 and Θ = 30° BTDC.

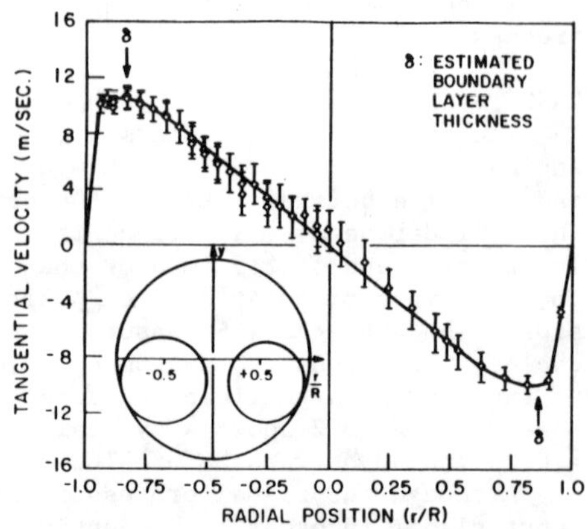

FIGURE 3. Swirl velocity profile at Θ = 30° BTDC. Vertical bars are the root-mean-square fluctuations.

The angular velocity of the swirling flow and its projected center of rotation were obtained as a function of crankangle Θ by weighted least squares fitting of a straight line to U vs r/R data in the range of $|r/R| < 0.75$. The weighting factor of 1/u' was used to reduce the influence of regions having large variability on the calculated angular velocity. Shown in Fig. 4 is the swirl ratio, defined as the angular velocity of the swirl normalized by that of the crankshaft, vs crank angle for various intake valve lifts. This figure shows that, within +-30 degrees around TDC of compression, the angular velocity of swirling flow increases as the maximum valve lift decreases. However, the difference between 100% and 75% lift is negligible. Predictions of the in-cylinder swirl ratio can be made using steady flow measurements of the dimensionless coefficients of discharge and angular momentum flux, C_f and N_f (21). Values of C_f and N_f for this intake port at two different valve lifts are shown in the following:

Valve lift(mm)	N_f	C_f	Nf/Cf~ swirl ratio
2.5	0.229	0.187	1.22
6.4	0.354	0.352	1.01

If constant values of N_f and C_f during intake were assumed, the swirl ratio would be proportional to the ratio Nf/Cf; see eqn. C1 of (21). Thus, even though both N_f and C_f decrease as the valve lift is decreased, the swirl ratio increases. This agrees, trendwise, with the LDA measurements. It should be noted however that depending on the flow geometry of the induction system, the swirl ratio may either increase or decrease for operation at reduced valve lift.

As shown in Fig. 4, the observed swirl decays during the expansion stroke. But, during the compression stroke the measured angular velocities behave in a complicated manner. This could be explained if swirl were stratified along the cylinder axis. As the piston moves upwards during compression, it would push regions of differing angular velocities into the measurement plane. Interactions between axial regions with different angular velocities would tend to diminish the swirl stratification. After the interactions were completed, the contents of the cylinder would behave as a single system with homogeneous angular velocity which decays in time due to wall shear. It appears from Fig. 4 that this occurs at about TDC.

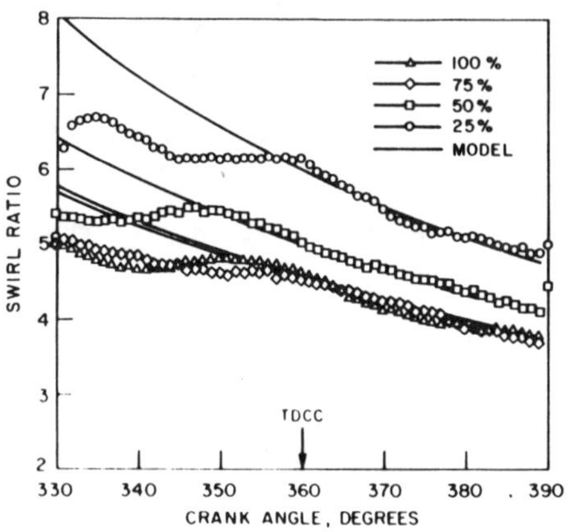

FIGURE 4. Temporal variation of the swirl ratio for various valve lifts.

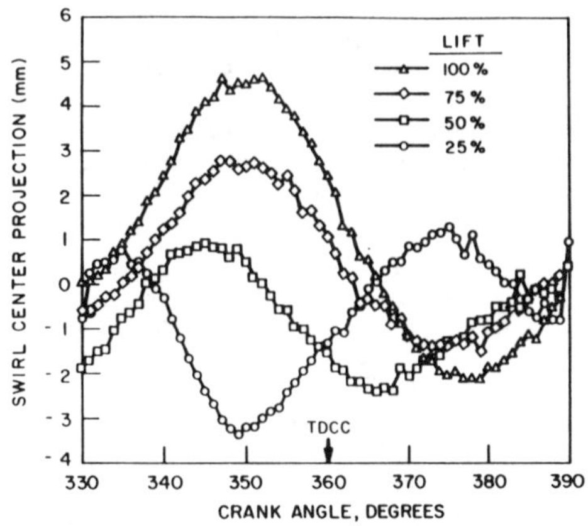

FIGURE 5. Temporal variation of the swirl center projection on a bore diameter for verious valve lifts.

The decay of angular velocity is caused by opposing torques resulting from wall shear at cylinder wall, cylinder head surface, and piston top. The wall shear torque may be written as

$$T = \lambda * \omega^{1.8}$$

where ω is the angular velocity and λ depends on cylinder bore diameter, instantaneous clearance height, and the density and viscosity of the fluid (21). Neglecting the small variation in clearance height that occurs within 30° of TDC, the value of λ can therefore be taken as a constant independent of valve lift. The time evolution of ω is governed by

$$d(I\omega)/dt = -T,$$

with I being the inertia of the cylinder contents, which does not vary due to piston motion (right circular cylinder of constant radius and mass) and is independent of intake valve lift (total air mass in cylinder is the same for all lifts within a few percent). The data after TDC in Fig. 4 were fitted to the solution of above differential equation to obtain the value of λ for each valve lift; all were within 15% of the value $\lambda = 0.213$. The solid lines in this figure are the result of these calculations. As can be seen, the agreement with data during the expansion stroke is reasonably good. This could be an indication that after TDC of compression, the contents of the cylinder rotate with a uniform ω which then decays due to wall shear torque. The higher values predicted by the model during the compression stroke suggests that the swirl motion is not homogeneous.

To consider this possibility further, we note that the rate of decay of ω is proportional to $\omega^{1.8}$. The nonlinear dependence on ω implies that the angular velocity would decay faster with axial swirl stratification than in the case of uniform angular velocity. Furthermore, some additional enhancement in the decay rate should occur due to the shear between layers of different angular velocities. These effects are not incorporated into the model, which makes plausible its overestimation of ω during the compression stroke.

In Fig. 5 the projection of the swirl center along the line of measurement is plotted versus crank angle for different valve lifts. Since the tangential velocities were measured only along a single bore diameter, it is not possible to locate the swirl center in the measurement plane. The variation of the swirl center projection with crank angle could result from either one or both of the following:

a- Processing of the swirl center at a given plane around the cylinder axis.

b- Variation of the swirl center along the cylinder axis; during compression, layers with different swirl centers enter the measurement plane.

If we assume that the variations of the swirl center with crank angle are solely due to precession, then the data indicates that the angular velocity of the precession is in the range of 3000 to 4000 RPM, or 6 to 8 times the engine RPM. Arcoumanis et al. (23) observed a similar result during the induction stroke for an on-axis valve and upstream swirl vanes. It is possible that precession alone might not account for such a high angular velocity. And we conjecture that the axial variation of the swirl center (b above) also contributes to the observed results.

Note that the variation of the projected swirl center is smaller after TDC than before. This could be an additional indication that the fluid motion becomes more homogeneous as the piston approaches TDC.

Figures 6a and 6b present the spatial variation of rms values at different crank angles for intake valve lifts of 100% and 25% respectively; the results for 75% and 50% lifts are similar to those for 100% lift. Figure 6 shows that at a given crank angle, the rms is almost constant for $|r/R| < 0.5$. Past this region, it decreases linearly with radius. Increased rms near the cylinder center for swirling flows has been generally observed (5, 23, 24). One possible explanation for this is that different turbulent length scales and decay rates could exist across the cylinder. In a swirling flow contained in a cylinder, the turbulence length scale in the central region of the vortex corresponds to the integral length of the free stream flow. In the outer region, the length scale is zero at a smooth wall and is proportional to the distance from the wall within the boundary layer. Since the turbulence decay is inversely

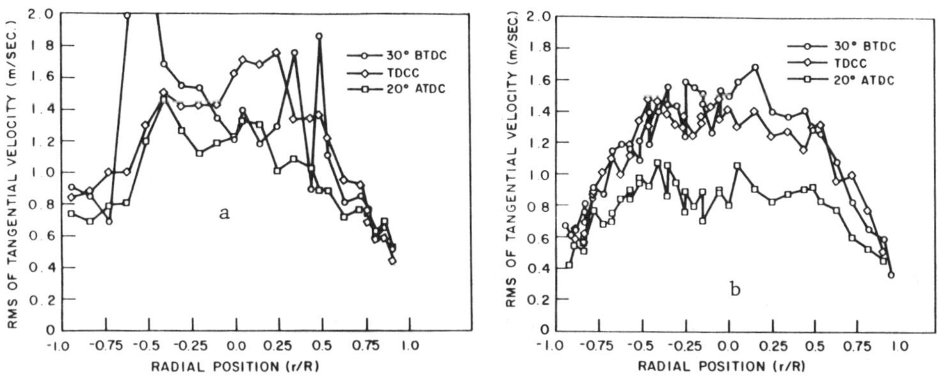

FIGURE 6. Spatial variation of root-mean-square fluctuations for (a) 25% and 100% (b) valve lifts.

proportional to the length scale (24), the turbulence generated during the intake process would tend to decay slower in the central region and faster in the wall region. Figures 6a and 6b support this possibility. However, if cycle-to-cycle variation of the swirl center occurred, this could account for some of the higher rms values in the central region.

In our experiment, the lowest rms values were observed near the cylinder wall. However, in references (5, 22), the rms increased sharply as the cylinder wall was approached. A large probe dimension used in a high velocity gradient, such as near the wall, would result in an overestimate of the rms. In our setup, this effect is negligible (probe width x mean velocity gradient << observed rms). The cylinder wall surface in our engine is smooth, which is typical of production engines. In the motored apparatus of reference (22), the cylinder surface within the clearance volume contains the intake and exhaust valves, which could act as roughness elements and increase the turbulence near the wall. Such surface disturbances might also be present in the apparatus of reference (5) in the form of a cylinder-mounted spark plug or side window. It is interesting to note that in Fig. 28 of reference (5), the rms did not increase as the (presumably smooth) surface of the piston was approached.

The observed effect of valve lift on the rms velocity at TDC is shown more clearly in Fig. 7. For both r/R = 0 and 0.94, the rms is constant or decreases slightly as the valve lift is reduced from full to 50% of full lift. This is in general agreement with the results of references (4 and 6). At 25% of full lift, the rms is somewhat larger than at full lift, which agrees with the hypothesis of references (15, 16), the LDA measurements of (4), and the modeling of (17).

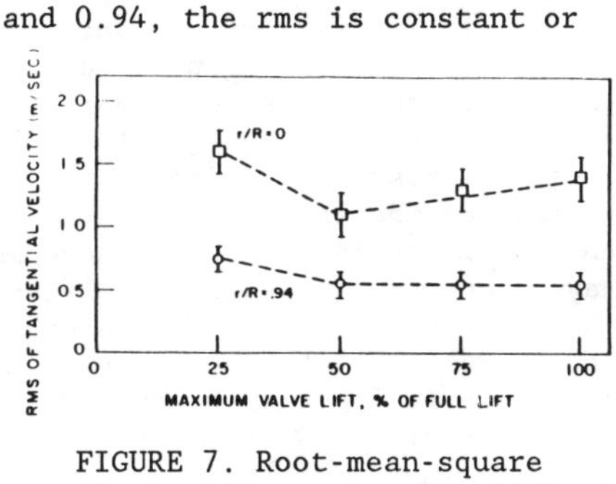

FIGURE 7. Root-mean-square fluctuations vs. valve lift at TDC for two radial locations.

Reducing the valve lift will increase the velocity of the mixture entering the cylinder for constant mass flow rate. This in turn will increase the production of turbulence during induction. If the turbulence dissipation rate were independent of u', the rms should increase as the valve lift is reduced. However, as seen in Fig. 8, the rms at TDC of compression is relatively insensitive to valve lift. This leads one to suspect that the dissipation rate may increase with u'. If this is the case, then the ability to increase turbulence at ignition via intake-generated turbulence is limited.

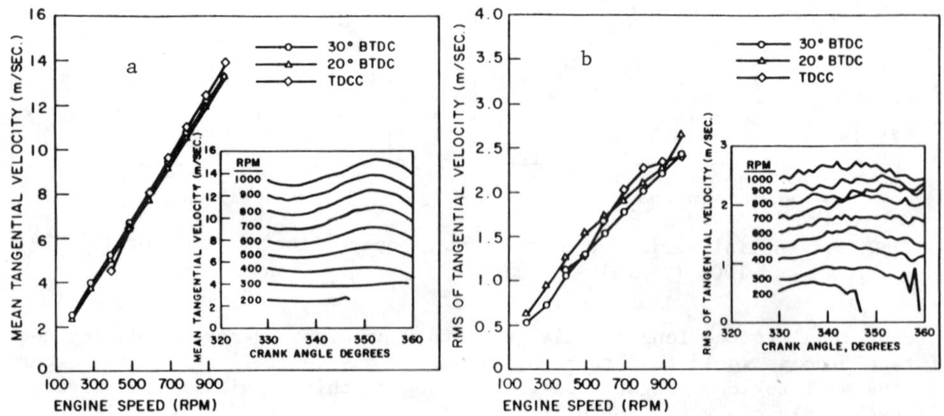

FIGURE 8. Mean tangential velocities (a), and root-mean-square fluctuations (b) at r/R = -0.5 vs RPM. The insets show the temperal variations at various engine speeds.

Finally in Figs. 8a and 8b, we plot the variation of the mean tangential velocity and its rms with respect to engine RPM. The data are for r/R = -0.5 and various crank angles for 100% valve lift. As can be seen, the mean and rms velocities scale linearly with RPM within the engine speeds investigated, 200 to 1000 RPM. Bopp et al. (20), also report linear scaling of mean velocities with RPM, but that the speed dependence of the rms depends on the location of measurements. And rms averaged over space increases faster than linear with RPM. However, their results for speeds up to 1000 RPM show an almost linear relation, in general agreement with our findings.

The insets in Figs. 8a and 8b show the measured velocities versus crank angles for various engine speeds. An interesting feature in inset of Fig. 8a is that the crank angle variation of the mean velocities increases as engine speed is increased. A possible explanation is that at the end of intake process, the swirl is nonhomogeneous within the cylinder and thereafter the interaction between different regions tends to homogenize the swirl. At lower speeds there may be less axial variation and/or there may be enough time between valve closing and the time of measurement so that the swirl becomes more homogenized. This would produce less variation in the velocity at a fixed point during compression. For higher speeds, the larger variation with crank angle could be due to larger axial stratification and/or too short a time interval for mixing to be completed.

CONCLUSIONS

In-cylinder velocity measurements were made using LDA in a motored engine with a high-swirl intake port. The data were taken along a single diameter at the middle of the clearance height for 30° on either side of TDC of compression. The main conclusions of this study are:

1. The swirl has rigid-body characteristics up to r/R = .75, or to within about 10 millimeters of the cylinder wall.

2. Prior to TDC, the swirl ratio and the swirl center vary with crank angle in a complicated way suggesting that the swirl could be stratified along the cylinder axis. However, after TDC the swirl ratio decays monotonically and is adequately modeled as solid-body rotation with a retarding torque due to wall shear.

3. The rms fluctuations in this engine are higher near the cylinder axis than near the cylinder wall. Cyclic variability of the swirl center could contribute to enhanced rms values near the axis.

4. Near TDC of compression, the mean and rms tangential velocities scale linearly with the engine speed in the range investigated (200 to 1000 RPM).

5. The swirl ratio increases as the intake valve lift is decreased. This confirms predictions based on steady flow discharge and angular momentum flux measurements for this intake port. It should be noted that different flow geometries can result in swirl ratio decreasing as intake valve lift is decreased.

6. A particularly interesting observation is that the rms velocity fluctuations are not affected significantly by reduction in the intake valve lift to 50% of full lift, in spite of a factor of two increase in the induction velocity. For 25% of full lift, some increase in rms was observed. This implies that increased production of turbulence during intake is accompanied by increasing dissipation rate.

ACKNOWLEDGEMENT

We wish to thank Dr. R. J. Tabaczynski for his interest, encouragement, and for stimulating discussions during the course of this investigation.

REFERENCES

1. Arcoumanis C, and Whitelaw J H: Fluid Mechanics of Internal Combustion Engines. A Review Intl. Symp. on Flows in Internal Combustion Engines: ASME: New York, 1985.
2. Tabaczynski R J: The Role of Fluid Mechanics in Spark Ignition Engine Design. Intl. Symp. on Flows in Internal Combustion Engines: ASME: New York, 1985.
3. Witze P O: The Effect of Spark Location on Combustion in a Variable-Swirl Engine. SAE Paper No. 820044, 1982.
4. Arcoumanis C, Bicen A F, and Whitelaw J H: Effect of Inlet Parameters on the Flow Characteristics in a Four-Stroke Model Engine. SAE Paper No. 820750, 1982.
5. Hall M J, and Bracco I V: A study of Velocities and Turbulence Instensities Measured in Firing and Motoring Engines. SAE Paper No. 870453, 1987.

6. Daneshyar H, and Fuller D E: Definition and Measurement of Turbulence Parameters in Reciprocating I.C. Engines. SAE Paper No. 861529, 1986.

7. Dent J C, and Salama N S: The Measurement of Turbulence Characteristics in an Internal Combustion Engine Cylinder. SAE Paper No. 750886, 1975.

8. Lancaster D R: Effects of Engine Variables on Turbulence in a Spark Ignition Engine. SAE Paper No. 760159, 1976.

9. Wakisaka T, Hammamoto Y, and Kinoshita S: Turbulence Characteristics in Internal Combustion Engines. Bul. ASME. Vol. 26, No. 212, 1983, p. 254.

10. Witze P O: A Critical Comparison of Hot-Wire Anemometry and Laser Doppler Velocimetry for I. C. Engine Applications. SAE Paper No. 800132, 1980.

11. Witze P O: Application of Laser Velocimetry to a Motored Internal Combustion Engine. Laser Velocimetry and Particle Sizing: Hemisphere Publishing Corporation, Washington D C: 1979, p. 239.

12. Haghgooie M, Kent J C, and Tabaczynski R J: Verification of LDA and Seed Generator Performance. Expts. Fluids, Vol. 4, 1986, p. 27.

13. Liou T M, and Santavicca: Cycle Resolved Turbulence Measurements in a Ported Engine With and Without Swirl. SAE Paper No. 830419, 1983.

14. Rask R B: Comparison of Window, Smoothed-Ensemble, and Cycle-by-Cycle Data Reduction Techniques for Laser Doppler Anemometer Measurements of In-Cylinder Velocity. Presented at the ASME Fluids Engineering Conference: Boulder, Colorado, June 22-23, 1981.

15. Stivender D L: Intake Valve Throttling (IVT) - A Sonic Throttling Intake Valve Engine. SAE Paper No. 680399, 1968.

16. Oforah O G: Effect of Valve Lift on Dynamic Pressure and Flame Propagation in a Spark Ignition Engine. M S Thesis, M.I.T., June, 1976.

17. Davis G C, Tabaczynski R J, and Belaire R C: The Effect of Intake Valve Lift on Turbulence Intensity and Burnrate in S.I. Engines - Model Versus Experiment. SAE Paper No. 840030, 1984.

18. Haghgooie M, Kent J C, and Tabaczynski R J: Turbulent Time Scale Measurements in an S. I. Engine Using Hot-Wire Anemometry and Fast Response Ion Probes. Presented at the ASME Winter Annual Meeting, Phoenix Arizona, November 14-19, 1982.

19. Rohsenow W M, and Choi H: Heat, Mass, and Momentum Transfer. Prentice-Hall, Inc., 1965, p. 76.

20. Bopp S, Vafidis C, and Whitelaw J H: The Effect of Engine Speed on the TDC Flowfield in a Motored Reciprocating Engine. SAE Paper No. 860023, 1986.

21. Davis G C, and Kent J C: Comparison of Model Calculations and Experimental Measurements of the Bulk Cylinder Flow Processes in a Motored PROCO Engine. SAE Paper No. 790290. 1979.

22. Johnston S C, Robinson C W, Rorke W S, Smith J R, and Witze P O: Application of Laser Diagnostics to an Injected Engine. SAE Paper No. 790092, 1979.

23. Arcoumanis C, Hadjiapostolou A, and Whitelaw J H: Swirl Center Precession in In-Cylinder Engine Flows. SAE Paper No. 870370, 1987.

24. Tennekes H, and Lumley J L: A First Course in Turbulence. The MIT Press, 1973, pp. 20-54.

MEASUREMENTS OF THE THREE-DIMENSIONAL TURBULENT FLOW IN THE CYLINDER OF AN I.C. ENGINE

M. PARSI and H. DANESHYAR

Department of Engineering, University of Cambridge

INTRODUCTION

It is now generally accepted that the fluid motion within the cylinder of an internal combustion engine has a major influence on the performance of the engine. Analysis of the combustion process requires an accurate knowledge of the fluid motion within the chamber; the ignition process, the burning rate and the cycle by cycle variations are all influenced by the motion.

Accurate measurements of the fluid motion are difficult to achieve due to the highly three-dimensional and turbulent nature of the flow. Reliable measurement techniques and data analysis procedures are yet to be developed. Presently the available techniques include (i) hot wire anemometry, (ii) laser doppler anemometry and (iii) flow visualisation. A combination of these methods, for example flow visualisation for understanding of the overall flow pattern, and hot wire anemometry or laser doppler anemometry, for detailed point measurements, provide a good compromise. This paper is concerned with the hot wire anemometry. Most workers in this field [1, 2, 3, 4, 5] have adopted a single sensor hot-wire probe. The measurements can not provide accurate values of the "mean" flow and the turbulence quantities since a single hot wire cannot resolve flow direction. An added difficulty is that the instantaneous flow structure may be different from cycle to cycle partly due to the random nature of breakdown of the jet inside the cylinder. It is therefore necessary to distinguish between fluctuations which arise because of cyclically different flow structures (cycle by cycle variations) and those fluctuations due to turbulence within any one of the structural flows, corresponding to a given cycle. There are basically two approaches to defining the flow characteristics, ensemble analysis and cycle by cycle analysis.

The ensemble analysis cannot give an accurate indication of the "mean" flow or bulk motion at a point in a given cycle if there are significant variations from cycle to cycle. For this case it will also overestimate the r m s value of the turbulent fluctuations for a given cycle. To overcome these shortcomings a cycle by cycle analysis needs to be carried out. Up to the present time no work has been published on the instantaneous measurements of the velocity vector within an engine cylinder by means of hot wire anemometry techniques. The absence of such measurements may be explained by problems due to unknown mean flow direction, high levels of turbulence and other difficulties which arise when a triple-sensor hot wire probe is used.

D. F. G. Durão et al. (eds.), Instrumentation for Combustion and Flow in Engines, 377–390.

This paper attempts to determine the magnitude of the instantaneous velocity vector within the combustion chamber of a 4-valve head by means of a triple-sensor hot wire probe. The magnitude of the components of the mean velocity vector and the associated turbulence quantities are estimated. It is considered that the 4-valve head used may provide a relatively good performance due to the influence of the associated flow field on the burn characteristics of the chamber.

EXPERIMENTAL APPARATUS AND PROCEDURE

The experiments were carried out on a Ricardo Hydra single cylinder research engine. The measurements were made in a 4-valve cylinder head. Figure 1 illustrates the general configuration of the experimental apparatus. The engine had a bore and a stroke of 80 mm and the compression ratio was 10:1. All the measurements were taken at 1000 r p m.

The hot-wire probe used in the experiments was a modified DISA 55P 91 triaxial probe (see Figure 2). The modification included the use of 90 % Pt - 10 % Rh wires of 3.4 mm length and 10 μm diameter, to allow high sensor operating temperatures (i.e. 600°C). Each hot-wire sensor was operated by a DISA 55 M constant temperature system. The hot wires were calibrated in the potential core of an air jet. The probe axis was aligned with the jet axis (i.e. each wire made an angle of 54.74 degrees with the jet axis). The effective cooling velocities were used in the calibration procedure. The heat transfer correlation used was a combination of those of Collis and Williams [6] and Davies and Fisher [7], both of which are modifications to the classical semi-empirical theory of King [8] known as the King's law. This correlation was proposed by Witze [9] as:

$$Nu_g = A + B\,Re_g^n$$

where A, B and n are calibration constants. Nu_g and Re_g are the Nusselt number and Reynolds number respectively in which all gas properties were calculated using the free-stream temperature. The calibration constants were determined for each wire individually.

The experimental measurements were made at the same location as the spark plug, with the probe axis parallel to the cylinder axis. The wire 1 was oriented in the plane YZ (see Figure 1) to determine the position of the probe coordinates relative to the cylinder coordinates. The hot-wire signals were directly digitized by a high speed data acquisition system at a rate of 13.3 KHz per channel. This corresponds to 0.45 degree crank-angle increments at 1000 r p m. Sampling was triggered by a TTL pulse every cycle (i.e. 225 degree BTDC to 45 degree ATDC of the compression stroke) to minimize the effect of engine speed variations. The cylinder temperature was measured using a piezo-electric pressure transducer connected to a charge amplifier. The values of temperature and pressure were averaged over 50 cycles which were used to determine the values of velocity. Nine hundred cycles were used to compute the ensemble-averaged velocity quantities.

ANALYSIS OF HOT-WIRE DATA

The interpretation of signals from hot-wire anemometer probes requires accurate description of the directional response of the individual wire sensors. This response is conventionally described in terms of an 'effective cooling velocity' that appears in the heat transfer law for convective cooling of the wire, and relates the cooling rate to the various components of velocity. In the following analysis the effective cooling velocity has been expressed according to the formula proposed by Jörgensen [10]:

$$V_e^2 = V_N^2 + k^2 V_T^2 + h^2 V_B^2 \tag{1}$$

where V_N, V_T and V_B are the normal, tangential, and binormal components of the velocity vector and k and h are the sensitivity coefficients (i.e. the yaw and the pitch coefficients respectively, see Figure 4).

The study by Adrian et al [11] on the directional response of single hot-wire probes indicates that the factors determining pitch response are due to aerodynamic disturbances created by the prongs, the probe body, and the curvature of the sensing wire while factors determining yaw response arise from aerodynamic disturbances, convective heat transfer due to flow parallel to the wire and the conductive end losses. The coefficients k and h depend basically on the length-to-diameter ratio, the yaw and the pitch angles (i.e. α and β, Figure 4), and the velocity magnitude itself. Previous investigations [10, 11, 12, 13] indicate that nominal averaged values of k vary between 0.1 and 0.3 and those of h vary between 1.1 and 1.2. The literature also indicates that the coefficient k is increased and h is decreased, with increasing velocity magnitude. Therefore the values of k and h were taken to be of the order of 0.1 and 1.2 respectively. This was justified by considering the velocity range expected in the engine cylinder (i.e. less than 15 m/s during the compression).

If the instantaneous velocity has components of magnitude $|V_{x1}|$, $|V_{x2}|$, $|V_{x3}|$ parallel to the respective axis x_1, x_2, x_3 in probe coordinates, then the effective velocity for each wire is given by the following:

$$V_{e1}^2 = k^2 |V_{x1}|^2 + |V_{x2}|^2 + h^2 |V_{x3}|^2 \tag{2}$$

$$V_{e2}^2 = h^2 |V_{x1}|^2 + k^2 |V_{x2}|^2 + |V_{x3}|^2 \tag{3}$$

$$V_{e3}^2 = |V_{x1}|^2 + h^2 |V_{x2}|^2 + k^2 |V_{x3}|^2 \tag{4}$$

It should be pointed out that the coefficients k and h are not necessarily the same for all wires. However it was assumed that the values of k and h are the same for all the wires.

In general a single hot-wire anemometer probe can not distinguish direction. In the case of a triple sensor probe, the direction as well as the magnitude of a velocity vector can be determined, provided we know the quadrant in which the velocity vector lies. One may have difficulty in determining the quadrant in complex situations. Generally in an engine cylinder the direction of the flow is unknown and the velocity vector does not lie within a specified quadrant throughout a cycle. The magnitude of the components of the instantaneous velocity vector in probe coordinates can always be obtained from:

$$\left[|V_{xi}|^2\right] = M^{-1}\left[V_{ei}^2\right] \tag{5}$$

where M is:

$$M = \begin{bmatrix} k^2 & 1 & h^2 \\ h^2 & k^2 & 1 \\ 1 & h^2 & k^2 \end{bmatrix}$$

Thus the magnitude of the instantaneous velocity vector (i.e. $|V| = \sqrt{|V_{x1}|^2 + |V_{x2}|^2 + |V_{x3}|^2}$) as well as the total kinetic energy per unit mass (i.e. $\frac{1}{2}\langle|V|^2\rangle$ = Mean K.E. + Turbulent K.E.) can always be determined. The magnitude of the individual components of the mean velocity vector (i.e. $|\langle V_{xi}\rangle|$) may be estimated for most of the crank angles assuming that the following condition is satisfied at each specified crank angle:

$$|\langle V_{xi}\rangle| = \langle|V_{xi}|\rangle$$

Thus the values of V_{xi} must have the same sign for all the number of cycles. This assumption is justified if at a specified crank angle the velocity vector lies within one of the eight possible quadrants for all the cycles. The r.m.s. values may then be obtained from:

$$\sqrt{\overline{v_{xi}^2}} = \left\{\frac{\sum_{j=1}^{n} \left(|V_{xi}|_j^2 - \langle|V_{xi}|\rangle^2\right)}{n}\right\}^{\frac{1}{2}}$$

where n is the number of cycles.

Finally the magnitude of the components of the mean velocity may be obtained for an arbitrary coordinate axes, xyz related to each of the eight quadrants (see Figure 3) from:

$$\begin{bmatrix} |\langle V_x\rangle| \\ |\langle V_y\rangle| \\ |\langle V_z\rangle| \end{bmatrix} = \left| N \begin{bmatrix} \langle|V_{x1}|\rangle \\ \langle|V_{x2}|\rangle \\ \langle|V_{x3}|\rangle \end{bmatrix} \right|$$

where N is:

$$N = \frac{1}{\sqrt{2}} \begin{bmatrix} 0 & -1 & 1 \\ \frac{2}{3} \cdot \sqrt{3} & -\frac{1}{3}\sqrt{3} & -\frac{1}{3}\sqrt{3} \\ \frac{\sqrt{6}}{3} & \frac{\sqrt{6}}{3} & \frac{\sqrt{6}}{3} \end{bmatrix}$$

ERROR ANALYSIS

In practice the values of V_{ei} are subject to errors. In the presence of the errors, $\pm \Delta V_{ei}$, Equation 5 may be written as:

$$\left[|V_{xi}|_m^2 \right] = M^{-1} \left[V_{ei}^2 \right] + M^{-1} \left[(\Delta V_{ei})^2 \pm 2V_{ei} \Delta V_{ei} \right] \qquad (6)$$

i.e.

$$\left[|V_{xi}|_m^2 \right] = \left[|V_{xi}|^2 \right] \pm [\varepsilon_i] \qquad (7)$$

where the subscript m denotes the measured values. The following conditions must therefore be satisfied:

$$\left[|V_{xi}|^2 \right] \geq [\varepsilon_i]$$

if the velocity vector lies nearly within either of the planes, $x_2 x_3$, $x_3 x_1$, $x_1 x_2$, at least one of the following cases may arise:

a) $|V_{x1}| \simeq 0$

b) $|V_{x2}| \simeq 0$

c) $|V_{x3}| \simeq 0$

The measured value of the corresponding component is then:

$$|V_{xi}|_m^2 = \pm [\varepsilon_i]$$

This indicates that the equation may not provide real roots for the corresponding component. However, the true magnitude of this component (i.e. approximately zero) may be substituted. This procedure has been carried out where necessary.

RESULTS AND DISCUSSION

Figure 5 shows the traces of the magnitude of the instantaneous velocity vector measured for operating with either of the inlet valves and the two valves. The figure shows the traces for 50 consecutive cycles. Significant cyclical variations may be observed. The break down of the swirl near the T D C of the compression is clearly shown for operation with valve 2 only. The use of valve 2 shows a high increase in the flow velocity due to the relatively high swirl imparted to the flow.

The characteristic dissipation time, t_d, for the turbulent kinetic energy is given by $L/_{v'}$ where L and v' denote the integral length scale and the r.m.s. of the velocity fluctuations respectively. For valve 2, at inlet valve closure, v' $\simeq$ 4 m/s (Figure 8) and L $\simeq$ 2 cm. Therefore $t_d \simeq$ 5 m s which is equivalent to 30 crank angle degrees at 1000 r.p.m. This means that in the absence of an ordered mean notion, the turbulent motion would decay rapidly relative to the time for an engine cycle, as in the case with the other two configurations shown in Figure 5.

Figure 6 shows the corresponding total kinetic energies measured directly from the values of $|V|$. The traces show a rapid dissipation of the kinetic energy near B D C where the inlet valves are nearly closed.

Figures 7 and 9 show the magnitude of the components of the mean velocity vector in the probe coordinate axes, x1 x2 x3, and in the coordinate axes, x y z, respectively. It may be seen that the swirl centre oscillates about the cylinder axis. In the coordinate axes, x y z, the magnitude of the component, $\langle V_z \rangle$, is much greater than the others. This implies from the consideration of the flow as well as the probe geometry and its location in the engine cylinder, that $|\langle V_z \rangle|$ may nearly be the magnitude of a tangential component of the mean velocity in the plane, XY.

Figure 8 shows the r.m.s. values in the probe coordinate axes. It is shown that the turbulence inside the cylinder is anisotropic at least during the intake stroke and the early part of the compression. However the turbulence tends to become isotropic towards the end of the compression stroke.

Figures 10 and 11 show the traces of the mean kinetic energies and the turbulent kinetic energies respectively. The turbulent kinetic energies generally follow the trends of the mean kinetic energies. In the case of valve 2, two distinct regions (i.e. - 225 to - 180 and - 45 to 0) may be observed where the mean kinetic energy reaches a peak and drops rapidly thereafter. Apparently the peaks correspond to the points of the break down of the swirl. The observed turbulent kinetic energy increases after the break down. The early peak (Figure 10) may be explained as follows. When the piston approached B D C the intake velocity is still appreciable. This can attribute to an increase in the swirl intensity. During the compression the swirl tends to spin up due to the conservation of angular momentum until the second break down occurs.

At least three mechanisms may be postulated for the break down of swirling flow into turbulence.

a) Instability associated with the growth of the vortex. Previous work, [14], has demonstrated that when the vortex reaches a critical size, small disturbances near the edge of the vortex grow and eventually cause break down of the swirling flow into random motion.

b) The second mechanism consists of the compression of the vortex into the clearance volume, which tends to change the geometry of the vortex flow. This process may influence the second break down near T D C.

c) Interaction between the roll up vortex [reference 14] and the barrel swirl.

Finally Figure 12 shows the comparison of the total kinetic energies for the direct measurements (i.e. from the values of $|V|$) and the indirect measurements (i.e. from the values of the mean K.E. and the turbulent K.E.). The comparison indicates good agreement of the methods thereby justifying the assumption made in the indirect measurements.

CONCLUSIONS

Velocity measurements are carried out with different inlet configurations using a triple-sensor hot wire probe. It is demonstrated that under conditions where the barrel swirl breaks down late in the compression stroke, the stored kinetic energy of the swirl can significantly affect the levels of the turbulence intensity long after the induction is over. The break down of the swirl would be advantageous in promoting a rapid propagation of an established flame.

ACKNOWLEDGEMENTS

The authors wish to thank the Ford Motor Company of the United Kingdom for the financial support provided for this work.

REFERENCES

1. Semenov, E. S., "Studies of turbulent gas flow in piston engines", Tech. Transl. F-97 NASA, pp. 122-147, 1963.

2. Daneshyar, H. and Fuller, D. E., "Definition and measurement of turbulence parameters in reciprocating i.c. engines". SAE Paper 861529, 1986.

3. Winsor, R. E. and Patterson, D. J., "Mixture turbulence - A key to cyclic combustion variation", SAE Paper 730086, 1973.

4. Woods, W.A. and Ghirlando, R., "Radial flow in an engine cylinder near the end of compression", I.Mech.E. Conf. Publ. 1975.

5. Witze, P. O., "Measurements of the spatial distribution and engine speed dependence of turbulent air motion in an i.c. engine", SAE Paper 770220, 1978.

6. Collis, D. C. and Williams, M. J., "Two-dimensional convection from heated wires at low Reynolds numbers", J. Fluid Mech., Vol. 6, pp. 357-384, 1959.

7. Davies, P. O. A. L. and Fisher, M. J., "Heat transfer from electrically heated cylinders", Proc. Roy. Soc. A., Vol. 280, pp.486-527, 1964.

8. King, L. V., "On the convection of heat from small cylinders in a stream of fluid: determination of the convective constants of small platinum wires with applications to hot-wire anemometry", Proc. Roy. Soc., Vol. 214A, No. 14, p. 373, 1914.

9. Witze, P. O., "A critical comparison of hot-wire anemometry and laser doppler velocimetry for I.C. engine applications", SAE Paper 800132, 1981.

10. Jörgensen, F. E., "Directional sensitivity of wire and fiber-film probes", DISA Information, No. 11, pp. 31-37, 1971.

11. Adrian, R. J., Johnson, R. E., Jones, B. G., Merati, P. and Tung, A. T. C., "Aerodynamic disturbances of hot-wire probes and directional sensitivity", J. Phys. E: Sci. Instrum., Vol. 17, pp. 62-71, 1984.

12. Champagne, F. H., Sleicher, C. A. and Wehrmann, O. H., "Turbulence measurements with inclined hot-wires", J. Fluid Mech., Vol. 28, Part I, pp. 153-175, 1967.

13. Hinze, J. O., "Turbulence", 1st ed. (New York: McGraw Hill), Chap. 2, 1959.

14. Daneshyar, H., Fuller, D. E. and Deckker, B. E. L., "Vortex motion induced by the piston of an internal combustion engine", Int. J. Mech. Sci., Vol. 15, pp. 381-390, 1973.

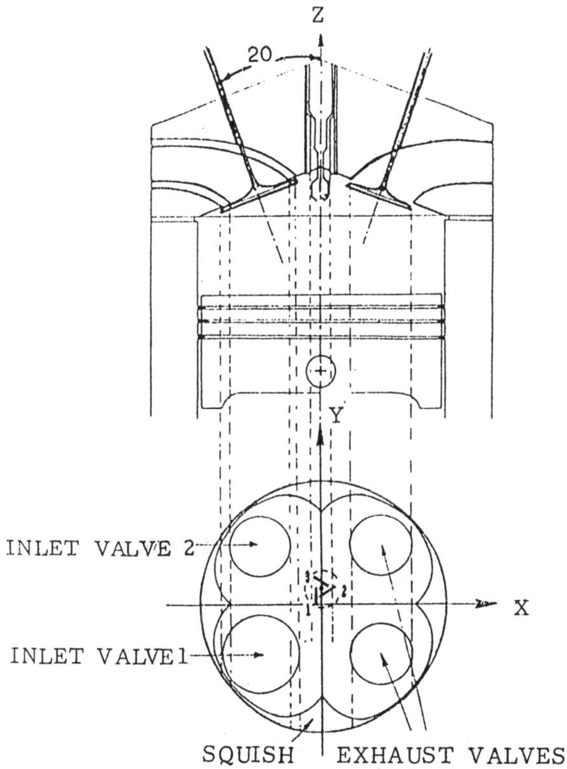

Fig. 1: General configuration of the engine, indicating the measurement location for the hot-wire sensors.

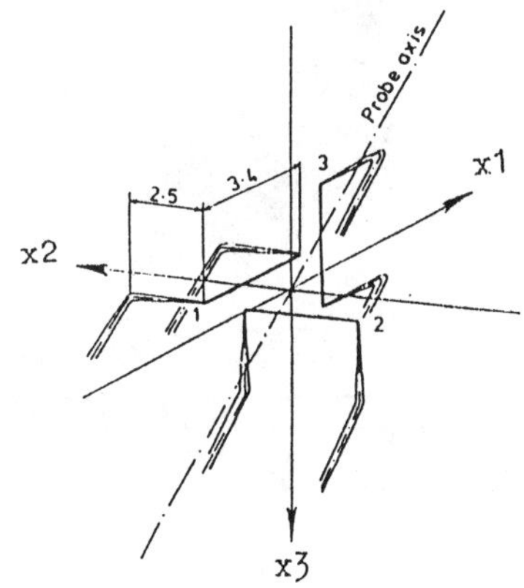

Fig. 2: Schematic of the triaxial hot-wire probe. All dimensions in mm.

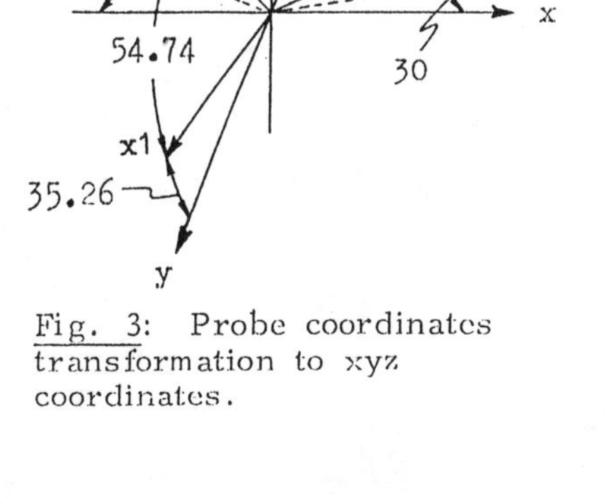

Fig. 3: Probe coordinates transformation to xyz coordinates.

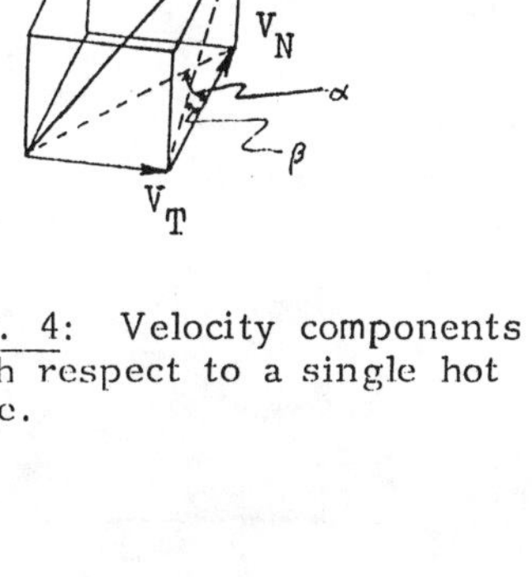

Fig. 4: Velocity components with respect to a single hot wire.

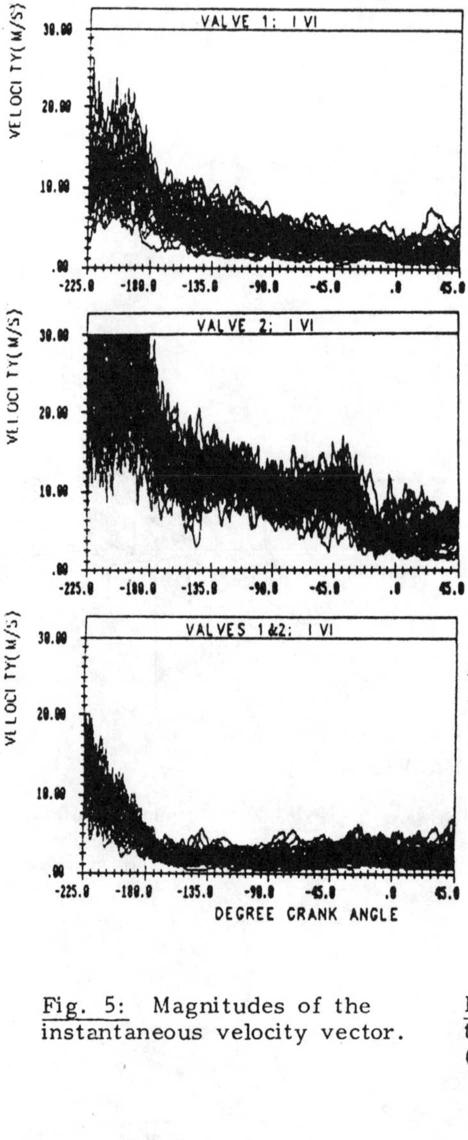

Fig. 5: Magnitudes of the instantaneous velocity vector.

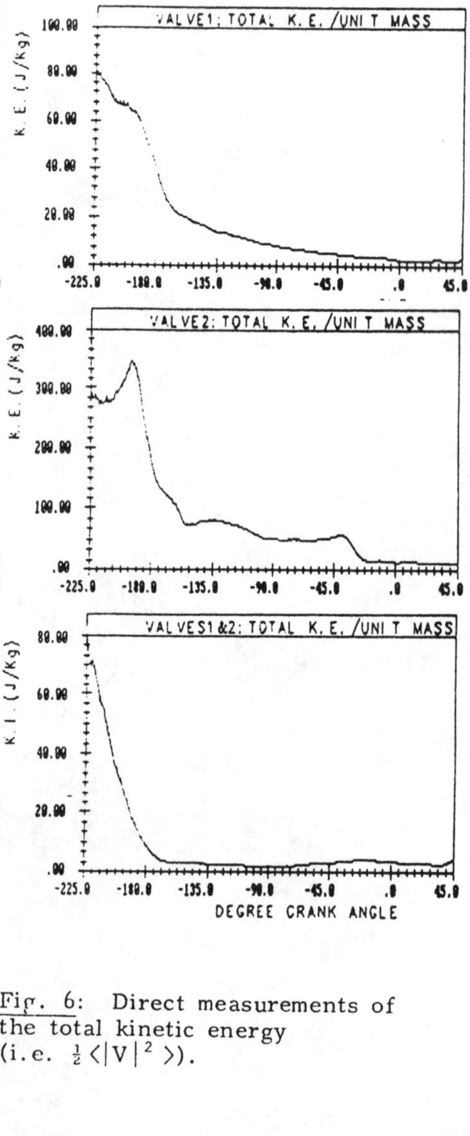

Fig. 6: Direct measurements of the total kinetic energy (i.e. $\frac{1}{2}\langle|V|^2\rangle$).

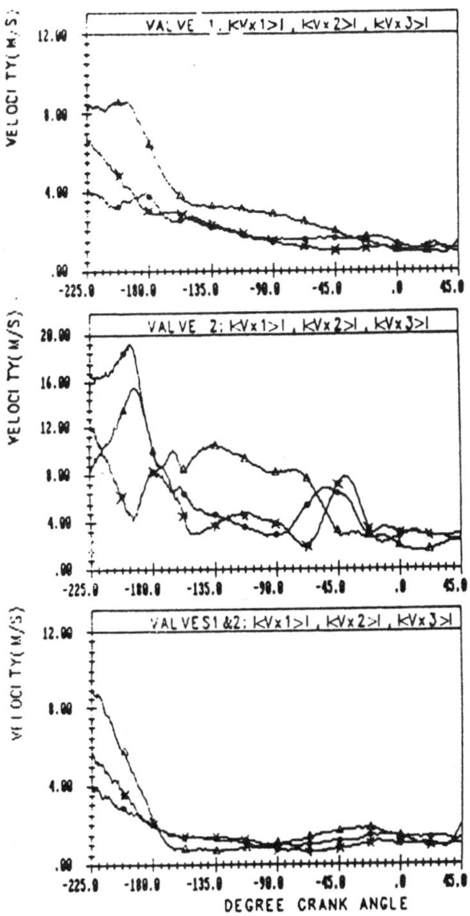

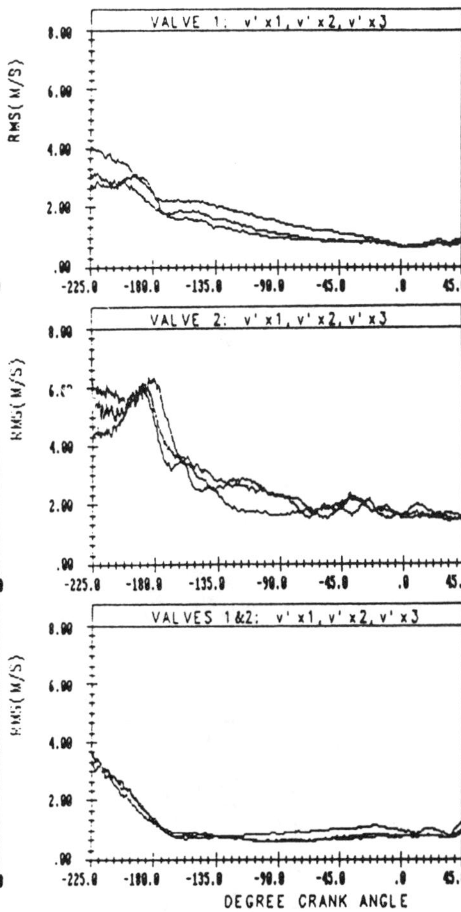

Figure 7 : Measurements of the magnitude of the components of the mean velocity vector in the probe coordinates.

x $|\langle V_{x_1} \rangle|$
• $|\langle V_{x_2} \rangle|$
Δ $|\langle V_{x_3} \rangle|$

Figure 8: The turbulence intensity relative to each component.

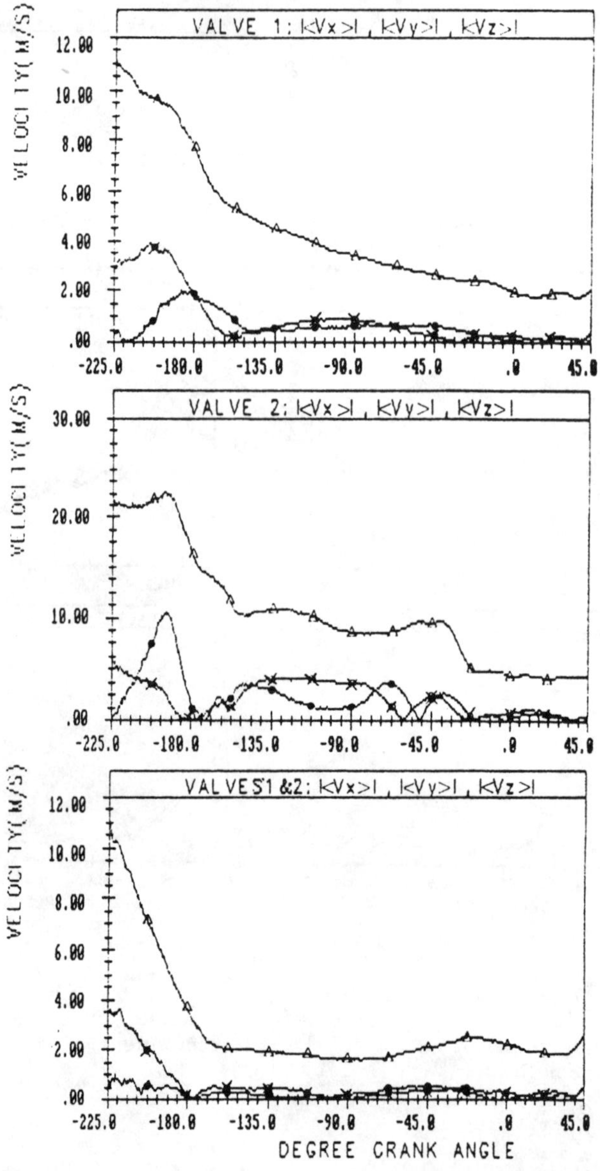

Figure 9: Magnitudes of the velocity vector in the xyz coordinates.

x $|\langle V_x \rangle|$

• $|\langle V_y \rangle|$

Δ $|\langle V_z \rangle|$

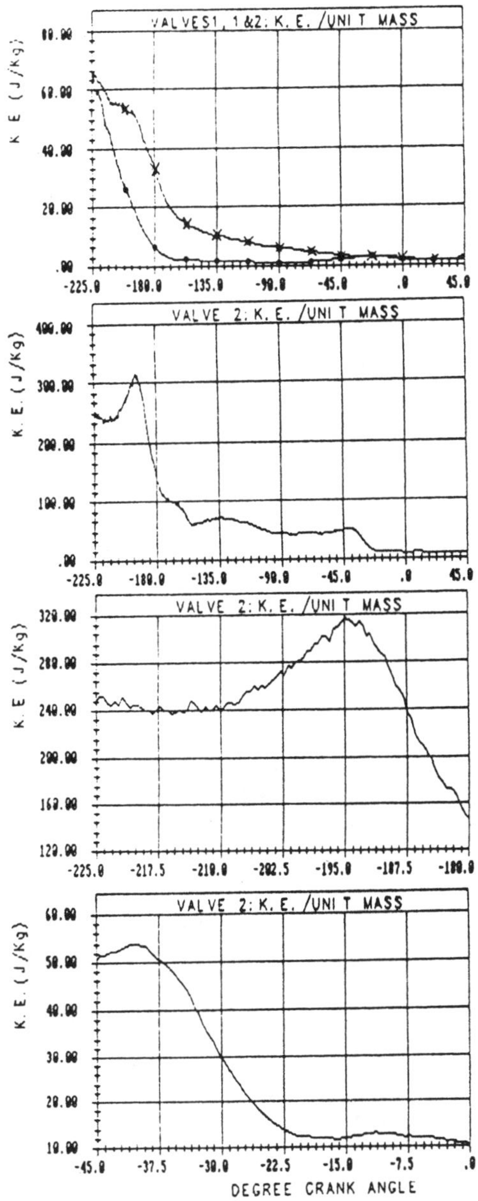

Figure 10: Mean kinetic energy

x INLET VALVE 1

• INLET VALVES 1 and 2

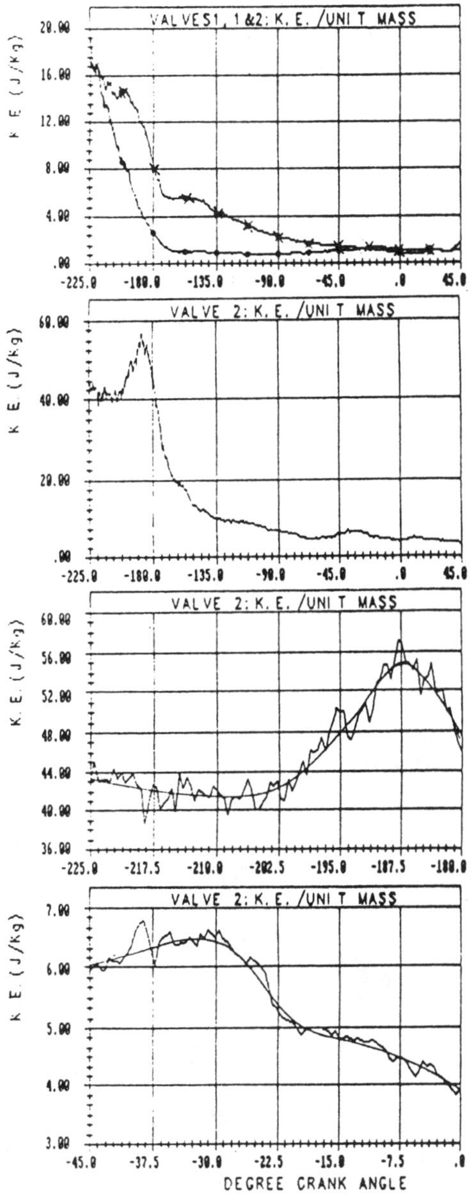

Figure 11: Turbulent kinetic energy.

x INLET VALVE 1

• INLET VALVES 1 and 2

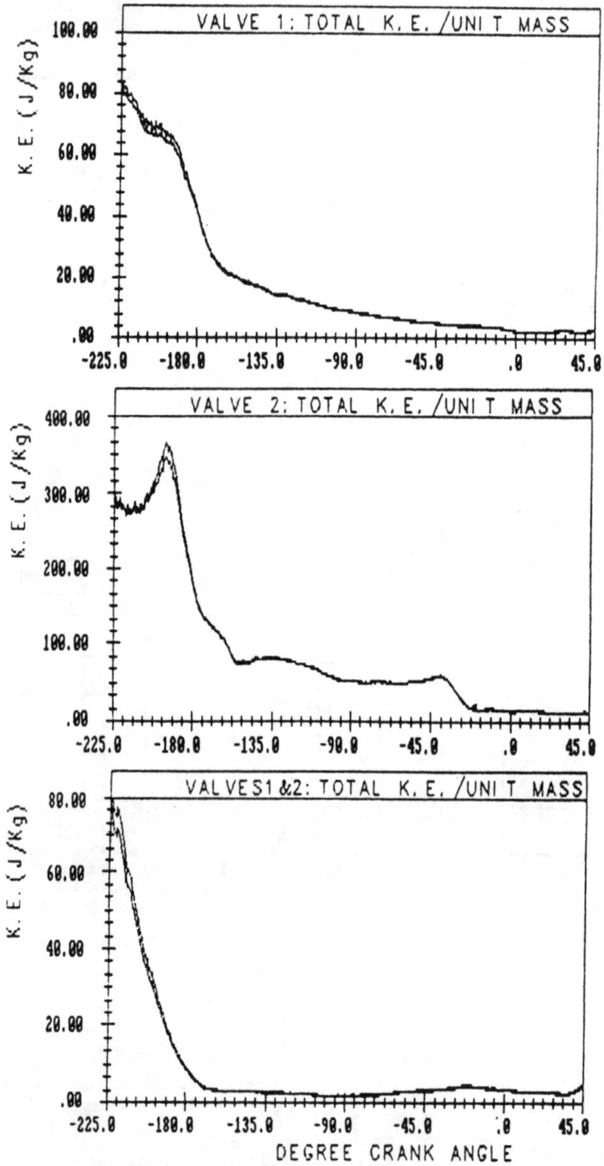

Figure 12: Comparison of the direct and the indirect measurements of the total kinetic energy.

SUBJECT INDEX